Alexander A. Kaminskii

Laser Crystals

Their Physics and Properties

Second Edition

With 89 Figures and 56 Tables

Springer-Verlag Berlin Heidelberg New York
London Paris Tokyo Hong Kong

Professor ALEXANDER A. KAMINSKII
Institute of Crystallography, Academy of Sciences of the USSR, Leninsky pr. 59,
SU-117333 Moscow, USSR

Title of the original Russian edition: ЛАЗЕРНЫЕ КРИСТАЛЛЫ (Laser Crystals)
© by „Nauka" Publishing House, Moscow 1975

ISBN 3-540-52026-0 2. Auflage   Springer-Verlag Berlin Heidelberg New York
ISBN 0-387-52026-0 2nd edition Springer-Verlag New York Berlin Heidelberg

ISBN 3-540-09576-4 1. Auflage Springer-Verlag Berlin Heidelberg New York
ISBN 0-387-09576-4 1st edition Springer-Verlag New York Heidelberg Berlin

Library of Congress Cataloging-in-Publication Data. Kaminskii, Aleksandr Aleksandrovich. [Lazernye kristally. English] Laser crystals : their physics and properties / Alexander A. Kaminskii. – 2nd ed. p. cm. – (Springer series in optical sciences ; v. 14) Translation of: Lazernye kristally. Includes bibliographical references. ISBN 0-387-52026-0 (U.S.) 1. Solid-state lasers. 2. Crystals. 3. Laser materials. I. Title. II. Series. TA1705.K3513 1990   621.36'61–dc20 89-26262 CIP

Printing: Zechnersche Buchdruckerei, Speyer. Bookbinding: J. Schäffer oHG, Grünstadt
2154/3130-543210 – Printed on acid-free paper

To the memory of my parents

Preface to the Second Edition

It was a greatest pleasure for me to learn that Springer-Verlag wished to produce a second edition of my book. In this connection, Dr. H. Lotsch asked me to send him a list of misprints, mistakes, and inaccuracies that had been noticed in the first edition and to make corresponding corrections without disturbing the layout or the typography too much. I accepted this opportunity with alacrity and, moreover, found some free places in the text where I was able to insert some concise, up-to-date information about new lasing compounds and stimulated emission channels. It was also possible to increase the number of reference citations. The reader of the second edition hence has access to more complete data on insulating laser crystals. However, sections on laser-crystal physics have not been updated, because a satisfactory description of the progress made in the last ten years in this field would have required the sections to be extended enormously or even a new book to be written.

Moscow, July 1989   ALEXANDER A. KAMINSKII

Preface to the First Edition

The greatest reward for an author is the feeling of satisfaction he gets when it becomes clear to him that readers find his work useful. After my book appeared in the USSR in 1975 I received many letters from fellow physicists including colleagues from Western European countries and the USA. Some of those letters, as well as official reviews of the book, made specific suggestions for improving the book. The satisfaction I derived from all those kind and warm responses gave me the determination to continue work on the book in order to fulfill these wishes in the next edition. This possibility arose when one of the scientific editors from Springer-Verlag, Heidelberg, H. Lotsch, who is the founder of the well-known series of quasi-monographs "Topics in Applied Physics", visited our Institute and suggested an English edition of my book. For all this, and for his subsequent help, I am sincerely thankful. I consider it my pleasant duty also to express my gratitude to the American physicist H. F. Ivey, who served as scientific editor of the translation.

The English version of the book retains the structure of the Russian edition, though it is supplemented with many new data in the tables and figures. It reflects trends in the development of the physics and spectroscopy of laser crystals in recent years. In the selection of material, special attention was paid to new research in such traditionally important problems as the theory of the crystal field, optical-transition intensity, electron-phonon interactions, concentration quenching of luminescence, laser garnets and high-concentration compounds. Finally, corrections were made and some more precise definitions given. The bibliography is also substantially larger and now includes work published up to the beginning of 1980. References to English translations, which are more-easily found in the West, are provided for nearly all Russian-language publications. I hope that they will help the reader to become acquainted with Russian work.

Many of the additions are based on the lectures I gave at the Physical Department of Kishinev State University in 1976, at the Annual Meeting of GDR Physical Society in Dresden in 1977, and at the International Schools on Luminescence of Inorganic Media in 1977 and on Radiationless Processes in 1979 in Erice, Italy. Such preliminary testing of the material was extremely useful for me.

Now that the book is to appear, I should like to express my gratitude for the help of many colleagues, both compatriots and foreign scientists. Their help took many forms, including reviewing various drafts of the manuscript;

communication of unpublished data; discussion and advice; critical comments; and donation of preprints, reprints, and also original figures from their publications.

I am specially thankful to P. P. Feofilov, Yu. E. Perlin, S. E. Sarkisov, A. A. Kaplyanskii, B. Z. Malkin, I. A. Shcherbakov, V. G. Dmitriev, and Yu. K. Danileiko; and also to B. DiBartolo, F. Williams, W. F. Krupke, M. J. Weber, L. G. DeShazer, R. G. Pappalardo, L. A. Riseberg, L. F. Mollenauer, R. C. Powell, and H. P. Jenssen from the U.S.A.; to D. Curie and F. Auzel from France; to H. G. Danielmeyer and H. Welling from Fed. Rep. of Germany; to G. Blasse from the Netherlands, to J. A. Koningstein from Canada; to H. P. Weber from Switzerland; to G. F. Imbusch from Ireland; to D. J. Newman from Australia, and to K. Washio and M. Sarawatari from Japan.

I am sincerely grateful to my closest co-workers and postgraduate students for their constant help and support. I am also cordially thankful to my wife Natasha and son Sasha for their understanding and patience through the many weekends that I took from them and devoted to the present book.

Moscow, July 1980   ALEXANDER A. KAMINSKII

Remarks by the Editor of the Original English Edition

When I first saw the original Russian version of this book, I was greatly impressed with the large amount of detailed information it contained, reflecting an obviously great investment of time by the author. I therefore suggested to Professor Kaminskii that an English version would be a useful addition to the Western literature. I was very happy later to learn that this would indeed be possible, through the cooperation of Dr. H. Lotsch of Springer-Verlag. Professor Kaminskii also decided in the English edition to enlarge certain sections and to bring the text, tables and references up to date.

I was provided a rough English version of the text. The task I set myself was to produce from this a readable and understandable text, as well as to correct (with the agreement of the author) some mistakes and typographical errors that had crept in. I hope that I have succeeded in this. However, I have made no attempt to make the material read as though I had written it. Rather, it seemed to me preferable that the text retain some of the flavor of the original version.

Monroeville, Pennsylvania, August 1980   HENRY F. IVEY[1]

---

[1] Westinghouse Electric Corporation Pittsburgh, PA 15235, USA

From the Editor of the Russian Edition

As is well known, the laser was first developed in 1960 on the basis of the single crystal of ruby, i.e., crystalline aluminum oxide activated by trivalent chromium ions. Notwithstanding the fact that subsequently gas and semiconductor lasers were developed, as well as lasers based upon glasses, liquids, and organic dyes; impure ionic crystals have retained their leading position, along with a variety of other active media regarded as promising for present-day lasers. Lasers based upon these crystals exhibit an exceptionally wide variety of properties due to the regularity of the crystal structure and the extraordinarily wide range of physical parameters of impure ionic crystals. The detailed and comprehensive study of all these properties has, in turn, made it realistic to search for new laser crystals that possess specified characteristics.

Recently, quantum electronics not only has benefited from the results of this search, but has also actively assisted in it. Owing to developments in quantum electronics, a new method of experimental investigation has been established: stimulated-emission spectroscopy. The study of the spectroscopic characteristics of laser emission by activated compounds furnishes an important supplement to conventional luminescence and absorption techniques. The importance of these spectroscopic studies for quantum electronics and solid-state physics can hardly be overestimated. Figuratively speaking, the field of quantum electronics begins just beyond the end faces of a laser crystal. From the viewpoint of quantum electronics, the crystal is an entity with definite properties. In contrast, stimulated-emission spectroscopy penetrates the interior of the crystal to investigate the relationship between the intrinsic (structural and other) properties of the crystal and the parameters that determine the operating characteristics of a laser.

The effectiveness of the direct search for new laser compounds depends directly on the amount of information accumulated in the field being studied. It is therefore not surprising that the rate of progress here has increased rapidly. An illustration of this is the fact that most laser crystals have been investigated only in the past few years. Up to now, the results of these studies have been available mostly as original papers or reviews published in a variety of journals; these results have not yet been generalized or systematized in a single publication. The present monograph by Alexander A. Kaminskii is an attempt to bridge this gap.

The field of laser crystals is one in which a number of major trends in quantum electronics, spectroscopy, and crystallography come together. In these rapidly advancing fields are engaged a large number of scientists, engineers, and technicians, each with his own scientific background and his own approach to this challenging problem. All these factors have considerably complicated the already extremely difficult task of writing such a book. Therefore, it is hardly worth enumerating the objectives pursued by the author in the limited space allowed by a monograph, especially since this has been done in the Author's Preface. I need only mention that, in my opinion, Alexander A. Kaminskii has succeeded in attaining his aims.

The main feature of the present monograph is that the techniques and some of the fundamental results of the stimulated-emission spectroscopy of activated crystals are described systematically for the first time. The author of this volume is a noted expert on these problems. Jointly with other Soviet investigators, and also independently, he has carried out a number of pioneering studies in the field whose development in our country was initiated by the outstanding work of Academicians N. G. Basov and A. M. Prokhorov. It does not seem superfluous here to mention that Alexander A. Kaminskii has discovered (and studied in detail) the stimulated-emission effect in over a third of the known insulating laser crystals.

The classification of crystals suggested in the book is both original and useful; it will be appreciated by both users of laser crystals and by physicists engaged in creating, and studying the properties of, new lasers. A high concentration of information in the monograph has been achieved by presenting most of the data in the form of tables. Rigorous selection of the most important theoretical results and the most reliable experimental data also constitutes a great merit of the book. The author decided not to treat many questions regarding crystalline lasers that have already been discussed at length in the literature; thus, the spectroscopic and emission properties of ruby, as well as a number of theoretical questions, are considered merely in outline.

I believe that the monograph will be found both interesting and helpful not only by specialists in the fields of quantum electronics, activated crystal spectroscopy, and crystallography, but also by those engaged in studies in related fields of solid-state physics.

B. K. VAINSHTEIN, Academician
Director of the A. V. Shubnikov Institute of Crystallography
Academy of Sciences of the USSR
Moscow

Contents

Folder (Table 2.1) at the end of the book

1. Brief Review of the Development of Laser Crystal Physics

Quantum electronics has developed from the concept of stimulated emission introduced by *Einstein* [1]. Yet it did not become a field of scientific research in its own right until the appearance of the well-known investigations by *Basov* and *Prokhorov* [2], as well as *Townes* et al. [3], in which this phenomenon was used for the amplification and generation of electromagnetic oscillations in the microwave region. In 1958 *Schawlow* and *Townes* [11] analyzed the possibility of generating and amplifying electromagnetic oscillations over the optical and near-optical range by stimulated emission. They considered the basic theory, and put forward some concrete suggestions concerning the choice of active media and methods for their excitation. In particular, they showed that for this purpose, one can use materials activated by lanthanide (Ln) ions. In the summer of 1960, a year and half later, *Maiman* announced in the journal *Nature* the development of a pulsed synthetic ruby laser ($Al_2O_3-Cr^{3+}$ crystal) emitting in the red region of the visible spectrum (0.6943 μm) [12]. These reports presented an obvious extension of quantum electronics to the optical range.

After the advent of the ruby laser, also in 1960, *Sorokin* and *Stevenson* described a CaF_2-U^{3+} crystal laser [13]. Whereas the ruby laser was a three-level system, this was a four-level one, laser action being accomplished at the temperature of liquid helium and at a wavelength near 2.6 μm. In 1961 the same authors observed stimulated emission in the visible (≈ 0.7 μm) by adding Sm^{2+} ions to a CaF_2 crystal [14]. Basic to the experiments of *Sorokin* and *Stevenson* were the results obtained by *Feofilov* et al. [15, 16] in detailed spectroscopic studies of these crystals. Thus, bivalent samarium proved to be the first laser Ln ion. Anticipating a little, we note that subsequently, Ln ions, especially trivalent ones, have taken a dominant position as laser-crystal activators responsible for stimulated-emission processes. This was very much due to the fundamental investigations of the spectroscopic behavior of Ln ions in various media, carried out by several scientists early in the "prelaser" period. The efforts of members of the spectroscopic schools headed by *Vavilov* in the Soviet Union, *Dieke* in the United States, and *Hellwege* in Western Europe were especially meritorious.

Activated crystals, however, best revealed their properties only after *Johnson* and *Nassau* developed a $CaWO_4-Nd^{3+}$ laser emitting near 1 μm at room temperature with an extremely low excitation threshold [17]. This behavior was attributed to the fact that the terminal level for an induced transition in the

Nd^{3+} ion lies about 2000 cm^{-1} above the ground level and is practically unpopulated at operating temperatures. Currently, the Nd^{3+} ion is the most widely used laser-crystal activator (see Tables 1.1–5)[1].

In 1962 crystal lasers were vigorously developed. Along with the discovery of the stimulated-emission effect in a great many crystals activated by other Ln ions, there appeared new types of solid-state lasers, for example continuous-wave (CW) [18] and Q-switched lasers [19]. Later they found extensive practical application.

In all the above crystal lasers, stimulated emission is attributed to transitions between electronic levels of activator ions. In 1963 *Johnson, Dietz,* and *Guggenheim* [20] showed that several crystals activated by certain bivalent ions of the iron group can also undergo induced emission in phonon-assisted transitions.

In subsequent years, scientists have searched for new laser crystals and for ways of improving their efficiency. Thus, in order to improve the generation characteristics, the principle of sensitization was employed [21]. *Voronko, Kaminskii, Osiko,* and *Prokhorov* [22, 23, 360] used another method to increase

---

[1] Information on the following laser crystals is also available: simple fluoride crystals-LiYF$_4$ with Ce^{3+} and Tm^{3+} ions, K$_5$NdLi$_2$F$_{10}$-Nd^{3+}, TbF$_3$-Sm^{3+} and BaYb$_2$F$_8$-Ho^{3+}; simple oxide crystals-KY(WO$_4$)$_2$ with Ho^{3+} and Er^{3+} ions, KEr(WO$_4$)$_2$-Er^{3+}, KLu(WO$_4$)$_2$ with Nd^{3+}, Ho^{3+} and Er^{3+} ions, Ca$_3$Ga$_2$Ge$_3$O$_{12}$-Nd^{3+}, Sc$_2$SiO$_5$-Nd^{3+}, PrP$_5$O$_{14}$-Pr^{3+}, Gd$_3$Ga$_5$O$_{12}$ with Ho^{3+} and Er^{3+} ions, Lu$_2$SiO$_5$-Nd^{3+}, Bi$_4$Ge$_3$O$_{12}$ with Ho^{3+} and Er^{3+} ions; disordered oxide crystal Ca$_2$MgSi$_2$O$_7$-Nd^{3+}; and Na$_2$Nd$_2$Pb$_6$(PO$_4$)$_6$Cl$_2$-Nd^{3+}.

Table 1.1. Simple fluoride laser crystals with ordered structure

Crystal	Activator ion							
	Ce^{3+}	Pr^{3+}	Nd^{3+}	Tb^{3+}	Dy^{3+}	Ho^{3+}	Er^{3+}	Tm^{3+}
LiYF$_4$	+	+	+	+	+	+	+	+
Li(Y,Er)F$_4$						+	+	
LiHoF$_4$						+		
LiErF$_4$							+	
MgF$_2$								
KMgF$_3$								
CaF$_2$			+			+	+	
CaF$_2$*^{a}			+				+	+
MnF$_2$								
ZnF$_2$								
SrF$_2$			+					+
BaF$_2$			+					
BaY$_2$F$_8$		+	+		+	+	+	
Ba(Y,Er)$_2$F$_8$					+			
Ba(Y,Yb)$_2$F$_8$		+				+	+	
LaF$_3$	+	+	+				+	
CeF$_3$			+					
HoF$_3$						+		

^{a} Crystals CaF$_2$* contained oxygen.

the efficiency of crystal lasers: it used mixed disordered systems (solid solutions) as the crystal matrix rather than simple crystals (see Tables 1.1, 3). A specific feature of this new promising class of laser media is the fact that Ln^{3+} ions are distributed over many activator centers with different structures. Hence, the absorption spectra of these materials consist of strong broad lines which provide much better utilization of pump power. Migration of energy over excited states of many activator centers, which takes place in these media, also contributes to the laser operating parameters. Mixed-crystal systems are similar, in certain spectroscopic properties, to such laser media as glasses and inorganic liquids doped with Ln^{3+} ions. A great number of mixed fluoride crystals with Ln^{3+} ions have been grown and used in about thirty laser systems (see Table 1.2). Some of these crystals impart unique properties to lasers. For instance, an α-NaCaYF$_6$-Nd^{3+} laser can operate at 1000 K [24]. The idea of disordered crystal systems has been successfully extended to crystals containing oxygen (see Table 1.4).

During the emergence of quantum electronics, spectroscopic studies of activated crystals suggested their possible use in laser systems. Later, experience accumulated in studies of stimulated-emission parameters allowed, in many cases, a much deeper and more complete analysis of the characteristics of these materials, and helped uncover their properties, adding to the information from such conventional spectroscopic methods as luminescence and absorption analyses. Thus, a new trend of spectroscopy originated at the junction of quantum electronics and classical spectroscopy, namely, the stimulated-emission

Yb^{3+}	Sm^{2+}	Dy^{2+}	Tm^{2+}	U^{3+}	V^{2+}	Co^{2+}	Ni^{2+}
					+	+	+
						+	+
+	+	+	+	+			
	+						
							+
						+	
	+	+		+			
				+			

Table 1.2. Mixed fluoride laser crystals with disordered structure

Crystal	Activator ion			
	Nd^{3+}	Ho^{3+}	Er^{3+}	Tm^{3+}
$5NaF \cdot 9YF_3$	+			
CaF_2-SrF_2	+			
CaF_2-YF_3	+	+	+	
$Ca_2Y_5F_{19}$	+			
CaF_2-LaF_3	+			
CaF_2-CeO_2	+			
CaF_2-CeF_3	+			
CaF_2-GdF_3	+			
CaF_2-ErF_3		+	+	+
SrF_2-YF_3	+	+		
$Sr_2Y_5F_{19}$	+	+		
SrF_2-LaF_3	+			
SrF_2-CeF_3	+			
SrF_2-GdF_3	+			
SrF_2-LuF_3	+			
CdF_2-YF_3	+			
CdF_2-LaF_3	+			
BaF_2-YF_3	+			
BaF_2-LaF_3	+			
BaF_2-CeF_3	+			
BaF_2-GdF_3	+			
BaF_2-LuF_3	+		+	
LaF_3-SrF_2	+		+	
$\alpha-NaCaYF_6$	+			
$\alpha-NaCaCeF_6$	+			
$\alpha-NaCaErF_6$		+		+
$CaF_2-YF_3-NdF_3$	+			
$CaF_2-ErF_3-TmF_3$			+	+
$SrF_2-CeF_3-GdF_3$	+			
$CdF_2-YF_3-LaF_3$	+			
$CaF_2-ErF_3-TmF_3-YbF_3$		+		
$CaF_2-SrF_2-BaF_2-YF_3-LaF_3$	+			

Table 1.3. Simple oxide laser crystals with ordered structure

Crystal	Activator ion								
	Pr^{3+}	Nd^{3+}	Eu^{3+}	Ho^{3+}	Er^{3+}	Tm^{3+}	Yb^{3+}	Ni^{2+}	Cr^{3+}
$LiNbO_3$		+		+		+			
$LiNdP_4O_{12}$		+							
$BeAl_2O_4$									+
MgO								+	
$NaNdP_4O_{12}$		+							
Al_2O_3									+
$KY(MoO_4)_2$		+							
$KY(WO_4)_2$		+		+	+				
$K(Y,Er)(WO_4)_2$				+	+				
$KNdP_4O_{12}$		+							
$KGd(WO_4)_2$		+		+	+				
$CaMg_2Y_2Ge_3O_{12}$		+					+		

Table 1.3 (*continued*)

Crystal	Activator ion								
	Pr^{3+}	Nd^{3+}	Eu^{3+}	Ho^{3+}	Er^{3+}	Tm^{3+}	Yb^{3+}	Ni^{2+}	Cr^{3+}
$CaAl_4O_7$		+			+				
$CaAl_{12}O_{19}$		+							
$CaSc_2O_4$		+							
$Ca_3(VO_4)_2$		+							
$Ca(NbO_3)_2$	+	+		+	+	+			
$CaMoO_4$		+		+		+			
$CaWO_4$	+	+		+	+	+			
$SrAl_4O_7$		+							
$SrAl_{12}O_{19}$		+							
$SrMoO_4$	+	+							
$SrWO_4$		+							
Y_2O_3		+	+						
$YAlO_3$	+	+		+	+	+			
$Y_3Al_5O_{12}$		+		+	+	+	+		+
Y_2SiO_5		+		+					
YP_5O_{14} [a]	+	+							
$Y_3Sc_2Al_3O_{12}$		+		+	+				+
$Y_3Sc_2Ga_3O_{12}$		+		+	+	+			+
YVO_4		+	+	+		+			
$Y_3Fe_5O_{12}$				+					
$Y_3Ga_5O_{12}$		+		+	+		+		+
$(Y,Er)AlO_3$				+	+	+			
$(Y,Er)_3Al_5O_{12}$				+	+	+			
$(Y,Yb)_3Al_5O_{12}$							+		
$(Y,Lu)_3Al_5O_{12}$		+							
$Ba_{0.25}Mg_{2.75}Y_2Ge_3O_{12}$		+							
$La_2Be_2O_5$		+							
La_2O_3		+							
$LaAlO_3$		+							
LaP_5O_{14}	+	+							
$LaNbO_4$		+		+					
CeP_5O_{14}		+							
$NdAl_3(BO_3)_4$		+							
NdP_5O_{14}		+							
Gd_2O_3		+							
$GdAlO_3$		+		+	+	+			
GdP_5O_{14}		+							
$GdScO_3$		+							
$Gd_3Sc_2Al_3O_{12}$		+					+		+
$Gd_3Sc_2Ga_3O_{12}$		+		+	+	+	+		+
$Gd_3Ga_5O_{12}$		+		+	+		+		+
$Gd_2(MoO_4)_3$		+							
$Ho_3Al_5O_{12}$				+					
$Ho_3Sc_2Al_3O_{12}$				+					
$Ho_3Ga_5O_{12}$				+					
Er_2O_3				+		+			
$ErAlO_3$				+	+	+			
$Er_3Al_5O_{12}$				+	+	+			
Er_2SiO_5				+					
$Er_3Sc_2Al_3O_{12}$				+					
$ErVO_4$				+					

[a] Some stable compositions with different structure from in $Y_xNd_{1-x}P_5O_{14}$ systems [249].   (*Continued*)

Table 1.3 (*continued*)

Crystal	Activator ion								
	Pr^{3+}	Nd^{3+}	Eu^{3+}	Ho^{3+}	Er^{3+}	Tm^{3+}	Yb^{3+}	Ni^{2+}	Cr^{3+}
$(Er,Tm,Yb)_3Al_5O_{12}$				+					
$(Er,Yb)_3Al_5O_{12}$						+			
$(Er,Lu)AlO_3$				+		+			
$(Er,Lu)_3Al_5O_{12}$					+	+			
$Yb_3Al_5O_{12}$				+	+				
$(Yb,Lu)_3Al_5O_{12}$								+	
$LuAlO_3$	+	+		+	+				
$Lu_3Al_5O_{12}$		+		+	+	+	+		
$Lu_3Sc_2Al_3O_{12}$		+					+		
$Lu_3Ga_5O_{12}$		+					+		
$PbMoO_4$		+							
$Bi_4Si_3O_{12}$		+							
$Bi_4Ge_3O_{12}$		+		+	+				

spectroscopy of activated crystals which is currently being rapidly developed [4, 5]. Among the most significant advances in this field are the creation of solid-state lasers with cascade operating schemes [25], with combined active media [26–28], with "photon injection" [29, 30] where some excitation of high-energy metastable states is due to various conversions of pump infrared quanta [31], and those with stimulated-emission frequency tuning (continuous or stepped) by variation of the temperature of the emitting crystal [24, 32–34] or of the parameters of the optical cavity [35]. Of interest also are crystal lasers that can be operated only at high temperatures [24, 36]. Multibeam lasers, suggested in [464], exhibit some new peculiar features. One of their most significant properties is the possibility of controlling the populations of excited metastable states during stimulated emission.

Of great concern for laser-crystal physics, quantum electronics, and integrated optics are high-concentration materials in which the emitting ions are dominant components of the crystalline lattice (self-activated or stoichiometric laser crystals). The first observation of stimulated emission with such crystals was by *Varsanyi* [76]. Using so-called surface lasers, he succeeded in demonstrating a highly efficient laser pump which excites stimulated emission from self-activated rare-earth crystals ($PrCl_3$ and $PrBr_3$) of size of order 1 μm. This is, naturally, important to those searching for new laser materials.

A good deal of impetus to research on the properties of activated crystals, particularly in the concentration quenching of the Ln^{3+} ion luminescence, was given by *Danielmeyer* et al. [84, 507]. They investigated self-activated NdP_5O_{14} crystals, which are nowadays used as a model for studying weak concentration quenching of the Nd^{3+} ion luminescence. Researchers in several laboratories in the Soviet Union have recently discovered new properties of $Y_3Al_5O_{12}$, $KY(WO_4)_2$, and $YAlO_3$ crystals containing high percentages (sometimes up to

Table 1.4.   Mixed oxide laser crystals with disordered structure

Crystal	Activator ion			
	Nd^{3+}	Ho^{3+}	Er^{3+}	Tm^{3+}
$LiLa(MoO_4)_2$	+			
$Li(Nd,La)P_4O_{12}$	+			
$Li(Nd,Gd)P_4O_{12}$	+			
$LiGd(MoO_4)_2$	+			
$NaLa(MoO_4)_2$	+	+		
$NaLa(WO_4)_2$	+			
$Na_3Nd(PO_4)_2$	+			
$Na_5Nd(WO_4)_4$	+			
$NaGd(WO_4)_2$	+			
$Na(Nd,Gd)(WO_4)_2$	+			
$KLa(MoO_4)_2$	+	+	+	
$K_3Nd(PO_4)_2$	+			
$K_3(Nd,La)(PO_4)_2$	+			
$K_5Nd(MoO_4)_4$	+			
$K_5Bi(MoO_4)_4$	+			
$CaY_4(SiO_4)_3O$	+	+		
$Ca_{0.25}Ba_{0.75}(NbO_3)_2$	+			
$CaLa_4(SiO_4)_3O$	+			
$Ca_4La(PO_4)_3O$	+			
$CaGd_4(SiO_4)_3O$	+			
$SrY_4(SiO_4)_3O$		+		
$SrLa_4(SiO_4)_3O$	+			
$YScO_3$	+			
$ZrO_2-Y_2O_3$	+			
$ZrO_2-Er_2O_3$		+	+	+
$Ba_2NaNb_5O_{15}$	+			
$Ba_2MgGe_2O_7$	+			
$Ba_2ZnGe_2O_7$	+			
$(Nd,Sc)P_5O_{14}$	+			
$(Nd,In)P_5O_{14}$	+			
$(Nd,La)P_5O_{14}$	+			
$(Nd,Gd)Al_3(BO_3)_4$	+			
$LuScO_3$	+			
$HfO_2-Y_2O_3$	+			
$Bi_4(Si,Ge)_3O_{12}$	+			

Table 1.5.   Other laser crystals

Crystal	Activator ion		
	Pr^{3+}	Nd^{3+}	Ho^{3+}
$Ca_5(PO_4)_3F$		+	+
$Sr_5(PO_4)_3F$		+	
La_2O_2S		+	
$LaCl_3$	+		
$LaBr_3$	+		
$CeCl_3$		+	
$PrCl_3$	+		
$PrBr_3$	+		
$Pb_5(PO_4)_3F$		+	

100%) of Er^{3+} and Ho^{3+} ions which emit efficiently at room temperature at about 2.6–3 μm—a range with good prospects for many applications [490, 546, 547, 572, 595, 641]. A peculiar feature of the latest lasers is their use of "self-saturating" transitions, where the terminal laser state has a longer lifetime than the initial one. In order to improve the characteristics of crystal lasers emitting on self-saturating transitions, excitation flow from the longer-lived terminal laser state was accelerated by the addition to the crystal of deactivator ions, which, during emission, take over the residual excitation, as was first reported in [546]. In this connection, we should mention that as long as 22 years ago, *Robinson* and *Devor* [111] observed laser action of Er^{3+} ions in a fluoride system, on such a transition. This relatively short list of advances leading to new crystal lasers confirms the fact that activated crystals open up practically inexhaustible possibilities for stimulated emission.

The methods of stimulated-emission spectroscopy involving the principles basic to these types of lasers help solve one of the main problems in quantum electronics, namely, extending and joining frequency ranges over which laser action can be established. High-temperature stimulated-emission spectroscopy [23, 32] provides a means of analyzing the behavior of induced transitions over a wide temperature range, which is of particular importance in studying electron–phonon interactions in activated crystals. The first encouraging studies along these lines have already been carried out with crystals of LaF_3, $YAlO_3$, $Y_3Al_5O_{12}$, and $Lu_3Al_5O_{12}$ doped with Nd^{3+} and Yb^{3+} ions [37, 38, 483, 548, 593].

The stimulated-emission effect has been observed so far in 425 insulating crystals doped with transition-element ions. Most of them are oxide laser crystals with ordered structures in which the dopant ions form mainly type-one activator centers (see Table 1.3). Here trivalent Ln ions are predominant, whereas simple fluoride laser crystals do not exhibit such selectivity (see Table 1.1). In addition to the trivalent chromium ion, the latter also use all types of known activator ions. At present roughly equal numbers of synthesized mixed fluoride and oxide disordered laser crystals are known (see Tables 1.2, 1.4).

As was mentioned earlier, the most extensively used activator ion in laser crystals is trivalent neodymium. It acts as a laser ion in almost one hundred and thirty media. In the second place is holmium, followed by erbium and thulium. It is interesting to note that of all the Ln ions, only thulium and dysprosium have laser properties in both the bivalent and the trivalent states.

From the great many known laser crystals, only those are widely used that best satisfy the latest requirements of quantum electronics. Besides ruby lasers which still preserve their position, we can mention, in the first place, $Y_3Al_5O_{12}$ and $YAlO_3$ crystals doped with Nd^{3+} ions. Pulsed and CW stimulated emission at room temperature was observed in the former crystal by *Geusic*, *Marcos*, and *Van Uitert* [9] in 1964 and in the latter by *Bagdasarov* and *Kaminskii* [10] in 1969. Recent efforts of *Prokhorov*, *Osiko* et al. have extended the range of active media that have good prospects for high-power lasers by including the CaF_2-Dy^{2+}

crystal [39], the laser effect in which was discovered as early as 1962 almost simultaneously by *Kiss* and *Duncan* [40], *Johnson* [41], and *Yariv* [42].

The development of laser-crystal physics has been considerably aided by new purely spectroscopic methods: polarized luminescence (developed in detail by *Feofilov* [43]); piezospectroscopy or the method of directional elastic deformation, suggested by *Kaplyanskii* [44]; and the concentration-series method [45] proposed by *Voronko*, *Osiko*, and the author of the present monograph. Along with conventional magneto-optics, luminescence, and absorption techniques, these new methods greatly helped in obtaining insight into the nature of activator-center formation in laser crystals. The extension of magneto-optics to Ln ions in crystals was largely due to the work of the research groups of *Dieke* [46, 47, 549] and *Zakharchenya* [48]. Of great importance for laser-crystal physics are the fundamental experimental studies of nonradiative relaxation in Ln^{3+} ions by *M. J. Weber* [49, 50] and by *Riseberg* and *Moos* [51–53].

The investigations of *DeShazer* [29] and the author of the present book [26, 28] were the earliest studies of the effect of narrow-band laser emission on processes that occur in excited disordered laser media. These studies were followed by investigations that gave rise to a highly informative laser spectroscopic method for studying the nature of homogeneously broadened spectral lines masked by inhomogeneously broadened lines in activated disordered systems (the laser-induced fluorescence-line narrowing method) [550–555, 626]. In studying absorption properties of activated crystals less than one millimeter in cross section, the laser derivative-spectroscopy method proves very helpful [556]. Of particular interest in recent studies of activated (in particular, laser) crystals is resonance Raman scattering. This phenomenon is the sudden increase in the intensity of scattered light observed when the frequency of the exciting light approaches that of an electronic transition of the crystal activator. Resonance Raman scattering spectra provide information about electronic-transition polarizations otherwise unavailable for isotropic crystals. The present state of this problem is reviewed by *Koningstein* [557], who has personally made great contributions to the development of this kind of spectroscopy.

Significant advances in the theoretical aspects of laser-crystal physics were made by the investigations of *Wybourne* [558] on the theory of crystal fields, and general theoretical problems of Ln^{3+} ion spectroscopy; of *Judd* [559] and *Ofelt* [560], on theoretical aspects of Ln^{3+} ion spectral intensity; of *Carnall*, *Fields* and *Rajnak* [728–730, 745], on theoretical analysis of energy levels of Ln^{3+} ions; of *Rebane* [54] and *McCumber* [55], on the theory of electronic-vibrational transitions; of *Dexter* [56], *Förster* [57], and *Galanin* [58], on the theory of transfer of electronic excitation energy; and of *Perlin* [59], on the theory of multiphonon nonradiative transitions. Chemical aspects of crystal field theory are analyzed in detail by *Ballhausen* [503] and *Jörgensen* [35]. It is impossible to overestimate the significance of the research of *Schawlow* [196, 198, 258], *Krupke* [267, 448, 561], *Geusic* [140, 408], *DeShazer* [562, 563], *Auzel* [466, 498, 771], *Yariv* [69, 335], *DiBartolo* [72], *Melamed* [130, 429], *Yen* [258, 626], *Kushida* [302], *Soffer* [146, 139],

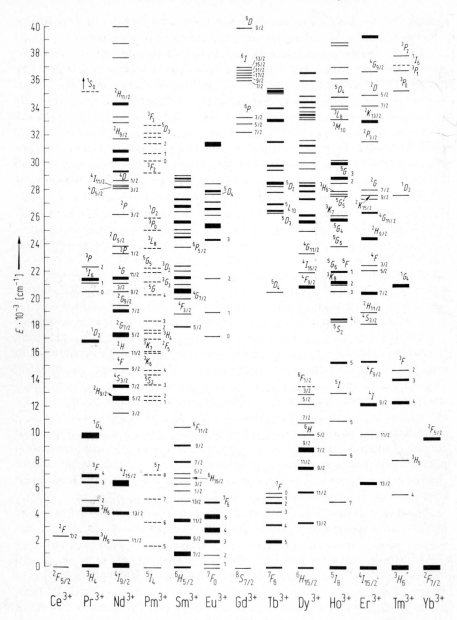

Fig. 1.1. Energy-level diagram of multiplets of Ln^{3+} ions in crystals. From data on $LaCl_3$-Ln^{3+} crystals in [46]

Powell [772, 773], and *Pappalardo* [174] for laser-crystal physics. A series of works by *Stepanov* et al. [60] deals with a detailed kinetic theory of emission in activated media. Investigating generation at electron-vibrational transitions of Ni^{2+} and Co^{2+} ions in fluorides using lamp pumping, *Moulton* and *Mooradian* established new possibilities for the performance of broad-band tunable crystalline lasers. An important step in this direction was the work of *Morris, Shand, Jenssen,* and *Walling* on laser action of Cr^{3+} ions in alexandrite and emerald. By the development of broad-band stimulated emission of Ti^{3+} ions in sapphire, *Moulton* revived specialist interest in this wonderful crystal. A notable contribution to solving tunable laser problems was made by the groups of *Huber* and *Dürr.* A large family of crystals with disordered structure of Ca-gallogermanate type with record broad-band stimulated-emission tuning of Cr^{3+} ions was discovered and investigated by *Kaminskii, Mill,* and *Shkadarevich* with co-workers. *Kenyon, Andrews, McCollum,* and *Lempicki* made an essential contribution to the understanding of the broad-band generation nature of this first laser activator. In the groups of *Chase, Esterowitz, Antipenko, Barnes, Shcherbakov,* and others, Cr^{3+} ions are used with success as sensitizing ions for creating highly efficient crystalline lasers. For calculation of Stark-structure levels of laser trivalent lanthanides in crystals the fundamental work of *Morrison* on angular momentum theory has cardinal meaning [860].

A valuable contribution to laser-crystal physics was made by specialists in crystal growth. New growth methods have been developed, and existing ones modified; important physicochemical problems have been solved; and several new laser crystals of practical interest have been produced. In particular, we should mention the pioneers of laser-crystal growth—*Nassau, Guggenheim, Morris,* and *Van Uitert.* Their work will be referred to frequently in the present monograph. The list of achievements might, of course, be continued on and on, because many scientists are currently engaged in laser-crystal physics and crystal growth, producing results no less important than those discussed above. Some of their work will be referred to in the following sections.

Our brief account of the main steps in the development of activated laser-crystal physics and of its most important results is concluded with the schematic energy-level diagram (without Stark component structure) for Ln^{3+} ions (the most widely used activators) constructed by *Dieke* and *Crosswhite* [46] (see Fig. 1.1). Their notation for the manifolds will be used throughout this book.

A list of activated insulating crystals in which stimulated emission has been excited over the last ten years can be found at the end of Table 5.1.

2. Spectral and Laser Characteristics of Activated Crystals

An impurity ion doped into a crystalline matrix has its own characteristic system of discrete (Stark) energy levels which serves to connect physically various quantum processes occurring in the activated medium with the radiation field. Basic to this connection are the energy transitions between different activator levels, which determine whether electromagnetic energy is absorbed or emitted by the medium. In his famous paper [1] in terms of thermodynamics, *Einstein* postulated the existence of the following elementary processes in a quantum system with discrete states: spontaneous emission, induced or stimulated absorption and emission. The latter two processes are only possible in the case of incident electromagnetic radiation. Since all transitions between energy levels are random, *Einstein* introduced three coefficients that characterize the transition probability per unit time, namely: the probability of spontaneous transition A_{ji}, and the probabilities of induced transitions for absorption $B_{ij} U(v_{ji})$, and for emission $B_{ji} U(v_{ji})$. Here $U(v_{ji})$ is the density of emission energy at the frequency

$$v_{ji} = |E_j - E_i|/h, \tag{2.1}$$

corresponding to the energy difference between the initial and terminal states of a system. In (2.1), E_j, E_i are the energies of the upper and lower levels, respectively, and $h = 6.6256 \times 10^{-27}$ erg·s is Planck's constant.

2.1 Basic Concepts of the Theory of the Absorption and Emission of Light

The basic laws governing the interaction of light with matter can be established by considering a simple quantum system with just two energy levels, the upper E_2 and the lower E_1. At equilibrium, such a system satisfies

$$\frac{N_2}{N_1} = \frac{g_2}{g_1} \exp\left(-\frac{E_2 - E_1}{kT}\right) \tag{2.2}$$

where N_1, N_2 are the numbers (populations) of activators with energies E_1, E_2, respectively; T is the temperature; $k = 1.38 \times 10^{-16}$ erg·K^{-1} is Boltzmann's constant; and g_1, g_2 are the degeneracies, or statistical weights of the levels, i.e., the number of independent quantum states with the same energy level.

When irradiated by light of frequency v_{21}, such a system gets out of thermodynamic equilibrium through the $1 \to 2$ transition, whose probability is $B_{12} U(v_{21})$. We make the simplification $g_1 = g_2$. The return of the system to the initial equilibrium, i.e., the $2 \to 1$ transition, can occur though two processes. Its total probability is thus the sum $A_{21} + B_{21} U(v_{21})$, and the condition for equilibrium of the system is

$$B_{12} U(v_{21}) N_1 = A_{21} N_2 + B_{21} U(v_{21}) N_2, \tag{2.3}$$

or

$$U(v_{21})(B_{12} N_1 - B_{21} N_2) = A_{21} N_2. \tag{2.4}$$

Using (2.1, 2), expression (2.4) takes the form

$$U(v_{21})[B_{12} \exp(hv_{21}/kT) - B_{21}] = A_{21}.$$

Hence, a simple relationship is derived between the spectral flux density and the Einstein coefficients

$$U(v_{21}) = \frac{A_{21}}{B_{12} \exp(hv_{21}/kT) - B_{21}}. \tag{2.5}$$

Comparison of this expression with the classical Planck formula for equilibrium flux density inside a closed cavity,

$$U(v) = \frac{8\pi h v^3}{c^3} \frac{1}{\exp(hv/kT) - 1},$$

readily reveals a universal relation between the Einstein coefficients (when the medium is not a vacuum),

$$B_{12} = B_{21},$$

$$B_{21} = \frac{c^3}{8\pi h v_{21}^3 n^3} A_{21}$$

or generally

$$B_{ij} g_i = B_{ji} g_j, \tag{2.6}$$

$$B_{ji} = \frac{c^3}{8\pi h v_{ji}^3 n^3} A_{ji}, \tag{2.7}$$

where c is the velocity of light in vacuum and n is the refractive index of the medium. The factor before A_{21} represents the number of radiative oscillators (oscillation modes) per unit volume emitting over a single frequency range. In (2.6, 7), there are no functions of particle distribution over energy levels or parameters describing external-field properties and they are not related to temperature. Relations between the coefficients depend only upon the energy gap between the interacting E_j and E_i levels, and upon their statistical weights. If (2.6) and (2.7) were invalid, this would contradict condition (2.3) and, hence, the thermodynamic equilibrium attained between matter and electromagnetic radiation. Thus, these equations are universal because they apply to any quantum system with any energy-level combination and exposed to any external field.

Let us revert to the condition for equilibrium of the system. As was already pointed out, the left-hand side of (2.3) describes absorption, while the right-hand side, consisting of two physically different terms, represents emission. Spontaneous emission does not depend on an external field and is produced by an ensemble of centers with random phases. The emission characteristics are determined solely by the properties of the activated medium. In the case of impurity media, this process appears as luminescence. The induced emission processes represented by the second term on the right of (2.3) are of paramount importance for us, since they represent laser radiation. Unlike luminescence, laser radiation is coherent; its frequency and phase are governed by those of the inducing electromagnetic field.

In considering the main laser properties, it is better to characterize induced radiative transitions by a parameter called the cross section, σ_e, rather than by probability. In order to pass from probability to σ_e, we introduce the notion of radiation intensity $J(v)$ which determines the total number of photons passing through a square centimeter of the surface of a medium per unit time. Then

$$B_{ji}U(v_{ij}) = \sigma_e J(v_{ij}) \tag{2.8}$$

or

$$\sigma_e = B_{ji}U(v_{ij})/J(v_{ij}).$$

If an emission spectrum consists of a single narrow line with the form factor $g(v)$, the parameter σ_e can have various forms. We shall use

$$\sigma_e = B_{ji}\frac{hv_{ij}n}{c}g(v) = A_{ji}\frac{c^2}{8\pi v_{ij}^2 n^2}g(v). \tag{2.9}$$

Hence

$$J(v_{ij}) = U(v_{ij})\frac{c}{hv_{ij}n}\frac{1}{g(v)}. \tag{2.10}$$

For a Lorentz luminescence line of width Δv_{lum}, expression (2.9) takes the form

$$\sigma_e = A_{ji} \frac{c^2}{4\pi^2 v_{ij}^2 n^2 \Delta v_{lum}} = A_{ji} \frac{\lambda_{ij}^2}{4\pi^2 n^2 \Delta v_{lum}}. \qquad (2.11)$$

Here λ_{ij} is the emission wavelength. Expression (2.11) is one of the various forms of the well-known Füchtbauer–Ladenburg formula[1].

Now we shall consider a notion of negative temperature which is often used in quantum electronics to characterize a loss of thermodynamic equilibrium by a quantum system. Let us write (2.2) as

$$T = -\frac{E_2 - E_1}{k \ln (N_2/N_1)}. \qquad (2.12)$$

The natural logarithm of the level population ratio in the denominator is negative under equilibrium conditions because $N_1 > N_2$. Consequently, at thermodynamic equilibrium, the temperature is positive. Radiative induced transitions are possible only when $N_2 > N_1$. According to the above equation, such a system is formally characterized by a negative temperature. To achieve such a state of a system and maintain it for a significant time is one of the problems to be solved in establishing laser generation. It follows from (2.3) that transition of a system from the state with $N_2 > N_1$ to equilibrium is attributed not only to induced transitions. Over the optical range, spontaneous emission usually predominates over induced emission since its weight in the decay channel of the upper level increases directly proportional to the cube of the transition frequency (2.7). That is why the second objective of a laser designer is to make induced transitions prevail over spontaneous transitions.

Having briefly considered the basic properties of quantum transitions, we shall now proceed to the conditions for laser or stimulated-emission generation in an impurity material. Let a parallel monochromatic light beam of frequency v_{21} and intensity J_0 be incident on an activated medium with negative temperature and length l. The interaction of weak incident radiation with matter is described by the expression

$$J = J_0 \exp (-k_v l)$$

or

$$J/J_0 = \exp (-k_v l), \qquad (2.13)$$

---

[1]For a Gaussian luminescence line, (2.11) can be represented as

$$\sigma_e = A_{ji} \frac{\lambda_{ij}^2}{8\pi n^2 \Delta v_{lum}}.$$

where J is the transmitted intensity, and

$$k_v = (k_1 + k_2) + \rho$$

is the total absorption coefficient per unit length of the medium. In the latter expression, the coefficient k_1 characterizing the $1 \rightarrow 2$ transition is proportional to σ_e and N_1. It is always positive because this process involves energy flow from the field to the matter. The coefficient k_2 characterizes the intensity of the $2 \rightarrow 1$ transitions. It is negative because in this case energy flows in the opposite direction. This coefficient is proportional to σ_e and N_2. Finally, the coefficient ρ represents losses in the medium other than those associated with the quantum system itself. These may arise from scattering by foreign particles, bubbles, etc. If $\rho = 0$, the total absorption coefficient will be

$$k_v = \sigma_e N_1 - \sigma_e N_2 = \sigma_e(N_1 - N_2) = -\sigma_e(N_2 - N_1) = -\sigma_e \Delta N. \qquad (2.14)$$

In a negative-temperature system, $\Delta N > 0$, so that $k_v < 0$. This means an increasing intensity of the incident radiation. This is evident if (2.14) is inserted into (2.13): the ratio J/J_0 exceeds unity. The value of ΔN is known as the number of active particles; it characterizes the degree of inversion involved in level transitions of a quantum system. The product $\sigma_e \Delta N = \alpha_v$ is referred to as the gain coefficient. Rewriting expression (2.13) using the new terminology, we obtain

$$J/J_0 = \exp\left[(\alpha - \rho)l\right]_v,$$

$$= \exp\left[(\Delta N \sigma_e - \rho)l\right]_v. \qquad (2.15)$$

In actual media, $\rho \neq 0$. According to (2.15), high gain can be achieved in a system in two ways—by increasing either ΔN or l. The second way is more common in practice, l being substantially increased by multiple passes of a beam through an active medium. In most cases, this involves a Fabry-Perot interferometer formed by two plane-parallel mirrors. As a rule, one mirror is totally reflecting, $R = 100\%$, while the other is partially reflecting. At a sufficiently high ΔN, a once-amplified flux from the totally reflecting mirror may be reamplified so that the medium and the mirror losses will not only be compensated but exceeded by a value comparable to the magnitude of the original flux. If this flux is returned to the medium by the partially reflecting mirror, the amplifier becomes a generator, since its operation no longer depends on external radiation. The condition for exciting such a generator, taking account of the losses on double beam passage through the medium and on the partially reflecting mirror of reflectance R, is

$$J/J_0 = R \exp\left[2(\Delta N \sigma_e - \rho)l\right]_v = 1. \qquad (2.16)$$

We have considered the most common way of achieving predominance of induced transitions over spontaneous ones. Note that in actual laser systems based on impurity media, stimulated emission is usually initiated by spontaneous emission of activator centers rather than by external irradiation. The familiar activated media are characterized by very complex systems of energy levels[2]. Their properties are governed by the type of impurity ion and the properties of the matrix as well as by the degree of their coupling. These properties, in turn, determine an operating scheme and type of laser system, i.e., the conditions for excitation and realization of $\Delta N > 0$, frequency range, generation mode, and temperature range of stimulated emission. Clearly, only a complex detailed investigation of many phenomena based on quantum transitions between energy levels (absorption, luminescence, and stimulated emission) can provide all the necessary information about an activator ion and the matrix into which it has been doped.

We now proceed to the characteristics of activated laser media, which are initially measured in spectroscopic studies and which serve as the basic parameters in a theoretical treatment of laser operation.

2.2 Spectroscopic Characteristics of Activated Crystals

Consider a parallel monochromatic laser beam of frequency v_{21} and density J_0 which is incident on a two-level isotropic medium with positive temperature (2.12) and length l. Taking (2.10) into account, the equilibrium condition (2.3) becomes

$$A_{21}N_2 = J_0 \frac{hv_{21}n}{c}(B_{12}N_1 - B_{21}N_2)g(v).$$

Unlike induced emission, spontaneous emission is omnidirectional and, for flux propagating along l, may prove to be a loss. Disregarding luminescence in the direction of propagation, the flux attenuation in a two-level medium can be represented as

$$-\frac{1}{J_0}\frac{dJ}{dl} = \frac{hv_{21}n}{c}B_{12}N_1\left(1 - \frac{N_2}{N_1}\right)g(v). \tag{2.17}$$

Integrating (2.17) over the whole length, we obtain

$$J = J_0 \exp(-k_v l) = J_0 \exp(-D), \tag{2.18}$$

----

[2]The locations of energy levels in familiar activated laser crystals will be treated in Chapter 4.

which is the formula (2.13) given earlier, where

$$k = D/l = \frac{h\nu_{21}n}{c} B_{12}N_1\left(1 - \frac{N_2}{N_1}\right)g(\nu), \qquad (2.19)$$

or, taking into account (2.7),

$$k_\nu = \frac{\lambda_{21}^2 A_{21} N_1}{8\pi n^2}\left(1 - \frac{N_2}{N_1}\right)g(\nu). \qquad (2.20)$$

This expression represents the so-called reduced absorption coefficient which is usually measured experimentally. As the case $N_1 \gg N_2$ is more frequent in practice, induced emission can be ignored. Then, integrating (2.20) over the entire line gives

$$\int k_\nu d\nu = \frac{\lambda_{21}^2}{8\pi n^2} A_{21}N_1. \qquad (2.21)$$

Expression (2.21) describes an integrated absorption coefficient and is one of the basic equations of absorption spectroscopy. Equation (2.18) contains the product $D = k_\nu l$. This parameter is also used in describing absorption and is referred to as optical density or optical thickness.

Comparison of expressions (2.19), (2.9) shows that the absorption coefficient is proportional to the lower-level population and to the transition cross section, i.e.,

$$k_\nu = \sigma_a(\nu)N_1, \qquad (2.22)$$

or

$$\sigma_a(\nu) = k_\nu/N_1.$$

Hence, knowing the number of particles at the lower level (in case of a large energy gap ΔE between the levels, N_1 is equal to C, the total concentration of activator ions) we can readily find the transition cross section by measuring an absorption coefficient. The above considerations lead to an important corollary. A transition cross section can be determined in two independent ways: from the spectral characteristics of spontaneous emission, and from absorption, i.e., using expressions (2.11), (2.22):

$$\sigma_e(\nu_{21}) = \frac{\lambda_{21}^2 A_{21}}{4\pi^2 n^2 \Delta\nu_{\text{lum}}} = k_\nu/N_1 = \sigma_a(\nu_{12}). \qquad (2.23)$$

Experimentally, the former method is usually employed when the terminal level of the transition concerned is rather high and practically unpopulated (say, in a four-level laser), because, in this case, the absorption method would require high-temperature devices or very long samples.

The coefficients A_{21} and B_{21} related by (2.7) can be written in terms of a transition of electric or magnetic moment. The analogs of these quantities in classical electrodynamics are the amplitudes of time variations in the electric and magnetic moment of a system. Two types of transitions are distinguished: electric-dipole and magnetic-dipole transitions. Impurity ions currently used in existing laser media primarily undergo electric dipole transitions. Here, the probability of spontaneous emission is [61]

$$A_{21} = \frac{64\pi^4 v_{21}^3}{3hc^3} P_0^2 \qquad (2.24)$$

where P_0^2 is the squared amplitude of a matrix element of the dipole moment P. As the theories of quantum and classical oscillators are in good agreement, in place of transition probability (2.24), a specific characteristic can be introduced, the oscillator strength f_{21}

$$f_{21} = \frac{8\pi^2 m v_{21}}{3he^2} P_0^2 \qquad (2.25)$$

where m and e are the electron mass and charge, respectively. By substituting (2.25) into the expression for spontaneous transition probability (2.24), we derive

$$A_{21} = \frac{8\pi^2 e^2 v_{21}^2}{mc^3} f_{21}. \qquad (2.26)$$

We now express A_{21} in terms of the integrated absorption coefficient using (2.21)

$$A_{21} = \frac{1}{N_1} \frac{8\pi v_{21}^2}{c^2} \int k_v dv. \qquad (2.27)$$

From (2.26) and (2.27), we obtain

$$f_{21} = \frac{1}{N_1} \frac{mc}{\pi e^2} \int k_v dv, \qquad (2.28)$$

which is the well-known Kravetz formula for the oscillator strength. In contrast to A_{21}, f_{21} is dimensionless. As a result, by measuring the absorption coefficient k_v for an absorption line as a function of frequency, we can easily determine both the spontaneous transition probability and the oscillator strength. Knowing A_{21}, we can calculate the Einstein coefficient for the induced transition probability.

The probability of spontaneous transition is associated with one of the most important spectroscopic parameters, the lifetime of the excited state, which characterizes the rate of luminescence decay after excitation has been discontinued. It is significant in laser physics because it is used to predict whether a given luminescence medium is suitable for various laser types or not.

We now analyze, within the frame of our two-level model, the decrease of the number of excited particles due to luminescence, and consider the relationship of the luminescence lifetime τ_{lum} of excited state 2 to other quantum parameters. At the instant $t = 0$, corresponding to the time at which excitation of the system was discontinued, let the population of the excited level be N_2^0. Some particles will be removed from level 2 by the spontaneous $2 \rightarrow 1$ transition during the period from t to $t + dt$. The number of the particles removed is

$$-dN_2 = Mdt = A_{21}N_2dt, \tag{2.29}$$

since the number of photons M emitted per unit time equals the product N_2A_{21}. In (2.29), N_2 corresponds to the population of the excited state at time t. From (2.29) we obtain

$$A_{21} = -\frac{1}{N_2}\frac{dN_2}{dt}. \tag{2.30}$$

In the general case, in which more than one luminescence channel originates from the excited state, expression (2.29) should be rewritten as

$$-dN_j = \sum_i A_{ji}N_j dt. \tag{2.31}$$

We introduce the notion of overall probability of spontaneous transitions $\Sigma_i A_{ji}$ the expression for which, according to (2.31), has the form

$$\sum_i A_{ji} = -\frac{1}{N_j}\frac{dN_j}{dt}.$$

The solution of this differential equation is

$$N_j = N_j^0 \exp\left(-\sum_i A_{ji}t\right). \tag{2.32}$$

Hence, it follows that, in case of several spontaneous-emission channels, the population of the excited state decreases exponentially with time, i.e., just as is the case with a single transition.

According to our model of a quantum system, the luminescence intensity is directly proportional to the number of excited particles. Therefore,

$$\frac{I_{\text{lum}}}{I_{\text{lum}}^0} = \frac{N_j}{N_j^0} = \exp\left(-\sum_i A_{ji}t\right). \tag{2.33}$$

At time $t = 1/\sum_i A_{ji}$,

$$\frac{I_{\text{lum}}}{I_{\text{lum}}^0} = \frac{1}{e}. \tag{2.34}$$

The quantity $1/\Sigma_i A_{ji}$ is used to characterize the mean lifetime of a particle at the level j and is usually referred to as the radiative lifetime of the excited state,

$$1/\sum_i A_{ji} = \tau_{\text{rad}}. \tag{2.35}$$

Coming back to a two-level system, whose de-excitation is attributed to luminescence alone, (2.35) is better rewritten

$$1/A_{21} = \tau_{\text{rad}}^{21}. \tag{2.36}$$

In this case, $1/A_{21}$ characterizes the lifetime of state 2 determined by emission on only one transition. However, because in actual laser systems excited states are most often associated with several decay channels, it is rather difficult to determine a specific partial τ_{rad} for a single transition. Experimental determination of decay usually involves oscillographic measurements of the intensity of a signal from a photoreceiver that records variations of luminescence intensity at the frequency of one of the transitions (Fig. 2.1) after an exciting pulse. As the sweep on the oscillograph tube is linear (horizontal, with respect to time; vertical, with respect to input-signal intensity), an exponential is displayed on the screen. Then, the value of the lifetime can be found as $t_0 - t_e$, where t_0 is the reference instant corresponding to the signal intensity I_0, and t_e is the time corresponding to the signal intensity $I_e = I_0/e$.

Besides purely radiative channels of deactivation, nonradiative transitions[3], which tend to reduce τ_{rad} can also occur. Indeed, taking account of

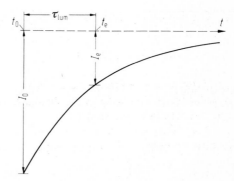

Fig. 2.1. Luminescence decay curve

---

[3]Phonon-assisted transitions, associated with emission of both photons and phonons, are also possible. These transitions are not treated here.

nonradiative transitions characterized by the probabilities d_{ji}, the number of particles removed from the excited state j is, according to (2.31),

$$-dN_j = \sum_i (A_{ji} + d_{ji})N_j dt. \cdot$$

(2.37)

From this expression, the law of variation of N_j due to two decay channels is derived in the same way as (2.32):

$$N_j = N_j^0 \exp\left[-\sum_i (A_{ji} + d_{ji})t\right].$$

(2.38)

This law of excited-state deactivation corresponds to the lifetime

$$\tau_{\text{lum}} = 1/\sum_i (A_{ji} + d_{ji}).$$

(2.39)

Equation (2.39) can be written in the form

$$\tau_{\text{lum}} = \frac{1}{\sum_i A_{ji}} \frac{\sum_i A_{ji}}{\sum_i (A_{ji} + d_{ji})},$$

or, taking (2.35) into account,

$$\tau_{\text{lum}} = \frac{\tau_{\text{rad}}}{1 + \sum_i d_{ji}/\sum_i A_{ji}}.$$

(2.40)

Thus, because of nonradiative channels, the τ_{rad} value is reduced by a factor of $1 + \sum_i d_{ji}/\sum_i A_{ji}$. These deactivation channels should, of course, also reduce the luminescence intensity by the same factor. This process, known as luminescence quenching, is easily characterized by another spectroscopic parameter, the quantum yield of luminescence. This is the ratio of the number of emitted photons to the number of systems that reach the state j, and is expressed by

$$\eta = \frac{1}{1 + \sum_i d_{ji}/\sum_i A_{ji}}.$$

(2.41)

For $d_{ji} = 0$, the quantum yield is equal to unity.

2.3 Homogeneous and Inhomogeneous Line Broadening

One of the fundamental problems in laser physics is the determination of the spectral character of the luminescence line at the frequency of which stimulated emission occurs. The shape of the luminescence line considerably affects the energy parameters of a laser system; in particular, such a significant characteristic as the generation threshold. All the known laser crystals fall into two types with regard to the spectroscopic properties of activator centers: simple materials and mixed systems, or solid crystalline solutions. (These may also include activated glasses and inorganic liquids.) By an activator center we mean a local formation, conventionally isolated from the crystal volume, whose radius is about the linear size of the elementary cell of the crystal, and which consists of a dopant ion and the surrounding ions of the host matrix lying in its immediate vicinity. Center localization implies the idea of isolation of activator ions from each other. It is a rather arbitrary concept, because the statistical distribution of impurities over a crystal, even in the case of very low concentrations, always means that some centers are located in the immediate vicinity of others, i.e., very closely spaced pairs, or more complex associates, occur. As a criterion we take the relative number of such formations. Numerous investigations indicate that the localization condition holds fairly well at activator ion concentrations up to $10^{19}-10^{20}$ cm^{-3}.

An isomorphic distribution of an activator impurity over an ordered-structure material, which places every ion of one type in exactly equivalent positions of the matrix, is an ideal simple single-center crystal. In it, all the elementary centers are identical, and the pertinent spectral line consists of many overlapping lines exactly alike in position and shape. In this case, physical processes that involve the whole ensemble of particles completely coincide with interactions that occur at each particular center. That is why the spectral lines of such crystals are called homogeneously broadened. Inhomogeneous line broadening of activator ions comes about by natural broadening of the corresponding states due to spontaneous radiative and nonradiative transitions. In actual simple crystals, because of microdefects in the matrix itself, activator centers may differ somewhat from one another even with isomorphic impurity distributions. This behavior is especially noticeable at low temperatures, where level broadening due to nonradiative transitions becomes slight. In this case, the observed line is a superposition of several frequency-separated lines that belong to separate centers. At the same time, the theory predicts that at $T\to 0$ the electronic linewidth should approach the natural linewidth, i.e., the quantity

$$\Delta v_{\rm e} \sim \frac{1}{2\pi} A_{ji}.$$

For the impurity ions used in laser crystals, this value ranges from 10^{-2} to 10^{-4} cm^{-1}. Broadening due to crystal defects is known as inhomogeneous broadening. As a rule, the width of such lines depends only slightly on temperature. Their shape is usually gaussian. This situation is responsible for the fact that departure of some frequencies from the mean value is random, and is governed by random distortions of the intercrystalline field. Experimental results on the temperature dependence of spectral line shape for the activator ions used in simple laser crystals have revealed that, at about 150 K and higher temperatures, inhomogeneous broadening is negligible, and the line shape may be considered lorentzian.

These considerations show that the inhomogeneous line width Δv_{inh} of an electronic line may be used at very low (helium) temperatures to estimate internal crystal homogeneity. Some examples of single-center simple crystals with isomorphic and homovalent impurity distributions are $YAlO_3$ and $Y_3Al_5O_{12}$, where Y^{3+} ions are replaced by Ln^{3+} ions. In simple crystals homovalent isomorphism may also create slightly different centers characterized by peculiar spectroscopic parameters. This situation is possible only if some ions to be replaced in the crystal occupy different spatially inequivalent positions. For example, in the sesquioxides Y_2O_3 and Er_2O_3, cations occupy two inequivalent positions with C_2 and C_{3i} symmetries. In such a case, we anticipate, on Ln^{3+} ion distribution, formation of two types of activator centers.

Similar situations occur in heterovalent activation of simple crystals, where the substituting and the substituted ions are of the same size but differ in charge. In this case, center formation is determined by the way in which charge is compensated in order to preserve local electrical neutrality. Heterovalent isomorphism may lead both to single-center systems (for example, $Ca_5(PO_4)_3F$ crystals with Ln^{3+} ions) and to multicenter systems. Typical examples are fluorite crystals (CaF_2) activated by Ln^{3+} ions [63].

The number of possible varieties of simple crystals that differ in activator-center formation is not limited to the examples above. Numerous laser materials include much more complex systems, for example, $LiNbO_3$ crystals with Ln^{3+} ions, in which the isomorphism principle [which dictates the allowable difference in radii of an impurity and a substituted ion (shown in the Table 2.1 on folder at the end of this book) and the desirable coincidence of their electric charges] completely fails.

We now discuss briefly some properties of mixed-crystal systems. Since their structure is disordered, there are always several differing positions that may be taken by activator centers. As a consequence, numerous kinds of centers are formed. In contrast to simple multicenter crystals, where each type of center has very clearly resolved lines in the optical spectra, mixed systems do not exhibit good line resolution even at very low temperatures; on the contrary, their spectra are characterized by broad bands formed by superposition of a great many lines. In other words, spectral lines for activator impurities in mixed crystals are always inhomogeneously broadened. Examples of these materials are fluoride crystals such as $CaF_2 - YF_3$-Nd^{3+} [23], $CaF_2 - SrF_2 - BaF_2 - YF_3 - LaF_3$-$Nd^{3+}$ [64], α-$NaCaCeF_6$-Nd^{3+} [65], $5NaF \cdot 9YF_3$-Nd^{3+} [66] and oxygen-containing systems, such as $NaLa(MoO_4)_2\text{-}Nd^{3+}$ [67, 68], and $CaY_4(SiO_4)_3O: Er^{3+}, Tm^{3+} - Ho^{3+}$ [130, 429].

Thus, in identifying the type of an actual impurity crystal, the temperature behavior of the shape of an activator electronic line (band) can be taken as a reliable criterion. If, at low temperatures, a homogeneous shape changes to an inhomogeneous one, the crystal in question is simple. Such a change does not take place in mixed systems, and the line (band) shape is always inhomogeneous. Sometimes the type of material can be identified from the character of the disorder of its crystalline structure.

Studies of stimulated emission in simple crystals doped with Ln^{3+} ions (in which activator centers are mostly structurally degenerate) were based on vast amounts of spectroscopic information, in contrast with those of multicenter disordered-structure media. For multicenter-crystal laser systems, the situation is that there are results concerning their laser characteristics, but very little information is available on their spectroscopic properties. This may retard efforts to find new materials of this class of active media, which are very promising for solid-state lasers.

In order to gain a better understanding of the physics of the processes that occur in a laser crystal during stimulated emission, it is important to construct a schematic diagram of (Stark) energy levels and to identify the observed induced transitions. There are over a hundred observed induced transitions in multicomponent fluoride systems (see Table 5.1). Some lag in studying the spectroscopic properties of these media appears to be caused by the difficulties involved in analysis of broad, mostly structureless, bands of absorption and luminescence spectra. Realizing the significance of spectroscopic studies of this new class of laser crystals, several laboratories in the Soviet Union have been actively engaged for several years in complex investigations of their laser and spectroscopic properties. Both conventional and new methods of analysis are used in this work [512, 513, 551, 736].

While studying the spectroscopic properties of various laser crystals, the author of the present monograph found that in the case of some doubly disordered fluoride systems, at certain (usually high) concentrations of an activator or of a second component, the spectra become insensitive to concentrations. At the same time, at low concentrations (3–5%) of these components, absorption and luminescence spectra are rather saturated, and individual lines exhibit intensity distributions which depend on concentration [92, 515]. Similar behavior has been observed by the authors of [516] with triply disordered fluoride crystals (gagarinites). These experimental observations led to the operational concept of a quasicenter [92]. An attempt was made to analyze in terms of this concept the band structure of two crystals that are of interest from the viewpoint of laser applications, namely $BaF_2 - LaF_3\text{-}Nd^{3+}$ [92] and $CaF_2 - YF_3\text{-}Er^{3+}$ [106]. By a quasicenter we mean some arbitrary structure that generalizes properties of several activator centers that differ in structure, but have very similar Stark splitting of energy states. The structure and number of separate centers depends on the permissible disorder [517] of a given mixed system and on the concentration of its components, including the activator. If the quasicenter "state" splitting is small compared with their Stark-component broadening, this model yields featureless shapes in absorption and luminescence spectra. At this point, it is worthwhile mentioning two studies using X-ray analysis [518, 519]. According to [518], in the cubic disordered CaF_2-CeF_3 system at high CeF_3 concentrations (about 40 wt.%) complex coordination cation polyhedrons are formed, in which fluorine atoms may occupy positions other than the (1/4, 1/4, 1/4) positions usual in CaF_2 crystals. In CaF_2-CeF_3 crystal (about 40 wt.%) the number of F^- ions which are in the immediate vicinity of the cation varies from 6 to 12. According to [518], the statistically most probable polyhedron is a deformed cube, with one fluorine atom replaced by two shifted a little bit from the positions (1/2, 1/2, 1/2). This symmetry may be considered monoclinic [517]. The same conclusion is reached in [519] which describes neutronographic studies of CaF_2-YF_3 (about 17 wt.%). Finally, it is appropriate to mention detailed investigations of the formation peculiarities of Er^{3+} ion activator centers in CaF_2 crystals, carried out by *Tallant* et al. [548], using up-to-date experimental techniques.

A systematization and analysis of the accumulated knowledge of the spectroscopic properties and temperature behavior of stimulated-emission parameters in activated condensed media has been made, as well as an investigation of these properties in connection with the crystal chemistry of these compounds. This has led to an understanding of the fundamental regularities of crystal-field disorder phenomena of Ln^{3+} activators in insulating laser crystals [861]. One of the regularities discovered in this work — structural-dynamic disorder — is connected with the formation of Ln-polyhedra statistics, the variety of which is dependent, in particular, on the concentration of activator ions. Such a situation occurs in nonstoichiometric laser fluoride crystals with fluorite-type structure. The other — structural-static disorder — is connected with the statistics of filling up the similar crystallographic positions by cations with different valency, it is generally independent of the concentration of Ln^{3+} laser ions in the crystal (for example, in $La_3Ga_5SiO_{14}$, $Ca_3(Nb, Ga)_2Ga_3O_{12}$, $LaSr_2Ga_{11}O_{20}$, and $Ca_2Ga_2SiO_7$).

3. Operating Schemes and Types of Lasers Based on Activated Crystals

3.1 Excitation Threshold for Stationary Single-Mode Emission

Operating schemes for lasers based on activated crystals are divided into three- and four-level systems, the latter having several variants (Fig. 3.1). In one four-level scheme, the terminal level for the induced transition belongs to the ground state of an activator ion and is rather low (Fig. 3.1c); this scheme of operation usually requires deep cooling of the crystal. In another variant of the four-level scheme, the induced transition terminates at the Stark component of one of the excited multiplets (Fig. 3.1b). In lasers based on crystals doped with Cr^{3+}, Ho^{3+}, V^{2+}, Co^{2+}, Ni^{2+}, or Sm^{2+} ions, stimulated emission at certain frequencies also occurs via phonon-assisted transitions, which terminate on "phonon" levels connected with certain vibronic modes of the host crystal (Fig. 3.1d).

These laser operating schemes have the following specific features in common: stimulated emission involves one metastable state of an activator ion

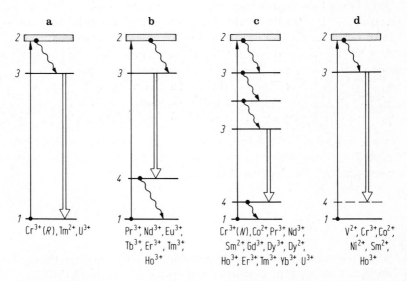

Fig. 3.1. Three- (a) and four- (b–d) level laser operating schemes. Double arrows indicate induced transitions; wavy arrows, nonradiative transitions. The dashed line indicates the position of a "phonon" level

and considerable losses in nonradiative transitions, particularly for pumping channels with nonselective optical excitation. The properties of these schemes have received rather detailed study, both experimental and theoretical, and are outlined in a number of comprehensive reviews and monographs (see, for example, [60, 69–74]. For this reason, the following discussion will be confined to the final equations for threshold excitation energy in the case of single-mode stationary laser action of three- and four-level systems, so that in further descriptions of these lasers the relationship between the most important spectroscopic properties of laser crystals and one of the main parameters of stimulated emission can be made clear.

In the case of a three-level operating scheme, the excitation threshold for lasing at the wavelength of a homogeneously broadened luminescence line is

$$P_{\text{thr}} = \frac{h v_{\text{exc}} V}{2} \frac{b_2}{b_x} \left(\frac{4\pi^2 c n \Delta v_{\text{lum}} \rho'}{\eta \lambda_g^2 \beta_x} + \frac{b_x'}{b_1} \frac{N_0}{\tau_{\text{lum}} \eta_1} \right). \tag{3.1}$$

Similar expressions for the two variants of four-level lasers are

$$P_{\text{thr}} = \frac{h v_{\text{exc}} V}{\eta} \frac{b_3}{b_x} \left[\frac{4\pi^2 c n \Delta v_{\text{lum}} \rho'}{\lambda_g^2 \beta_x} + \frac{b_x'}{b_1} \frac{N_0}{\tau_{\text{lum}}} \exp\left(-\frac{\Delta E_{21}}{kT} \right) \right] \tag{3.2}$$

for a low terminal level, and

$$P_{\text{thr}} = \frac{h v_{\text{exc}} V}{\eta} \frac{b_3}{b_x} \frac{4\pi^2 c n \Delta v_{\text{lum}} \rho'}{\lambda_g^2 \beta_x} \tag{3.3}$$

for a negligible population of the terminal state.

In all the above expressions, N_0 is the total concentration of activator ions for a given type of generating center; V is the crystal volume; $h v_{\text{exc}}$ is the effective energy of the exciting quantum; β_x is the spectral density, or luminescence branching ratio, determining the fraction of photons emitted in a given Stark transition with respect to the total number of photons emitted from a given state to all lower levels; $b_1, b_2, b_3 = \Sigma_i b_i$, where b_i is the Boltzmann factor of the ith Stark component with respect to the lowest component in this state; b_x is the same Boltzmann factor for the initial level of the induced transition x; and ρ' represents the total optical losses, equal to

$$\rho + \frac{1}{l} \ln \frac{1}{\sqrt{R_1 R_2}}$$

where ρ represents the losses in the medium; l is the length of the crystal; R_1, R_2 are reflectances of the mirrors of the optical cavity. In (3.1–3)

$$\eta = \eta_1 \eta_2$$

is the quantum yield of luminescence, where

$$\eta_1 = d_{23}/p_2,$$

$$\eta_2 = A_{31}/p_3$$

for a three-level scheme; and

$$\eta_1 = d_{23}/p_2,$$

$$\eta_2 = A_{34}/p_3$$

for a four-level scheme. Here p_i is the total probability of radiative and nonradiative transitions from the ith state, the d's are nonradiative transition probabilities, and the A's are spontaneous-emission probabilities.

Expressions (3.1–3) assist in the analysis of the influence of different spectroscopic parameters of an activated medium upon the excitation threshold for stimulated emission. Thus, it follows from (3.3) that, in the case of a four-level medium, the laser threshold is independent of τ_{lum}. It is affected mostly by the luminescence linewidth, branching ratio, quantum yield, and Q factor (i.e., $1/\rho'$). In the case of three-level media, such characteristics as $\Delta\nu_{\text{lum}}$, β_x, and Q factor have, on the contrary, a very slight bearing on P_{thr} because the second term of (3.1) is substantial; here the dominant parameter becomes τ_{lum}. The longer the lifetime of the metastable state in a three-level laser crystal, the easier it is to excite stimulated emission.

Expressions for the excitation threshold for the long-pulse mode (pulse duration comparable to that of pumping) and the Q-switched mode (sometimes referred to as the giant-pulse mode, or simply monopulse mode) are given in the review by *Mak*, *Ananev*, and *Ermakov* [70].

3.2 Operating Schemes of Crystal Lasers with One Metastable State

The strongest lines in the luminescence spectra of activated crystals are associated with transitions between the Stark components of a multiplet pair. For example, these transitions for Nd^{3+} ions are $^4F_{3/2} \rightarrow {^4I_{11/2}}$ and for Ho^{3+} ions, $^5I_7 \rightarrow {^5I_8}$. It is on these transitions (arbitrarily called "principal") that, under ordinary conditions (broad-band mirrors in an optical cavity and broad-band pumping), stimulated emission is excited. In most known laser activator ions, the principal transitions are connected with the lowest metastable states in their energy spectra. In order for laser action to occur at other line frequencies due to transitions connected with high metastable states or other multiplet pairs

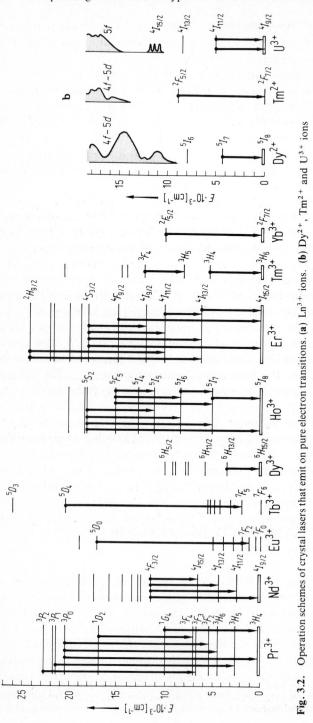

Fig. 3.2. Operation schemes of crystal lasers that emit on pure electron transitions. (**a**) Ln^{3+} ions. (**b**) Dy^{2+}, Tm^{2+} and U^{3+} ions

(such transitions will be hereafter called "additional", and usually have smaller cross sections), the amplifying contour of the system should be specially designed to provide maximum gain at the excited-line frequency.

In stimulated-emission spectroscopy [4, 5], the amplifying contour can be modified in two ways. The first, most common, way is to introduce into the cavity "selective losses," which suppress amplification at the frequency of the principal transition line and provide the highest Q factor values at the frequency of the desired line. Selective mirrors are more often used for this purpose instead of cavities with various dispersion elements. The second way is to introduce "selective gain" into the cavity, so as to increase the effective gain coefficient at the frequency of the excited line to a desired value.

In all the cases above, the operating schemes of crystal lasers involve one metastable state of an activator ion. Such schemes will be called simple, in contrast to other more complex cases. The main disadvantage of the simple schemes for broad-band optical excitation is the great energy losses in the nonradiative transitions that occur between the states above the metastable multiplet or below the terminal laser level.

All known simple operating schemes of crystal lasers that generate on purely electronic transitions are given in Fig. 3.2[1]. For the sake of simplicity, only induced transitions and the multiplets pertaining to the stimulated-emission process are represented in the diagrams. Note that Nd^{3+}, Eu^{3+}, Tb^{3+}, Dy^{3+}, Yb^{3+}, Dy^{2+}, Tm^{2+}, and U^{3+} ions have only one laser metastable state, whereas the Er^{3+} ion has five and Ho^{3+} has four. One metastable state can be connected with several laser channels. Thus, the $^4F_{3/2}$ manifold of the Nd^{3+} ion, $^5S_2$ of the Ho^{3+} ion, and the $^4S_{3/2}$ manifold of the Er^{3+} ion are assigned to four stimulated-emission channels, covering rather a wide range of wavelength. Figure 3.2 indicates that there may be various operating schemes based on crystals activated by trivalent Ln ions.

Information on additional stimulated-emission channels indicating crystal type, laser wavelength range, and optical-cavity type is listed in Table 3.1. To conclude the present section, we note that detailed theoretical treatments of the main generating parameters of simple three- and four-level laser operating schemes, using spectroscopic data, can be found in [60, 72–74].

Table 3.1 see next page.

---

[1] The operating schemes of CaF_2-Sm^{2+}, SrF_2-Sm^{2+}, and Al_2O_3-Cr^{3+} lasers which can emit also on phonon-assisted transitions will be presented later. References to more recent information on new laser transitions can be found at the end of this chapter.

Table 3.1. Induced transitions of laser crystals (*Footnotes at the end of the table*)

Activator ion	Principal and additional transitions	Laser wavelength [μm]	Crystal	Reference
Pr^{3+}	$^3P_0 \rightarrow {}^3H_6$ [a]	≈ 0.6		
	1. $^3P_0 \rightarrow {}^3H_4$	≈ 0.48	LiYF$_4$	[606]
			LaCl$_3$	[116]
			PrCl$_3$	[116]
	2. $^3P_1 \rightarrow {}^3H_5$	≈ 0.53	LaCl$_3$	[116]
			LaBr$_3$	[564]
			PrCl$_3$	[76]
	3. $^3P_2 \rightarrow {}^3F_3$	≈ 0.63	LaBr$_3$	[564]
	4. $^3P_0 \rightarrow {}^3F_2$	≈ 0.65	LaCl$_3$	[116]
			LaBr$_3$	[564]
			PrCl$_3$	[116]
			PrBr$_3$	[76]
	5. $^1D_2 \rightarrow {}^3F_4$	≈ 1.04	CaWO$_4$	[385, 386]
	6. $^1G_4 \rightarrow {}^3H_4$	≈ 1.04	Ca(NbO$_3$)$_2$	[383]
			SrMoO$_4$	[69]
Nd^{3+}	$^4F_{3/2} \rightarrow {}^4I_{11/2}$	≈ 1.06		
	1. $^4F_{3/2} \rightarrow {}^4I_{9/2}$	≈ 0.9	KY(WO$_4$)$_2$	[77]
			CaWO$_4$	[78]
			YAlO$_3$	[79]
			Y$_3$Al$_5$O$_{12}$	[80, 81] [b]
			Lu$_3$Al$_5$O$_{12}$	[191]
	2. $^4F_{3/2} \rightarrow {}^4I_{13/2}$	≈ 1.35	LaF$_3$	[87, 88]
			CeF$_3$	[89]
			5NaF · 9YF$_3$	[88]
			CaF$_2$–YF$_3$	[87, 90] [b]
			Ca$_2$Y$_5$F$_{19}$	[90]
			CaF$_2$–LaF$_3$	[536]
			CaF$_2$–CeF$_3$	[88]
			CaF$_2$–GdF$_3$	[536]
			SrF$_2$–YF$_3$	[88]
			Sr$_2$Y$_5$F$_{19}$	[317]
			SrF$_2$–LaF$_3$	[88]
			SrF$_2$–CeF$_3$	
			SrF$_2$–GdF$_3$	[91]
			SrF$_2$–LuF$_3$	[536]
			CdF$_2$–YF$_3$	[89]
			BaF$_2$–YF$_3$	[487]
			BaF$_2$–LaF$_3$	[89, 92]
			LaF$_3$–SrF$_2$	[93]
			α-NaCaYF$_6$	[89]
			α-NaCaCeF$_6$	[89]
			LiNbO$_3$	[89]
			LiNdP$_4$O$_{12}$	[752]
			NaNdP$_4$O$_{12}$	[752]
			KY(MoO$_4$)$_2$	[93]
			KY(WO$_4$)$_2$	[77, 93, 94]
			KNdP$_4$O$_{12}$	[752]
			KGd(WO$_4$)$_2$	[595]
			CaAl$_4$O$_7$	[88, 95]
			CaSc$_2$O$_4$	[488]
			Ca(NbO$_3$)$_2$	[93]

Table 3.1 (*continued*)

Activator ion	Principal and additional transitions	Laser wavelength [μm]	Crystal	Reference
			$CaWO_4$	[78, 93]
			$SrAl_4O_7$	[88, 96]
			$SrAl_{12}O_{19}$	[489]
			$SrMoO_4$	[97]
			Y_2O_3	[770]
			$YAlO_3$	[93, 98]
			$Y_3Al_5O_{12}$	[99, 100][b]
			Y_2SiO_5	[82]
			$Y_3Sc_2Al_3O_{12}$	[488, 529]
			$Y_3Sc_2Ga_3O_{12}$	[536]
			YVO_4	[88]
			$Y_3Ga_5O_{12}$	[504]
			$La_2Be_2O_5$	[633]
			$Gd_3Sc_2Al_3O_{12}$	[84, 488, 529]
			$Gd_3Ga_5O_2$	[504, 594]
			$LuAlO_3$	[171, 565]
			$Lu_3Al_5O_{12}$	[89]
			$Lu_3Sc_2Al_3O_{12}$	[86, 488]
			$Lu_3Ga_5O_{12}$	[506]
			$PbMoO_4$	[97]
			$Bi_4Si_3O_{12}$	[566]
			$Bi_4Ge_3O_{12}$	[567]
			$LiLa(MoO_4)_2$	[97]
			$LiGd(MoO_4)_2$	[97]
			$NaLa(MoO_4)_2$	[97]
			$NaLa(WO_4)_2$	[97]
			$KLa(MoO_4)_2$	[428]
			$CaLa_4(SiO_4)_3O$	[627]
			$ZrO_2-Y_2O_3$	[90]
			$HfO_2-Y_2O_3$	[90]
			$Ca_5(PO_4)_3F$	[90]
	3. $^4F_{3/2} \rightarrow {}^4I_{15/2}$	≈ 1.85	$Y_3Al_5O_{12}$	[101][c]
Ho^{3+}	$^5I_7 \rightarrow {}^5I_8$	≈ 2.0		
	1. $^5S_2 \rightarrow {}^5I_8$	≈ 0.55	CaF_2	[102]
			$Ba(Y,Yb)_2F_8$	[31]
	2. $^5S_2 \rightarrow {}^5I_7$	≈ 0.75	$LiYF_4$	[568]
	3. $^5F_5 \rightarrow {}^5I_7$	≈ 0.98	$LiYF_4$	[568]
			$LiHoF_4$	[545]
	4. $^5S_2 \rightarrow {}^5I_6$	≈ 1.01	$LiYF_4$	[568]
	5. $^5S_2 \rightarrow {}^5I_5$	≈ 1.4	$LiYF_4$	[568]
	6. $^5F_5 \rightarrow {}^5I_6$	≈ 1.5	$LiHoF_4$	[545]
	7. $^5F_5 \rightarrow {}^5I_5$	≈ 2.36	$LiHoF_4$	[545]
			BaY_2F_8	[532]
	8. $^5I_6 \rightarrow {}^5I_7$	$2.9-3$	$KGd(WO_4)_2$	[641]
			$YAlO_3$	[547, 565]
			$Y_3Al_5O_{12}$	[546, 547, 565]
			$LaNbO_4$	[607]
			$Lu_3Al_5O_{12}$	[546, 547, 565]
Er^{3+}	$^4I_{13/2} \rightarrow {}^4I_{15/2}$	≈ 1.6		
	1. $^2H_{9/2} \rightarrow {}^4I_{13/2}$	≈ 0.56	BaY_2F_8	[103]
	2. $^2H_{9/2} \rightarrow {}^4I_{11/2}$	≈ 0.7	BaY_2F_8	[103]
	3. $^4S_{3/2} \rightarrow {}^4I_{15/2}$	≈ 0.55	BaY_2F_8	[103]

(*Continued*)

Table 3.1 (*continued*)

Activator ion	Principal and additional transitions	Laser wavelength [μm]	Crystal	Reference
	4. $^4S_{3/2} \rightarrow {}^4I_{13/2}$	≈ 0.85	LiYF$_4$	[104]
			CaF$_2$	[105]
			CaF$_2$-YF$_3$	[25, 106]
			CaF$_2$-HoF$_3$-ErF$_3$	[25]
			CaF$_2$-ErF$_3$-TmF$_3$	[25]
			KGd(WO$_4$)$_2$	[641]
			YAlO$_3$	[107]
			Y$_3$Al$_5$O$_{12}$	[546, 547, 565]
			Lu$_3$Al$_5$O$_{12}$	[546, 547, 565]
	5. $^4S_{3/2} \rightarrow {}^4I_{11/2}$	≈ 1.3	CaF$_2$*^{d}	[108]
	6. $^4S_{3/2} \rightarrow {}^4I_{9/2}$	≈ 1.7	LiYF$_4$	[545]
			CaF$_2$*^{d}	[108]
			KGd(WO$_4$)$_2$	[641]
			YAlO$_3$	[109, 110]
			Y$_3$Al$_5$O$_{12}$	[537]
			LuAlO$_3$	[571]
			Lu$_3$Al$_5$O$_{12}$	[271, 546]
	7. $^4F_{9/2} \rightarrow {}^4I_{15/2}$	≈ 0.67	BaY$_2$F$_8$	[31, 103]
			Ba(Y,Yb)$_2$F$_8$	[31, 103]
	8. $^4I_{11/2} \rightarrow {}^4I_{13/2}$	2.7–2.95	CaF$_2$-ErF$_3$	[469]
			CaF$_2$-ErF$_3$-TmF$_3$	[111]
			KY(WO$_4$)$_2$	[595]
			KGd(WO$_4$)$_2$	[641]
			YAlO$_3$	[547, 565]
			Y$_3$Al$_5$O$_{12}$	[490]
			(Y,Er)$_3$Al$_5$O$_{12}$	[490, 572]
			Er$_3$Al$_5$O$_{12}$	[572]
			(Er,Lu)$_3$Al$_5$O$_{12}$	[546, 547, 565]
			Lu$_3$Al$_5$O$_{12}$	[546, 547, 565]
Tm^{3+}	$^3H_4 \rightarrow {}^3H_6$	≈ 1.9		
	1. $^3F_4 \rightarrow {}^3H_5$	≈ 2.35	YAlO$_3$	[112]
			Y$_3$Al$_5$O$_{12}$	[563]

[a] Due to the absence of data in the literature on the luminescence intensity of the Pr^{3+} ion in crystals, $^3P_0 \rightarrow {}^3H_6$ is taken for the present as the principal transition.
[b] Both selective dielectric interference mirrors and dispersion resonators were used.
[c] Only the dispersion resonator was used. In all other cases, selective interference mirrors were used for stimulated-emission excitation of additional transitions.
[d] Crystals CaF$_2$* contained oxygen. During investigations of them in laser experiments, broad-band silver mirrors were used.

3.3 Phonon-Terminated Crystal Lasers

In the preceding section, we considered the operating schemes of insulating crystal lasers whose stimulated emission was attrributed to electronic transitions between the Stark levels of impurity ions. New possibilities for activated crystals were disclosed by *Johnson* et al. [20] in developing the so-called phonon-terminated laser. The operating schemes of this type of laser are shown in Fig. 3.3. In a phonon-terminated laser, an induced transition causes changes not only in the electronic energy of an activator ion but also in the energy of one of the vibrational modes of the host crystal.

Stimulated emission has been observed with a laser based on the MgF_2-Ni^{2+} crystal over a temperature range 20–77 K in the infrared at 1.6223 μm (about 6164 cm^{-1})[20]. The peculiarity of this laser operation is illustrated in Fig. 3.4 by the polarized luminescence spectrum for MgF_2-Ni^{2+} and the crystal-field splitting scheme for the $^3T_2$ and $^3A_2$ manifolds of Ni^{2+} ions, the vibrational level of the MgF_2 crystal being represented by a dashed line. Generally speaking, although a laser transition terminates at the phonon level, stimulated emission in such a system is determined in practice by the properties of the electronic levels of the activator impurity. This is because the spectral intensity and shape of an electronic-vibrational (or vibronic) band, which represents, in this case, the shape of the gain curve, depend on the electron–phonon coupling in the crystal and, hence, on the individual properties of electronic states of the emitting ion. It seems difficult to build up stimulated emission at the frequency of a purely electronic line of the MgF_2-Ni^{2+} crystal, because the $^3A_2$ ground state has such a small splitting ($\Delta E \approx 6$ cm^{-1}).

The narrow lines around 1.53 μm in the luminescence spectra correspond to purely electronic transitions, while the broad bands extending to 1.85 μm are electronic-vibrational. In the π spectrum, the arrow shows the frequency of the observed laser line. The solid arrow in the energy-level scheme represents an induced transition. The levels involved are indicated by the notation: 6506 cm^{-1} $^3T_2 \rightarrow {}^3A_2 + (h\nu_{ph} \approx 340$ cm$^{-1})$.

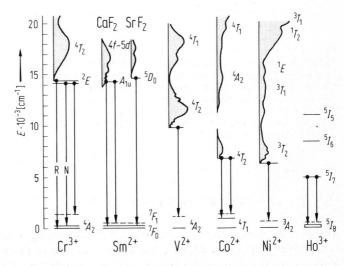

Fig. 3.3. Operation schemes of "phonon-terminated" lasers. The dashed lines indicate the position of "phonon" levels, connected with electron-vibrational transitions

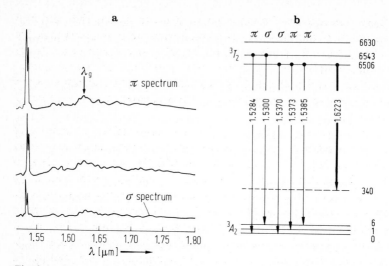

Fig. 3.4. Spectra of polarized luminescence, due to the $^3T_2 \rightarrow {}^3A_2$ transition (**a**) and crystal-field splitting scheme of the $^3T_2$ and $^3A_2$ manifolds of Ni^{2+} ions in the MgF_2 crystal at 20 K (**b**) [20]. The dashed line indicates the position of a "phonon" level. The positions of the levels are given in cm^{-1}, and the wavelengths of transitions between them are given in μm. The induced transition is shown by the thick arrow

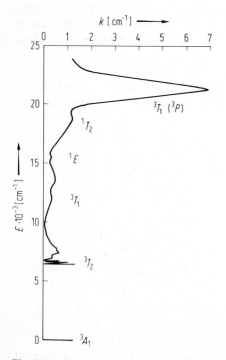

Fig. 3.5. Absorption spectrum of an MgF_2-Ni^{2+} (0.5%, $\boldsymbol{E} \| c$) crystal at 20 K [113]

An attractive feature of phonon-terminated lasers is that they not only provide information about the spectral-emission behavior of impurity ions, but also open up new ways for studying an important phenomenon in solid-state physics, the electron-phonon interaction. The possibility of laser-frequency tuning over a wide wavelength range by means of temperature variations or by introduction of selective losses or gain into the optical cavity may prove to be of practical value.

The original work on MgF_2-Ni^{2+} [20] was followed by a number of communications concerned with the properties of phonon-terminated lasers based on other crystals [113–115, 117, 118], and with the theoretical analysis of these [54, 55, 119]. As this type of laser has been virtually ignored in quantum electronics literature, we shall now briefly outline its main properties and specific features, using the example of an MgF_2-Ni^{2+} crystal laser [113]. The polarized absorption spectrum for a MgF_2-Ni^{2+} crystal at 20 K is given in Fig. 3.5. It consists of broad and rather strong bands (the absorption coefficient scale is on the upper horizontal axis) which allow effective pumping of this crystal by the broad-band excitation sources employed in laser techniques. The spectral bands have been identified from the results of *Tanabe* and *Sugano* [120, 121].

Figure 3.6 shows four luminescence spectra for the MgF_2-Ni^{2+} (1.5%) crystal; they correspond to the $^3T_2 \rightarrow ^3A_2$ transition and were taken at higher temperatures than the π spectrum of Fig. 3.4. It is evident that, as the temperature rises, the stimulated-emission frequency is changed stepwise to different side-band maxima which extend toward the longer-wavelength portion of the spectrum.

Temperature-tuning curves of the phonon-terminated MgF_2-Ni^{2+} crystal laser are plotted in Fig. 3.7. The temperature dependence of threshold generation energy, $E_{thr}(T)$, is evident. The observations show that, for a tempera-

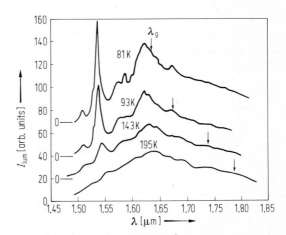

Fig. 3.6. Polarized luminescence spectra of Ni^{2+} ions in an $MgF_2(E \| c)$ crystal, due to the $^3T_2 \rightarrow ^3A_2$ transition [113]. The arrows indicate wavelengths of stimulated emission

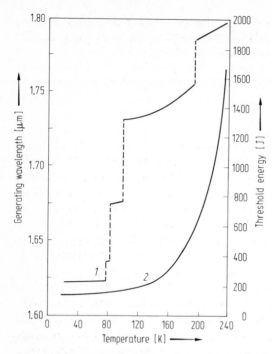

Fig. 3.7. Temperature-tuning curves of stimulated-emission wavelength for MgF_2-Ni^{2+} (1.5%) crystal laser (1) and $E_{thr}(T)$ dependence (2) [113]. The emission threshold energies were measured under pulse excitation by an FT-524 type xenon lamp

ture rise from 20 to 240 K, laser emission occurs in five discrete frequency ranges from 1.62 to 1.8 μm. Analysis has revealed that these ranges correspond to the side-band peaks of the luminescence spectrum ($^3T_2 \rightarrow {}^3A_2$). The widest range is 1.731–1.756 μm (continuous tuning range of about 0.025 μm or about 82 cm^{-1}).

To describe the temperature tuning of the π polarized emission of this laser, we shall use the results of theoretical investigations into the properties of phonon-terminated lasers conducted by *Rebane* and *Sild* [119] and *McCumber* [55]. According to these works, the gain in a two-level approximation is

$$\alpha_\pi(\omega) = \frac{g_2}{g_1}\{N_2 - N_1 \exp[\hbar(\omega - \omega_0)/kT]\}\,\sigma_e^\pi(\omega),$$

where g_1, g_2 are the statistical weights, and N_1, N_2 are the populations of the lower and upper levels of the whole electronic-vibrational system; and $\hbar(\omega - \omega_0)$ is the energy measured from the zero-phonon line[2] (in our case, the

---

[2] It would be more correct to take the Stark structure of the $^3A_2$ manifold into account, but because of its small splitting (about 6 cm^{-1}), the ground state may be assumed in the discussion to be a single level.

energy of the transition assigned to a purely electronic line is $\hbar\omega_0 \approx 6500 \text{ cm}^{-1}$).
A cross section for an induced transition with π polarization $\sigma_e^\pi(\omega)$ can be derived
as usual from the luminescence spectrum:

$$\sigma_e^\pi(\omega) = f_\pi(\omega)\left[\frac{2\pi c}{\omega n_\pi(\omega)}\right]^2 .$$

Here $f_\pi(\omega)$ describes the shape of a luminescence band (a zero-phonon line plus a
side-band structure) and represents the density of photons emitted per unit solid
angle per second over a single frequency range at the frequency ω, and $n_\pi(\omega)$ is
the refractive index of the medium for a given emission polarization. Thus,
knowing the shape of the whole luminescence band and the level populations, we
can define the gain function in a phonon-terminated laser and determine the
spectral range over which laser action will take place. The results of such an
analysis for the MgF_2-Ni^{2+} crystal are plotted in Fig. 3.8. The maxima of the
curves agree with the emission frequencies shown in Fig. 3.6.

A similar but somewhat more detailed analysis was performed by the
authors of [118] who constructed a phonon-terminated laser based on the CaF_2-
Sm^{2+} crystal. A distinctive feature of their experiment, compared to that just
described, is a new method of excitation: they used monochromatic pumping
(ruby laser light) which permitted more-accurate measurement of the crystal
temperature and luminescence line shape. This led to better agreement with the
calculated gain function and spectral composition of stimulated emission. The
results obtained are shown in Fig. 3.9, and the energy-level scheme for the Sm^{2+}
ion in the CaF_2 crystal with identified induced transitions is presented in Fig.
10. Moreover, this method of excitation makes possible much higher (by many
orders of magnitude) spectral flux and, hence, stimulated emission in laser
materials characterized by short lifetime of the metastable state and low

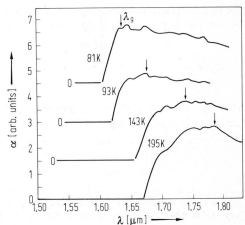

Fig. 3.8. Theoretical dependences of the gain function α on wavelength for an MgF_2-Ni^{2+} crystal laser [113]. The arrows indicate wavelengths of stimulated emission

quantum yield of luminescence. This method of excitation enabled the authors of [118] to achieve a wider range of frequency tuning (about 700 cm^{-1}). A significant advantage of their laser is that it emits in the red region of the spectrum i.e., in a range where sensitive photoreceivers are available.

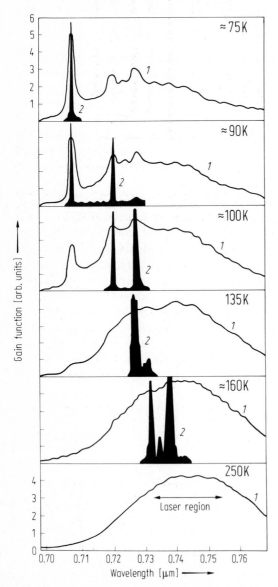

Fig. 3.9. Wavelength dependences of gain function α (1) and stimulated-emission spectral composition (photography of generation spectra) (2) of the phonon-terminated CaF_2-Sm^{2+} crystal laser at different temperatures [118]

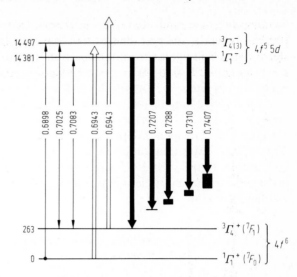

Fig. 3.10.   Crystal-field splitting scheme of Sm^{2+} ions levels in CaF$_2$ crystal, showing levels directly connected with the stimulated emission process [118]. The thick arrows denote induced transitions, and the double arrows are laser-excitation channels. The other notation is as in Fig. 3.4

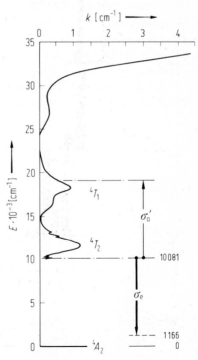

Fig. 3.11.   Absorption spectrum of an MgF$_2$-V^{2+} (0.4%, $l = 0.4$ cm) crystal at 2 K [114]. The dashed line indicates the position of a "phonon" level, and the thick arrow indicates the induced transition

The threshold and frequency of lasers that emit on phonon-assisted transitions can be strongly affected by absorption from the excited metastable state. Figure 3.11 is an absorption spectrum for the MgF_2-V^{2+} (0.4%) crystal taken at about 2 K [114]. With sufficiently powerful pumping and long τ_{lum} (for MgF_2-V^{2+} crystals, it is about 2.3 ms), an undesired absorption ($^4T_2 \rightarrow {}^4T_1$) associated with a broad band at about 18 000 cm^{-1} can occur during stimulated emission. Figure 3.12 gives the luminescence spectrum of V^{2+} ions in the MgF_2 crystal, which corresponds to the $^4T_2 \rightarrow {}^4A_2$ transition, as well as the $\sigma_e(v)$ curve. The induced transition observed at 77 K is represented by an arrow. It corresponds to $\sigma_e = 8 \times 10^{-21}$ cm^2. If the $^4T_2 \rightarrow {}^4T_1$ absorption happens to occur at the frequency of the laser transition, the expression for the gain function will have the form

$$\alpha(\omega) = \frac{g_2}{g_1} \{N_2 - N_1 \exp\left[\hbar(\omega - \omega_0)/kT\right]\} \left[\sigma_e(\omega) - \sigma_a(\omega)\right], \qquad (3.4)$$

where σ_a is the cross section for absorption from the $^4T_2$ state. Unfortunately, the absorption spectrum of the excited MgF_2-V^{2+} crystal is not yet known.

For quantitative considerations of the effect of this absorption channel, the absorption spectrum associated with the $^4T_2 \rightarrow {}^4T_1$ transition can be assumed to be close to that of the $^4A_2 \rightarrow {}^4T_1$ transition. This assumption was made in [114], where the $\sigma_a'(v)$ dependence was calculated and compared with the experimental $\sigma_e(v)$ curve. The results of this comparison are presented in Fig. 3.13. It follows from (3.4) that stimulated emission is possible at a frequency where the σ_a'/σ_e ratio is minimum, i.e., where an absorption channel can be disregarded. If $\sigma_a'(v)$

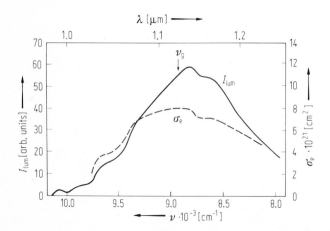

Fig. 3.12. Luminescence spectrum of V^{2+} ions in the MgF_2 crystal, due to the $^4T_2 \rightarrow {}^4A_2$ transition, and the $\sigma_e(v)$ dependence, shown by the dashed line [114]. The pure electron luminescence line lies in the region of 10080 cm^{-1}. The laser frequency is indicated by an arrow

corresponded to the absorption channel concerned, stimulated emission would be expected not at the frequency of about 8915 cm^{-1} (1.1217 μm) but at a higher frequency (about 9040 cm^{-1}) where the ratio σ_a'/σ_e is minimum. Because the MgF$_2$-V^{2+} crystal exhibits laser action at 1.1217 μm, the actual value of σ_a is, probably, less than 3×10^{-21} cm^2. Thus, studies of lasing on phonon-assisted transitions may also provide information about vibronic modes of the host matrix and about spectra of absorption from excited states. Data concerning the luminescence-emission characteristics of phonon-terminated lasers are listed in Table 3.2.

The possibility of CW operation of phonon-terminated lasers is also of interest. If the absorption spectra of the MgF$_2$-Ni^{2+} and MgF$_2$-V^{2+} crystals (see Figs. 3.5, 11) are examined, it is evident that these crystals are very suitable for excitation purposes. In the case of pumping into the low-lying bands, the quantum yield of luminescence is close to unity. The luminescence decay of the metastable states is also very favorable for the desired inversions. With the MgF$_2$-Ni^{2+} crystal, the $\tau_{lum}(^3T_2)$ value is 3.7, 11.5, and 12.8 ms for 295, 77, and 20 K, respectively. These properties allow CW operation of an MgF$_2$-Ni^{2+} crystal laser at a sufficiently low pumping threshold and an output power of about one watt, excitation being accomplished by a tungsten-iodine filament lamp.

In lasers emitting on phonon-assisted transitions, the emitted wavelengths can be varied by introducing dispersion elements into the optical cavity. The results of the experimental realization of this possibility, using an MgF$_2$-Ni^{2+} crystal, are represented by the tuning curves of Fig. 3.14. The use of dispersive optical cavities at a fixed crystal temperature makes it possible, in the course of laser experiments, also to study the spectral composition of the side-band

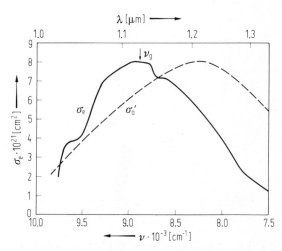

Fig. 3.13. The dependence of $\sigma_e(v)$ due to the $^4T_2 \to {}^4A_2$ transition, and $\sigma_a'(v)$ for the (proposed) $^4T_2 \to {}^4T_1$ transition in an MgF$_2$-V^{2+} crystal [114]

Table 3.2. Phonon-terminated lasers based on activated crystals

Activator ion	Crystal	Induced transition	Laser wavelength [μm]	Temperature [K]	Phonon terminal level [cm^{-1}]	Reference
V^{2+}	MgF$_2$	$^4T_2 \rightarrow {}^4A_2$	1.1217	77	1166	[113, 114]
Co^{2+}	MgF$_2$[a]	$^4T_2 \rightarrow {}^4T_1$	≈ 1.99	77	1780	[113, 117]
			≈ 2.05	77	1930	
	KMgF$_3$	$^4T_2 \rightarrow {}^4T_1$	1.821	77	1420	[113]
	ZnF$_2$	$^4T_2 \rightarrow {}^4T_1$	2.165	77	1895	[113, 117]
Ni^{2+}	MgO	$^3T_2 \rightarrow {}^3A_2$	1.3144	77	398	[113]
	MgF$_2$[b]	$^3T_2 \rightarrow {}^3A_2$	1.623	77	340	[20, 113]
			1.636	77–82	390	
			1.674–1.676	82–100	526–533	
			1.731–1.756	100–192	723–805	
			1.785–1.797	198–240	898–935	
	MnF$_2$	$^3T_2 \rightarrow {}^3A_2$	1.865	20	560	[113]
			1.915	77	580	
			1.922	77	600	
			1.929	85	620	
			1.939	85	650	
Sm^{2+}	CaF$_2$	$5d \rightarrow {}^7F_1$	≈ 0.7207	85–90	506	[118]
			≈ 0.7287	110–130	658	
			≈ 0.7310	155	700	
			≈ 0.745	210	958	
Cr^{3+}	Al$_2$O$_3$	$^2E \rightarrow {}^4A_2$	0.7670	300	—	[122]
Ho^{3+}	BaY$_2$F$_8$	$^5I_7 \rightarrow {}^5I_8$	2.171	295	—	[532]

[a] In [832] continuous tuning from 1.63 to 2.08 μm at 80 K was carried out.
[b] In [832] continuous tuning from 1.61 to 1.74 μm at 80 K was carried out.

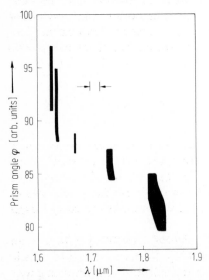

Fig. 3.14. Prism tuning of stimulated emission in an MgF$_2$-Ni^{2+} (1.5%) crystal laser at 85 K [113]. The horizontal width of the darkened areas indicates the region over which emission is observed for a given prism angle

continuum, which is structureless. Comparison of it with the actual phonon spectrum of the host matrix may give valuable information on electron-phonon interactions in a laser medium.

3.4 Sensitized-Crystal Lasers

Several methods have been suggested for improving the efficiency of crystal lasers. Among them the sensitizing method is of particular interest. Along with the main activator ions, ions of another type are also introduced into a crystal, the so-called sensitizers or donors (D). Their function is to absorb excitation energy and transfer it to the main ions, acceptors (A). This process can be represented by the scheme:

$$D_{exc} + A \rightarrow D + A_{exc},$$

where the subscript "exc" represents an excited state of a given ion. Thanks to the additional absorption bands of the sensitizing ions, the effective pumping band will be extended in the case of broad-band optical excitation. Hence, the excitation threshold will decrease, while the emitting-ion intensity increases (sometimes rather greatly).

Dexter [56] seems to have been the first to carry out a detailed theoretical analysis of excitation-energy transfer in activated crystals. He succeeded in establishing some general patterns. In particular, he showed that the most effective energy transfer from donors to acceptors occurs when the donor emission spectrum is in resonance with the acceptor absorption spectrum. How much their spectra overlap is shown by a parameter known as the overlap integral,

$$\int \frac{A_D(E)\sigma_A(E)}{E^4} dE.$$

Here, $\int A_D(E) dE = 1/\tau_{rad}^D$ is the spontaneous-emission probability of the donor ions, where E is the photon energy, τ_{rad}^D is the radiative lifetime measured at low concentrations of the sensitizing ions, and $\int \sigma_A(E) dE = Q_A$ is the integrated cross section, proportional to the area covered by the absorption curve, which is due to transitions between the acceptor-ion states involved in excitation-energy transfer.

The excitation-energy-transfer probability depends, among other factors, on the nature of the interacting transitions [56, 124]. Thus, in the case of electric dipole transitions, which are most typical for the activator and sensitizing ions (Ln^{3+}) used in laser crystals, this probability (for equal statistical weights of the levels involved in the process and broad spectral bands), takes the form [56]:

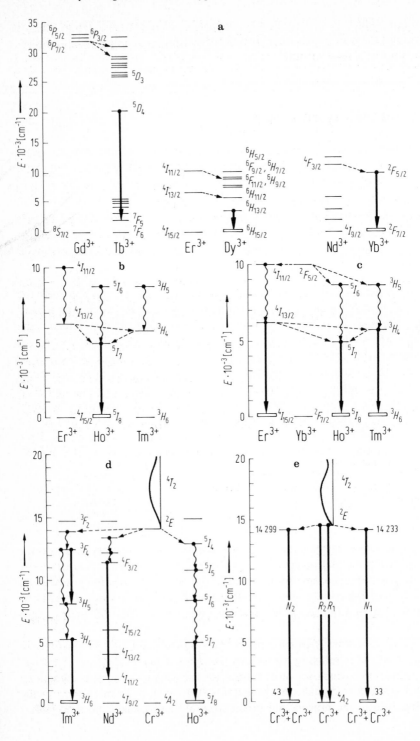

◀ **Fig. 3.15.** Simplified schemes of excitation energy transfer in sensitized laser crystals. (**a**) From Gd^{3+} to Tb^{3+} ions, from Er^{3+} to Dy^{3+} ions and from Nd^{3+} to Yb^{3+} ions. (**b**) From Er^{3+} and Tm^{3+} to Ho^{3+} ions. (**c**) From Yb^{3+} to Er^{3+} ions, from Yb^{3+} and Er^{3+} to Tm^{3+} ions and from Er^{3+}, Yb^{3+}, and Tm^{3+} to Ho^{3+} ions. (**d**) From Cr^{3+} to Tm^{3+}, Nd^{3+}, and Ho^{3+} ions. (**e**) From simple to dimeric activator centers of Cr^{3+} ions in ruby crystals. The thick arrows denote induced transitions and the wavy arrows denote nonradiative transitions

$$P_{ed} = \frac{3\hbar^3 c^4}{4\pi n^4} \frac{1}{R^6} \frac{Q_A}{\tau_{rad}^D} \int \frac{F_A(E)f_D(E)}{E^4}\, dE.$$

Here R is the donor-ion–acceptor-ion distance; $\hbar = h/2\pi$, where h is Planck's constant; $\int F_A(E)dE = 1$, because it is assumed that $\sigma_A(E) = Q_A F_A(E)$; and $\int f_D(E)dE = 1$ under the condition $1/\tau_{rad}^D = A_D f_D(E)$.

More often, in practice, however, the transitions involved in energy transfer are not in resonance; nevertheless, experiments show this process to be rather effective. Investigations have revealed that, in this case, the energy-transfer process also involves phonons of the host crystal.

Possible mechanisms responsible for excitation-energy transfer in laser crystals have been outlined in a number of comprehensive reviews [124, 125]. For this reason, we shall only consider in this book the most frequent sensitization schemes (Fig. 3.15). For the sake of simplicity, broken arrows connect only those multiplets that participate in excitation-energy transfer. Solid arrows represent, as before, induced transitions. The schemes presented show that, besides Ln ions, ions of the iron group (Cr^{3+}) may also serve as sensitizing ions in laser crystals. Sometimes, sensitizing ions belong to the crystalline matrix or are color centers. All known donor–acceptor pairs and more complex combinations that are used in laser insulating crystals are listed in Table 3.3.

Table 3.3.   Sensitized laser crystals

Laser activator ion	Induced transition	Sensitizing ions	Crystal	Energy transfer scheme on figure	Reference
Nd^{3+}	$^4F_{3/2} \to {}^4I_{11/2}$	Cr^{3+}	$YAlO_3$	3.15d	[127]
		Cr^{3+}	$Y_3Al_5O_{12}$	3.15d	[126]
		Ce^{3+}	CeF_3	—	[345]
Tb^{3+}	$^5D_4 \to {}^7F_5$	Gd^{3+}	$LiYF_4$	3.15a	[83]
Dy^{3+}	$^6H_{13/2} \to {}^6H_{15/2}$	Er^{3+}	$Ba(Y,Er)_2F_8$	3.15a	[479]
Ho^{3+}	$^5I_7 \to {}^5I_8$	Cr^{3+}	$Y_5Al_5O_{12}$	3.15d	[128]
		Cr^{3+}	$Ca_5(PO_4)_3F$	3.15d	[129, 130]
		Fe^{3+}	$Y_3Fe_5O_{12}$	—	[573]
		Er^{3+}	$LiYF_4$	3.15b	[131, 132]
		Er^{3+}	CaF_2-ErF_3	3.15b	[133]
		Er^{3+}	α-$NaCaErF_6$	3.15b	[134]
		Er^{3+}	Y_2SiO_5	3.15b	[638]

(*Continued*)

Table 3.3 *(continued)*

Laser activator ion	Induced transition	Sensitizing ions	Crystal	Energy transfer scheme on figure	Reference
		Er^{3+}	$CaMoO_4$	3.15b	[21]
		Er^{3+}	$LaNbO_4$	3.15b	[406]
		Er^{3+}	$(Y,Er)_3Al_5O_{12}$	3.15b	[128, 135]
		Er^{3+}	Er_2O_3	3.15b	[136]
		Er^{3+}	$ErAlO_3$	3.15b	[137]
		Er^{3+}	Er_2SiO_5	3.15b	[271, 638]
		Er^{3+}	$(Er,Lu)AlO_3$	3.15b	[421]
		Er^{3+}	$Yb_3Al_5O_{12}$	3.15b	[422]
		Er^{3+}	$NaLa(MoO_4)_2$	3.15b	[406]
		Er^{3+}	$ZrO_2-Er_2O_3$	3.15b	[431]
		Er^{3+},Tm^{3+}	$Li(Y,Er)F_4$	3.15b	[132]
		Er^{3+},Tm^{3+}	BaY_2F_8	3.15b	[532]
		Er^{3+},Tm^{3+}	$K(Y, Er) (WO_4)_2$	3.15b	[595]
		Er^{3+},Tm^{3+}	Y_2SiO_5	3.15b	[638]
		Er^{3+},Tm^{3+}	YVO_4	3.15b	[538, 660]
		Er^{3+},Tm^{3+}	$Er_3Sc_2Al_3O_{12}$	3.15b	[529]
		Er^{3+},Tm^{3+}	$Y_3Fe_5O_{12}$	3.15b	[138, 139]
		Er^{3+},Tm^{3+}	$(Y,Er)AlO_3$	3.15b	[107, 141]
		Er^{3+},Tm^{3+}	$(Y,Er)_3Al_5O_{12}$	3.15b	[140]
		Er^{3+},Tm^{3+}	$ErVO_4$	3.15b	[660]
		Er^{3+},Tm^{3+}	$Lu_3Al_5O_{12}$	3.15b	[143]
		Er^{3+},Tm^{3+}	$CaY_4(SiO_4)_3O$	3.15b	[130, 429]
		Er^{3+},Tm^{3+}	$SrY_4(SiO_4)_3O$	3.15b	[130]
		Er^{3+},Tm^{3+},Yb^{3+}	$CaF_2-ErF_3-TmF_3-YbF_3$	3.15c	[111, 145]
		Er^{3+},Tm^{3+},Yb^{3+}	$(Y,Er)_3Al_5O_{12}$	3.15c	[140]
		Er^{3+},Tm^{3+},Yb^{3+}	$(Er,Tm,Yb)_3Al_5O_{12}$	3.15c	[86, 486]
		Er^{3+},Tm^{3+},Yb^{3+}	$Lu_3Al_5O_{12}$	3.15c	[143]
Er^{3+}	$^5S_2 \to {}^5I_8$	Yb^{3+}	$Ba(Y,Yb)_2F_8{}^a$	3.19b	[31]
	$^4S_{3/2} \to {}^4I_{13/2}$	Ho^{3+}	$CaF_2-ErF_3-HoF_3$	3.16d	[25]
	$^4I_{13/2} \to {}^4I_{15/2}$	Color centers	CaF_2	—	[146]
		Yb^{3+}	$Y_3Al_5O_{12}$	3.15b	[147]
		Yb^{3+}	$Yb_3Al_5O_{12}$	3.15b	[86, 486]
	$^4F_{9/2} \to {}^4I_{15/2}$	Yb^{3+}	$Ba(Y,Yb)_2F_8{}^a$	3.19a	[31]
Tm^{3+}	$^3F_4 \to {}^3H_5$	Cr^{3+}	$YAlO_3$	—	[112, 525]
		Cr^{3+}	$Y_3Al_5O_{12}$	—	[563]
	$^3H_4 \to {}^3H_6$	Cr^{3+}	$YAlO_3$	3.15b	[525]
		Cr^{3+}	$Y_3Al_5O_{12}$	3.15b	[128]
		Er^{3+}	CaF_2-ErF_3	3.15b	[111, 148]
		Er^{3+}	$\alpha-NaCaErF_6$	3.15b	[134]
		Er^{3+}	$CaMoO_4$	3.15b	[21]
		Er^{3+}	$YAlO_3$	3.15b	[107]
		Er^{3+}	$(Y,Er)AlO_3$	3.15b	[107]
		Er^{3+}	Er_2O_3	3.15b	[149]
		Er^{3+}	$ErAlO_3$	3.15b	[225]
		Er^{3+}	$Er_3Al_5O_{12}$	3.15b	[150]
		Er^{3+}	$(Y,Er)_3Al_5O_{12}$	3.15b	[128, 135]
		Er^{3+}	$(Er,Lu)AlO_3$	3.15b	[421]
		Er^{3+}	$ZrO_2-Er_2O_3$	3.15b	[431]
		Er^{3+},Yb^{3+}	$(Er,Yb)_3Al_5O_{12}$	3.15b	[86, 486]

^{a} Cooperative processes.

(Continued)

Table 3.3   (*continued*)

Laser activator ion	Induced transition	Sensitizing ions	Crystal	Energy transfer scheme on figure	Reference
Yb^{3+}	$^2F_{5/2} \to {}^2F_{7/2}$	Nd^{3+}	CaF_2	3.15a	[152]
		Nd^{3+}	$Y_3Al_5O_{12}$	3.15a	[86]
		Nd^{3+}	$Y_3Ga_5O_{12}$	3.15a	[486]
		Nd^{3+}	$Gd_3Ga_5O_{12}$	3.15a	[486]
		Nd^{3+}	$Lu_3Al_5O_{12}$	3.15a	[86, 483]
		Nd^{3+}	$Lu_3Sc_2Al_3O_{12}$	3.15a	[488]
		Nd^{3+}	$Lu_3Ga_5O_{12}$	3.15a	[86]
		Cr^{3+},Nd^{3+}	$Y_3Al_5O_{12}$	—	[86, 483]
		Cr^{3+},Nd^{3+}	$Lu_3Al_5O_{12}$	—	[86, 483]
Ni^{2+}	$^3T_2 \to {}^3A_2$	Mn^{2+}	MnF_2	—	[113]

3.5 Cascade Operating Schemes for Crystal Lasers

The recent successful spectroscopic and emission studies of laser media carried out by many groups of authors have stimulated the development of new types of laser that reveal further spectroscopic properties of activated crystals. Thus, the author of the present monograph has suggested and realized in practice operating schemes of crystal lasers called "cascade schemes" [25, 106], by analogy with gas lasers operating on several successive transitions. As Fig. 3.16 shows, unlike the simple case, cascade stimulated emission involves several metastable states that belong to one or more activator-ion types.

We distinguish between direct cascade operating schemes involving at least two metastable states of similar ions, and those with an intermediate transfer of excitation energy [25]. The latter case is possible in crystals with two or more impurities. The former type of cascade generation scheme has been realized in practice with the $CaF_2 - YF_3$-Er^{3+} crystal. Figure 3.16a shows that this scheme is characterized by a single common long-lived activator $^4I_{13/2}$ state, induced transitions being initiated and terminated at its Stark components. For Er^{3+} ions, such stimulated emission channels are $^4S_{3/2} \to {}^4I_{13/2} \to {}^4I_{15/2}$. The second cascade scheme has been realized by using CaF_2-HoF_3-ErF_3 and CaF_2-ErF_3-TmF_3 fluoride mixed systems (Figs. 3.16c, d). For an Er^{3+}-Ho^{3+} pair, the emission process consists of stimulated emission on the $^4S_{3/2} \to {}^4I_{13/2}$ transition (Er^{3+}: $\lambda_g \approx 0.84$ μm), excitation-energy transfer between the $^4I_{13/2}$ and $^5I_7$ states by the simultaneous transitions $^4I_{13/2} \to {}^4I_{15/2}$ and $^5I_8 \to {}^5I_7$, and stimulated emission on the $^5I_7 \to {}^5I_8$ transition (Ho^{3+}: $\lambda_g \approx 2.03$ μm). A similar scheme represents a CaF_2-ErF_3-TmF_3 laser (Fig. 3.16c).

Besides the cascade schemes discussed above, many variations are possible in principle, which use the media employed in [25, 106], as well as other fluoride

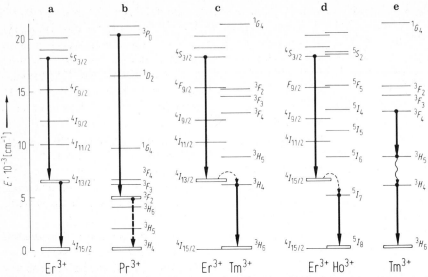

Fig. 3.16. Simplified schemes of cascade laser operation. (**a, b, e**) Direct cascade schemes. (**c, d**) With intermediate excitation-energy transfer. The dashed arrow indicates a proposed laser transition (**b**). Only induced transitions are indicated in these schemes

systems. For instance, the following scheme is readily realizable in practice with a $CaF_2-ErF_3-TmF_3$ crystal laser (see Fig. 3.16c): excitation-energy transfer between the $^3F_4$ and the $^4I_{9/2}$ states by the simultaneous transitions $^3F_4 \rightarrow {}^3H_6(Tm^{3+})$ and $^4I_{15/2} \rightarrow {}^4I_{9/2}(Er^{3+})$; nonradiative relaxation $^4I_{9/2} \rightsquigarrow {}^4I_{11/2}(Er^{3+})$; stimulated emission on the $^4I_{11/2} \rightarrow {}^4I_{13/2}$ transition $(Er^{3+}: \lambda_g \approx 2.69~\mu m^3)$; excitation-energy transfer between the $^4I_{13/2}$ and $^3H_4$ states by the simultaneous transitions $^4I_{13/2} \rightarrow {}^4I_{15/2}$ and $^3H_6 \rightarrow {}^3H_4(Tm^{3+})$; and, finally, stimulated emission on the $^3H_4 \rightarrow {}^3H_6$ transition $(Tm^{3+}: \lambda_g \approx 1.86~\mu m)$.

Recent investigations [546, 547, 525, 565] have shown the practical importance of cascade emission in oxygen-containing crystal lasers in broadening the prospects for further extension of the infrared range, with λ_g above 2 μm. In [525] a cascade $YAlO_3: Cr^{3+}-Tm^{3+}$ laser is described that operates by the direct cascade scheme $^3F_4 \rightarrow {}^3H_5 \rightsquigarrow {}^3H_4 \rightarrow {}^3H_6$ (see Fig. 3.16e). The authors of [546] have discussed the possibility of cascade emission in garnet-structure crystals ($Y_3Al_5O_{12}$ and $Lu_3Al_5O_{12}$) activated by Er^{3+} and Tm^{3+} ions (or by Er^{3+} and Ho^{3+} ions), by a scheme with an intermediate transfer of excitation energy. In addition to the cascade-laser schemes represented in Fig. 3.16, other variants are also possible, including those using $YAlO_3-Ln^{3+}$ crystals.

---

[3]Emission at 2.69 μm due to the $^4I_{11/2} \rightarrow {}^4I_{13/2}$ transition of Er^{3+} ions in $CaF_2-ErF_3-TmF_3$ crystals has been observed [111].

Stimulated emission in a $YAlO_3$-Er^{3+} crystal laser at about 1.663 μm wavelength at 300 K has been observed by *Weber*, *Bass*, and *De Mars* [109]. The emission was assigned, however, to the $^4S_{3/2} \rightarrow {}^4I_{9/2}$ transition rather than to the $^4I_{13/2} \rightarrow {}^4I_{15/2}$ transition, which is most common in many media with Er^{3+} ions (see Table 3.1). The Er^{3+} ion is also known to be an effective sensitizer for such ions as Ho^{3+} and Tm^{3+} (see Table 3.3). Thus, there should be no doubt that the scheme with an intermediate transfer of excitation energy is readily achievable in practice at certain concentrations of Er^{3+} and Ho^{3+} ions in $YAlO_3$ crystals at suitable operating temperatures. A possible transition sequence in such a laser is: stimulated emission on the $^4S_{3/2} \rightarrow {}^4I_{9/2}$ transition (Er^{3+}: $\lambda_g \approx 1.663$ μm); nonradiative relaxation $^4I_{9/2} \rightsquigarrow {}^4I_{11/2} \rightsquigarrow {}^4I_{13/2}$ (see Fig. 3.16d); excitation-energy transfer between the $^4I_{13/2}$ and $^5I_7$ states (Ho^{3+}) by the simultaneous transitions $^4I_{13/2} \rightarrow {}^4I_{15/2}$ and $^5I_8 \rightarrow {}^5I_7$ or by some other channels; and stimulated emission on the $^5I_7 \rightarrow {}^5I_8$ transition (Ho^{3+}: $\lambda_g \approx 2$ μm).

It follows from the considerations above that lasers with cascade operating schemes are expected to have far smaller excitation-energy losses due to nonradiative transitions, because stimulated emission involves levels that spread over wider energy range. Indeed, the overall efficiency of a CaF_2-HoF_3-ErF_3 crystal laser [25] was observed to increase considerably when the emission of the $^4S_{3/2} \rightarrow {}^4I_{13/2}$ transition of Er^{3+} ions was introduced. Here we also turn attention of the readers to the recently made cascade crystal lasers on the base of $Gd_3Ga_5O_{12}$, $Y_3Al_5O_{12}$, and $Lu_3Al_5O_{12}$ crystals with Ho^{3+} ions on the $^5S_2 \rightarrow {}^5I_5 \rightarrow {}^5I_6 \rightarrow {}^5I_8$ transitions [774, 775]; $Y_3Al_5O_{12}$ with Ho^{3+} ions on the $^5I_6 \rightarrow {}^5I_7 \rightarrow {}^5I_8$ transitions [794]; $LiYF_4$ with Er^{3+} ions on the $^4S_{3/2} \rightarrow {}^4I_{11/2} \rightarrow {}^4I_{13/2}$ transitions [776]; and with Ho^{3+} ions on the $^5S_2 \rightarrow {}^5I_5 \rightarrow {}^5I_7$ and $^5S_2 \rightarrow {}^5I_5 \rightarrow {}^5I_6$ transitions [777].

In lasers that operate by the scheme with intermediate transfer of excitation energy, "second" emitting ions (Ho^{3+} and Tm^{3+} ions in Fig. 3.16c, d), which take up the residual excitation energy, serve concurrently as deactivators of the long-lived terminal $^4I_{13/2}$ state of Er^{3+} ions. This new function of Ln^{3+} ions in crystals extends their potential for the development of new types of laser crystals and the improvement the characteristics of the existing ones. Thus, deactivator Ln^{3+} ions in high-concentration crystal lasers that emit on self-saturating transitions [546, 547, 565] considerably improve their output parameters (see Sect. 3.6). Their presence is particularly beneficial in giant-pulse lasers.

We now discuss the requirements for materials to serve as active media for lasers with cascade emission schemes. In [25] these were formulated as follows: activator ions should have several states sufficiently distant from one another and characterized by highly probable spontaneous emission; the crystalline matrix should exhibit natural vibrations so that its active modes in electron-phonon interactions should provide sufficient metastability of these states. The most suitable activator impurities for cascade crystal lasers excited by broad-band emission are Ho^{3+}, Er^{3+}, and Tm^{3+} ions [25, 106]. For narrow-band laser excitation (see Sect 3.15 below), which provides the necessary conditions for stimulated emission from practically any metastable state of Ln^{3+} ions, a list of active media and activator ions can be extended, for example, by including activated crystals with a narrow-impurity active-phonon spectrum, say, trichlorides or tribromides of yttrium, lanthanum or the rare earths. Evidence of this is the observation by *Varsanyi* [76] of stimulated emission with $PrCl_3$-Pr^{3+} and $PrBr_3$-Pr^{3+} crystals on the $^3P_0 \rightarrow {}^3F_2$ transition (see Fig. 3.16b). Failure of the laser output to saturate due to the long lifetime of the $^3F_2$ state suggested to him simultaneously occurring infrared transition between the $^3F_2$ and, say, $^3H_4$ manifolds (indicated by a broken arrow in the figure) in the $PrCl_3$-Pr^{3+} system. Unfortunately, this channel has not been investigated. The requirements for the cascade-laser operation regime of activated crystls with laser excitation (see Sect. 3.16) are here simplified. The possibility of using nanosecond (and even picosecond) pumping pulses broadens both the assortment of Ln^{3+} ions and the list of potential initial laser states. The latter must be characterized only by spontaneous radiative transitions which are strong enough.

3.6 Self-Saturating Laser Transitions

Self-saturating laser transitions are those between the Stark components of two activator-ion states, the lower having a longer lifetime ($\tau_{lum}^{term} \gtrsim \tau_{lum}^{init}$). Stimulated emission on such transitions has been observed in a great many crystal lasers (listed in Table 3.4). Table 3.4 shows that in some laser materials the ratio $\tau_{lum}^{term}/\tau_{lum}^{init} \gg 10$.

One of the difficulties involved in exciting emission on self-saturating transitions lies in the production of sufficient inversions of the operating levels. In actual crystal systems, this problem is aggravated because of a low quantum yield of luminescence from the initial level. Compared to the case discussed previously, we have here an extraordinary spectroscopic situation as regards providing the conditions for achieving and studying stimulated emission. Although lasers of this type are in the early stages of development, the main aspects of their fundamental features are reasonably clear. This is largely due to the work on establishing and improving the parameters of cascade crystal lasers considered above.

If the problem of low quantum yield of luminescence from the initial state can be avoided by proper choice of the excitation regime $\tau_{exc} < \tau_{lum}^{init}$ [546, 569, 590], there may be several solutions of the saturation-effect problem, depending on the crystal type and the given transition. Three methods have been suggested, so far, for reducing this undesirable (saturating) effect; two have already been employed in actual lasers. The first method is to establish conditions for exciting stimulated emission by the direct-cascade scheme [25]. Unfortunately, this method, as well as the other two to be discussed later, suffers from some weaknesses. If the second operating induced channel is directly connected with the ground state, an

Table 3.4. Self-saturation laser transitions of Ln^{3+} ions in crystals observed using broad-band flash-lamp excitation

Ion	Transition	Laser wavelength [μm]	Crystal	Temperature [K]	τ_{lum}^{init} [ms]	τ_{lum}^{term} [ms]	Reference
Ho^{3+}	$^5S_2 \rightarrow {}^5I_7$	≈ 0.75	$LiYF_4$	90	0.05	≈ 10	[568]
	$^5F_5 \rightarrow {}^5I_7$	≈ 0.98	$LiYF_4$	90	0.1	≈ 10	[568]
			$LiHoF_4$	90	0.04	—	[545]
	$^5I_6 \rightarrow {}^5I_7$	2.9–3	$KGd(WO_4)_2$	300	—	—	[641]
			$YAlO_3$	300	0.8–1.0	6.6	[547, 565]
			$Y_3Al_5O_{12}$	300	≈ 0.05	≈ 7.5	[547, 565]
			$LaNbO_4$	300	—	—	[607]
			$Lu_3Al_5O_{12}$	300	≈ 0.05	≈ 7.5	[546, 547, 565]
Er^{3+}	$^2H_{9/2} \rightarrow {}^4I_{13/2}$	≈ 0.56	BaY_2F_8	77	0.025	—	[103]
	$^4S_{3/2} \rightarrow {}^4I_{13/2}$	≈ 0.85	$LiYF_4$	300	0.2	—	[104]
			CaF_2	77	1.0	≈ 15	[105]
			$CaF_2 - YF_3$	77	≈ 0.5	—	[25, 106]
			$CaF_2 - HoF_3 - ErF_3$	77	≈ 0.5	—	[25]
			$CaF_2 - ErF_3 - TmF_3$	77	≈ 0.5	—	[25]
			$YAlO_3$	77, 300	≈ 0.13	≈ 5	[107, 565]
			$Y_3Al_5O_{12}$	77, 300	≈ 0.14	≈ 6.5	[546, 547, 565]
			$Lu_3Al_5O_{12}$	77, 300	≈ 0.13	≈ 6.4	[546, 547, 565]
	$^4I_{11/2} \rightarrow {}^4I_{13/2}$	2.7–2.94	$CaF_2 - ErF_3$	300	8.8	9	[469]
			$CaF_2 - ErF_3 - TmF_3$	300	4.5	15–20	[111]
			$KY(WO_4)_2$	300	—	—	[595]
			$KGd(WO_4)_2$	300	—	—	[641]
			$YAlO_3$	300	0.9	≈ 5	[547, 565]
			$Y_3Al_5O_{12}$	300	0.11	6.5	[490, 572]
			$(Y,Er)_3Al_5O_{12}$	300	0.08	≈ 5	[490, 572]
			$Er_3Al_5O_{12}$	300	0.07	≈ 2	[572]
			$(Er,Lu)_3Al_5O_{12}$	300	0.09	≈ 6	[546, 547, 565]
			$Lu_3Al_5O_{12}$	300	0.11	6.4	[546, 547, 565]

increase of activator concentration at room temperature (which is desired, say, for higher gain in the first emission channel) may stop laser action completely, owing to strong reabsorption. This can be the case for Er^{3+} ions, with generation by the direct cascade $^4I_{11/2} \rightarrow {}^4I_{13/2} \rightarrow {}^4I_{15/2}$ scheme. As for Ho^{3+} ions in $Y_3Al_5O_{12}$ crystal, direct-cascade laser-emission regime by the $^5I_6 \rightarrow {}^5I_7 \rightarrow {}^5I_8$ scheme has been observed [794].

The second method involves a well-known phenomenon, transfer of electronic excitation energy (resonant or nonresonant) between Ln^{3+} ions. This method has proved to be highly effective in improving the generation efficiency of many sensitized-crystal systems (see Sect. 3.4). In the case of lasers that emit on self-saturating transitions, emitting ions function as donors, while acceptor ions serve as deactivator ions (see Fig. 3.17a).

This scheme hardly needs further comment. Deactivator ions were first used to accelerate the decay of residual excitation from the terminal laser state [546, 547]. Here stimulated emission was observed at about $\lambda_g \approx 2.9$ μm on the

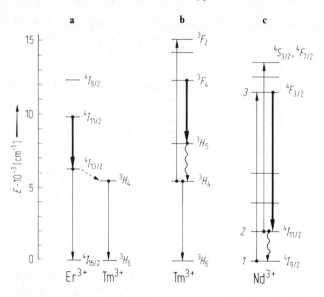

Fig. 3.17. Simplified diagrams of deactivation schemes for some terminal laser states in doped crystals. (a) Energy transfer from Er^{3+} to Tm^{3+} ions. (b) "Absorptional" deactivation with preliminary nonradiative transition. (c) "Absorptional" deactivation with simultaneous nonradiative transition

additional $^4I_{11/2} \rightarrow {}^4I_{13/2}$ transition in a $Lu_3Al_5O_{12}$-Er^{3+} crystal with Tm^{3+}, or Tm^{3+} and Ho^{3+}, as deactivator ions (this laser is discussed further in Sect. 7.2). In certain crystals, deactivator ions may function, in turn, also as emitting ions. As shown in [25], this case has a cascade laser scheme with intermediate excitation-energy transfer. *Johnson* and *Guggenheim* [532] suggest Eu^{3+} or Pr^{3+} ions as deactivators of the $^5I_7$ state, in order to excite stimulated emission of Ho^{3+} ions on the self-saturating $^5I_6 \rightarrow {}^5I_7$ transition in a BaY_2F_8 crystal at $\lambda_g \approx 2.9$ μm.

In analyzing the feasibility of a CW $YAlO_3$-Tm^{3+} crystal laser operating at 300 K, involving an additional $^3F_4 \rightarrow {}^3H_5$ transition, *Caird*, *DeShazer*, and *Nella* [563] expressed apprehension that the long-lived $^3H_4$ state may prevent this regime of emission. Excitation "accumulation" at levels of the $^3H_4$ state due to the large energy gap of about 2000 cm^{-1} between the $^3H_5$ and $^3H_4$ manifolds (see Fig. 3.17b), may seriously deteriorate "four-level" conditions of generation on the $^3F_4 \rightarrow {}^3H_5$ transition ($\lambda_g \approx 2.35$ μm). These authors point to a possible effect of indirect saturation of the terminal operating $^3H_5$ state by Boltzmann repopulation via the next-lower $^3H_4$ state. To avoid this effect, which is detrimental to CW generation, they suggest that the long-lived $^3H_4$ state be strongly deactivated by another external selective emission that provides absorption transitions from its Stark levels in the upward direction. In particular, for a $YAlO_3$-Tm^{3+} crystal, they propose neodymium-laser emission at $\lambda_g \approx 1.06 \mu m$ (about 9435 cm^{-1}). Here deactivation of the long-lived $^3H_4$ state by absorption would be accomplished through the $^3H_4 \rightarrow {}^3F_4$ transition (see Fig. 3.17b). The authors of [563] also discuss the first two methods for deactivating the $^3H_4$ state, namely, use of crystals with deactivator ions (Pr^{3+}, Eu^{3+}, and Tb^{3+}) or laser operation on two successive $^3F_4 \rightarrow {}^3H_5 \rightsquigarrow {}^3H_4 \rightarrow {}^3H_6$ transitions. The latter has already been realized in a pulsed laser [525].

To conclude this section, we mention another useful application of this method of active absorption de-activation of the lower operating state to improve the characteristics of self-activated neodymium laser crystals. At high concentration of Nd^{3+} ions and large cross section for the $^4F_{3/2} \leftrightarrow {}^4I_{11/2}$ transition, the generation efficiency may be adversely affected by reabsorption (see Fig. 3.17c) due to a high absorption coefficient at the frequency ν_{23}

$$k(\nu_{23}) = \sigma_{e.23} N_0 \exp(-\Delta E_{21}/kT).$$

In neodymium self-activated crystals $\sigma_{e.23} \approx 2 \times 10^{-19} cm^2$ for a concentration of activator ions $N_0 \approx 3.5 \times 10^{21} cm^{-3}$ (see Table 7.2). At 300 K this would give $k(\nu_{23}) \approx 7 \times 10^{-2} cm^{-1}$. Such a high absorption coefficient is, of course, undesirable, particularly for the CW regime. Currently, to pump neodymium-doped self-activated laser crystals, semiconductor light-emitting diodes are used [580, 581], pumping being effected into the absorption band corresponding to the $^4I_{9/2} \rightarrow {}^4F_{3/2}$ transition. The simplified scheme of Fig. 3.17c shows that the $^4I_{11/2} \rightarrow {}^2S_{3/2}, {}^4F_{7/2}$ transitions are resonant to this transition. Thus, laser emission at 0.87 μm performs two functions: simultaneously exciting the crystals and deactivating their lower operating $^4I_{11/2}$ state due to the absorption $(^4I_{11/2} \rightarrow {}^4S_{3/2}, {}^4F_{7/2})$. This can also, of course, be applied to improve parameters of neodymium lasers with other active media.

3.7 Infrared-Pumped Visible Crystal Lasers

Every year, a variety of cooperative phenomena is discovered and investigated in activated systems. These phenomena are also of importance for crystal-laser physics. For example, simple cooperative processes, such as migration in mixed multicenter media and excitation-energy transfer in sensitized crystals, have yielded much-higher laser efficiencies. Some cooperative phenomena open up ways of developing lasers based on new excitation principles. These will be treated briefly in the present section.

In studying cumulative processes in activated systems, *Feofilov* and *Ovsyankin* [153–155] observed a phenomenon which they called "cooperative luminescence sensitization". The nature of this phenomenon is clearly seen in Fig. 3.18, which is the scheme of visible luminescence of Tm^{3+} ions in a CaF_2-YbF_3

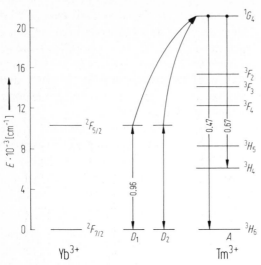

Fig. 3.18. Scheme of cooperative sensitization of Tm^{3+} ion luminescence by Yb^{3+} ions under infrared excitation [155]

crystal excited by the infrared absorption of Yb^{3+} ions $(^2F_{7/2} \to {}^2F_{5/2})$. The process of cooperative sensitized luminescence proceeds by the scheme [155]

$$D + D + A + 2h\nu_D \to D_{exc} + D_{exc} + A \to D + D + A_{exc}$$
$$\to D + D + A + h\nu_A,$$

i.e., the Yb^{3+} ions (D), upon absorption of photons with a wavelength of about 0.96 μm $(h\nu_D)$, are excited to the energy of the $^2F_{5/2}$ level. Subsequent interaction of two excited Yb^{3+} ions (D_{exc}) leads to energy accumulation on one of the Yb^{3+} ions, followed by its transfer to the Tm^{3+} ion. As a result, the Tm^{3+} ion (A_{exc}) is excited to the energy of the $^1G_4$ state (about 21275 cm^{-1}) and exhibits luminescence at 0.47 and 0.67 μm $(h\nu_A)$ when it returns to the ground state.

Higher-lying states of Ln^{3+} ions can be excited by another mechanism. Thus, *Auzel* [466] proposes successive energy excitation: the first infrared quantum brings a system into some intermediate metastable state, from which, upon absorption of a second quantum, it goes to the upper level. Some aspects of the nature of up-conversion of infrared emission in activated crystals have also been analyzed in [498, 582, 583]. These investigations were probably very helpful to *Johnson* and *Guggenheim*, who were the first to develop lasers based on $Ba(Y,Yb)_2F_8$ crystals doped with Er^{3+} and Ho^{3+} ions operating in the visible as a result of infrared excitation processes [31]. Operating schemes of these lasers are given in Fig. 3.19. Visible emission was excited by infrared radiation from a xenon flash lamp. For this purpose, a visible filter was placed between the cryostat containing the laser element and the lamp. In the case of Er^{3+} ions, the filter passed emission at $\lambda > 0.67$ μm, and in the case of Ho^{3+}

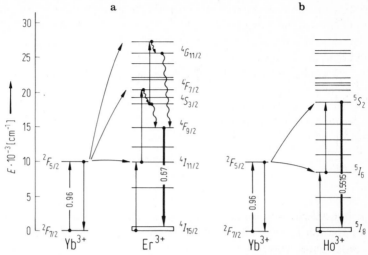

Fig. 3.19. Infrared-to-visible conversion by $Yb^{3+}-Er^{3+}$ and $Yb^{3+}-Ho^{3+}$ ions in $Ba(Y,Yb)_2F_8$ crystals [31]. (**a**) With Er^{3+} ions, (**b**) with Ho^{3+} ions. Generation in the visible spectral region is due to cooperative sensitization processes. Notation is the same as in Fig. 3.15

ions, at $\lambda > 0.61$ μm. In the former case (Fig. 3.19a), the excitation threshold for generation at 0.67 μm ($^4F_{9/2} \rightarrow {}^4I_{15/2}$ transition) was observed to be 195 J; in the latter case (Fig. 3.19b), 635 J ($\lambda_g = 0.5515$ μm, $^5S_2 \rightarrow {}^5I_8$ transition). Unfortunately, no concrete excitation mechanism for the high-energy states was established in [31].

To conclude this section, we note that excitation of luminescence in activated crystals by cumulative processes may turn out to be very effective also in the case of ultraviolet lasers. The use of the high-power infrared emission of semiconductor lasers is most attractive in this case.

3.8 Crystal Lasers with Self-Doubling Frequency

Nowadays, there are several physical devices that use stimulated emission with various nonlinear optical dispersive substances: generators of optical harmonics (frequency multiplication), devices for modulation and deflection of laser emission, parametric generators, etc. Information has lately become available that permits development of crystal lasers with frequency self-doubling, i.e., quantum generators in which the active medium itself serves as a nonlinear element. The first laser of this type was made by *Johnson* and *Ballman* [156] on the base of the ferroelectric $LiNbO_3$ crystal activated by Tm^{3+} ions. Its fundamental laser emission, corresponding to the $^3H_4 \rightarrow {}^3H_6$ transition at $\lambda_g = 1.8532$ μm, is converted by the $LiNbO_3$ crystal itself into the second harmonic at $\lambda_d = 0.9266$ μm. It is not our purpose here to consider in detail nonlinear optical phenomena and their specific features that result in the generation of optical harmonics. These questions are treated elsewhere [157–160]. Keeping to our topic of lasers, we shall discuss only those aspects necessary for understanding the operation of lasers with frequency self-doubling.

In nonlinear crystals exposed to laser-produced electromagnetic fields (particularly, to the very strong fields of giant laser pulses), nonlinear effects may occur and generate harmonics. These effects can be phenomenologically described in terms of nonlinear susceptibility tensors derived by expanding the susceptibility tensor in terms of the field strength:

$$\kappa(E) = \kappa^0 + \chi_k E_k + \chi_{kl} E_k E_l, \tag{3.5}$$

where κ^0 is the linear susceptibility tensor; χ_k is the nonlinear susceptibility tensor responsible for second-harmonic generation; χ_{kl} is the same tensor responsible for the third harmonic; and E_k, E_l are the field-strength components of the fundamental wave.

Because of general regularities, crystals with a center of symmetry have zero coefficients of the χ_k tensor [161] and, hence, they are unsuitable for second-harmonic generation. For this purpose crystals with $\chi_k \neq 0$ must be used; for example, piezoelectric crystals such as quartz, KH_2PO_4 (KDP), $LiNbO_3$ [160, 161].

The intensity of second-harmonic generation is characterized by a conversion ratio which can be expressed as χ_k / κ^0. It follows from (3.5) that the efficiency of conversion of the fundamental-wave energy into a second harmonic depends both upon the χ_k value and upon the field strength E of the fundamental wave. High field intensities are known to involve considerable difficulties associated with self-breakdown and destruction of crystals. Therefore, crystals with a high conversion ratio are most desirable materials. Investigations have revealed that by proper choice of propagation directions of the fundamental and the second harmonic waves with respect to the optic axes in nonlinear crystals we can, under certain conditions, increase the conversion ratio. It is known from theory [157, 158] that

$$I(2\omega) \approx a_s \chi_k^2 l^2 I^2(\omega) \frac{\sin^2(\Delta l/2)}{(\Delta l/2)^2}. \tag{3.6}$$

Here $I(\omega)$ and $I(2\omega)$ are the intensities of the fundamental and the second harmonics, respectively; $\Delta = \omega(n_\omega - n_{2\omega})/c$ is the wave mismatch due to dispersion; n_ω and $n_{2\omega}$ are the refractive indices of the medium at the frequencies ω and 2ω, respectively; c is the velocity of light; l is the crystal length; and a_s is a proportionality coefficient that depends on the crystal symmetry. At $\Delta = 0$, the intensity $I(2\omega)$ is proportional to l^2. In piezoelectric crystals there are certain directions in which n_ω and $n_{2\omega}$ are equal (i.e., the phase velocities of electromagnetic-wave propagation at frequencies of the fundamental and the second harmonic are equal). It is known that, in a uniaxial crystal, the phase velocities of two monochromatic waves of the same frequency may be different because of polarization. Thus, the phase velocity of an extraordinary wave depends on its direction of propagation through the crystal.

Figure 3.20a shows a slab of KDP crystal cut so that the major optical axis of the crystal is oriented along z. The letter O stands for a point source of light. In the case of monochromatic emission at the frequency ω, two waves will propagate through the crystal. The front of an ordinary wave is spherical, i.e., phase velocities are equal in all directions. The front of an extraordinary wave is the surface of an ellipsoid of revolution. Figure 3.20 also shows wave surfaces from a monochromatic source located at the same point O but with frequency 2ω. Along the direction θ_0, $n_o(\omega) = n_e(2\omega)$ and, hence, the phase velocities are equal: $v_{ph}^o(\omega) = v_{ph}^e(2\omega)$.

Expression (3.6), which is the condition for synchronism, shows that for frequency-doubling experiments, a slab of single crystal should be cut as shown in Fig. 3.20b. In this case, the conditions for converting the fundamental-wave energy into a second harmonic are most favorable. If this direction is chosen at random, the second-harmonic intensity will depend on the crystal length and may prove to be zero at the output. A laser with self-doubling frequency operates in exactly the same way as a conventional laser does, except that it has further losses due to energy transfer from the fundamental wave to a harmonic. It is desirable, therefore, to grow crystals for this purpose with orientations such that the direction of synchronism coincides with the geometric axis.

As was pointed out before, the first laser with frequency doubling was based on an optically negative ($n_e < n_o$) ferroelectric crystal of $LiNbO_3$-Tm^{3+} (space group C_{3v}^6–$R3c$) [156]. In this crystal, generation is possible only at low temperatures, because it proceeds by a four-level operating scheme with a low-lying terminal level (5667 cm^{-1} $^3H_4 \rightarrow {}^3H_6$ 271 cm^{-1} transition). A second harmonic in this crystal is possible only for the emission with the σ polarization ($E \perp c$) and is associated with an ordinary wave. Figure 3.21 shows luminescence

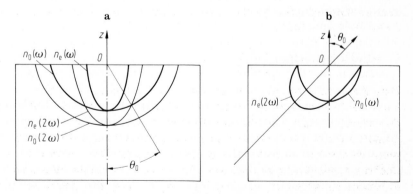

Fig. 3.20. Wave surfaces in a KDP crystal. (**a**) Along the optical axis, (**b**) along the direction of synchronism

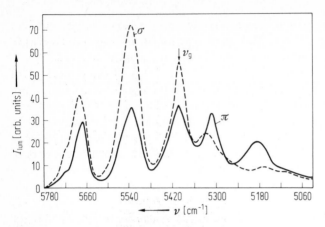

Fig. 3.21. Polarized luminescence spectra of Tm^{3+} ions in a $LiNbO_3$ crystal, due to the $^3H_4 \rightarrow {}^3H_6$ transition at 77 K [156]. The arrow indicates the wavelength of stimulated emission

spectra for the $LiNbO_3$-Tm^{3+} crystal corresponding to the $^3H_4 \rightarrow {}^3H_6$ transition and taken at 77 K for two polarizations. An arrow represents the line of observed stimulated emission. As is seen, it belongs to the σ spectrum, so the authors of [156] were able to achieve a second harmonic simultaneously.

Of most interest are frequency-doubling lasers that use a $LiNbO_3$-Nd^{3+} crystal. First, they can be operated at room [34, 162, 163] and even higher temperatures; second, the emission, attributed to the $^4F_{3/2} \rightarrow {}^4I_{11/2}$ transition, is at about 1.08 μm and, upon conversion to a second harmonic, corresponds to the green region of the visible spectrum. The first $LiNbO_3$-Nd^{3+} laser with a low excitation threshold at 300 K was developed by the authors of [162]. As is shown by Fig. 3.22, its emission corresponded to the π polarization ($\boldsymbol{E} \| c$) and is associated with the 11254 cm^{-1} $^4F_{3/2} \rightarrow {}^4I_{11/2}$ 2034 cm^{-1} transition (A line at $\lambda_g = 1.0846$ μm). In [163] use was made of a $LiNbO_3$-Nd^{3+} crystal with another orientation. It provided stimulated emission at the wavelength of the B line at $\lambda_g = 1.0933$ μm, assigned to the 11254 cm^{-1} $^4F_{3/2} \rightarrow {}^4I_{11/2}$ 2107 cm^{-1} transition. Emission in this line corresponds to an ordinary wave, which is a necessary condition for second-harmonic generation.

In recent years, much progress has been made in producing $LiNbO_3$-Nd^{3+} crystals of high optical quality [584] and with high activator concentrations [530]. As a result, the characteristics of lasers based on these crystals are greatly improved. In order to increase the distribution ratio of Nd^{3+} ions, the authors of [530] took advantage of *Nassau*'s idea [588] of simultaneously coactivating $LiNbO_3$ crystals with Nd^{3+} and Mg^{2+} ions. Because of the relatively good spectral-generation properties of $LiNbO_3$-Nd^{3+} crystals, *Kaminow* and *Stulz* [584] estimated the possibility of using them for a thin-film wave-guide laser.

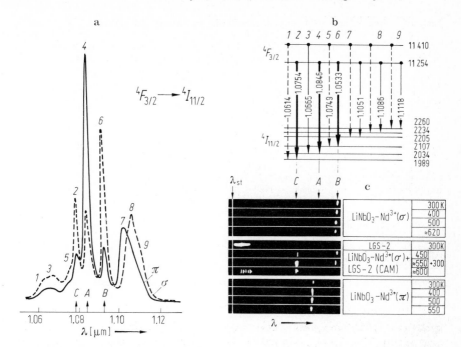

Fig. 3.22. Polarized luminescence spectra of Nd^{3+} ions in a $LiNbO_3$ crystal $(^4F_{3/2} \rightarrow {}^4I_{11/2}$ transition) at 300 K (**a**) and the crystal-field splitting scheme of the $^4F_{3/2}$ and $^4I_{11/2}$ manifolds (**b**) as well as high-temperature generation spectra obtained with conventional and CAM lasers (**c**) [5]. The level positions are given in cm^{-1}, and the wavelength of transitions between them are given in μm. The thick arrows denote induced transitions. Lines in the spectra and corresponding transitions in the energy scheme are denoted by the same numeration. The arrow indicates the wavelength of a standardizing line, $\lambda_{st} = 1.0651 \ \mu m$

Some spectroscopic and emission properties of laser-activated crystals without a center of symmetry, suitable for frequency-doubling lasers, are listed in Table 3.5 [164, 165] and Table 3.6.

Recent theoretical studies by *Dmitriev* and *Zenkin* [586, 587] have shown that the process of second-harmonic generation in nonlinear activated crystals has some specific features. Since their results relate directly to the problem considered in this section, it is interesting to discuss them briefly here. The linear susceptibility tensor κ^0 used in expression (3.5) is written [586] in the following form

$$\kappa^0 = \kappa_1^0 - i\left(\kappa_2^0 - \frac{\kappa_3^0}{1 + \alpha_{ij}EE^*}\right). \tag{3.7}$$

In (3.7), the starred quantity is the complex conjugate of the field strength of the fundamental wave; κ_1^0 is the real part of the κ^0 tensor; in its imaginary part, $\kappa_2^0 \sim \rho$ characterizes nonresonance losses (see Sect. 3.1), while κ_3^0 is the initial

Table 3.5. Spectroscopic and laser properties of activated ferroelectric crystals

Crystal	Space group	Activator ion	Temperature [K]	Laser wavelength [μm]	Laser rod orientation	Polarization of emission	Reference
LiNbO$_3$	C_{3v}^6–$R3c$	Nd^{3+}	300	1.0846 $c \perp F$		π	[162, 530, 584]
			300	1.0933 $c \parallel F$		σ	[163, 530]
			300	1.3745 $c \perp F$		—	[89]
			300	1.3870 $c \parallel F$		—	[89, 530]
		Ho^{3+}	77	2.0786 $c \parallel F$		σ	[156]
		Tm^{3+}	77	1.8532 $< cF \approx 43°\,\sigma^a$			[156]
Ca$_{0.25}$Ba$_{0.75}$(NbO$_3$)$_2$	D_{4h}^{14}–$P4_2/mnm$	Nd^{3+}	300	1.062 —		—	[156]
Ba$_2$NaNb$_5$O$_{15}$	C_{2v}^{11}–$Cmm2$	Nd^{3+}	300	1.0613 $c \parallel F$		σ	[585]
Gd$_2$(MoO$_4$)$_3$	C_{2v}^8–$Pba2$	Nd^{3+}	300	1.0701 $c \parallel F$		—	[164, 166]

[a] Second harmonic obtained with $\lambda = 0.9266\ \mu$m.

Table 3.6. Acentric activated laser crystals

Crystal	Space group	Activator ion	Laser wavelength [μm]	Laser rod orientation	Polarization of emission	Temperature [K]	Reference
KNdP$_4$O$_{12}$	C_2^2–$P2_1$	Nd^{3+}	≈ 1.051	$[101]\parallel F$	$[010]\parallel \mathbf{E}$	300	[579, 581]
			≈ 1.32	—	—	300	[752]
NdAl$_3$(BO$_3$)$_4$	D_3^7–$R32$	Nd^{3+}	≈ 1.065	$c \perp F$	$c \perp \mathbf{E}$	300	[579, 581, 589]

(with no field at the frequency ω) gain in a nonlinear activated crystal, and α_{ij} is the spectroscopic nonlinearity parameter (see expression (1.34) on p. 22 of Ref. [60]) which determines saturation of the gain coefficient due to excitation redistribution over activator operating levels for high-power emission.

In [586, 587], two regimes of interaction between the fundamental ω frequency field and nonlinear activated crystals are analyzed. In the first, the "amplification" regime, a wave of frequency ω is passed through the excited crystal (by any optical-pumping method) where it is amplified and concurrently converted to a harmonic at frequency 2ω. The second "generation" regime corresponds to the operation of a laser with frequency self-doubling, discussed earlier, where an excited nonlinear activated crystal inside a cavity interferometer (Fabry–Perot, for example) simultaneously serves as a generator and a converter. Following [586, 587] we shall first consider the main peculiarities of the "amplification" regime. In the case of complete synchronism ($n_\omega = n_{2\omega}$ or $\Delta = 0$) and with no nonresonance losses ($\kappa_2^0 = 0$), the field amplitudes for the fundamental ω and the second-harmonic 2ω frequencies of an excited nonlinear laser crystal will increase. This behavior is clearly shown in Fig. 3.23, for reduced field amplitudes. Analysis of the "amplification" regime of second-harmonic generation [586] has indicated that at some (conventionally called threshold) length of a nonlinear crystal $l = l_{\text{thr}}$, where

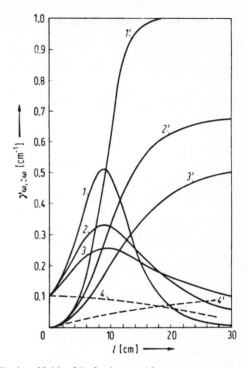

Fig. 3.23. Dependences of the reduced amplitudes of fields of the fundamental frequency γ_ω (curves 1–4) and of the harmonic $\gamma_{2\omega}$ (curves 1'–4') upon the length l, for a nonabsorbing ($\rho = 0$) and nondispersive ($\varDelta = 0$) nonlinear crystal [586]: $K_0 = 0$ (curves 4 and 4') and $K_0 = 0.5$ cm^{-1} (curves 1–3, 1'–3'), $\gamma_{2\omega}(0) = 0$, $\gamma_\omega(0) = 0.1$ cm^{-1}. The parameter for curves 1–3, 1'–3' is the normalized saturation parameter Ξ, which is equal to 0, for curves 1 and 1'; 7.5 cm^2, for curves 2 and 2'; and 25 cm^2, for curves 3 and 3'

$$l_{\mathrm{thr}} = -\frac{2}{\zeta_0}\tanh^{-1}\left(\frac{K_0 - \gamma_{2\omega}(0)}{\zeta_0}\right), \tag{3.8}$$

the field amplitude for the fundamental frequency ω approaches a maximum. The physical significance of the threshold length l_{thr} is quite simple: it is the length of the excited nonlinear crystal at which the gain at the fundamental frequency ω (due to inversions between activator levels) compensates for both nonresonance (since generally $\kappa_2^0 \neq 0$) and nonlinear losses associated with power flow from the fundamental wave to the second harmonic, of frequency 2ω. A peculiar property of the threshold length is that at $l > l_{\mathrm{thr}}$, the pump power that excites a nonlinear laser crystal is consumed in "amplification", primarily of the harmonic rather than of the fundamental wave. Thus, theory [586, 587] predicts a possible way of improving the efficiency of harmonic generators by using nonlinear laser crystals as converters. This point is corroborated by the plots of Fig. 3.23. Broken curves 4 and 4' represent $\gamma_{\omega,2\omega}(l)$ plots that correspond to an unexcited nonlinear activated crystal or to one without an activator ($\kappa_3^0 = 0$).

At present there are several laser insulating crystals (see Tables 3.5 and 3.6) that can function simultaneously as nonlinear and as amplifying media. Available experience [89, 156, 530, 584, 585] indicates that, among them, such crystals as $LiNbO_3$ and Ba_2NaNbO_{15} doped with Ln^{3+} ions can be rather large in size.

The behavior of the $\gamma_{\omega,2\omega}(l)$ curves in Fig. 3.23 is very well described by the parameters of expression (3.8),

$$K_0 = 2\pi\kappa_3^0 \frac{\omega n_\omega}{c}, \tag{3.9}$$

$$\gamma_{\omega,2\omega}(0) = \frac{q}{n_\omega}\left[\frac{32 n_{\omega,2\omega}}{c} J_{\omega,2\omega}(0)\right]^{1/2}, \tag{3.10}$$

$$\xi_0 = \{[K_0 - \gamma_{2\omega}(0)]^2 + [\gamma_\omega(0)]^2\}^{1/2}, \tag{3.11}$$

and

$$q = \frac{4\pi\omega}{c}\chi_k \sin\theta_0.$$

In expression (3.9), K_0 is the power-gain coefficient in a nonlinear laser crystal with no field of frequency ω. It depends on the κ_3^0 tensor, which is proportional to $\Delta N\,\sigma_e$. The parameters $J_{\omega,2\omega}(0)$ in (3.10) represent the initial power densities (at the input end of an amplifying nonlinear crystal) for the fundamental and harmonic frequencies, respectively. The coefficient ξ_0 in expression (3.11) can be treated as a generalized gain coefficient dependent on the spectroscopic properties of the host matrix itself and on the reduced field amplitudes at the frequencies ω and 2ω. The parameter q in expression (3.10) is the coefficient of nonlinear coupling between waves of frequencies ω and 2ω [157].

Referring again to Fig. 3.23, if at $l > l_{thr}$, the amplitudes γ_ω decrease monotonically, the reduced amplitudes of the second-harmonic field tend to saturate; saturation depends upon the magnitude of the normalized parameter of gain-coefficient saturation [586],

$$\Xi = \frac{\alpha_{ij}cn_\omega}{32\pi q^2}.$$

From the shape of the $\gamma_{\omega,2\omega}(l)$ curves, we can see that the higher the $\gamma_\omega(0)$ value, the faster the increase of the second-harmonic field strength. Thus every medium has its own optimal crystal length. Even a very brief analysis of the amplification regime clearly displays the advantage of excited nonlinear laser crystals for converting the fundamental frequency to a second harmonic.

These advantages are even more likely to apply in the generation regime, with a nonlinear laser medium in the optical cavity. As was mentioned before, this

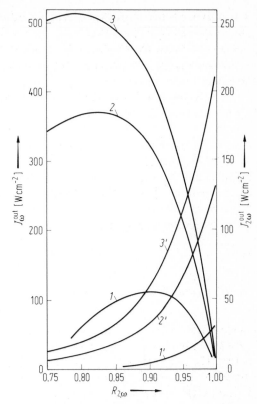

Fig. 3.24. Dependences of the density of output power at the fundamental frequency J_ω^{out} (curves 1–3) and the harmonic $J_{2\omega}^{out}$ (curves 1'–3') upon the reflection coefficient of the output mirror of an optical resonator at the fundamental frequency ω for a self-doubling frequency laser based on a LiNbO$_3$-Nd^{3+} crystal with the following parameters [587] $R_{1,\omega} = R_{1,2\omega} = 1$; $R_{2,2\omega} = 0$; $\rho_\omega = \rho_{2\omega} = 0.01$ cm^{-1}; $\Delta = 0$; $\alpha_{ij} = 5 \times 10^{-4}$ cm$^2 \cdot$W^{-1} and $q(8\pi/cn)^{1/2} = 3.7 \times 10^{-4}$ cm$^{-1/2}$. The parameter for the curves is the initial gain coefficient equal to 0.04 cm^{-1} for curves 1, 1'; 0.08 cm^{-1} for curves 2, 2'; and 0.1 cm^{-1} for curves 3, 3'

regime was analyzed in great detail in [587]. Figure 3.24 shows the most-interesting results of this paper from our point of view. The authors derived theoretically the output power density for the emission at the fundamental frequency J_ω^{out} and at the harmonic $J_{2\omega}^{out}$ as a function of the reflectance $R_{2,\omega}$ of the output mirror of the interferometer for the LiNbO$_3$-Nd^{3+} crystal. The $J_{\omega,2\omega}^{out}(R_{2,\omega})$ plots are evidently equivalent to those of conventional ($\kappa_3^0 = 0$) lasers (see, for example, [60]), with the exception that the optimum reflectances of the output mirror at the frequency ω are a little bit lower. This is attributed to the existence of both linear (κ_2^0) and nonlinear losses associated with generation of the second harmonic. As is to be expected, maximum output at the second harmonic is observed only in the case of totally reflecting mirrors ($R_{1,\omega} = R_{2,\omega} = 1$) of the interferometer.

3.9 Lasers with Combined Active Media

The idea of controlling the emission spectra of solid-state lasers and of studying excitation migration in disordered laser crystals by use of selective amplification introduced into the optical cavity occurred independently to the author of the present monograph and *DeShazer* [26–30]. On the basis of this principle, the former has suggested and realized in practice lasers with combined active media (CAM) [26–28]. *DeShazer* studied generators with internal[4] and external injection of narrow-band emission into the laser cavity [29, 30]. In case of lasers with CAM, the Q factor at a desired frequency is provided by a proper amplification contour formed by combining several selected media with correspondingly different luminescence spectra. In the new type of laser, this is achieved by inserting into the laser cavity several different active elements with their own pumping systems (Fig. 3.25a).

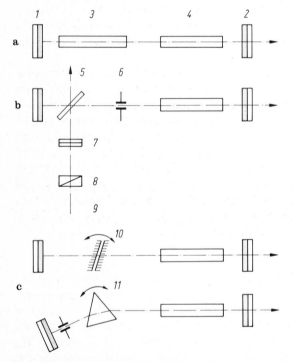

Fig. 3.25. Block diagrams of spectroscopically controlled lasers with combined active media (CAM) (**a**), with "photon injection" (**b**), and with dispersive optical resonator (**c**). 1, 2: resonator mirrors; 3, 4: active media of different types; 5: beam splitter; 6: diaphragm; 7: attenuator; 8: polarizer; 9: narrow-band external emission source. 10, 11: discriminating interferometer-selector or prism

---

[4]*DeShazer* [29] called generators with internal injection of a narrow-band emission into the laser cavity "composite oscillators".

Combined media for such lasers can be not only crystals and glasses with like and unlike ions, but also inorganic liquids (e.g., $POCl_3$ [167]) and other types of laser-active media. Of special concern are lasers with CAM based on simple crystals where activator ions form one type of center (for example, $Y_3Al_5O_{12}$-Nd^{3+}, $YAlO_3$-Nd^{3+} and others [9, 10]) or laser media with a variety of luminescence centers (for example, mixed fluoride crystals, glasses, or inorganic liquids [6–8, 22, 70, 73, 74]). The former type of laser crystals usually have rather narrow ($3–15$ cm^{-1} at 300 K) and homogeneously broadened luminescence bands (at 300 K and higher temperatures). The latter type of media have inhomogeneously broadened luminescence bands that result from superposition of lines of numerous centers with close structures. These materials exhibit wide luminescence bands, sometimes of the order of several hundreds of cm^{-1}. Though lasers with CAM and composite generators with injection of narrow-band emission, as was mentioned before, involve one and the same principle of "introduced selective amplification" [4, 5], they are rather different, primarily in the properties of their active media. In CAM lasers the amplification contour generalizes the luminescent properties of various materials, whereas the active media in composite generators preserve [29, 30] their individual luminescence properties.

The principle of operation of CAM lasers is similar to that of conventional lasers, so we shall not treat it in detail but consider instead only the condition for exciting stimulated emission in them. For the sake of simplicity, we analyze a CAM laser with an optical cavity formed by two plane-parallel mirrors with reflectances R_1 and R_2, filled with a combined medium consisting of two crystals of the same cross section s and of lengths l_1 and l_2 placed in series (see Fig. 3.26). Assume that the system has no other reflecting surfaces and that the crystals exhibit different luminescence characteristics. The properties of such a system will be analyzed by the method outlined in [60]. Let $J_1(z)$ and $J_2(z)$ stand for the fluxes per unit area propagating from R_1 to R_2, and vice versa, respectively. Then, the total fluxes per cross section of the combined rod will be $sJ_1(z)$ and $sJ_2(z)$.

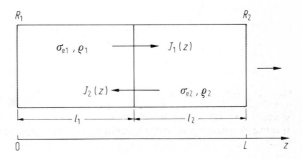

Fig. 3.26. Axial energy flows in a rod with plane-parallel end faces, composed of two laser media of different types

The differential equations describing variations in $sJ(z)$ over a path dz for the fluxes propagating through the amplifying medium in two opposite directions are

$$dJ_1(z) = \{[\alpha_1(z) - \rho_1'(z)] + [\alpha_2(z) - \rho_2'(z)]\} J_1(z) \, dz,$$

$$dJ_2(z) = -\{[\alpha_1(z) - \rho_1'(z)] + [\alpha_2(z) - \rho_2'(z)]\} J_2(z) \, dz. \tag{3.12}$$

When integrating (3.12), the conditions associated with flux reflection from the mirrors should be taken into account. These conditions, for the case in which there are no external emission sources at the generation frequency, are

$$J_1(0) = R_1 J_2(0),$$

$$J_2(L) = R_2 J_1(L). \tag{3.13}$$

Integrating (3.12) from $z = 0$ to $z = L$ gives

$$J_1(L) = J_1(0) \exp \int_0^L \{[\alpha_1(z) - \rho_1'(z) + [\alpha_2(z) - \rho_2'(z)]\} \, dz,$$

$$J_2(L) = J_2(0) \exp \int_0^L (-\{[\alpha_1(z) - \rho_1'(z)] + [\alpha_2(z) - \rho_2'(z)]\}) \, dz. \tag{3.14}$$

In the second equation, the minus sign before the left brace indicates that the directions of the flux J_2 and of the z axis are opposite. Now write

$$P = \exp \int_0^L \{[\alpha_1(z) - \rho_1'(z)] + [\alpha_2(z) - \rho_2'(z)]\} \, dz. \tag{3.15}$$

Expression (3.14) can then be recast in the simpler form

$$J_1(L) = J_1(0) P,$$

$$J_2(L) = J_2(0)/P. \tag{3.16}$$

In the cavity under consideration, with $R_1 = R_2$, amplification of the two fluxes determined by P should be equal. In fact,

$$\frac{J_1(L)}{J_1(0)} = \frac{J_2(0)}{J_2(L)} = P.$$

It has been shown that in the stimulated-emission case, amplification does not depend on the parameters of (3.12), but is completely determined by the boundary conditions, which impose severe limitations on the field inside the generating medium [60]. Expression (3.14) together with the independent expressions (3.13) form a system of equations with respect to four unknown quantities

$$J_1(0) - R_1 J_2(0) = 0,$$

$$J_1(L) - P J_1(0) = 0,$$

$$J_2(L) - R_2 J_1(L) = 0,$$

$$J_2(L) - J_2(0)/P = 0.$$

(3.17)

A nonzero solution of (3.17) is possible only for

$$P = (R_1 R_2)^{-1/2},$$

(3.18)

which, taking (3.15) into account, becomes

$$(R_1 R_2)^{1/2} \exp \int_0^L \{[\alpha_1(z) - \rho_1'(z)] + [\alpha_2(z) - \rho_2'(z)]\}\, dz = 1$$

or

$$(R_1 R_2)^{1/2} \exp \int_0^{l_1} [\alpha_1(z) - \rho_1'(z)]\, dz + \int_{l_1}^{l_2} [\alpha_2(z) - \rho_2'(z)]\, dz = 1$$

(3.19)

When the mean gain and loss coefficients are introduced,

$$\bar{\alpha}_1 = \frac{1}{l_1} \int_0^{l_1} \alpha_1(z)\, dz,$$

$$\bar{\rho}_1 = \frac{1}{l_1} \int_0^{l_1} \rho_1'(z)\, dz,$$

$$\bar{\alpha}_2 = \frac{1}{l_2} \int_{l_1}^{l_2} \alpha_2(z)\, dz,$$

$$\bar{\rho}_2 = \frac{1}{l_2} \int_{l_1}^{l_2} \rho_2'(z)\, dz,$$

(3.19) can be written in the form

$$R_1 R_2 \exp 2[(\bar{\alpha}_1 - \bar{\rho}_1)l_1 + (\bar{\alpha}_2 - \bar{\rho}_2)l_2] = 1.$$

(3.20)

When $R_1 = 1$ under steady-state conditions, the final form of the condition for exciting stimulated emission of frequency v_g in CAM lasers is

$$R \exp 2[(\sigma_{e1} \Delta N_1 - \rho_1)l_1 + (\sigma_{e2} \Delta N_2 - \rho_2)l_2]_{v_g} = 1.$$

(3.21)

If the losses in the media are similar and depend only slightly on wavelength, the shape of the amplification curve for these lasers will be determined by superposition of their luminescence spectra. Intensity in the net luminescence spectrum is governed by the wavelengths of the active elements and by their relative excitation energies. In other words, by proper choice of lengths and pump energies amplification can be attained at any desired frequency, within the frequency range of the luminescence spectra of the combined media, and generation can be excited at it.

Operation of CAM lasers will be illustrated by a combination of a simple $Y_3Al_5O_{12}$-Nd^{3+} crystal with a mixed fluoride α-$NaCaYF_6$-Nd^{3+} system, neodymium-doped glass of the LSG-1 type, or inorganic $POCl_3$-Nd^{3+} liquid. Figure 3.27 represents the luminescence spectra of these media, corresponding to the principal $^4F_{3/2} \rightarrow {}^4I_{11/2}$ transition. The same figure also shows an empirical scheme of the crystal-field splitting of the $^4F_{3/2}$ and $^4I_{11/2}$ manifolds of Nd^{3+} ions in the $Y_3Al_5O_{12}$ crystal, including the luminescent and induced transitions. For convenience, all the lines of the $Y_3Al_5O_{12}$-Nd^{3+} crystal luminescence spectrum and transitions in the energy-level scheme are numbered (this procedure will be employed in subsequent similar cases). As seen, the spectral lines of the garnet are narrow, while those of $POCl_3$-Nd^{3+}, α-$NaCaYF_6$-Nd^{3+}, and neodymium-doped glass are relatively wide. In conventional laser systems, laser action in the $Y_3Al_5O_{12}$-Nd^{3+} crystal under moderate pumping energies and at room temperature is observed at 1.06415 μm (9398 cm^{-1}) corresponding to the A line

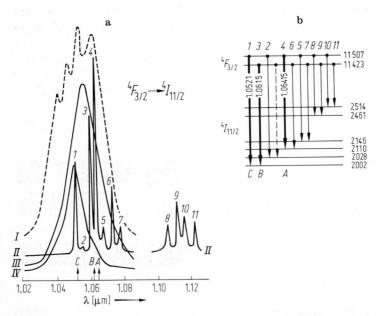

Fig. 3.27. Luminescence spectra ($^4F_{3/2} \rightarrow {}^4I_{11/2}$ transition) at 300 K (**a**) and crystal-field splitting scheme of the $^4F_{3/2}$ and $^4I_{11/2}$ manifolds of Nd^{3+} ions (**b**) [28]. (**a**) I—mixed fluoride α-$NaCaYF_6$-Nd^{3+} system; II—$Y_3Al_5O_{12}$-Nd^{3+} crystal; III—neodymium glass (LGS-1 type); IV—$POCl_3$-Nd^{3+} inorganic liquid. (**b**) $Y_3Al_5O_{12}$-Nd^{3+} crystal. The other notation is the same as in Fig. 3.22

[168] which is assigned to a transition from the upper Stark component of the $^4F_{3/2}$ state. In a CAM laser based on $Y_3Al_5O_{12}$-Nd^{3+} (300 K) + α-$NaCaYF_6$-Nd^{3+} (300 K), stimulated emission occurs at 1.0615 μm (9421 cm^{-1}), the B line associated with the 11413 cm^{-1} $^4F_{3/2} \rightarrow {}^4I_{11/2}$ 2002 cm^{-1} transition. If a combined medium consists of $Y_3Al_5O_{12}$-Nd^{3+} + $POCl_3$-Nd^{3+}, generation may be possible under certain conditions at a wavelength of 1.0521 μm (9505 cm^{-1}), the C line (the 11507 cm^{-1} $^4F_{3/2} \rightarrow {}^4I_{11/2}$ 2002 cm^{-1} transition) as well [167]. Analysis of the spectra emitted by these lasers provided precise identification of the crystal-field splitting of the metastable $^4F_{3/2}$ state of Nd^{3+} ions in $Y_3Al_5O_{12}$ crystals, which at 300 K may be taken to be 84 $\pm$ 0.3 cm^{-1}.

Before discussing other applications of CAM lasers in spectroscopic research on activated crystals, it should be noted that the power parameters of this type of laser (see Fig. 3.28) will be discussed later.

Interaction of the dopant ions with crystalline-lattice vibrations manifests itself in thermal shifts and broadening of spectral lines, in variations of the probability for nonradiative transitions, and in the appearance of side-band satellites in the optical spectra. Study of these manifestations of the electron–phonon interaction is particularly significant for quantum electronics, since it determines many of the spectroscopic characteristics of crystalline laser media. Among the many elaborate methods of stimulated-emission spectroscopy [4, 5], that using CAM lasers occupies a special place. In this connection, let us return to the results of investigations on $Y_3Al_5O_{12}$-Nd^{3+} crystals, both in the case of conventional lasers and in the case of CAM lasers. Figure 3.29 shows stimulated-emission spectra for a $Y_3Al_5O_{12}$-Nd^{3+} crystal, obtained in the wide temperature range 77–850 K. It is seen that the spectra emitted by a conventional laser contain information on the behavior of only two lines of stimulated emission—A from 130 to about 850 K, and B from 77 to about

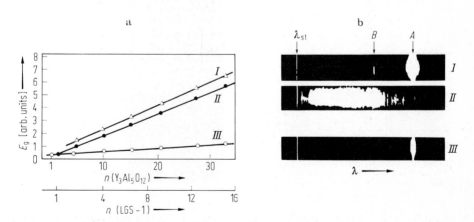

Fig. 3.28. Dependences $E_g(n)$, where $n = E_{exc}/E_{thr}$ (**a**) and stimulated emission spectra of lasers (**b**) [28]. I—CAM laser; II—glass laser (LGS-1 type); III—$Y_3Al_5O_{12}$-Nd^{3+} crystal laser. Wavelength of standard line $\lambda_{st} = 1.0561$ μm

300 K. On the other hand, the spectra emitted by a CAM laser contain much more information concerning a wider temperature range for the A and B lines and additional data on their temperature behavior. It is noteworthy that the luminescence method usually employed in such experiments is not applicable at

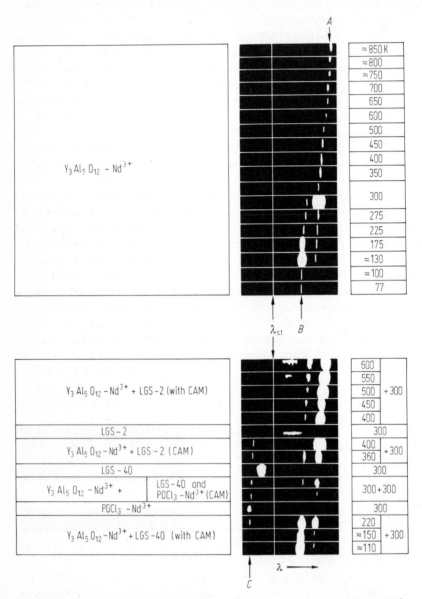

Fig. 3.29. Stimulated-emission spectra ($^4F_{3/2} \rightarrow {}^4I_{11/2}$ transition) of a $Y_3Al_5O_{12}$-Nd^{3+} crystal, obtained in a wide temperature range, with conventional and CAM lasers. Wavelength of standard line $\lambda_{st} = 1.0561\ \mu m$

high temperatures because of thermal broadening and superposition of lines [170]. Use of the above results permitted interpretation of the thermal shift of these emitted lines, which, in turn, led to conclusions about the vibrational modes of the $Y_3Al_5O_{12}$ crystal active in electron–phonon interactions [38].

The CAM method has been also applied to excite emission of new lines (not observed previously with conventional lasers) of YVO_4-Nd^{3+} [167, 171], LaF_3-Nd^{3+} [37], $YAlO_3$-Nd^{3+} [10], $LiNbO_3$-Nd^{3+} [34] and other crystals. This method was indispensable in analyzing the spectral properties of the YVO_4-Nd^{3+} crystal, since the crystal-field splitting of the metastable $^4F_{3/2}$ state at 300 K is only about 14 cm^{-1}. This characteristic relates directly to the parameter A_2^0 of the crystal field $|A_2^0| = |E/6\alpha|$, where α is the constant of equivalent operators. Since the inaccuracy of conventional spectroscopic methods is of the order of 2–5 cm^{-1}, at low absolute values of ΔE ($^4F_{3/2}$) the parameter A_2^0 will exhibit great variations and seriously disturb determination of other constants of the crystal-field potential. The use of a CAM laser made it possible to determine the splitting of the $^4F_{3/2}$ manifold within an accuracy of 0.2 cm^{-1}.

Similar spectral-emission investigations have been carried out on the ferroelectric and ferroelastic $Gd_2(MoO_4)_3$-Nd^{3+} laser crystal [164–166]. Its luminescence spectra associated with the $^4F_{3/2} \rightarrow {}^4I_{11/2}$ transition at 300 K are shown in Fig. 3.30. Since activated $Gd_2(MoO_4)_3$ crystals are characterized by a domain structure, stimulated emission in conventional lasers may be achieved only with zero-oriented ($F\|c$) samples. This case corresponds to the luminescence spectrum labeled I in Fig. 3.30. As was expected, in a conventional laser at 300 K, stimulated emission was observed at the strongest luminescence line—the A line, at $\lambda_g = 1.0701$ μm (9345 cm^{-1}) corresponding to the 11355 cm^{-1} $^4F_{3/2} \rightarrow {}^4I_{11/2}$ 2010 cm^{-1} transition. The spectra emitted by this crystal are also represented in Fig. 3.30. It is still a problem to produce homogeneous $Gd_2(MoO_4)_3$-Nd^{3+} samples with 90° orientation ($F \perp c$). This circumstance prevents laser action on the B line, which is the strongest in the luminescence spectrum. This difficulty was eliminated in [166] by utilizing a CAM laser based on $Gd_2(MoO_4)_3$-Nd^{3+} ($F\|c$, 300 K) and neodymium silicate glass of the LGS-2 type. Here, laser action was achieved on the B line at the wavelength $\lambda_g = 1.0606$ μm (9429 cm^{-1}) assigned to the 11355 cm^{-1} $^4F_{3/2} \rightarrow {}^4I_{11/2}$ 1926 cm^{-1} transition.

As shown by Fig. 3.31, the CAM method was very effective in studying stimulated emission from the $YAlO_3$-Nd^{3+} crysral [10]. Yttrium aluminate combined with LGS-2 silicate glass emits at 77 K at the frequencies of two lines—C and D. At room temperature the stimulated emission spectrum of $YAlO_3$-Nd^{3+} combined with the fluoride LaF_3–SrF_2-Nd^{3+} system contains four lines—A, B, C, and D.

The author of the present monograph employed a combination of high-temperature stimulated-emission spectroscopy [32, 34] and the CAM method to examine the laser characteristics of the nonlinear $LiNbO_3$-Nd^{3+} crystal [34]. Figure 3.22 shows in addition to polarized luminescence spectra, the stimulated-

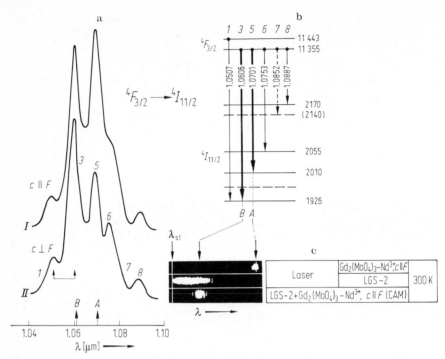

Fig. 3.30. Polarized luminescence spectra of Nd^{3+} ions ($^4F_{3/2} \rightarrow {}^4I_{11/2}$ transition) in a $Gd_2(MoO_4)_3$ crystal at 300 K (**a**) and the crystal-field-splitting scheme of the $^4F_{3/2}$ and $^4I_{11/2}$ manifolds (**b**), as well as stimulated emission spectra (**c**) [164]. Generation was investigated in conventional and CAM lasers, using a neodymium glass (LGS-2 type) laser rod. The square brackets in the spectra show the $^4F_{3/2}$ state splitting. The dashed level and the level with energy in brackets require more accurate definition of their positions. The other notation is the same as in Fig. 3.22

emission spectra and an empirical scheme of the crystal-field splitting of the $^4F_{3/2}$ and $^4I_{11/2}$ manifolds of the activator ion. It is evident that in a CAM laser based on $LiNbO_3$-Nd^{3+} and LGS-2 glass (300 K), stimulated emission has also been observed on the C line at the wavelength $\lambda_g = 1.0794$ μm (9265 cm^{-1}). As a rule, this line, associated with the 11254 cm^{-1} $^4F_{3/2} \rightarrow {}^4I_{11/2}$ 1989 cm^{-1} transition, is not excited with conventional lasers.

The CAM method is also very effective in studying the laser properties of simple crystals with inhomogeneously broadened luminescence lines at low temperatures, for example, $Ca(NbO_3)_2$ crystals doped with Ln^{3+} ions. If Ca^{2+} ions are replaced with Ln^{3+} ions, several different activator centers with a similar structure may form in the crystal, because the excess ionic charge can be compensated in various ways. Indeed, at low temperatures the luminescence spectra of $Ca(NbO_3)_2$-Nd^{3+} crystals grown by the melt technique without adding a compensating impurity contain weak satellites near the strong lines corresponding to the main center. These satellites belong to centers that are difficult to analyze. In conventional lasers, stimulated emission of the

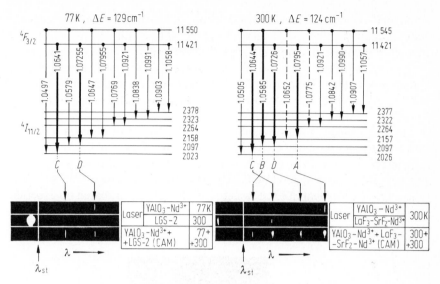

Fig. 3.31. Crystal-field-splitting schemes of the $^4F_{3/2}$ and $^4I_{11/2}$ manifolds of Nd^{3+} ions in a $YAlO_3$ crystal with corresponding stimulated-emission spectra [5]. Generation was investigated in conventional and CAM lasers, at 77 and 300 K. Notation is the same as in Fig. 3.22

$Ca(NbO_3)_2$-Nd^{3+} crystal [172, 173] is assigned to the line of the main center ($^4F_{3/2} \rightarrow {}^4I_{11/2}$ transition) at the wavelength $\lambda_g = 1.0615$ μm (9421 cm^{-1}) at 300 K and at $\lambda_g = 1.0612$ μm (9423 cm^{-1}) at 77 K (Fig. 3.32). Other activator centers do not show up in stimulated emission. This may be because they transfer energy to the main center, whose conditions for generation are more favorable. Use of a CAM laser based on the $Ca(NbO_3)_2$-Nd^{3+} crystal at 77 K and the LGS-2 glass at 300 K or on the mixed fluoride BaF_2-LaF_3-Nd^{3+} system (300 K) allowed excitation of three new induced transitions N_1-N_3 which cannot be assigned to the main center. Their emission wavelengths are 1.0614 μm (9421 cm^{-1}), 1.0626 μm (9411 cm^{-1}) and 1.0588 μm (9444 cm^{-1}). These lines are clearly seen in the spectra. This example illustrates very well the possibilities of the CAM method for studying the emission properties of such activator centers that do not appear spectroscopically under usual conditions.

The CAM method proves to be useful also in studying excitation-energy migration over the activator centers of similar ions in laser crystals characterized by inhomogeneously broadened lines of the optical spectra [28–30]. For a long time, this phenomenon has attracted the attention of many researchers, since it determines some basic parameters of lasers and laser amplifiers using such materials. On the other hand, this phenomenon is interesting in itself. Migration rates (or probabilities) are mostly experimentally estimated in the stimulated-emission regime. These estimates are subject to great uncertainties. Indeed, in lasers based on such media, the accumulated excitation energy may go to the stimulated-emission channel through "induced" deactivation of the metastable state, because for the kth center, the probability of induced emission at the frequency v_g will be determined by the flux density $J(v_g)$ and by the value σ_e^k, i.e., by $J(v_g^-)\sigma_e^k$. It is thus quite possible that the probability of induced emission is comparable to that of migration. Such a situation is corroborated by the results of numerous investigations on ultrashort-pulse laser amplifiers and by the results of [175].

Our description of the possibilities provided by CAM lasers will conclude with an account of some of their applications. One of the problems encountered in quantum electronics and in determining the role of crystal lasers in science and applications is that of extending the emission-frequency range. This problem is generally solved by search for new active media. However, based on experience

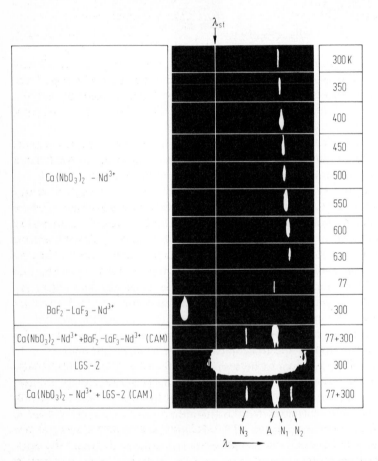

Fig. 3.32. Stimulated-emission spectra, due to the $^4F_{3/2} \rightarrow \,^4I_{11/2}$ transition, of a Ca(NbO$_3$)$_2$-Nd^{3+} crystal, obtained in a wide temperature range [172]. Conventional and CAM lasers were used. Wavelength of standard line $\lambda_{st} = 1.0561$ μm

accumulated in this area for many years, we have to recognize that this is a thorny path. The difficulty lies in finding a medium with high output-energy parameters rather than one with new spectroscopic properties. Tables 1.1–5 show there are now nearly three hundred crystal lasers, covering a wide range of emission frequencies. Unfortunately, only a few of them enjoy wide practical use (ruby, and $Y_3Al_5O_{12}$ and $YAlO_3$ doped with Nd^{3+} ions), the others being merely the objects of laboratory investigations because of their unsatisfactory energy parameters. Creation of CAM lasers [26–28] would facilitate solution of this problem since one emission frequency can be extended substantially by using crystals with the best available energy characteristics.

To see this more clearly, consider the properties of a CAM laser based on $Y_3Al_5O_{12}$-Nd^{3+} (300 K) and LGS-1 neodymium-doped glass. Figure 3.28 shows three $E_g(E_{exc}/E_{thr})$ plots. The line of lowest slope corresponds to excitation of the garnet crystal alone. In this case, the emission efficiency is relatively low. Excitation of the glass alone yields the middle line. Stimulated-emission spectra corresponding to each plotted case of $E_g(E_{exc}/E_{thr})$ are shown for clarity. The garnet emits one narrow line, while the glass emission spectrum is a broad band about 70 cm^{-1} wide, and the emission efficiency is much greater. The slope of the $E_g(E_{exc}/E_{thr})$ plot is steepest for excitation of the combined medium, the emitted energy being concentrated mostly in a very narrow band (about 1 cm^{-1}) of the A line of the $Y_3Al_5O_{12}$-Nd^{3+} crystal.

Similar results have been obtained with many other combined-media pairs [26–28]. This leads to the conclusion that CAM lasers exhibit, on the one hand, a spectral composition of emission and an excitation threshold which are characteristic of simple crystals and, on the other hand, rather high efficiency typical of media with a variety of activator centers. A comparison of these properties with the parameters of conventional lasers reveals that higher efficiencies are obtained than with simple lasers, and much stronger spectral lines and lower excitation thresholds than with various activator centers. Thus, in CAM lasers, the substance with various activator centers is a kind of accumulator of excitation energy. Note that this principle for stimulating emission may be applied also in high-power lasers used experimentally to create high-temperature plasmas.

In all the cases above, the combined medium consists of a simple crystal and a laser material with a disordered structure and a variety of activator centers. Here, the overall crystallo-optical properties of the CAM as a whole are determined only by the parameters of the simple crystal, because glasses, mixed fluoride systems[5] and inorganic liquids are optically isotropic. Explorations of CAM lasers based on similar anisotropic crystals, with different crystallographic orientations (sometimes such lasers are called composite) with respect to the geometric laser axis, have opened new possibilities of activated crystals that lead to lasers with temperature-controlled spectral and polarization characteristics

---

[5]Not every mixed fluoride system has cubic symmetry (see [6–8]).

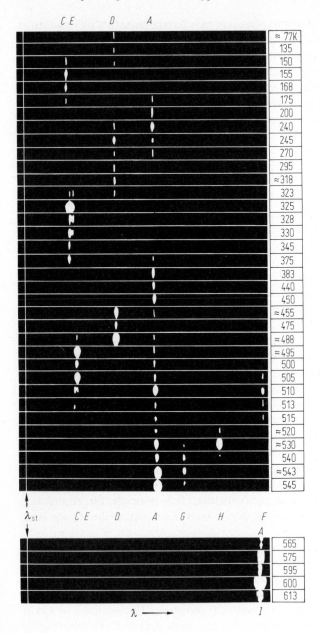

Fig. 3.33. Stimulated-emission spectra of a $YAlO_3$-Nd^{3+} crystal, due to the $^4F_{3/2} \rightarrow {}^4I_{11/2}$ transition, obtained in a wide temperature range [178]. Wavelength of standard line $\lambda_{st} = 1.0561$ μm

[176]. Some properties of composite laser media are treated also in [177]. A specific feature of such a combined medium is that it has complex crystallo-optical properties that change when one or all the crystals are heated or cooled[6]. In the case of simple crystals doped with Ln^{3+} ions, which have many strong luminescence lines, this may lead to temperature pulsations and multifrequency emission [176]. Investigations show that these effects may be observed with a crystal composed of several disoriented blocks. The latter case is exemplified by the stimulated-emission spectra of a laser based on the orthorhombic $YAlO_3$-Nd^{3+} crystal (space group D_{2h}^{16}—$Pbnm$) shown in Fig. 3.33. When heated from 77 to 613 K, this crystal, with the above properties, may generate on eight transitions. The results of identifying the stimulated-emission lines observed in this sample with Stark transitions between the $^4F_{3/2}$ and the $^4I_{11/2}$ manifolds, will be presented below. Along with purely scientific interest concerning the thermal behavior of activator-ion levels, lasers based on media with complex crystallo-optical properties are expected to be useful for applications.

An obstacle to the wide use of the CAM method for analyzing Stark structures of activator-ion levels is the limited choice of emitting materials, which makes it rather difficult to produce a combined medium with all the desired spectral parameters. This difficulty can be eliminated, under certain conditions, by use of a laser with injection of narrow-band external radiation, as shown in the schematic diagram Fig. 3.25b [29, 30]. The properties of such a laser will be illustrated by an example of *DeShazer* [29, 30]. The active medium in these experiments was neodymium-doped glass. Its stimulated emission is not polarized and Δv_g is about 100 cm^{-1}. When a beam splitter 5 (see Fig. 3.25) is inserted into the cavity to introduce a beam from an external laser, the emission from the glass becomes partially polarized, with its spectral composition preserved. Finally, if completely polarized emission from an external laser based on an anisotropic crystal (say, $CaWO_4$ or LaF_3 crystals doped with Nd^{3+} ions) is injected into the optical cavity, the neodymium-doped glass acquires practically all the emission properties of the external laser [29, 30]. As the strongest field in the optical cavity corresponds to the frequency of the external laser, lasers with injection of a narrow band, as well as CAM lasers, can emit all the accumulated energy in a narrow band. Effective injection into some lasers also seems to be possible from a conventional source, say, from a pulsed-xenon lamp together with a narrow-band filter and a polarizer.

---

[6]Thermal rotation of the plane of polarization is described by the familiar expression $\delta\varphi = 2\pi l \delta n(T)/\lambda$, where l is the length of the optical medium, λ is the emission wavelength, and $\delta n(T)$ is the thermal variation of the refractive index. In order to rotate the polarization plane through the angle $\delta\varphi = \pi$ for $l = 0.2$ cm, $\lambda = 10^{-4}$ cm and $\delta n(T) = 5 \times 10^{-6}$, the crystal temperature should be changed by about 100 K.

3.10 Tunable Crystal Lasers with Dispersive Cavities

The work of *Soskin* et al. [35, 178] shows how methods that employ solid-state lasers with dispersive cavities can be highly effective in the spectroscopy of the stimulated emission of activated media. Schematic diagrams of two simple lasers of this type are presented in Fig. 3.25c. An optical cavity employing a prism proved to be most convenient. In the geometric-optics approximation, standing waves may arise only at the frequency of emission that propagates normal to the plane mirrors. At this frequency the Q factor of the cavity will be maximum. Because of dispersion by the prism, all other frequencies will involve great losses. Such a cavity can be tuned by turning one of the mirrors. If the dispersive cavity contains an active medium with an inhomogeneously broadened luminescence line, its total Q is determined by the Q of the cavity itself, $Q_{cav} = v_{tun}/\Delta v_{cav}$ and by the Q of the luminescence line, $Q_{lum} = v_0/\Delta v_{lum}$ (here v_{tun} is the tuning frequency, Δv_{cav} is the half-width of the Q curve for the dispersive cavity, and v_0 is the frequency corresponding to the maximum of the luminescence line). When v_{tun} and v_0 coincide, emission is stimulated at the frequency of the luminescence-band maximum. Otherwise, the emission frequency is determined by [179]

$$v_g = \frac{v_{tun} + v_0(\Delta v_{cav}/\Delta v_{lum})}{1 + \Delta v_{cav}/\Delta v_{lum}}.$$

Hence, the emission frequency will be effectively tuned if

$$\Delta v_{cav}/\Delta v_{lum} < 1. \tag{3.22}$$

Values of Δv_{cav} of about 10 cm^{-1} can be obtained with disordered media with a variety of activator centers, for example, for mixed fluoride systems, where Δv_{lum} is about several times this value.

Dispersive optical cavities are also useful for exciting stimulated emission at the frequencies of individual homogeneously broadened luminescence lines of activator ions in simple crystals, and for tuning emission frequencies within the width of these lines. With media with ordered crystalline structures, condition (3.22) is harder to satisfy, because their luminescence line widths at 300 K are about 10 cm^{-1}. In this case, a Fabry-Perot interferometer can be used as the cavity dispersion element (see Fig. 3.25c), in which case Δv_{cav} may be under 1 cm^{-1}. Use of dispersive optical cavities has already provided valuable information about the nature of inhomogeneously broadened luminescence bands for media with disordered structures and various types of center. In particular, the widths of the homogeneous components (Δv_{hom}) in these bands have been estimated for a number of materials and the duration of excitation-energy migration over a band have been determined [179]. For example, according to [180], the value of Δv_{hom} for a mixed $CaF_2 - YF_3$-Nd^{3+} crystal was found to be (30 ± 5) cm^{-1}.

We now consider some results on emission-frequency tuning and spectral composition of the stimulated emission from activated crystals with homogeneously broadened luminescence lines. To tune continuously the emission frequency of a ruby laser, a Fabry–Perot interferometer was used in [179], with transparency maximum (v_k) and angle of orientation with respect to the laser axis related by

$$v_k = k/2d \cos \varphi,$$

where d is the distance between the interferometer mirrors and k is the order of interference. It follows from this expression that by varying the angle φ, a frequency-dependent transmission curve can be obtained for a given direction. If the gain contour of the active medium encompasses several maxima of interferometer transparency, the tuning limit will coincide with the interferometer period,

$$\Delta v_{\text{int}} = 1/2d \cos \varphi.$$

Figure 3.34 shows a luminescence spectrum for the Al_2O_3-Cr^{3+} crystal and a range of continuous tuning of the emission frequency within the shape of the R_1 and R_2 lines, obtained in [179]. In the experiment, a ruby laser rod 86 mm long and 8 mm in diameter was used. The cavity was formed by plane dielectric mirrors with about 99% reflectance at the wavelength λ_g and a Fabry-Perot interferometer which was placed between one mirror and the active element. The interferometer had 68% transmittance and $d = 0.1$ mm. As a glance at Fig. 3.34 shows, the continuous-tuning range at the R_2 line was about 14 cm^{-1} [the $^2E(2\bar{A}) \rightarrow {}^4A_2$ transition][7]. A further variation in the angle φ would stop

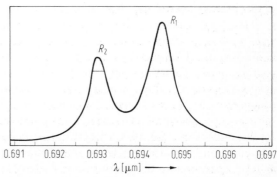

Fig. 3.34. Luminescence spectra of an Al_2O_3-Cr^{3+} crystal, due to the $^2E \rightarrow {}^4A_2$ transition, at 300 K [179]. The shaded regions correspond to the regions of continuous tuning of the stimulated-emission frequency inside the R_1 and R_2 line contours

---

[7] A tuning range of about 16 cm^{-1} was observed by *Karpushko, Sinitsin, Zheltov* and *Rubanov*, at the Minsk Institute of Physics.

stimulated emission on the R_2 line and cause emission on the R_1 line [the $^2E(\bar{E}) \rightarrow {}^4A_2$ transition]. It has been proposed [179] that an interferometer can also serve as a selector, which helps stabilize the frequency with respect to the thermal shift of the luminescence line that occurs at high pump powers. In this example, the maximum value of Δv_g was 0.1 cm^{-1} over the whole tuning range, while in a laser with a conventional cavity the same sample yielded $\Delta v_g = 0.6$ cm^{-1}.

In some laser crystals, for example, $CaWO_4$, heterovalent replacement of Ca^{2+} ions by Ln^{3+} ions, with no other impurity to compensate for the excess charge, leads to the formation of several activator centers with similar structures. They show up very well in optical spectra at low temperatures [182], whereas at high temperatures, they are concealed by homogeneously broadened strong lines of the main center[8]. Introduction of a special charge compensator (Na^+, Nb^{5+}, or other ions) may lead to centers with an energy-level splitting quite different from the main one. Stimulated-emission experiments with lasers based on such crystals with nondispersive cavities have revealed emitting transitions that belong to additional centers [183–185]. Subsequent experiments [178] that use multicenter simple crystals and lasers with dispersive cavities have yielded further information on the emission properties of the additional centers. In [178], an interferometer with $\Delta v_{int} = 50$ cm^{-1} was inserted into the cavity and use was made of $CaWO_4$-Nd^{3+} crystals (about 1.8 at.%) with compensating Na^+ and Nb^{5+} ions. By turning the interferometer, the 1.0582 μm (300 K, $^4F_{3/2} \rightarrow {}^4I_{11/2}$ transition) line was scanned. The experiment revealed several new induced transitions that had not been excited in nondispersive cavity lasers. The results of these measurements are represented in Fig. 3.35.

A more-complicated dispersive cavity was fabricated in [591] for a tunable Q-switched laser based on the $YAlO_3$-Nd^{3+} crystal emitting on the lines of the additional $^4F_{3/2} \rightarrow {}^4I_{13/2}$ transition (λ_g is about 1.34 μm). The schematic diagram of this laser is shown in Fig. 3.36. A laser crystal (50 mm long, 3 mm in diameter, $C \approx 1$ at.%, $F \| c$) was excited by two xenon flash lamps of type ISP-1200 in a water-cooled irradiation chamber (6). The crystal ends, one of which was at an angle of about three degrees with respect to the laser axis, had an antireflection coating at wavelength 1.34 μm. To tune the emission frequency, a dispersive cavity (7) was used, consisting of a 5 × prism telescope, an interferometer selector (2 mm base, 60% mirror reflectance) and a 60° prism of TF-5 glass. One of the mirrors of the interferometer selector was mounted on the piezoceramic element for fine tuning of the laser frequency. Coarse tuning was effected by the 60° prism. The emission linewidth is of great importance in laser spectroscopy and laser isotope separation. In the laser under consideration, to produce emission in one transverse mode (see Fig. 3.37), two diaphragms (5) with an opening of 1 mm were placed in the cavity. The output interference mirror (3) transmitted about 40% of the $\lambda \approx 1.34$ μm wavelength. For the nanosecond regime of generation, use was made of an electro-optical shutter using an $LiNbO_3$ crystal.

---

[8]For $CaWO_4$-Nd^{3+} crystals, this center is of the L type [186].

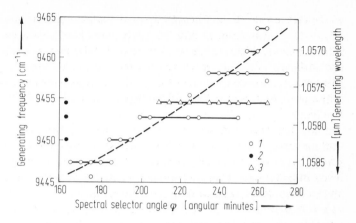

Fig. 3.35. Dependence of λ_g in the region of the 1.0582 μm luminescence line $({}^4F_{3/2} \to {}^4I_{11/2}$ transition) on dispersive-resonator tuning in a $CaWO_4$-$Nd^{3+}(Na^+)$ crystal laser [178]. (1) Experimental values of v_g in dispersive resonator at 300 K; (2) values of v_g, corresponding to one stimulated-emission pulse of a laser with plane-parallel mirrors, at 100 K; (3) values of v_g in an unselective and nondispersive resonator, at 300 K. The dashed line corresponds to the theoretical v_g tuning curve of the laser

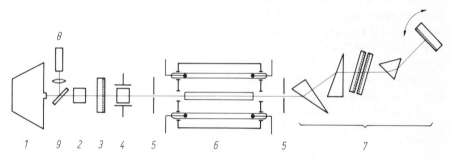

Fig. 3.36. Schematic diagram of a laser with frequency doubling and tuning at the lines of the additional ${}^4F_{3/2} \to {}^4I_{13/2}$ stimulated-emission channel in a $YAlO_3$-Nd^{3+} crystal (for notation, see text) [591]

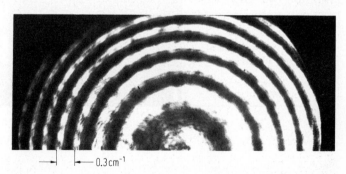

Fig. 3.37. Stimulated-emission-spectrum interferogram of a $YAlO_3$-Nd^{3+} crystal at the 1.3413 μm line (the linewidth of harmonic generation is approximately 0.08 cm^{-1}) [591]

In measuring the wavelength λ_g of the laser-emission lines assigned to the $^4F_{3/2} \rightarrow {}^4I_{13/2}$ transition of Nd^{3+} ions, conventional photoelectric methods are subject to large uncertainties, of the order of 0.001 μm. For this reason, in [591], the nonlinear optics method was used for second-harmonic generation, with subsequent photographic recording of the converted emission by use of a diffraction spectrograph (1) with a dispersion of about 0.0093 μm mm^{-1}. The nonlinear element (2) was a 18 $\times$ 8 $\times$ 8 mm $LiIO_3$ crystal ($\angle Fc \approx 24°$). A standard wavelength was provided by the emission from lamp (8) with a hollow lithium cathode (resonant doublet 0.67078 μm) which was directed by beam splitter (9) to the spectrograph input slit. Under these conditions, the excitation threshold of stimulation of emission of the $^4F_{3/2} \rightarrow {}^4I_{11/2}$ transition lines was about thirty joules, in the long-pulse-generation case ($\tau_{exc} \approx 200$ ms); and about one hundred joules, in the Q-switched case ($\tau_g \approx 70$ ns). The generation linewidths of the emission were measured by means of a 16 mm base Fabry-Perot interferometer. Such detailed description of the experiment is given here because several original experimental procedures were used in [591] that are of interest both for specialists engaged in developing commercial crystal lasers, and for physicists studying the spectroscopic peculiarities of laser crystals.

The measurements in [591] permitted more-accurate definition of the wavelengths emitted by the $YAlO_3$-Nd^{3+} crystal on the $^4F_{3/2} \rightarrow {}^4I_{13/2}$ transition at 300 K. According to these measurements, in this case, the laser emits at 1.3413 and at 1.3393 μm (with a measurement accuracy of $\pm$0.0001 μm). The emission in these lines is practically orthogonally polarized. For the 1.3413 μm line, the **E** vector is perpendicular to the crystallographic c direction (σ polarization). The results of the identification of these lines are presented in Fig. 6.17. The spectra

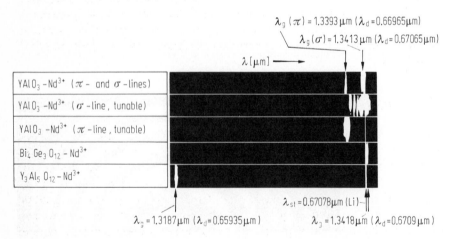

Fig. 3.38. Stimulated-emission spectra (second harmonic) of $YAlO_3$, $Bi_4Ge_3O_{12}$, and $Y_3Al_5O_{12}$ crystals with Nd^{3+} ions (the two latter crystals for comparison), due to the $^4F_{3/2} \rightarrow {}^4I_{13/2}$ transition, at 300 K [591]. Wavelength of standard line $\lambda_{st} = 0.67078$ μm

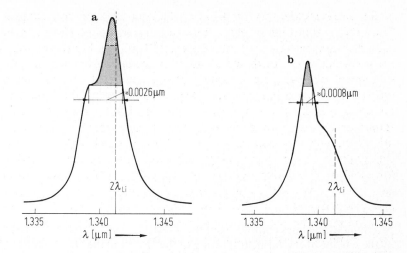

Fig. 3.39. Fragments of the polarized luminescence spectra ($^4F_{3/2} \rightarrow {}^4I_{13/2}$ transition) of a YAlO$_3$-Nd^{3+} ($c \| F$) crystal at 300 K [591]. (**a**) Corresponds to σ polarization. (**b**) Corresponds to π polarization. The light shading shows the continuous-tuning region for the stimulated-emission wavelength inside the luminescence 1.3393 and 1.3413 μm line contours in the long-pulse regime, and the darker shading shows the tuning region in the Q-switching regime

(Fig. 3.38) corresponding to second-harmonic generation in the case of a YAlO$_3$-Nd^{3+} crystal, and fragments of its spectra (Fig. 3.39) for polarized luminescence ($^4F_{3/2} \rightarrow {}^4I_{13/2}$), demonstrate tuning ranges in the red region of the spectrum, near the resonance doublet of lithium ($ns^2S_{1/2} \rightarrow np^2P^\circ_{1/2,3/2}$), and in the infrared region, within the luminescence line shapes of an additional emission channel. At $\tau_g = 100$ μs the spectral tuning range is about 0.0026 μm for the long-wave line and about 0.0008 μm for the short-wave line. It is evident that in the Q-switched case, the tuning range is somewhat narrower: it is about 0.001 μm for the σ line at $E_{\text{exc}} = 200$ J.

3.11 Temperature-Tunable Crystal Lasers

In Sects. 3.9, 10 we described "external" ways of controlling the emission frequencies of lasers based on activated crystals, whose emissions are attributed to transitions between electronic states of impurities. Now we shall consider the behavior of lines emitted when the temperature of the active medium is varied.

The thermal effects observed in stimulated-emission spectra are gradual change of emitted frequency, and sudden switching of induced-emission channels (rather frequent with certain crystals) that connect different pairs of Stark levels of two multiplets of the activator ion[9]. The thermal behavior of

---

[9]Increase of the crystal temperature leads also to broadening of the emission line.

stimulated-emission lines over a wide temperature range have been studied most closely for crystals activated with Nd^{3+} ions [4, 5, 32], since in most cases Nd^{3+} ions impart good spectroscopic properties to laser crystals and provide a practical four-level operating scheme on the $^4F_{3/2} \to ^4I_{11/2}$ and $^4F_{3/2} \to ^4I_{13/2}$ transitions over a range from the lowest possible temperatures to temperatures that exceed room temperature by a factor of two or three [24, 32, 34].

Figure 3.29 shows the temperature dependence of the spectra emitted by the $Y_3Al_5O_{12}$-Nd^{3+} crystal. It is evident from this figure that, along with a gradual frequency change of stimulated emission on the A and B lines, switching from one to the other takes place at about 225 K. At low temperatures, the laser emits at the frequency of the B line (the 11427 cm^{-1} $^4F_{3/2} \to ^4I_{11/2}$ 2002 cm^{-1} transition), while at room temperature and higher, emission is at the frequency of the A line (the 11507 cm^{-1} $^4F_{3/2} \to ^4I_{11/2}$ 2110 cm^{-1} transition). Thermal switching of the lines emitted from the $Y_3Al_5O_{12}$-Nd^{3+} crystal is shown clearly by the scheme of crystal-field splitting of the $^4F_{3/2}$ and the $^4I_{11/2}$ manifolds in Fig. 3.27b. The A transition is evidently associated with the upper Stark component of the metastable state, and the B transition with the lower component. With a higher probability for the spontaneous A transition above a certain temperature, according to the Boltzmann thermal distribution, the population of the upper Stark level increases and the A line luminescence becomes stronger than that of the B line. This is indeed just the case. According to [271], $A_A = 1250$ s^{-1} and $A_B \approx 520$ s^{-1}. The luminescence spectrum of Fig. 3.27a (II) also shows that at 300 K the A line is already the strongest[10].

As a further argument in favor of such an explanation, we cite results on the thermal behavior of the induced transition ($^4F_{3/2} \to ^4I_{11/2}$) of Nd^{3+} ions in $Lu_3Al_5O_{12}$ which has the same structure as the $Y_3Al_5O_{12}$ crystal. The crystal fields of these compounds produce very similar spectroscopic properties of Nd^{3+} ions. The character of the splitting of most terms is practically the same (see Table 4.3 below) and the intensities of the corresponding transitions are identical. But they differ in a very important respect, the magnitude of the ΔE splitting of the $^4F_{3/2}$ state. For Nd^{3+} ions in the $Lu_3Al_5O_{12}$ crystal, it is about 20% less; at 77 K it is approximately 67 cm^{-1}. Therefore, emission-channel switching from the B line (at 77 K, $\lambda_g = 1.0605$ μm) to the A line (at 300 K, $\lambda_g = 1.06425$ μm) [210] occurs at about 177 K rather than at about 225 K, as is the case with the $Y_3Al_5O_{12}$-Nd^{3+} crystal. Since the ratio of probabilities for spontaneous A and B transitions are equal in the two crystals, the following equality should hold:

$$\frac{225 \text{ K } (T_{AB} \text{ } Y_3Al_5O_{12})}{177 \text{ K } (T_{AB} \text{ } Lu_3Al_5O_{12})} = \frac{85 \text{ cm}^{-1}(\Delta E^4F_{3/2} \text{ } Y_3Al_5O_{12})}{67 \text{ cm}^{-1}(\Delta E^4F_{3/2} \text{ } Lu_3Al_5O_{12})}.$$

---

[10] According to [187] $A_A = 1440$ s^{-1}.

In order to illustrate the effect of stimulated-emission channel switching in the $Lu_3Al_5O_{12}$-Nd^{3+} crystal, Fig. 3.40 demonstrates the thermal behavior of the emitted spectra over a temperature range from 77 to about 750 K (for comparison, see Fig. 3.29).

Switching of the stimulated-emission lines with temperature has also been observed in lasers based on anisotropic crystals doped with Nd^{3+} ions. In these cases, however, the picture is more complicated. Here, the switching phenom-

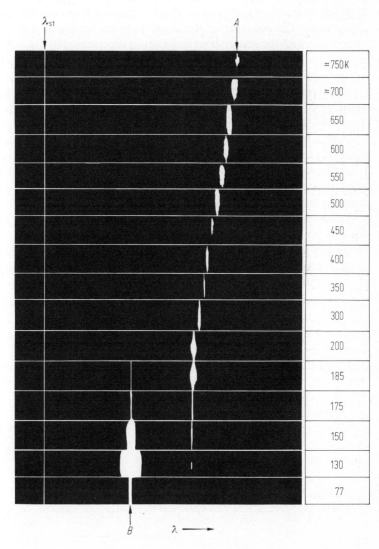

Fig. 3.40. Stimulated-emission spectra ($^4F_{3/2} \rightarrow ^4I_{11/2}$ transition) of $Lu_3Al_5O_{12}$-Nd^{3+} crystal obtained in a wide temperature range [271]. Wavelength of standard line $\lambda_{st} = 1.0561$ μm

enon is also influenced by polarization effects, which render the emission-line intensities dependent upon the directions of the geometric and the crystallographic axes of the laser element; this, in turn, leads to an orientation dependence of the point of equality of the excitation thresholds for the two induced transitions that participate in the switching.

As an example, consider some results obtained with a laser based on the hexagonal LaF_3-Nd^{3+} crystal by *Vylegzhanin* and *Kaminskii* [37]. Figure 3.41 shows $E_{thr}(T)$ for three active elements at various orientations (various angles ζ between the F and c axes). Among all the experimental data in this figure, we shall discuss only the behavior of the $E_{thr}(T)$ curves for the A and B lines. Their connection with the energy-level diagrams is represented in Fig. 3.42. In contrast to the $Y_3Al_5O_{12}$-Nd^{3+} case, the A and B lines emitted by the LaF_3-Nd^{3+} crystal are connected only with the lower level of the $^4F_{3/2}$ manifold. Hence, depending on the orientations of the F and c axes of the crystal, at low temperatures, stimulated emission may occur either on the A line with wavelength $\lambda_g = 1.0400\,\mu m$ ($\zeta = \angle Fc = 0$) or on the B line with wavelength $\lambda_g = 1.0630\,\mu m$ ($\zeta = 90°$). Investigations have shown this switching to be due to a resonance effect caused by the thermal shift of activator-ion energy levels. This resonance results from the overlapping of the B and F lines. It is evident from Fig. 3.42 that the F line is connected to the upper level of the $^4F_{3/2}$ state; its intensity increases with temperature. As a result, the gain on the B line also increases with temperature, and its excitation threshold becomes less than that of the A line. The threshold equality point, i.e., the temperature at which the emission channel switches from the A line to the B line, shifts with changing crystal orientation because the angular dependence of the intensity of the F line differs greatly from those of the A and B lines.

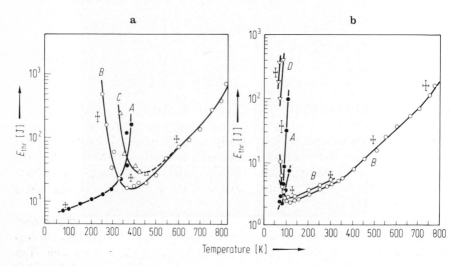

Fig. 3.41. Dependences $E_{thr}(T)$ of generation A, B, C and D lines in an LaF_3-Nd^{3+} crystal laser [37]. **(a)** $\zeta = \angle cF = 20°$. **(b)** $\zeta = 63, 70°$

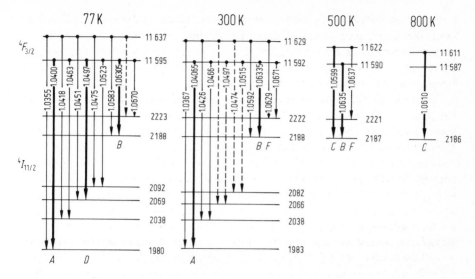

Fig. 3.42. Crystal-field-splitting schemes of the $^4F_{3/2}$ and $^4I_{11/2}$ manifolds of Nd^{3+} ions in LaF_3 crystal [37]. The dashed arrows in the schemes correspond to insufficiently resolved lines in the luminescence spectra. Other notation is the same as in Fig. 3.22

The condition for equality of thresholds of the A and B lines of the LaF_3-Nd^{3+} crystal is [37]

$$(\Delta v_{lum}/\Delta v_{lum}^A)\Phi_A(\zeta)A_A = \Phi_B(\zeta)A_B + b_i\varphi(T)\Phi_F(\zeta)A_F, \qquad (3.23)$$

where

$$\varphi(T) = (\Delta v_{lum}^B/2)^2(v_B - v_F)^{-2} + (\Delta v_{lum}/2)^2$$

or

$$\Phi_x(\zeta) = 1.5[0.5 \sin^2\theta_x (1 + \cos^2\zeta) + \cos^2\theta_x \sin^2\zeta]$$

determines the dependence of line intensity on the direction of observation [188] (θ_x is a parameter determined experimentally for each transition x; b_i is the Boltzmann factor for the upper level of the $^4F_{3/2}$ state). Condition (3.23) has been derived by assuming $\Delta v_{lum}^F \approx \Delta v_{lum}^B$ and that the F luminescence line has a Lorentzian shape.

An analysis of the experimental $E_{thr}(T)$ plots of Fig. 3.41 using (3.23) is shown in Fig. 3.43. This model of the effect of emission-channel switching from the A to the B line in a LaF_3-Nd^{3+} laser describes the experiment very well. The experimental points represented by open circles in Fig. 3.43 have been taken from [189], which also indicates that the multifrequency emission of a LaF_3-Nd^{3+}

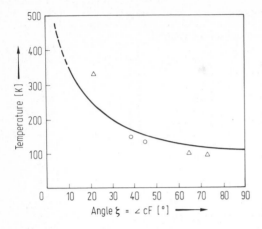

Fig. 3.43. Angular and temperature dependences of the point of equality of $E_{thr}(T_{AB})$ of the A and B lines in a LaF_3-Nd^{3+} crystal laser [37]. The continuous line is the result of calculation; triangles represent the experimental data obtained in [37] and circles, data from [189]

laser corresponds to a low probability of nonradiative transitions from levels of the $^4I_{11/2}$ state. This is evidence that analysis of thermal effects of emission-channel switching in lasers based on activated insulating crystals may yield many interesting results concerning the physics of processes that determine laser properties.

This section will be concluded with a consideration of continuously changing emission-line frequencies. Rather extensive experimental information is now available about the thermal shift of stimulated-emission lines for a great many crystals activated with Nd^{3+} and Cr^{3+} ions [4, 5, 32–34, 37, 38, 169, 170, 190]. The main conclusion of these studies is to attribute continuous change of emission-line frequencies to shift of activator-ion Stark levels, due to the lattice effect. In most experimentally known cases, this shift makes the emitted lines shift to longer wavelengths with increasing temperature (Table 3.7).

In the harmonic approximation of crystalline-lattice behavior, the energy of impurity-ion levels is changed by the electron–phonon interaction. Frequently, this factor is predominant [170, 192, 193] and will be treated in the following discussion. Among the effects associated with anharmonic crystalline vibrations, thermal expansion of the lattice may contribute most to the level shift. The greater the distance from neighboring atoms, the lower the potential of the crystal field of an impurity ion and the smaller (in some cases) the Stark splitting of its multiplets. In the case of rather complex crystalline structures, this effect may be accompanied by a change in the symmetry of the immediate environment, which is responsible not only for quantitative but also for qualitative changes in the pattern of Stark splitting. The latter effect may be noticeable, say, near phase transitions. A maximum effect of the lattice expansion on level shift is

Table 3.7. Temperature-tunable crystal lasers

Laser-tuning region[a] [μm]	Temperature [K]	Activator ion	Crystal	Reference
0.6943→0.6952	300→550	Cr^{3+}	Al_2O_3	[32, 190]
0.7207→0.745	85→210	Sm^{2+}	CaF_2	[118]
1.0293→1.0298	77→210	Yb^{3+}	$Y_3Al_5O_{12}$	[86, 483]
1.0294→1.0297	77→175	Yb^{3+}	$Lu_3Al_5O_{12}$	[86, 483]
1.0370→1.0395	300→530	Nd^{3+}	SrF_2	[36]
1.04065→1.0410	300→430	Nd^{3+}	LaF_3	[32]
1.0461→1.0468	300→530	Nd^{3+}	CaF_2	[32]
1.0534→1.0563	300→920	Nd^{3+}	BaF_2-LaF_3	[32]
1.0539→1.0549	300→550	Nd^{3+}	α-$NaCaYF_6$	[32]
1.0582→1.0587	300→700	Nd^{3+}	$CaWO_4$	[32]
1.0595→1.0613	380→820	Nd^{3+}	LaF_3	[32, 37]
1.0597→1.0583	300→800	Nd^{3+}	SrF_2-LaF_3	[32]
1.0610→1.0627	77→600	Nd^{3+}	$Y_3Al_5O_{12}$	[32, 169]
1.0615→1.0625	300→650	Nd^{3+}	$Ca(NbO_3)_2$	[34]
1.0623→1.0585	300→700	Nd^{3+}	$CaF_2-SrF_2-BaF_2-YF_3-LaF_3$	[32]
1.0628→1.0623	300→560	Nd^{3+}	CaF_2	[32]
1.0629→1.0597	300→ ≈1000	Nd^{3+}	α-$NaCaYF_6$	[32]
1.0632→1.0642	400→700	Nd^{3+}	LaF_3	[32, 37]
1.0632→1.0603	300→950	Nd^{3+}	CaF_2-YF_3	[32]
1.06335→1.0638	300→650	Nd^{3+}	LaF_3	[32, 37]
1.0637→1.0670	170→900	Nd^{3+}	$Y_3Al_5O_{12}$	[32, 169]
1.06375→1.0672	130→900	Nd^{3+}	$Lu_3Al_5O_{12}$	[191]
1.06405→1.0654	77→500	Nd^{3+}	$YAlO_3$	[176]
1.06425→1.0644	77→300	Nd^{3+}	$Bi_4Ge_3O_{12}$	[567]
1.0652→1.0659	310→500	Nd^{3+}	$YAlO_3$	[176]
1.0653→1.0633	300→920	Nd^{3+}	α-$NaCaCeF_6$	[32]
1.0653→1.0665	300→700	Nd^{3+}	$NaLa(MoO_4)_2$	[32]
1.0656→1.0629	300→600	Nd^{3+}	CdF_2-YF_3	[34]
1.0657→1.0640	300→750	Nd^{3+}	CaF_2-CeF_3	[32]
1.0664→1.0672	300→690	Nd^{3+}	YVO_4	[34]
1.0671→1.0687	77→550	Nd^{3+}	$LuAlO_3$	[565]
1.0687→1.0690	77→600	Nd^{3+}	$KY(WO_4)_2$	[94]
1.07255→1.0730	77→490	Nd^{3+}	$YAlO_3$	[176]
1.0787→1.0783	450→590	Nd^{3+}	$LiNbO_3$	[34]
1.07955→1.0802	77→700	Nd^{3+}	$YAlO_3$	[176]
1.0796→1.0803	600→700	Nd^{3+}	$YAlO_3$	[176]
1.0831→1.0834	120→550	Nd^{3+}	$LuAlO_3$	[565]
1.0885→1.0889	300→420	Nd^{3+}	CaF_2*[b]	[32]
1.0933→1.0922	300→620	Nd^{3+}	$LiNbO_3$	[34]
1.674→1.0676	82→100	Ni^{2+}	MgF_2	[113]
1.731→1.756	100→192	Ni^{2+}	MgF_2	[113]
1.785→1.797	198→240	Ni^{2+}	MgF_2	[113]

[a] Spectral composition and laser-tuning region of anisotropic crystals depend on the mutual orientation of geometric and crystallographic axes.
[b] Crystal CaF_2* contained oxygen.

expected in "soft" crystals[11] characterized by moderate Debye temperatures and rather high coefficients of thermal expansion. In the case of "hard" crystals, such as $YAlO_3$ and $Y_3Al_5O_{12}$, it is commonly thought (see, for example, [170, 196]) that the effect of thermal expansion upon level shift in Ln^{3+} ions is negligibly small compared with the electron-phonon interaction.

Electron-phonon interactions perturb various electronic states, changing their wave functions and energies. The first theory of this phenomenon was proposed by *McCumber* and *Sturge* [197], who considered both single-phonon and two-phonon Raman processes of electron-phonon interaction. In this theory, single-phonon processes yield the following temperature dependence of the energy of each of the two interacting levels i and j of an activator ion:

$$E_i(T) - E_i(0) = \beta_{ij} \left(\frac{T}{T_D}\right)^2 \mathscr{P} \int_0^{T_D/T} \frac{x^3}{e^x - 1} \cdot \frac{1}{x^2 - (\Delta E_{ij}/kT)^2} dx, \qquad (3.24)$$

where $\mathscr{P}$ signifies the principal value of the integral.

In the case of closely spaced levels ($\Delta E_{ij} < kT_D$, most often these are Stark levels of the same multiplet of Ln^{3+} ion) the interaction between them at low temperatures ($kT < \Delta E_{ij}$) leads to their mutual repulsion, and at high temperatures ($kT > \Delta E_{ij}$) to mutual attraction. When $\Delta E_{ij} \gg kT_D$, (3.24) is much simpler[12]:

$$E_i(T) - E_i(0) = \beta'_{ij} \left(\frac{T}{T_D}\right)^4 \int_0^{T_D/T} \frac{x^3}{e^x - 1} dx; \qquad (3.25)$$

at all practicable temperatures, the levels repel one another. Such an effect will be called an intermultiplet interaction, since it involves distant levels. The diagonal quadratic term of the electron-vibrational interaction operator gives the dependence $E_i(T)$, which differs from (3.25) only by a constant factor. Therefore, expression (3.25) is generally used to describe the total contribution of intermultiplet interactions, the symbol α_i standing for the coefficient β'_{ij}.

A particular state of an activator ion may be connected through electron-phonon interaction with a great many levels separated from it by various energy gaps. For this reason, it is not easy to establish which interactions are responsible for its shift, the more so because the observed line shift is a sum of the shifts of two levels.

A commonly employed method for interpreting experimental data on thermal line shifts is to choose the parameters α, β and T_D contained in expressions (3.24, 25). The values of α and β indicate what elementary processes of the electron–phonon interaction–single- or two-phonon Raman–are responsible for the shift; T_D is the effective Debye temperature characterizing the impurity-active portion of the lattice vibrations of the host matrix. This

---

[11] Theoretical calculations of the thermal lattice-expansion effect are rather easy only for relatively simple structures such as CaF_2 [194]. Elegant procedures for its evaluation, in some cases, from experimental results are demonstrated in [192, 195].

[12] The integral of expression (3.25) is tabulated in the book by *DiBartolo* [72].

interpretation is rather simple and is widely used. It suffers, however, from some disadvantages. First, the theory [197] assumed a long-wave approximation for the acoustic-phonon spectrum of the crystal matrix, which limits its validity to low temperatures compared with T_D. Hence, in particular, use of this method is questionable in interpreting the results obtained by high-temperature spectroscopy, with which shifts of some lines up to the Debye and even higher temperatures can be determined. Second, expressions (3.24) and (3.25) are only rough approximations. By choosing suitable values of α, β, and T_D, they can be made to fit experimental curves over the whole temperature range observed; however, the "Debye temperature" T_D found in this way is considerably underestimated.

The actual phonon density distribution in crystals may deviate substantially from the Debye distribution. As will be shown later [see (3.26)], different frequencies can make contributions of opposite signs to the thermal shifts of levels, so this point should be taken into consideration when interpreting experimental results. A source of information about phonon-density distributions may be (along with results, say, on slow-neutron scattering) data on the side bands of zero-phonon lines. To the best of our knowledge, not a single example has been reported of impurity ions with vacant d and f shells, that would support the Debye model and, in particular, eliminate the need to assume interaction with optical vibrations. An example indicating the insufficiency of the Debye model is the electronic-vibrational spectrum of the $MgO-Cr^{3+}$ crystal [198].

It is rather difficult to derive any reliable information about the electron–phonon interaction from the coefficients α and β. In [199, 508] analytical expressions for the coefficients of [197] were derived in terms of the point model of an activator center and the electron–phonon interaction. It was then shown in [592] that a single-term expression such as (3.25) does not suffice to interpret the observed thermal shift of the R line in ruby, $LaAlO_3$, etc., because this effect is influenced not only by the distant $^2T_{2g}$ level of the Cr^{3+} ion but also by the near-lying $^2T_{1g}$ level as well. Thus, going from (3.24) to (3.25) is not justified, even for ions of the iron group in which the crystal field is strong and the Stark splitting is large. This is even more evident when expressions (3.24) and (3.25), which were first applied to the iron-group ions, are also used to interpret the Ln^{3+} ion experiments[13]. The latter are known to differ from the iron group by their weak interaction with the crystal field, which manifests itself in smaller Stark splitting and weaker electron–phonon interactions.

This situation has stimulated detailed theoretical studies of the thermal shift of the lines due to Ln^{3+} ions as well as other manifestations of the electron–phonon interaction. Such studies were initiated by *Perlin* et al. [38, 200, 540, 593], who developed a theory of impurity–phonon interaction in crystals

---

[13]The unconvincing interpretation of data by the other manifestation of electron-phonon interaction in Ln^{3+} ions—line broadening—was given (taking into account only Raman processes) by *Gourley* [201].

doped with Ln^{3+} ions. The objects of these investigations were two laser crystals most commonly used and thoroughly studied—$Y_3Al_5O_{12}$ and $YAlO_3$ activated by Nd^{3+} ions (see Chap. 6). The observed thermal shift of the stimulated-emission lines associated with the principal $^4F_{3/2} \rightarrow {}^4I_{11/2}$ transition of these crystals consists of:

1) changed distance between the centers of gravity of the $^4F_{3/2}$ and $^4I_{11/2}$ manifolds;

2) shifted Stark levels of these states relative to their centers of gravity.

The former effect has not yet had a quantitative theoretical interpretation, since intermultiplet vibrational admixing is not expected to cause a noticeable shift, because of weak electron-phonon interaction [508]. This effect may be accounted for by modulation of a portion of the spin–orbit interaction, depending on the ion–ligand distance, by vibrations of the crystalline lattice.

The second effect has been investigated in the studies mentioned above. The theory suggested by them is not limited by a long-wave approximation and takes into consideration interaction of $4f$ electrons with all of the branches of crystalline vibrations. Unfortunately, no systematic data are currently available concerning the detailed phonon spectra of laser crystals, and such information cannot be provided by means of calculations at present. Therefore, in order to calculate various manifestations of the electron-phonon interaction, it is reasonable to employ the method of effective phonon density outlined in [200]. This method provides the following expression for the temperature shift of the ith Stark level of a multiplet:

$$E_i(T) - E_i(0) = \sum_i B_{ij}F_{ij}(T).$$

(3.26)

Here

$$F_{ij}(T) = \mathscr{P} \int_0^{\omega_{max}} \rho_{eff}(\omega)\bar{n}(\omega) \frac{\Delta E_{ij}}{(\Delta E_{ij})^2 - (\hbar\omega)^2} d\omega,$$

(3.27)

where ρ_{eff} is the phonon density averaged over the directions of the Brillouin zone and the branches of crystalline vibrations, as well as over the constants of interaction with various symmetry vibrations; $\bar{n}(\omega)$ is the equivalent average density of phonons of the frequency ω; $\mathscr{P}$ indicates the principal value of the integral. The values of (3.26) contain the squared matrix elements of the interaction in question; methods for the calculations of these in terms of the point-ion model of the crystal field are outlined in [38, 540]. An example of the interpretation of experimental results is given in Fig. 3.44b, which shows temperature dependences of shifts of the 2002 and 2110 cm^{-1} Stark levels of the $^4I_{11/2}$ manifold (see Fig. 3.27) of the Nd^{3+} ion in the $Y_3Al_5O_{12}$ crystal relative to the center of gravity of this state. The induced B transition is terminated at the first level, and the A transition at the second level; the temperature behavior of the corresponding luminescence lines is given in Fig. 3.44a.

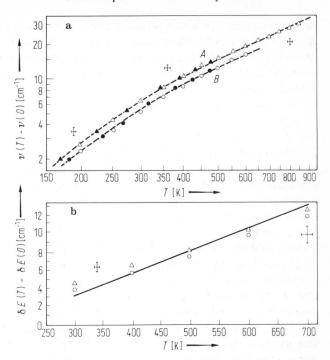

Fig. 3.44. Temperature shift of the A and B lines [169, 170] (a) and temperature shift of the $^4I_{11/2}$ manifold levels, which are the terminal levels for the A and B induced transitions, in a $Y_3Al_5O_{12}$-Nd^{3+} crystal (b). The experimental points from luminescence measurements [170] are shown by dark dots; those from stimulated-emission measurements [169] by open dots (a); the continuous line denotes the result of calculations by formula (3.28) with $x = 18$ (b)

Without entering into details about calculation of the matrix elements contained in B_{ij} (this special theoretical question is beyond the scope of the present book) it should be noted that, according to (3.26), contributions of different sign are made to the observed shift of the 2002 and 2110 cm^{-1} levels by the other five components of the same multiplet. The relative size of these contributions is determined not only by the corresponding matrix elements of the operator of the electron–phonon interaction but also by the actual phonon density distribution (dispersion law) of crystalline vibrations. The level shift of the $^4I_{11/2}$ manifold associated with the A and B lines is satisfactorily fitted by a single-term formula derived from (3.26) by assuming that the maximum phonon density for the $Y_3Al_5O_{12}$ crystal is at $\omega \approx 380$ cm^{-1}

$$\delta E(T) - \delta E(0) = \frac{x}{\exp{(\hbar\omega/kT)} - 1} \quad [\text{cm}^{-1}], \tag{3.28}$$

where x is a proportionality constant. Since this consideration is rather approximate, the agreement reached between the theoretical and experimental values for thermal shifts of these levels may be considered relatively good. According to [86, 483, 484], a sharp peak is observed at $\omega \approx 400$ cm^{-1} of the side-band spectra of the Yb^{3+} ion in the Y$_3$Al$_5$O$_{12}$ crystal. This vibrational spectrum was used in [200] to calculate an effective phonon density for the Y$_3$Al$_5$O$_{12}$ crystal by

$$\rho_{eff}(\omega) = A\omega^3 N_s(\omega). \tag{3.29}$$

In this expression, $N_s(\omega)$ is the form function of the spectrum of single-phonon side bands of the zero-phonon line at $T = 0$; the frequency ω is measured from the maximum of the zero-phonon line; the constant A is found from the normalizing condition. The calculated results are given in Fig. 3.45, from which one can see that the dependence $\rho_{eff}(\omega)$ has high peaks over a range in ω of from 350 to approximately 450 cm^{-1}.

The dependence $\rho_{eff}(\omega)$ for the YAlO$_3$ crystal has been similarly calculated [593]. The results obtained are presented in Fig. 3.46. In order to determine an effective phonon density, use was made of the side-band spectrum of the R line of Cr^{3+} ions given by M. J. Weber [477]. Using expression (3.26) and the established dependence $\rho_{eff}(\omega)$, [593] analyzed the thermal shift of the emitted A line (the $^4F_{3/2} \rightarrow {}^4I_{11/2}$ transition) of the YAlO$_3$ crystal (see Fig. 6.16). It has been shown that, just as in case with Y$_3$Al$_5$O$_{12}$-Nd^{3+} crystal, the thermal shift of the A line is attributed to the thermal shift of the lower Stark component of the $^4F_{3/2}$ manifold, which is due to single-phonon intramultiplet interaction. This is corroborated by the data tabulated in Table 3.8.

Thus, the physical basis of the thermal shift of lines is rather well understood. The thermal shift of the Stark levels of Ln^{3+} ions is mostly caused by direct single-phonon intermultiplet interactions. Nevertheless, quantitative interpretations of experimental results involve serious difficulties because, on the

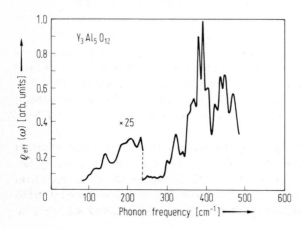

Fig. 3.45. Effective phonon density of the Y$_3$Al$_5$O$_{12}$ crystal [200]

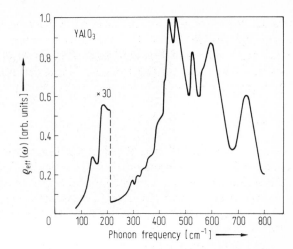

Fig. 3.46. Effective phonon density of the $YAlO_3$ crystal [593]

Table 3.8. Lower Stark-level thermal shift relative to the center of gravity of the $^4F_{3/2}$ manifold of Nd^{3+} ions in $YAlO_3$ crystals [593]

Temperature [K] $E_i(T) - E_i(0)$ [cm^{-1}]	200	300	400	500	600
Theory	0.6	1.7	3	4.5	6
Experiment	—	2.3	2.8	3.8	4.8

one hand, there are no reliable methods for separating the contributions of thermal lattice expansion and of electron-phonon interaction and, on the other hand, the theoretical models are, to some extent, idealized.

3.12 High-Temperature Induced Transitions of Nd^{3+} Ions in Crystals

Some information about high-temperature induced transitions (transitions excited only at temperatures higher than room temperature) has been already presented in Sect. 3.9, in connection with the properties of a CAM laser based on the $YAlO_3$-Nd^{3+} crystal consisting of several disordered blocks, and in Sect. 3.11 in connection with the emission parameters of the LaF_3-Nd^{3+} crystal. Now we shall briefly discuss two more examples.

Figure 3.47 shows the high-temperature spectra of stimulated emission ($^4F_{3/2} \to {^4I_{11/2}}$ transition) from two cubic (O_h^5—$Fm3m$) CaF_2 and SrF_2 crystals activated by Nd^{3+} ions [24, 36]. It is evident that at about 500 K new laser lines arise in these spectra. Analysis of the spectroscopic properties of these materials [5, 338] attributed the high-temperature lines of stimulated emission to

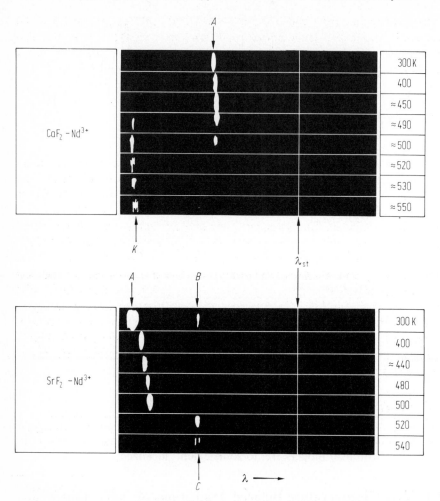

Fig. 3.47. Stimulated emission spectra ($^4F_{3/2} \rightarrow {}^4I_{11/2}$ transition) of CaF_2 and SrF_2 crystals with Nd^{3+} ions obtained in a wide temperature range [5, 36]. Wavelength of standard line $\lambda_{st} = 1.0561$ μm

transitions characterized by the highest probabilities of spontaneous emission[14] associated with the upper Stark component of the metastable $^4F_{3/2}$ state. For such cases, the condition of equal excitation thresholds for low- and high-temperature emission lines, which describe the effect of induced-transition switching in anisotropic media, according to (3.3), can be written as

$$\frac{\beta^l}{\varDelta v_{lum}^l} = \frac{\beta^h b^h}{\varDelta v_{lum}^h},$$

---

[14] In the case of the SrF_2-Nd^{3+} crystal, the B transition can be excited also, owing to other resonant transitions at close frequencies [338].

where the superscripts l and h stand for low and high temperature, β is the branching ratio, and b^h is the Boltzmann factor $\exp\left[-\Delta E(^4F_{3/2})/kT\right]$. It is also assumed that $\lambda_g^l \approx \lambda_g^h$ and $n^l \approx n^h$, and that these characteristics are independent of temperature.

3.13 Continuous-Wave Crystal Lasers

Continuous-wave (CW) laser operation is of great importance in solving many scientific and applied problems. For activated-crystal lasers, this is one of the most technically difficult modes, because it requires low excitation thresholds and relatively good thermophysical characteristics of the crystal. The latter are particularly significant for crystals that emit at room temperature. So far, about thirty activated crystals capable of CW generation are known (see Table 3.9). Their laser frequencies cover a spectral range from 0.6943 to 2.613 μm. The $Y_3Al_5O_{12}$ and $YAlO_3$ crystals doped with Nd^{3+} ions exhibit the best emission characteristics. At low temperatures (77 K), the CaF_2-Dy^{2+} and $(Y, Er)_3Al_5O_{12}$: Tm^{3+}-Ho^{3+} crystals excel. For communication systems employed in space explorations, CW lasers excited by solar radiation are of special interest. The properties of such lasers (those based on the $CaWO_4$-Nd^{3+}, $Y_3Al_5O_{12}$-Nd^{3+}, and CaF_2-Dy^{2+} crystals) are described elsewhere. Some of their parameters are included in Table 3.10.

Table 3.9. CW lasers based on activated crystals with broad-band excitation

Laser wavelength [μm]	Activator ion	Crystal	Induced transition	Temperature [K]	Power output [W]	Efficiency [%]	Reference
0.6934	Cr^{3+}	Al_2O_3	$^2E(\bar{E}) \to {}^4A_2$	77	—	—	[203, 204]
0.6943	Cr^{3+}	Al_2O_3	$^2E(\bar{E}) \to {}^4A_2$	300	2.37	0.1	[205, 206]
1.04065	Nd^{3+}	LaF_3	$^4F_{3/2} \to {}^4I_{11/2}$	300	—	—	[207]
1.0521	Nd^{3+}	$Y_3Al_5O_{12}$	$^4F_{3/2} \to {}^4I_{11/2}$	300	—	—	[99]
1.0582	Nd^{3+}	$CaWO_4$	$^4F_{3/2} \to {}^4I_{11/2}$	300	<0.1	0.01	[208, 209]
1.0559	Nd^{3+}	$LiGd(MoO_4)_2$	$^4F_{3/2} \to {}^4I_{11/2}$	300	—	—	[210]
1.060	Nd^{3+}	$Gd_3Sc_2Al_3O_{12}$	$^4F_{3/2} \to {}^4I_{11/2}$	300	0.55	—	[531]
1.061	Nd^{3+}	$CaMoO_4$	$^4F_{3/2} \to {}^4I_{11/2}$	300	—	—	[211]
1.0610	Nd^{3+}	$Y_3Al_5O_{12}$: Cr^{3+}	$^4F_{3/2} \to {}^4I_{11/2}$	77	—	—	[126]
1.0612	Nd^{3+}	$CaLa_4 (SiO_4)_3O$	$^4F_{3/2} \to {}^4I_{11/2}$	300	—	—	[474]
1.0612	Nd^{3+}	$Ca(NbO_3)_2$	$^4F_{3/2} \to {}^4I_{11/2}$	77	0.1	0.05	[212]
1.0615	Nd^{3+}	$Ca(NbO_3)_2$	$^4F_{3/2} \to {}^4I_{11/2}$	300	1	0.1	[172, 212, 213]
1.0615	Nd^{3+}	$Y_3Al_5O_{12}$	$^4F_{3/2} \to {}^4I_{11/2}$	300	—	—	[99]
1.0621	Nd^{3+}	$Gd_3Ga_5O_{12}$	$^4F_{3/2} \to {}^4I_{11/2}$	300	—	—	[594]
1.0629	Nd^{3+}	$Ca_5(PO_4)_3F$	$^4F_{3/2} \to {}^4I_{11/2}$	300	≈ 10	≈ 2	[130, 214, 215]
1.06415	Nd^{3+}	$Y_3Al_5O_{12}$	$^4F_{3/2} \to {}^4I_{11/2}$	300	to 1100	2–4	[9, 216, 217]
1.06415	Nd^{3+}	$Y_3Al_5O_{12}$: Cr^{3+}	$^4F_{3/2} \to {}^4I_{11/2}$	300	—	—	[216]

(Continued)

Table 3.9 (*continued*)

Laser wave-length [μm]	Acti-vator ion	Crystal	Induced transition	Temper-ature [K]	Power output [W]	Effi-ciency [%]	Reference
1.06425	Nd^{3+}	Lu$_3$Al$_5$O$_{12}$	$^4F_{3/2} \rightarrow {}^4I_{11/2}$	300	0.35	0.01	[210]
1.0644	Nd^{3+}	YAlO$_3$	$^4F_{3/2} \rightarrow {}^4I_{11/2}$	300	35	≈ 0.8	[98]
1.0644	Nd^{3+}	YAlO$_3$: Cr^{3+}	$^4F_{3/2} \rightarrow {}^4I_{11/2}$	300	6.5	0.3	[127]
1.0653	Nd^{3+}	NaLa(MoO$_4$)$_2$	$^4F_{3/2} \rightarrow {}^4I_{11/2}$	300	—	—	[218]
1.0672	Nd^{3+}	KGd(WO$_4$)$_2$	$^4F_{3/2} \rightarrow {}^4I_{11/2}$	300	—	—	[595]
1.0673	Nd^{3+}	CaMoO$_4$	$^4F_{3/2} \rightarrow {}^4I_{11/2}$	77	—	—	[211]
1.0688	Nd^{3+}	KY(WO$_4$)$_2$	$^4F_{3/2} \rightarrow {}^4I_{11/2}$	300	0.55	≈ 0.03	[210]
1.0698	Nd^{3+}	La$_2$Be$_2$O$_5$	$^4F_{3/2} \rightarrow {}^4I_{11/2}$	300	9	≈ 3	[596]
1.0737	Nd^{3+}	Y$_3$Al$_5$O$_{12}$	$^4F_{3/2} \rightarrow {}^4I_{11/2}$	300	—	—	[99]
1.075	Nd^{3+}	La$_2$O$_2$S	$^4F_{3/2} \rightarrow {}^4I_{11/2}$	300	—	—	[219]
1.0796	Nd^{3+}	YAlO$_3$	$^4F_{3/2} \rightarrow {}^4I_{11/2}$	300	100	2	[10, 98, 220]
1.1119	Nd^{3+}	Y$_3$Al$_5$O$_{12}$	$^4F_{3/2} \rightarrow {}^4I_{11/2}$	300	—	—	[99]
1.1158	Nd^{3+}	Y$_3$Al$_5$O$_{12}$	$^4F_{3/2} \rightarrow {}^4I_{11/2}$	300	—	—	[99]
1.1160	Tm^{2+}	CaF$_2$	$^2F_{5/2} \rightarrow {}^2F_{7/2}$	4.2–27	—	—	[221]
1.1225	Nd^{3+}	Y$_3$Al$_5$O$_{12}$	$^4F_{3/2} \rightarrow {}^4I_{11/2}$	300	—	—	[99]
1.3187	Nd^{3+}	Y$_3$Al$_5$O$_{12}$	$^4F_{3/2} \rightarrow {}^4I_{13/2}$	300	0.03	0.01	[99]
1.3370	Nd^{3+}	CaWO$_4$	$^4F_{3/2} \rightarrow {}^4I_{13/2}$	300	—	—	[222]
1.3387	Nd^{3+}	Lu$_3$Al$_5$O$_{12}$	$^4F_{3/2} \rightarrow {}^4I_{13/2}$	300	—	—	[222]
1.3393	Nd^{3+}	YAlO$_3$	$^4F_{3/2} \rightarrow {}^4I_{13/2}$	300	—	—	[98, 222]
1.3413	Nd^{3+}	YAlO$_3$	$^4F_{3/2} \rightarrow {}^4I_{13/2}$	300	14	0.4	[98]
1.3525	Nd^{3+}	KY(WO$_4$)$_2$	$^4F_{3/2} \rightarrow {}^4I_{13/2}$	300	—	—	[222]
1.674	Ni^{2+}	MgF$_2$	$^3T_2 \rightarrow {}^3A_2$	82–100	1	0.2	[113]
1.731	Ni^{2+}	MgF$_2$	$^3T_2 \rightarrow {}^3A_2$	100–192	1	0.2	[113]
1.860	Tm^{3+}	CaF$_2$–ErF$_3$	$^3H_4 \rightarrow {}^3H_6$	65	—	—	[145]
1.929	Ni^{2+}	MnF$_2$	$^3T_2 \rightarrow {}^3A_2$	85	—	—	[113]
1.934	Tm^{3+}	Er$_2$O$_3$	$^3H_4 \rightarrow {}^3H_6$	77	—	—	[149]
1.939	Ni^{2+}	MnF$_2$	$^3T_2 \rightarrow {}^3A_2$	85	—	—	[113]
2.0132	Tm^{3+}	Y$_3$Al$_5$O$_{12}$	$^3H_4 \rightarrow {}^3H_6$	85	—	—	[128]
2.0132	Tm^{3+}	Y$_3$Al$_5$O$_{12}$: Cr^{3+}	$^3H_4 \rightarrow {}^3H_6$	77	—	—	[128]
2.014	Tm^{3+}	(Y,Er)$_3$Al$_5$O$_{12}$	$^3H_4 \rightarrow {}^3H_6$	85	—	—	[128]
2.0412	Ho^{3+}	YVO$_4$: Er^{3+}, Tm^{3+}	$^5I_7 \rightarrow {}^5I_8$	77	—	—	[660]
2.0416	Ho^{3+}	ErVO$_4$: Tm^{3+}	$^5I_7 \rightarrow {}^5I_8$	77	—	—	[660]
2.0746	Ho^{3+}	BaY$_2$F$_8$: Er^{3+}, Tm^{3+}	$^5I_7 \rightarrow {}^5I_8$	77	—	—	[532]
2.0866	Ho^{3+}	BaY$_2$F$_8$: Er^{3+}, Tm^{3+}	$^5I_7 \rightarrow {}^5I_8$	77	0.56	0.25	[532]
2.0975	Ho^{3+}	Y$_3$Al$_5$O$_{12}$: Cr^{3+}	$^5I_7 \rightarrow {}^5I_8$	85	—	—	[128]
2.0977	Ho^{3+}	(Y,Er)$_3$Al$_5$O$_{12}$: Tm^{3+}	$^5I_7 \rightarrow {}^5I_8$	77	50	6.5	[597]
2.098	Ho^{3+}	(Y,Er)$_3$Al$_5$O$_{12}$: Tm^{3+}	$^5I_7 \rightarrow {}^5I_8$	77	20.5	≈ 4	[223]
2.0990	Ho^{3+}	(Y,Er)$_3$Al$_5$O$_{12}$: Tm^{3+}	$^5I_7 \rightarrow {}^5I_8$	77	15	5	[140]
≈ 2.1	Ho^{3+}	CaF$_2$-ErF$_3$-TmF$_3$-YbF$_3$	$^5I_7 \rightarrow {}^5I_8$	65	—	—	[145]
2.1020	Ho^{3+}	Lu$_3$Al$_5$O$_{12}$: Er^{3+}, Tm^{3+}	$^5I_7 \rightarrow {}^5I_8$	77	—	—	[143]
2.121	Ho^{3+}	Er$_2$O$_3$	$^5I_7 \rightarrow {}^5I_8$	77	—	—	[136]
2.1223	Ho^{3+}	Y$_3$Al$_5$O$_{12}$: Cr^{3+}	$^5I_7 \rightarrow {}^5I_8$	85	—	—	[128]
2.1227	Ho^{3+}	(Y,Er)$_3$Al$_5$O$_{12}$: Tm^{3+}	$^5I_7 \rightarrow {}^5I_8$	85	—	—	[140]
2.1227	Ho^{3+}	(Y,Er)$_3$Al$_5$O$_{12}$: Tm^{3+}, Yb^{3+}	$^5I_7 \rightarrow {}^5I_8$	85	7.6	—	[140]
2.35867	Dy^{2+}	CaF$_2$	$^5I_7 \rightarrow {}^5I_8$	77	150	—	[39]
2.35867	Dy^{2+}	CaF$_2$	$^5I_7 \rightarrow {}^5I_8$	77	1.2	0.06	[40–42]
2.613	U^{3+}	CaF$_2$	$^4I_{11/2} \rightarrow {}^4I_{9/2}$	77	—	—	[18]

Table 3.10.   Solar-pumped CW crystal lasers

Crystal	Laser wavelength [μm]	Temperature [K]	Power output [W]	Reference
CaF_2-Dy^{2+}	2.35867	27	—	[226]
CaF_2-Dy^{2+}	2.35867	77	—	[227, 511]
$CaWO_4-Nd^{3+}$	1.0582	300	0.13	[511]
$Y_3Al_5O_{12}-Nd^{3+}$	1.06415	300	1–1.5	[228, 229]
$Y_3Al_5O_{12}-Nd^{3+}$	1.06415	300	$\approx 5.6^a$	[598]
$Y_3Al_5O_{12}:Cr^{3+}-Nd^{3+}$	1.0641	300	4.8	[381]

[a] Multimode and 2.05 W in TEM_{00} mode.

3.14 Q-Switched Solid-State Lasers

In order to generate high-power ultra-short laser pulses, in 1962, *McClung* and *Hellwarth* [19] made use of an optical cavity with a controlled Q factor, thereby developing a new type of laser. Its operation is based on accumulation of excitation energy in the metastable state due to introduction into the cavity of emission-preventing losses that are removed when the quantum system has attained population inversion greatly exceeding the threshold. Such lasers have wide application because they help solve many important problems in science and engineering. Their use ranges from communication systems [230] to investigations on production of high-temperature plasmas [231, 495]. A ruby laser can generate pulses of monochromatic light with a peak power of about 10^9 W at an energy of about 300 J ($\tau_g \approx 10^{-7}$ s). Because of such high-power pulses, these lasers are also called giant-pulse lasers.

The difficulties encountered in constructing Q-switched lasers have been thoroughly analyzed in many original papers and monographs on quantum electronics, so we shall not treat them here. We think it necessary to focus attention on the following three points which are, to our mind, extremely important: the stability of activated crystals under high-density fluxes of coherent radiation; limitations of pulsed emitted power imposed by the nonradiative relaxation processes that occur in a quantum system, i.e., between energy levels of activator ions; and the effect of amplified luminescence channels (superluminescence). Above a certain energy density (threshold density) of laser radiation, damage is inflicted first on the end surfaces of the emitting crystal and then in its bulk. According to *Danileiko* et al. [232], these phenomena are caused by thermal breakdown due to radiation absorbed by local defects (or inclusions) in the crystal. This breakdown may take the form of thermoelastic stresses or phase changes, such as melting, or thermal explosions generally accompanied by a high-temperature glow. Some of the available information on surface and volume damage of activated laser crystals is presented in Table 3.11. Note that the data tabulated here are by no means complete and should be considered only

Table 3.11. Surface and volume optical strength of laser crystals

Crystal	Light-damage thresholds[a] $[W\,cm^{-2}]$		τ_{pulse} [s]	Growth method	Laser wave-length $[\mu m]$	Reference
	Surface	Volume				
$LiNbO_3$	1.2×10^8	—	2×10^{-8}	Czochra-lski	≈ 1.06	[239, 240]
$Al_2O_3-Cr^{3+}$	10^8-10^9	—	$2\times10^{-9}-2.10^{-6}$	Verneuil	0.6943	[233, 234]
	—	$(0.6-1) \times 10^{10}$	$(2-3) \times 10^{-8}$			[235–237]
	$\approx 10^6$	—	$5\times10^{-6}-10^{-4}$			[232]
Al_2O_3	$(1-2)\times10^{10}$	—	2×10^{-8}	—		[235, 237]
$Al_2O_3-(Ti, V,$ Co, Ni, Fe, Mg)	10^8-10^{10}	—	3×10^{-8}	—		[238]
$CaLa_4(SiO_4)_3O-$ Nd^{3+}	2×10^9	2×10^9	—	Czochra-lski	≈ 1.06	[756]
$Y_3Al_5O_{12}-Nd^{3+}$	$(1-3)\times10^9$	$(6-15)\times10^8$	2×10^{-8}	Czochra-lski	≈ 1.06	—

[a] Measurements in single-mode emission.

Table 3.12. Nonlinear refractive index n_2 for several laser-crystal hosts [599, 600, 765, 831]

Crystal	$n_2[10^{-13}\,esu]$[a]
$LiYF_4$	≈ 0.6
MgF_2	≈ 0.3
CaF_2	≈ 0.65
SrF_2	≈ 0.6
BaF_2	≈ 1
LaF_3	≈ 1.5
CeF_3	≈ 1.6
Al_2O_3	≈ 1.4
$Y_3Al_5O_{12}$	≈ 4.08
$La_2Be_2O_5$	≈ 2.1

[a] Most measurements were carried out at a wavelength of 1.06 μm.

as relative or tentative, since damage thresholds depend strongly on the growth technique and the quality of the crystals.

Numerous experiments on the nature of optical damage show that this phenomenon, which limits the output power of crystal lasers, depends directly on the duration of the laser pulse, i.e., on the peak power. Thus, at power fluxes of the order of 10^2 $W\,cm^{-2}$, effects due to nonlinear polarization of the medium become substantial. Among these effects is the phenomenon of self-focusing, which relates to the problem dealt with in this section. The self-focusing effect is related to the nonlinear refractive index n_2 of a medium [599]:

$$n_2 \langle E^2 \rangle \sim \Delta n,$$

where Δn is the change in index induced by an intense optical (laser) beam, $n = n^0 + \Delta n$, where n^0 is the refractive index measured with a low-intensity light beam. In the nonlinear refractive index, Δn is expressed in terms of the time-averaged optical electric-field amplitude E in electrostatic units. The measured values of n_2 for some laser crystalline matrices are given in Table 3.12. Note the favorable prospects for mixed-fluoride Nd^{3+} activated crystals for use in high-power lasers and laser amplifiers, because their spectral-emission properties are competitive with the best neodymium-doped glasses, and their thermophysical and nonlinear parameters are much better [5–8].

Mechanisms determining the structural stability of crystal media exposed to powerful laser radiation are of special concern in connection with optical damage. It is reasonable to expect such stability from perfectly pure crystals without defects (i.e., crystals without any foreign impurities, structural defects, or inclusions). The most probable processes responsible for damage of highly pure and perfect crystals are avalanche-collision ionization (see, for example, [603]) and multiphoton ionization [604]. At intensities in the laser beam of the order of 10^9 W cm^{-2}, the electric field in the light wave is comparable with the fields within the atoms. Laser damage of optically transparant crystals begins at this point. Indeed, if a crystal contains several free electrons, their interactions with the wave field will rapidly increase their energies by classical brems-absorption. If the electron energy exceeds the ionization energy of the atoms in the crystal, the latter are ionized by collision of electrons. These additional electrons rapidly build up energy in the electric field of the laser beam and ionize more atoms. This mechanism of electron-avalanche breakdown continues until the electron density becomes so high that the crystal is no longer transparent to laser emission. The density of energy liberation is thereby so increased that practically every atom in the local region of interaction is ionized and the substance turns into a highly ionized plasma. Such breakdown causes structural damage to the adjacent regions.

With further increase in the peak power of the laser beam, the electric-field strength may exceed the field within the atoms, thereby causing direct multiphoton ionization. This effect is one of the main ionization mechanisms in the case of picosecond laser pulses. Here threshold intensities are about 10^{11}–10^{12} W cm^{-2}.

Damage thresholds observed with some highly perfect crystals are 10^{10}–10^{11} W cm^{-2}. Unfortunately, these thresholds are achievable only in some local regions (about 10^{-8} cm^{-3}) and cannot characterize the optical stability of the whole crystal.

The probabilities of nonradiative relaxation which determine the thermodynamic equilibrium of a quantum system of activator ions depend upon the nature of the process. Direct nonradiative relaxation, which generally occurs between Stark levels within a multiplet, is characterized by relaxation times of 10^{-12}–10^{-14} s [604]. If the terminal laser level is a Stark component of the ground state, it is deactivated in this time during giant-pulse emission. Therefore, in such lasers, it is better to use activated crystals using a three-level scheme or a four-level scheme with a low-lying terminal level.

The probabilities of intermultiplet, usually multiphonon, processes of nonradiative relaxation are many orders of magnitude lower. In some laser crystals these probabilities may amount to 10^8 s^{-1}; more often, however, they range between 10^4 and 10^7 s^{-1} [49, 53, 151, 169, 476, 477, 703]. As a result, in crystals that employ a four-level scheme with a high-lying terminal level, the lower level may saturate under certain conditions, thereby limiting the output power.

This undesirable effect may also occur if a crystal is characterized by a large cross section for laser transition. At considerable excitation energies, the gain $\Delta N \sigma_e^{eff}$ of the system may be so high that the metastable state is effectively deactivated through superluminescence channels. For this reason, the $Y_3Al_5O_{12}$-Nd^{3+} crystals, which have high value of σ_e^{eff} ($\approx 3.3 \times 10^{-19}$ cm^2) for the main laser A transition (1.06415 μm at 300 K), are not optimal. To achieve effective emission in this case, crystals with somewhat smaller cross sections for the operating transition, for example, $YAlO_3$-Nd^{3+} crystals, are preferable, other conditions being equal. It is worthwhile mentioning the paper [602], whose authors, after a special treatment of the side surface of an active element, succeeded in reducing the detrimental effect of superluminescence upon the gain characteristics of a $Y_3Al_5O_{12}$-Nd^{3+} crystal. They used a round rod of $Y_3Al_5O_{12}$-Nd^{3+} with small square grooves on the surface and achieved a gain higher by a factor of 2.4 compared with that with a ground surface. The effect of superluminescence on the emission parameters of solid-state lasers has been considered in detail [70, 601, 602].

Account should also be taken of possible absorption at the emission frequency, both in the operating transition and from the metastable state to higher-lying levels of activator multiplets. All these facts should be borne in mind when selecting a crystal for a Q-switched laser.

3.15 Nonmagnetic Pyrotechnically Pumped Crystal Lasers

Nonmagnetic crystal lasers pumped by special pyrotechnic flashlamps were first described in [241]. Potassium perchlorate ($KClO_4$) was the oxidizer, and zirconium was the fuel. When this mixture is ignited, the following chemical reactions take place:

$$KClO_4 \rightarrow KCl + 2O_2,$$

$$Zr + O_2 \rightarrow ZrO_2,$$

$$2KCl \rightarrow 2K + Cl_2.$$

The second reaction is highly exothermic; therefore, the reaction products are heated to high temperatures and intense radiation in the yellow-red portion of the spectrum (color temperature of 4800–5200 K) is observed, This type of laser is interesting for both research and applications [243]. For instance, a $Y_3Al_5O_{12}$-Nd^{3+} laser pumped with special explosive lamp can generate, at 300 K, a pulse with an energy of about 1 J [242]. Because pyrotechnic lasers do not involve electric current, they can be placed in strong pulsed and stationary magnetic fields, permitting studies of the Zeeman effect on electron-induced transitions. Hence, it can be expected that in the near future, we shall approach a solution of the problem of establishing the local symmetry of the crystal field of emitting ions. In contrast to other types of lasers, crystal lasers with pyrotechnic excitation are characterized by quasicontinuous emission, of a duration sometimes amounting to about 50 ms.

3.16 Crystal Lasers with Laser or LED Excitation

Table 3.13 contains some spectroscopic and emission characteristics of activated crystals with laser excitation. Formally, such arrangements should more properly be called radiation converters rather than lasers, because they convert laser pumping emission into laser (Stokes) emission of another frequency. In addition to the practical interest in this type of excitation (integrated optics, optical communication), it provides much additional information about the properties of stimulated emission from crystals. This is valuable in searching for new laser materials because, often at the initial stage of investigations, it is technically difficult and economically unjustified to produce large and optically perfect crystals. Some results have been briefly mentioned earlier, for example, when lasers that operate on phonon-assisted transitions were discussed. Others are given in the publications referred to in Table 3.13.

Though stimulated emission in minute crystals (of the order of a millimeter or less in size) is fairly well excited by lasers, the most promising way of exciting them seems to be light-emitting-diode (LED) pumping [248, 254, 621, 625]. The recent increase of interest in this type of excitation is due to the advent of several high-concentration laser crystals (see Chap. 7 and the reviews [574, 575, 581]). Table 3.14 contains some parameters of known crystals with LED pumping, Figs. 3.48, 49 show two characteristic types of optical geometry (end, or longitudinal, and side, or transversal) for their excitation with $LiNdP_4O_{12}$ and $Y_3Al_5O_{12}$ crystals doped with Nd^{3+} ions as examples [621, 622]. In Fig. 3.48a, an argon-ion laser was used to tune the system and also as an additional excitation source. For the detailed design of these devices, the original papers should be consulted. In conclusion of this section it is proper and necessary to say something about the $Y_3Al_5O_{12}$-Nd^{3+} crystal. It is used now in lasers with the activator-ion concentration of about 1–1.2 at.%. Such optimal values of concentration are determined by the maximum of luminescence (see Sect. 6.1)

Table 3.13. Laser converters based on activated crystals

Crystal	Induced transition	Laser wavelength [μm]	Pumping wavelength [μm]	Exciting-laser type	Reference
$LiYF_4-Pr^{3+}$	$^3P_0 \to ^3H_4$	0.479	0.444	Dye laser (coumarin)	[606]
	$^3P_0 \to ^3F_2$	≈ 0.64	0.444	Dye laser (coumarin)	[123]
CaF_2-Sm^{2+}	$5d(A_{1u}) \to ^7F_1$	0.7205–0.745	0.6943	Q-switched ruby laser	[118, 279]
CaF_2-Dy^{2+}	$^5I_7 \to ^5I_8$	2.35867	0.6943	Ruby laser	[244, 245]
LaF_3-Nd^{3+}	$^4F_{3/2} \to ^4I_{11/2}$	1.06335	0.6943	Q-switched ruby laser	[189]
$LiNbO_3-Nd^{3+}$	$^4F_{3/2} \to ^4I_{11/2}$	1.0846	0.7525	Kr CW laser	[584]
$LiNdP_4O_{12}-Nd^{3+}$	$^4F_{3/2} \to ^4I_{11/2}$	1.0477	0.4579,0.4765, 0.5145 and 0.5965	Ar pulse and CW dye (rhodamine 6G) lasers	[608–611]
			≈ 0.87	Semiconductor laser (double-heterostructure $Ga_xAl_{1-x}As$)	[580]
	$^4F_{3/2} \to ^4I_{13/2}$	≈ 1.317	0.5145	Ar CW laser	[752]
$NaNdP_4O_{12}-Nd^{3+}$	$^4F_{3/2} \to ^4I_{11/2}$	≈ 1.051	0.5145	Ar laser	[578]
	$^4F_{3/2} \to ^4I_{13/2}$	≈ 1.32	0.5145	Ar CW Laser	[752]
$Al_2O_3-Cr^{3+}$	$^2E \to ^4A_2$	0.7670	0.6943	Q-switched ruby laser	[122]
		0.6943	0.4880 and 0.5145	Ar laser	[246]
$KNdP_4O_{12}-Nd^{3+}$	$^4F_{3/2} \to ^4I_{11/2}$	≈ 1.052	≈ 0.58	Dye laser	[579]
	$^4F_{3/2} \to ^4I_{13/2}$	≈ 1.32	0.5145	Ar CW laser	[752]
$CaWO_4-Nd^{3+}$	$^4F_{3/2} \to ^4I_{11/2}$	—	0.6943	Q-switched ruby laser	[247]
		1.0649	0.5145	Ar CW laser	[246]
$Y_2O_3-Nd^{3+}$	$^4F_{3/2} \to ^4I_{11/2}$	≈ 1.07	0.7525	Kr CW laser	[770]
	$^4F_{3/2} \to ^4I_{13/2}$	≈ 1.31	0.7525	Kr CW laser	[770]
$YAlO_3-Nd^{3+}$	$^4F_{3/2} \to ^4I_{9/2}$	≈ 0.930	—	Xe pulse laser	[79]
$Y_3Al_5O_{12}-Nd^{3+}$	$^4F_{3/2} \to ^4I_{9/2}$	0.8910 0.8999 0.9385	0.49–0.54	Ar pulse laser	[81]
	$^4F_{3/2} \to ^4I_{11/2}$	1.06415	0.6943	Q-switched ruby laser	[247, 253]
			0.5145	Ar pulse laser	[480]
			≈ 0.87	Semiconductor (GaAs)	[251]
			0.745 and 0.805	Raman laser	[569]
$YP_5O_{14}-Nd^{3+}(C_{2h}^6)^a$	$^4F_{3/2} \to ^4I_{11/2}$	1.0525	—	CW laser	[249]
$YP_5O_{14}-Nd^{3+}(C_{2h}^5)^a$	$^4F_{3/2} \to ^4I_{11/2}$	≈ 1.051	0.581	Dye laser	[249, 252]
			0.4765,0.5017 and 0.5145	Ar pulse multimode laser	[249, 252]
YVO_4-Nd^{3+}	$^4F_{3/2} \to ^4I_{11/2}$	1.0641	0.5145	Ar CW laser	[605, 614]
	$^4F_{3/2} \to ^4I_{13/2}$	≈ 1.34	0.5145	Ar CW laser	[605, 614]
YVO_4-Ho^{3+}	$^5I_7 \to ^5I_8$	2.0412	0.4880 and 0.5145	Ar CW laser	[660]
$Y_3Fe_5O_{12}-Ho^{3+}$	$^5I_7 \to ^5I_8$	≈ 2.1	≈ 1.06	Nd glass laser	[573]
			0.6943	Ruby laser	[573]
$CeP_5O_{14}-Nd^{3+}$	$^4F_{3/2} \to ^4I_{11/2}$	≈ 1.051	0.4765,0.5017 and 0.5145	Ar pulse multimode laser	[615]
$NdAl_3(BO_3)_4-Nd^{3+}$	$^4F_{3/2} \to ^4I_{11/2}$	1.0635	≈ 0.58	CW dye laser	[579]
			≈ 0.8	Semiconductor (double-heterostructure $Ga_{1-x}Al_xAs$)	[624]
$NdP_5O_{14}-Nd^{3+}$	$^4F_{3/2} \to ^4I_{11/2}$	1.0512	≈ 0.58	Pulse dye laser (rhodamine 6G)	[84, 616]

(*Continued*)

Table 3.13 (*continued*)

Crystal	Induced transition	Laser wavelength [μm]	Pumping wavelength [μm]	Exciting laser type	Reference
			0.6764,0.7525 and 0.7993	Kr CW laser	[612]
			≈ 0.8	Semiconductor (double heterostructure Ga$_x$Al$_{1-x}$As)	[616, 624]
		1.0520–1.0530	0.5145	Ar CW laser	[617]
		1.0630	0.5145	Ar CW laser	[617]
GdP$_5$O$_{14}$–Nd^{3+}	$^4F_{3/2} \to {}^4I_{11/2}$	1.0512	0.4765,0.5017 and 0.5145	Ar pulse multimode laser	[615]
ErVO$_4$–Ho^{3+}	$^5I_7 \to {}^5I_8$	2.0416	0.4880 and 0.5145	Ar CW laser	[660]
Li(Nd,La)P$_4$O$_{12}$–Nd^{3+}	$^4F_{3/2} \to {}^4I_{11/2}$	1.0477	0.5145	Ar CW laser	[618]
Li(Nd,Gd)P$_4$O$_{12}$–Nd^{3+}	$^4F_{3/2} \to {}^4I_{112}$	1.0477	0.5145	Ar CW laser	[618]
Na$_3$Nd(PO$_4$)$_2$–Nd^{3+}	$^4F_{3/2} \to {}^4I_{11/2}$	≈ 1.05	≈ 0.58	Dye laser	[577]
Na$_5$Nd(WO$_4$)$_4$–Nd^{3+}	$^4F_{3/2} \to {}^4I_{11/2}$	≈ 1.063	0.586	Dye laser	[577]
K$_3$Nd(PO$_4$)$_2$–Nd^{3+}	$^4F_{3/2} \to {}^4I_{11/2}$	≈ 1.06	—	—	[619]
K$_3$(Nd,La)(PO$_4$)$_2$–Nd^{3+}	$^4F_{3/2} \to {}^4I_{11/2}$	≈ 1.06	—	—	[619]
K$_5$Nd(MoO$_4$)$_4$–Nd^{3+}	$^4F_{3/2} \to {}^4I_{11/2}$	1.0660	0.745 and 0.805	Raman laser	[569]
(Nd,Sc)P$_5$O$_{14}$–Nd^{3+}	$^4F_{3/2} \to {}^4I_{11/2}$	≈ 1.051	0.4727,0.4765, 0.4965,0.5017 and 0.5145	Ar CW laser	[533]
(Nd,In)P$_5$O$_{14}$–Nd^{3+}	$^4F_{3/2} \to {}^4I_{11/2}$	≈ 1.051	0.4765,0.5017 and 0.5145	Ar multimode laser	[615]
(Nd,La)P$_5$O$_{14}$–Nd^{3+}	$^4F_{3/2} \to {}^4I_{11/2}$	1.0511	0.7525	Kr CW laser	[475]
(Nd,Gd)Al$_3$(BO$_3$)$_4$–Nd^{3+}	$^4F_{3/2} \to {}^4I_{11/2}$	1.0635	≈ 0.58	Dye laser	[579]
LaCl$_3$–Pr^{3+}	$^3P_0 \to {}^3H_4$	0.4892	0.488	Dye laser (rhodamine 6G)	[116]
	$^3P_1 \to {}^3H_5$	0.5298	0.474		
	$^3P_0 \to {}^3H_6$	0.6164	0.488	Dye laser (rhodamine G6)	[116, 564]
		0.619			
	$^3P_0 \to {}^3F_2$	0.6452	0.488	Dye laser (rhodamine 6G)	[116]
LaBr$_3$–Pr^{3+}	$^3P_1 \to {}^3H_5$	0.532	—	Dye laser (rhodamine 6G)	[564]
	$^3P_0 \to {}^3H_6$	0.621	—		
	$^3P_2 \to {}^3F_3$	0.632	—		
	$^3P_0 \to {}^3F_2$	0.647	—		
CeCl$_3$–Nd^{3+}	$^4F_{3/2} \to {}^4I_{11/2}$	1.0647	0.5145	Ar CW laser	[620]
PrCl$_3$–Pr^{3+}	$^3P_0 \to {}^3H_4$	0.489	0.488	Dye laser	[116]
	$^3P_1 \to {}^3H_5$	0.531	0.474	(rhodamine 6G)	
	$^3P_0 \to {}^3H_6$	0.617	0.488	Dye laser	[116, 564]
		0.620		(rhodamine 6G)	
	$^3P_0 \to {}^3F_2$	0.647	0.488	Dye laser (rhodamine 6G)	[76, 116]
PrBr$_3$–Pr^{3+}	$^3P_0 \to {}^3H_6$	0.622	—	Dye laser	[564]
	$^3P_0 \to {}^3F_2$	0.649	—	Dye laser (rhodamine 6G)	[76, 564]

[a] Some stable compositions with different structure form in the Y$_x$Nd$_{1-x}$P$_5$O$_{14}$ system (see Table 5.1 and [249]).

Table 3.14.   Lasers based on activated crystals with LED pumping

Crystal	Induced transition	Laser wavelength [μm]	Pumping wavelength [μm]	LED type	Reference
$LiNdP_4O_{12}-Nd^{3+}$	$^4F_{3/2} \rightarrow {}^4I_{11/2}$	1.0477	0.78–0.82	$Al_xGa_{1-x}As$	[621]
$Y_3Al_5O_{12}-Nd^{3+}$	$^4F_{3/2} \rightarrow {}^4I_{11/2}$	1.06415	0.8 –0.82	$GaAs_xP_{1-x}$	[248, 250]
			0.808	Super-luminescent ($Al_xGa_{1-x}As$)	[622]
			0.79–0.82	Double-heterostructure (AlGaAs)	[623]
$Y_3Al_5O_{12}-Yb^{3+}$	$^2F_{5/2} \rightarrow {}^2F_{7/2}$	1.0296	0.93–0.96	GaAs: Si	[254]

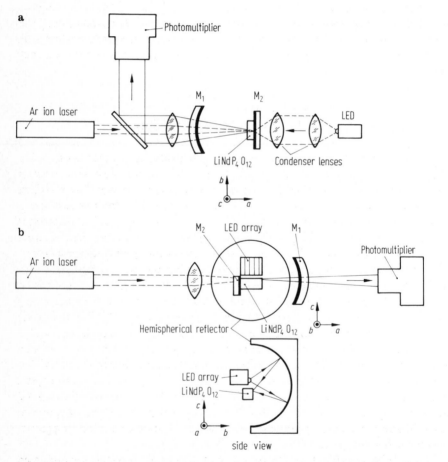

Fig. 3.48.   Block diagrams of $LiNdP_4O_{12}$ crystal lasers end (longitudinally) and side (transversely) pumped by LED emission [621]. (**a**) End-pumping configuration; (**b**) side-pumping configuration. M_1, M_2 are resonator mirrors

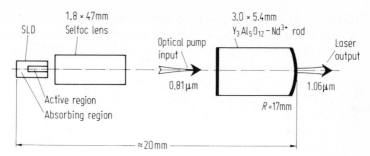

Fig. 3.49. Block diagram of a $Y_3Al_5O_{12}$-Nd^{3+} crystal laser end-pumped by a superluminescent diode [622]

and crystal growth conditions from the melt [6–8, 150, 438]. For lasers with laser or LED pumping (as well as for conventional lasers with lamp pumping), this crystal with higher concentration (up to 3–4 at.%) is of obvious interest. Such crystals of high optical quality can be synthesized by growing from solution in a melt [441].

3.17 Single-Crystal Film and Fiber-Crystal Lasers

The new generation of lasers based on activated insulating crystals fabricated in the form of single-crystal films or crystal fibers will, of course, be of great interest for integrated optics. The results of some investigations on such lasers will be briefly discussed here. The first single-crystal thin-film laser based on a multicomposite garnet doped with Ho^{3+} ions (the precise chemical formula is $Y_{1.25}Ho_{0.1}Er_{0.55}Tm_{0.5}Yb_{0.6}Al_5O_{12}$) was reported by *Van der Ziel, Bonner, Kopf,* and *Van Uitert* [290]. These authors observed stimulated emission of Ho^{3+} ions at 77 K at the 2.098 μm wavelength of the principal $^5I_7 \rightarrow {}^5I_8$ transition. The active element in their experiment was a $6.5 \times 3 \times 0.2$ mm film deposited from the liquid phase by the epitaxial method on a 0.8 mm thick substrate consisting of a high-quality $Y_3Al_5O_{12}$ crystal oriented in the plane (111). The operating ends of the film together with the substrate were thoroughly treated, then coated with a layer of gold that functioned as the mirrors of the optical cavity. With a spiral xenon lamp of type FT-524, the excitation threshold for emission was 72 J. Further crystal-growth studies [478] led to more perfect samples of this crystal and, hence, to lasers that emit at 77 K both in the pulsed mode with $E_{thr} \approx 1.7$ J and in the CW mode with $P_{thr} \approx 200$ W. High optical quality of thin-film crystal lasers has been achieved by matching the *a* lattice constant of the emitting garnet and the substrate garnet.

Pulsed CW thin-film lasers based on $Y_3Al_5O_{12}$-Nd^{3+} crystals are discussed in [478, 480]. *Stone* et al. [623] used the same crystal to develop a CW fiber laser operating at 300 K. The excitation source was a double-heterostructure AlGaAs

LED. A fiber element 5 mm long and 80 mm in diameter, inserted between the mirrors of the optical cavity, when pumped through the end, started to generate when a 45 mA current flowed through the LED. Pumping was into the absorption band corresponding to the $^4I_{9/2} \rightarrow {}^4F_{3/2}$ transition. *H. P. Weber* et al. [475] recently developed a fiber laser converter by use of a high-concentration $Nd_{0.5}La_{0.5}P_5O_{14}$ crystal, 0.68 mm in length and 12×12 mm^2 in cross section. The schematic diagram of this device, including some technical details, is given in Fig. 3.50. On absorbing about 10 mW power from a krypton-ion laser, the $Nd_{0.5}La_{0.5}P_5O_{14}$ crystal emitted at a wavelength of 1.051 μm. In a recent study [770], CW laser action of Y_2O_3-Nd^{3+} single-crystal fibers was obtained on both the emission channels $^4F_{3/2} \rightarrow {}^4I_{11/2}$ and $^4F_{3/2} \rightarrow {}^4I_{13/2}$, with krypton-ion laser excitation.

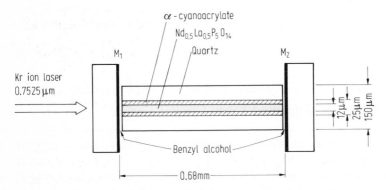

Fig. 3.50. Block diagram of an $Nd_{0.5}La_{0.5}P_5O_{14}$ crystal laser end-pumped by krypton-ion laser emission [475]. The light pipe is placed between two plane-parallel mirrors, M_1, M_2 with 99.8% transparency at 1.05 μm wavelength. The benzyl alcohol solution is used to decrease reflection and scattering losses

Despite the advances mentioned above in the construction and investigation of film and fiber crystal lasers, the problem of producing laser-crystal films and fibers of high optical quality is still very complicated. In this connection, of interest, at least for research, are certain crystals from which thin plates several microns thick can be cleaved or chipped off. Among such crystals is $KY(MoO_4)_2$-Nd^{3+} [85, 93, 377]. Figure 3.51 shows a Y^{3+} polyhedron and the structural motif of this compound. Yttrium ions are seen to be located inside the oxygen octahedron. They take particular positions on the two axes of rotation that are parallel to the a axis. The Y^{3+} polyhedron can be represented as an irregular cube whose two faces, which are approximately perpendicular to the direction [001], are inclined about 35° relative to one another. The adjacent Y^{3+} polyhedra have common edges; in the direction (010), they form continuous bands. The adjacent parallel bands spaced by $a/2$ are interconnected by K^+ ions

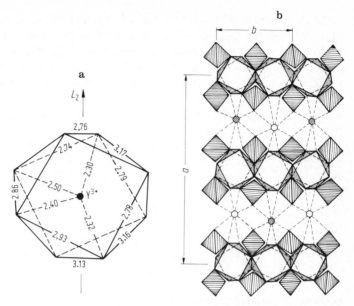

Fig. 3.51. Polyhedron of a Y^{3+} ion in $KY(MoO_4)_2$ crystal structure (**a**) and $KY(MoO_4)_2$ structure projected onto the xy plane (**b**) [85]. The figures show Y–O and O–O interatomic distances in Å, circles denote K^+ ions positions (**b**)

located in crystallographically independent positions. A comparatively weak K–O coupling provides perfect cleavage along the (100) plane. This compound is optically transparent over a range from 0.33 to about 5.5 μm.

3.18 Multibeam Crystal Lasers

In the previous sections, we have briefly described the operating principles of various types of laser based on activated insulating crystals with broad-band pumping. In these lasers, practically all the excitation energy that is accumulated by the active medium is generally emitted at the frequency of one induced Stark transition in a strictly definite (with respect to the geometry of the laser element) direction, determined by the type of optical cavity. Without reference to the mode structure of the field inside the active medium or the cavity as a whole, such lasers can be classified as single-beam.

We now consider lasers with multidirectional emission [464]. Such lasers differ from the single-beam class by having an active medium common to several separate cavities. The block diagrams of five simple designs with plane-parallel mirrors are given in Fig. 3.52. In multibeam lasers, each direction may correspond to stimulated emission of a different frequency and polarization, attributed to transitions between different levels of two states and between Stark

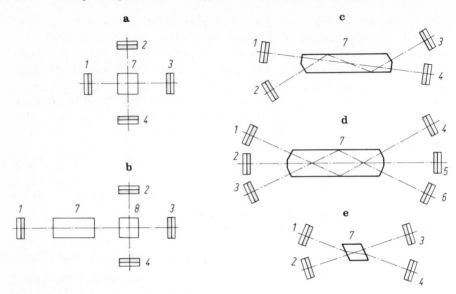

Fig. 3.52. Block diagrams of multibeam lasers [464]. 1–6—mirrors; 7, 8—laser media

components of several multiplet pairs of one and the same activator ion. Since these lasers can, in principle, be based on all known laser crystals (see Tables 1.1–5), as well as on activated glasses and inorganic liquids (or in general, on any laser media), they have familiar operating schemes, i.e., three- and four-level systems with one common or several metastable states (see Fig. 3.2a). For this reason, we shall not go into their operating principles but proceed to their peculiar features.

Consider, for example, a laser element of a neodymium-doped active material in a system of two cavities (say, as in Fig. 3.52a). One cavity is tuned to the emission wavelength of the principal $^4F_{3/2} \rightarrow {}^4I_{11/2}$ transition ($\lambda_g \approx 1.06$ μm) and the other to the wavelength of the additional $^4F_{3/2} \rightarrow {}^4I_{13/2}$ transition ($\lambda_g \approx 1.35$ μm). In this case emission can be excited by a four-level scheme with the metastable $^4F_{3/2}$ state in common. The "excitation-energy reservoir" established by pumping, which is referred to as inversion in conventional lasers, can feed either of the two emission channels or both simultaneously, if the gains are equal. The former case has been experimentally realized using simple $Y_3Al_5O_{12}$-Nd^{3+} and $KY(WO_4)_2$-Nd^{3+} crystals and the mixed fluoride CaF_2–YF_3-Nd^{3+} system [464]. The laser behavior was determined by the excitation conditions for stimulated emission in each beam.

The main peculiarity of multibeam lasers is the possibility of control over the accumulated excitation energy in any operating regime. Thus, one beam can be emitting and the other controlling; one can provide **CW** operation, the other, pulsed action. One channel may involve a CAM (Fig. 3.52b); the other may be used to inject the emission from an external laser. One can be a "generation" direction; the other, an "amplification" one. With multibeam lasers, within the laser medium practically any form and any distribution of excitation density along different directions is possible. Some variants of multibeam crystal lasers (see, for instance, Fig. 3.52c, d) will exhibit "waveguide" properties for certain beams, the wavefront of the exciting wave being preserved [462, 704]. The conception of a laser beam passing undistorted through a lasing medium thermally perturbed by pumping has recently found widespread use in so-called zigzag slab crystalline lasers. A fairly detailed treatment of this problem is presented in [855–859].

The multibeam principle can be basic to a variety of combinations and such lasers can be used for various purposes. For instance, in high-power lasers utilized for creation of high-temperature plasma, this principle is likely to help eliminate the detrimental effect on the active elements of amplification of laser energy reflected from the target [465, 495], by means of rapid deactivation (at the wavelength of the principal or the additional emission channel) of their residual excitation. The multibeam principle also permits development of peculiar "trigger" laser systems that have generation channels with sufficient separation of frequencies. This may facilitate faster response of devices in integrated optics.

In conclusion, we note that the recent advent of CAM lasers, of those with frequency self-doubling, with cascade operating schemes, with sensitization, with cumulative processes in excitation and, finally, of multibeam lasers is clear proof of the inexhaustible properties of active media represented by insulating laser crystals. Discovery of these possibilities was very much due to the development of new spectroscopic methods, particularly the stimulated-emission spectroscopy of activated crystals [271].

More recent information on new laser transitions can be found in the following references: $^5I_6 \rightarrow {}^5I_8$ [774, 775], $^5I_5 \rightarrow {}^5I_7$ [778], $^5I_5 \rightarrow {}^5I_6$ [777], and $^5S_2 \rightarrow {}^5F_5$ [833] of Ho^{3+} ions; $^1D_2 \rightarrow {}^1H_4$ [779], $^3F_4 \rightarrow {}^3H_4$ [834], $^1G_4 \rightarrow {}^3H_4$ and $^1G_4 \rightarrow {}^3F_3$ [835] of Tm^{3+} ions; $^4G_{5/2} \rightarrow {}^6H_{7/2}$ of Sm^{3+} ions [780]; $5d \rightarrow {}^2F_{7/2}$ [781] and $5d \rightarrow {}^2F_{5/2}$ [836] of Ce^{3+} ions; $^3P_1 \rightarrow {}^3H_4$ [837], $^3P_0 \rightarrow {}^3H_5$ [838], $^3P_0 \rightarrow {}^3F_3$ [839], $^3P_0 \rightarrow {}^3F_4$ [840], $^3P_0 \rightarrow {}^1G_4$ [841], $^1D_2 \rightarrow {}^3F_3$ [842], $^1G_4 \rightarrow {}^3F_4$ and $^1G_4 \rightarrow {}^3H_5$ [843] of Pr^{3+} ions; $5d \rightarrow {}^4I_{11/2}$ (?) [844], $^4D_{3/2} \rightarrow {}^4I_{11/2}$ [845], $^4I_{13/2} \rightarrow {}^4I_{11/2}$ [846], $^2P_{3/2} \rightarrow {}^2H_{9/2}$ ($^4F_{5/2}$) and $^2P_{3/2} \rightarrow {}^4I_{11/2}$ [847] of Nd^{3+} ions; $^4I_{9/2} \rightarrow {}^4I_{11/2}$ [848], $^4F_{9/2} \rightarrow {}^4I_{11/2}$ [849], $^4I_{9/2} \rightarrow {}^4I_{13/2}$ [850], and $^4F_{9/2} \rightarrow {}^4I_{13/2}$ [851] of Er^{3+} ions; $^4T_2 \rightarrow {}^4A_2$ of Cr^{3+} ions [852]; $^2E \rightarrow {}^2T_2$ of Ti^{3+} ions [853]; $^6H_{11/2} \rightarrow {}^6H_{13/2}$ of Dy^{3+} ions [854].

4. Stark Level Structure and Optical Transition Intensities of Activator Ions in Laser Crystals

The first and primary objective of spectroscopic investigations on activated laser crystals is to analyze the Stark structure of the spectra and to establish the scheme of energy levels of the activator ion. The second stage of investigation is associated with determining the point symmetry group of the activator, symmetry properties of the wave functions of the Stark levels, and selection rules for the optical transitions and their intensities. Purposeful synthesis and search for new materials with given spectral parameters requires, first of all, theoretical interpretation of dependences between the composition and the structural peculiarities of the base matrix, on the one hand, and the energy spectrum of the activator, on the other. The first problem is solved experimentally using the conventional methods of absorption and luminescence spectroscopy at low temperatures (see, for instance, [43, 46, 47, 61, 63, 497, 549, 722]) which have recently been facilitated by Raman light-scattering methods [501, 557] and stimulated-emission spectroscopy [4, 5, 271]; the latter provides accurate measurements at relatively high temperatures. Information on symmetry properties of the Stark levels and the structure of the activator-ion surroundings is largely obtained from polarization measurements and optical spectra in external fields (Zeeman effect, piezospectroscopy) as well as from EPR results [181, 500, 604, 663]. In recent years the newest spectroscopic methods have been successfully applied to different types of lasers [62, 722]. The principal method for theoretical interpretation of spectral measurements is crystal-field theory [255, 502, 503, 558], which actively employs the procedures of group theory [664].

4.1 Basic Concepts of Crystal-Field Theory

The present section contains experimental results on the energies of the Stark levels of activator ions for some laser crystals. However, prior to considering these results, we shall briefly discuss the basic concepts of crystal-field theory. The reader who would like a closer acquaintance with the procedure for calculating spectra of activated crystals is referred to several excellent monographs (for instance, [663–665]).

The main approximation of crystal-field theory is associated with introduction of the effective Hamiltonian of the dopant ion with an unfilled electron (d or f) shell in a crystal

$$\mathscr{H} = \mathscr{H}_0 + \mathscr{V}_{cr}, \tag{4.1}$$

where $\mathscr{H}_0$ is the Hamiltonian of a free ion, and $\mathscr{V}_{cr}$ is the energy of the ion interaction with the lattice, averaged over quantum-mechanical states of the latter so that $\mathscr{V}_{cr}$ can be represented as the energy of electrons on the activator ion in an effective crystalline electric field. The centrally symmetric self-consistent potential determining one-electron energies having been selected in $\mathscr{H}_0$, in order to find the energy levels of every electronic configuration governed by electron distribution over the states with fixed principal (n) and orbital (l) quantum numbers, it is necessary to calculate the eigenvalue of the operator

$$\mathscr{V} = \mathscr{V}_{ee} + \mathscr{V}_{so} + \mathscr{V}_{cr}, \tag{4.2}$$

which is treated as a perturbation. Here $\mathscr{V}_{ee}$ is the energy of the electrostatic interaction of the activator electrons and $\mathscr{V}_{so}$ is the operator of the spin–orbit interaction.

The Hamiltonian of the impurity-ion interaction with the crystal lattice is generally written down using one- and two-electron operators

$$\mathscr{V}_{cr} = \sum_i \mathscr{V}_1(r_i) + \sum_{i \neq j} \mathscr{V}_2(r_i, r_j), \tag{4.3}$$

where r_i is the radius vector of the ith electron of the activator ion. In the state space of the electron configuration $(nl)^N$, the $\mathscr{V}_1$ and $\mathscr{V}_2$ operators may be expanded in irreducible tensor operators $U_q^{(k)}$ of rank $k = 2m$ ($0 \leqslant m \leqslant l$) [666]

$$\mathscr{V}_1 = \sum_{kq} B_q^{(k)} U_q^{(k)}; \tag{4.4}$$

$$\mathscr{V}_2 = \sum_{pp'kq} B_q^{(k)}(pp') \{U^{(p)} U^{(p')}\}_q^{(k)}. \tag{4.5}$$

The braces represent the Kronecker product; $|p - p'| \leqslant k \leqslant |p + p'|$. The crystal field parameters $B_q^{(k)}$ and $B_q^{(k)}(pp')$ are the functionals of electron density in a crystal with parametric dependence upon the lattice structure. In the pairwise ion-interaction approximation (the "superposition" model), the crystal-field parameters $B_q^{(k)}$ are represented by the sum over the lattice

$$B_q^{(k)} = \sum_\lambda B^{(k)}(R_\lambda) C_q^{(k)}(\theta_\lambda, \varphi_\lambda), \tag{4.6}$$

where R_λ, θ_λ, and φ_λ are the polar coordinates of the λ ion and $C_q^{(k)}$ are functions of the angular coordinates of the λ ion (in the coordinate system centered on the impurity ion) which have the same transformation properties as the $U_q^{(k)}$ tensor operators. Various $B_q^{(k)}$ parameters are described by the equations resulting from the invariance of operator (4.3) under transformations belonging to the point symmetry group of the activator.

In the case of insulating crystals, the wave functions of the localized d and f electrons of activator ions have considerable overlap only with the wave functions of the outer electron shells of the immediate-neighbor ligands; therefore, the ligand coordinates completely determine the values of the $B_q^{(k)}$ (pp') parameters in the operator (4.5) governed by the direct-exchange interaction of activator and ligand electrons [666].

Numerous microscopic calculations of the energy spectra of activator centers (clusters) that contain the impurity ion and its immediate surroundings helped clarify a number of questions pertaining to the nature and relative role of various mechanisms of activator interaction with the lattice (see, for example, [667, 668]). In particular, the contributions of two-electron operators to the observed Stark splitting seem to be of the order of several per cent [666].

In interpreting the spectral results, the crystal-field parameters are found by fitting the spectra calculated with the Hamiltonian (4.4) to the experimental measurements. Since the approximation of $\mathscr{V}_{cr}$ by one-electron operators is known not to allow an accurate description of the spectral structure [note also that the observed Stark splittings are influenced (up to 10%) by the energy-level shifts induced by the electron–phonon interaction], this fitting procedure often leads to equivocal results that make no physical sense. Analysis of experimental data should be based on preliminary calculations (based on a microscopic model of the activator center) of initial parameters of the crystal field which will subsequently be clarified, their qualitative relationship being preserved by comparing the calculated and experimental Stark structures and the EPR spectra. This approach not only provides physically reasonable parameters of the crystal field but in some cases it also helps interpret the spectrum and choose the correct model of the center [669, 670].

In early works, the crystal-field parameters were calculated in terms of the point model of the ion lattice; the impurity ion was considered to be in an electric field of point ions with effective charges $Z_\lambda e$. The $B^{(k)}$ parameters in this case are proportional to

$$B^{(k)} (R_\lambda) \sim (1 - \sigma_k) \, e^2 Z_\lambda \left\langle r^k \right\rangle R_\lambda^{-(k+1)}, \tag{4.7}$$

where $\left\langle r^k \right\rangle$ is the mean value of the kth power of the radius of the optical electron, and σ_k are the constants of linear shielding, which allow consideration of the effect of the outer $5s^2 5p^6$ or $6s^2 6p^6$ shells deformed in the crystal field in the case of RE and actinide ions [671]. The crystal-field parameters calculated in terms of the point model using the $\left\langle r^k \right\rangle$ values obtained with one-electron Hartree–Fock radial functions of free ions [663] are usually of the same sign as the experimental results but have considerably different values. An account of electric dipole (for matrix ions at points without inversion) and quadrupole polarizations of ions in the crystal does not improve the agreement with experiment; the contributions to $B_q^{(k)}$ from the point-charge fields, dipole and quadrupole moments (found from the polarizability of free ions) summed over the lattice are close in value but may be of opposite sign [672]. Neglecting effects of short-range interactions—spatial distribution of electron density over the ligands—the relationship between various crystal-field parameters or their behavior in cases of hydrostatic and uniaxial lattice deformations cannot be correctly described in terms of the point model.

As quantum-mechanical calculations of activator-ion energy spectra are laborious and the numerical results obtained are subject to great uncertainties, the Stark structure of activator optical spectra has recently been interpreted with various models that involve semiphenomenological parametrization of the energies of the optical electrons. In particular, a simplified variant of the "superposition" model, suggested by *Newman* et al. (see review paper [668]) has been widely used. This model assumes complete mutual compensation of the contributions to the crystal-field parameters from electric multipoles localized on the matrix ion outside the cluster consisting of the activator ion and its immediate neighbors [673]. As a consequence, the sums over the lattice (4.6) include only the ligand contributions and, for a given geometry of the immediate surrounding, the magnitude of the Stark splittings are completely determined by the $B^{(k)}$ (R_λ) parameters. Assuming a power dependence of the $B^{(k)}$ parameters upon the distance R_λ between the ligand and the activator, $B^{(k)}(R_\lambda) = (R_0/R_\lambda)^{t_k} B^{(k)}(R_0)$, interaction of a fixed activator-ligand pair of ions is characterized by four parameters $(t_2, t_4, B^{(2)}, B^{(4)})$ in the d electron case and by six parameters $(t_2, t_4, t_6, B^{(2)}, B^{(4)}, B^{(6)})$ in the f electron case. The crystal-field parameters $B_q^{(4)}$ and $B_q^{(6)}$ used in the "superposition" model agree fairly well with the microscopic calculations. In some cases, considering the Stark structure of activated crystal spectra in terms of this model permits analysis of the local lattice deformations near dopant ions [674, 717], but artificial overestimation of the role of the short-range interactions does not yield physically reasonable powers t_2 for the quadrupole parameters of the crystal field [668].

The problem of parametrization of the effective Hamiltonian of the activator-ion interaction with the lattice has been further developed by *Malkin* et al. [718, 719]. They formulated the charge-exchange model which was then used in calculating static and dynamic spectral characteristics of both high-symmetry (cubic) and low-symmetry activator centers of impurity ions [669, 670, 718–720]. The charge-exchange model in crystal-field theory is based on just the same concept of the electron-density distribution over a crystal as the similar model in the theory of crystal-lattice dynamics—fictitious positive charges proportional to the squared overlap integrals of the wave functions are placed on the bonds between ions with overlapping electron shells [721]. Electrostatic valence-electron interactions with point charges and dipole moments on the matrix ions are explicitly treated, while only the part of the crystal field created by exchange charges is parametrized. Ion polarizability in a crystal should be distinguished from that of a free ion, and account should be taken of polarization due to deformation of electron shells of the neighboring ions with overlapping wave functions.

In any theory of the crystal field based on pairwise interactions, the dependence of individual ion contributions to the crystal-field parameters on the angular coordinates θ_λ, φ_λ remains exactly the same as in the point model [see (4.6)]; therefore, calculations of overlapping and nonorthogonality of the wave functions of activator and ligand electrons reduce to renormalizing the average values $\langle r^k \rangle$. In order to proceed from the point model to the charge-exchange model, it is sufficient to replace $\langle r^k \rangle$ in the $B^{(k)}(R_\lambda)$ parameters (λ is the ligand number) for the point model (4.7) by

$$\langle r^k \rangle - \frac{2(2k+1)}{(2l+1)} \frac{G_k}{Z_\lambda(1-\sigma_k)} S_k^{(l)}(R_\lambda),$$

where Z_λ is the ligand valence, l is the orbital moment of the activator electrons, G_k are the dimensionless parameters of the model, and $S_k^{(l)}(R_\lambda)$ are linear combinations of the squared overlap integrals which, in the case of ligands with filled outer $s^2 p^6$ shells, are given by

$$S_k^{(l)} = S_s^2 + S_\sigma^2 + P_k^{(l)} S_\pi^2,$$

with $P_2^{(3)} = -P_6^{(3)} = 3/2$,

$$P_4^{(3)} = 1/3$$

for f electrons; and

$$P_2^{(2)} = 1,$$

$$P_4^{(2)} = -4/3,$$

$$P_6^{(2)} = 0$$

for d electrons. Here S_s, S_σ, and S_π are two-center overlap integrals of the wave functions of f and d electrons with the s, p_z, and p_x ligand functions, respectively.

In a simple variant of this model, involving only one fitting parameter $G = G_k$ which characterizes the magnitude of exchange charges and is constant for a particular activator–ligand pair, it is possible to describe satisfactorily both the Stark structure of an activator spectrum in one matrix and its variations in homological crystals. In calculation of the crystal-field parameters, local lattice deformation near the activator should first of all be considered.

The various crystal fields that depend upon the relationship between $\mathscr{V}_{ee}$, $\mathscr{V}_{so}$, and $\mathscr{V}_{cr}$ are generally divided, following *Bethe* [502] into three types: weak fields ($\mathscr{V}_{ee}$, $\mathscr{V}_{so} \gg \mathscr{V}_{cr}$), medium fields ($\mathscr{V}_{ee} \gg \mathscr{V}_{cr} \gg \mathscr{V}_{so}$), and strong fields ($\mathscr{V}_{cr} \gg \mathscr{V}_{ee} \gg \mathscr{V}_{so}$). The weak-crystal-field case arises with RE and actinide ions whose inner unfilled $4f$ and $5f$ shells are compressed and shielded by outer filled

$5s^2p^6$ and $6s^2p^6$ shells, respectively. In the case of normal electron coupling ($\mathcal{V}_{ee} \gg \mathcal{V}_{so}$, Ln ions), electrostatic electron interaction creates terms that are characterized by certain total orbital and spin angular moments (L and S). The spin–orbit interaction $\mathcal{V}_{so} = \lambda \boldsymbol{L} \cdot \boldsymbol{S}$ partially reduces the $(2L + 1)(2S + 1)$ multiple-term degeneracy, and forms a multiplet structure of free-ion spectra. In considering Stark splittings in a weak crystal field, the eigenfunctions $|L, S, J, J_z\rangle$ of the total momentum $\boldsymbol{J} = \boldsymbol{L} + \boldsymbol{S}$ are used as a basis of the zero-order approximation. This case is seen in Fig. 4.1, which shows the scheme of energy-level splitting (exemplified by the $^4I_{9/2}$ ground state) for the Nd^{3+} ion in a $Y_3Al_5O_{12}$ crystal.

At medium crystal fields, we take the eigenfunctions $|L, L_z, S, S_z\rangle$ of the $\boldsymbol{L}, \boldsymbol{S}$ operators as the zero-order wave functions. In a medium crystal field, Stark splittings are smaller than the energy gaps between various terms of the same multiplicity (the same total spin) but larger than the fine-structure splitting in a free ion. Along with low-symmetry components of the crystal field, the spin–orbit interaction is generally treated as a perturbation that takes the cubic component of the crystal field into account. The medium-crystal-field case is not observed in its pure form. As a rule, iron-group ions are exposed to an intermediate crystal field ($\mathcal{V}_{ee} \approx \mathcal{V}_{cr}$), so the sum $\mathcal{V}_{ee} + \mathcal{V}_{cr}$ should be considered as a perturbation, or term interactions should be calculated by successive diagonalization of the $\mathcal{V}_{ee}$, $\mathcal{V}_{cr}$ matrices.

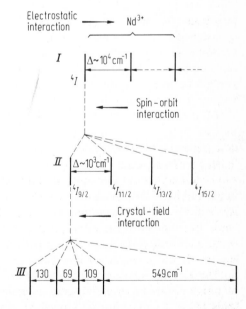

Fig. 4.1. Level splitting scheme for Nd^{3+} ions in a $Y_3Al_5O_{12}$ crystal. I—terms, characterized by L and S quantum numbers; II—manifolds, characterized by L, S and J quantum numbers; III—Stark levels, corresponding to irreducible representations of the site-symmetry group

Table 4.1. Splitting of Ln^{3+} ion manifolds in crystal fields of given symmetry[a]

Local symmetry	Schönflies designation	$J=0$	1	2	3	4	5	6	7	8
		$2J+1=1$	3	5	7	9	11	13	15	17
Cubic	O_h, T_d, O, T_h, T	1	1	2	3	4	4	6	6	7
Hexagonal	$D_{6h}, D_{3h}, C_{6v}, D_6,$ C_{6h}, C_{3h}, C_6	1	2	3	5	6	7	9	10	11
Trigonal	$D_{3d}, C_{3v}, D_3, C_{3i}, C_3$	1	2	3	5	6	7	9	10	11
Tetragonal	$D_{4h}, D_{2d}, C_{4v}, D_4,$ C_{4h}, S_4, C_4	1	2	4	5	7	8	10	11	13
Orthorhombic	D_{2h}, C_{2v}, D_2	1	3	5	7	9	11	13	15	17
Monoclinic	C_{2h}, C_s, C_2	1	3	5	7	9	11	13	15	17
Triclinic	C_i, C_1	1	3	5	7	9	11	13	15	17
		$J=1/2$	3/2	5/2	7/2	9/2	11/2	13/2	15/2	17/2
		$2J+1=2$	4	6	8	10	12	14	16	18
Cubic	O_h, T_d, O, T_h, T	1	1	2	3	3	4	5	5	6
Other lower symmetry types		1	2	3	4	5	6	7	8	9

[a] Possible site positions with indicated symmetry type in different crystal structures are given in Table 5.2.

A strong crystal field breaks the coupling between the orbital and spin momenta of electrons in the unfilled shell; an electron configuration in this case is determined by occupation numbers of one-electron Stark levels whose energies are found by diagonalization of the $\mathscr{V}_{cr}$ operator in the basis of one-electron functions $|ll_z\rangle$. In the strong crystal fields that act on the palladium- and platinum-group ions, the effects caused by the covalent bonds formed between paramagnetic ions and ligands are of particular importance.

The number of Stark levels in crystal fields of various symmetries depends on the quantum numbers l (strong field), L (medium field), and J (weak field), and is determined by the methods of the group theory. Table 4.1 lists the numbers of Stark components for various symmetries of the activator center for J running from 0 to 8 (integral values correspond to ions with an even number of electrons in the unfilled shell) and from 1/2 to 17/2. For example, in the case of the Nd^{3+} ion, which is the most common laser activator in fields of any symmetry except cubic, the number of electrons is odd and the number of Stark levels in a given multiplet is always the same. This greatly facilitates the interpretation of optical spectra in establishing a scheme of energy levels. In the case of ions with an even number of electrons, e.g., Ho^{3+} or Tm^{3+}, in fields of different symmetries, the number of Stark levels will also be different. In this case the point symmetry of the activator center must be known for the interpretation of optical spectra.

4.2 Stark Levels of Activator Ions in Laser Crystals (Experimental Data)

At present, about 280 activated insulating laser crystals (see Tables 1.1–5) are known, but the positions of the energy levels directly associated with the stimulated-emission process have been established reliably and sufficiently completely only for about one-third of this number. This situation is attributed, on the one hand, to the severe difficulties involved in interpreting the spectra of multicenter and, particularly, mixed crystals and, on the other hand, to the theoretical problems that arise in their interpretation (see Sect. 2.3). Unfortunately, the main key that has been found in the form of crystal-field theory for describing the observed spectral structure can be satisfactorily applied only to individual cases.

The results on the energies of Stark levels of laser-crystal activator centers so far available are listed in Tables 4.2–15. References have been kept to the minimum required to provide a sufficiently complete idea of the current state of spectroscopic studies on a given crystal. In these tables, the starred energies of the levels and complete splittings ΔE of the manifolds need further clarification. If the levels of the near-lying manifolds have not been identified, their total splitting is given in brackets. For most crystals, the positions of the Stark levels are represented to within an accuracy of 3–5 cm^{-1}, or better.

Note, finally, the order of the tables showing the positions of energy levels in laser crystals. The tables giving the Stark levels of simple fluoride crystals are first, followed by mixed fluoride crystals and so on. In this arrangement, crystals doped with lanthanide Ln^{3+} or Ln^{2+} ions are listed first, followed by actinide and iron-group ions.

Table 4.2.   Crystal-field splitting of Pr^{3+} ion manifolds

[SL]J manifold	Stark-level positions [cm^{-1}]	Number of components		ΔE [cm^{-1}]
		Theory	Exper.	
	LaF$_3$ crystal, 77 K [258, 259]			
$^3H_4$	0, 57, 76, 151, 195, 204, 299, 338, 508	9	9	508
$^3H_5$	2179, 2299, 2305, 2310, 2356, 2434	11	6	265*
$^3H_6$	4222, 4266, 4302, 4385, 4439, 4504, 4523, 4553, 4577, 4604, 4670, 4785	13	12	563*
$^3F_2$	5136, 5181, 5200, 5275, 5280	5	5	144
$^3F_3$	6453, 6495, 6499, 6587, 6602, 6622, 6722	7	7	269
$^3F_4$	6932, 6944, 6981, 6997, 7030, 7035, 7099*, 7103, 7164	9	9	232
$^1G_4$	9716, 9751, 9876, 9912, 10005, 10042, 10048, 10163, 10499	9	9	783
$^1D_2$	16873, 16893, 17083, 17183, 17204	5	5	331
$^3P_0$	20930	1	1	—
$^3P_1 + ^1I_6$	21458, 21475, 21479, 21484, 21487, 21516, 21522, 21529, 21536, 21554, 21567, 21580	16	12	(122*)

(Continued)

Table 4.2 (*continued*)

[SL]J manifold	Stark-level positions [cm^{-1}]	Number of components		ΔE [cm^{-1}]
		Theory	Exper.	
$^3P_2$	22691, 22714, 22734, 22772, 22819	5	5	128
$^1S_0$	46986	1	1	—
	LaCl$_3$ crystal, 4.2 K [372]			
$^3H_4$	0, 33, 97, 131	6	4	131*
$^3F_3$	6280, 6292, 6296, 6299, 6349	5	5	69
$^3F_4$	6773, 6782, 6785	6	3	12*
$^1G_4$	9741, 9772, 9777, 9813	6	4	72*
$^1D_2$	16636*, 16736, 16784	3	3	148*
$^3P_0$	20483	1	1	—
$^3P_1$	21075, 21101	2	2	26
$^3P_2$	22215, 22234, 22254	3	3	39
	PrCl$_3$ crystal, 4 K [260]			
$^3H_4$	0, 33, 96	6	3	96*
$^3H_5$	2115, 2134, 2169, 2188, 2202, 2222	7	6	107*
$^3H_6$	4231, 4250, 4275, 4295, 4331	9	5	102*
$^3F_2$	4922, 4950	3	2	28*
$^3F_3$	6303, 6308, 6352, 6367	5	4	64*
$^3F_4$	6700, 6751, 6772, 6785, 6804	6	5	104*
$^1G_4$	9763, 9772	6	2	9*
	LaBr$_3$ crystal, liquid-air temperature [767]			
$^3H_4$	0, 18, 86, 156	6	4	156*
$^3H_5$	2138, 2165, 2186	7	3	46*
$^3H_6$	4214*, 4226*, 4250, 4274*, 4298	9	5	84*
$^3F_2$	4919	3	1	—*
$^3F_3$	6253, 6281, 6284, 6312, 6329	5	5	76
$^3F_4$	6688, 6735, 6750, 6763, 6779	6	5	91*
$^1G_4$	9732, 9745, 9783	6	3	51*
$^1D_2$	16572, 16672, 16718	3	3	46
$^3P_0$	20378	1	1	—
$^3P_1$	20971, 20997	2	2	76
$^1I_6$	21236, 21339, 21352	9	3	16*
$^3P_2$	22115, 22129, 22153	3	3	38

Table 4.3. Crystal-field splitting of Nd^{3+} ion manifolds

[SL]J manifold	Stark-level positions [cm^{-1}]	Number of components		ΔE [cm^{-1}]
		Theory	Exper.	
	LiYF$_4$ crystal, 77 K [261, 262, 493]			
$^4I_{9/2}$	0, 132, 182, 249, 528	5	5	528
$^4I_{11/2}$	1998, 2042, 2079, 2228, 2264	6	5	266*
$^4I_{13/2}$	3948, 3976, 3995, 4026, 4205, 4228, 4238	7	7	290
$^4I_{15/2}$	5851, 5912, 5947, 6026, 6315, 6347, 6388, 6432	8	8	581
$^4F_{3/2}$	11538, 11597	2	2	59

(*Continued*)

Table 4.3 (*continued*)

[SL]J manifold	Stark-level positions [cm^{-1}]	Number of components		ΔE [cm^{-1}]
		Theory	Exper.	
	CaF$_2$ crystal (type I), 77 K [45]			
	Stark levels of L center			
$^4I_{9/2}$	0, 82, 202, 744	5	4	744*
$^4I_{11/2}$	1977, 2084*, 2094, 2103, 2462	6	5	485*
$^4F_{3/2}$	11594, 11707	2	2	113
$^4F_{7/2} + {}^4S_{3/2}$	13541, 13594, 13638, 13761	6	4	(220*)
$^2P_{1/2}$	23480	1	1	—
$^4D_{3/2}$	28106, 28280	2	2	174
	Stark levels of M center			
$^4I_{9/2}$	0, 36	5	2	36*
$^4I_{11/2}$	2028, 2052, 2191	6	3	163*
$^4F_{3/2}$	11582, 11622	2	2	40
$^4F_{7/2} + {}^4S_{3/2}$	13468, 13706, 13712, 13770	6	4	(302*)
$^2P_{1/2}$	23416	1	1	—
$^4D_{3/2}$	28272, 28288	2	2	16
	Stark levels of N center			
$^4I_{9/2}$	0, 42	5	2	42*
$^4I_{11/2}$	2031, 2061, 2085	6	3	54*
$^4F_{3/2}$	11602, 11641	2	2	39
$^4F_{7/2} + {}^4S_{3/2}$	13486, 13602, 13727, 13732	6	4	(246*)
$^2P_{1/2}$	23441	1	1	—
$^4D_{3/2}$	28281*, 28320	2	2	39*
	CaF$_2$ crystal (type II), 77 K [316, 496]			
$^4I_{9/2}$	0, 85, 440, 475, 925	5	5	925
$^4I_{11/2}$	1986, 2022, 2304, 2336, 2584, 2618	6	6	632
$^4I_{13/2}$	4175, 4255, 4525, 4585, 4619	7	5	444*
$^4I_{15/2}$	6112, 6158, 6211, 6623, 6671, 6765	8	6	653*
$4F_{3/2}$	11398, 11494	2	2	96
$^4F_{5/2} + {}^2H_{9/2}$	12394, 12503, 12525, 12687, 12729, 12867, 12920, 13046	8	8	(652)
$^4F_{7/2} + {}^4S_{3/2}$	13369, 13519, 13580, 13680, 13624, 13646	6	6	(277)
$^4F_{9/2}$	14637, 14747, 14769, 14874, 15024	5	5	387
$^2H_{11/2}$	15855, 16033, 16067, 16098, 16170, 16292	6	6	437
$^4G_{5/2} + {}^2G_{7/2}$	17007, 17082, 17167, 17218, 17370, 17409, 17790	7	7	(783)
$^2K_{13/2} + {}^4G_{7/2} + {}^4G_{9/2}$	18829, 18882, 19048, 19346, 19410, 19558, 19665, 19700, 19728, 19860, 20235	16	11	(1406*)
$^2P_{1/2}$	23277	1	1	—
	LaF$_3$ crystals, 4.2 and 77 K[a] [37, 263, 264]			
$^4I_{9/2}$	0, 44, 140, 297, 502	5	5	502
$^4I_{11/2}$	1980, 2038, 2069, 2092, 2188, 2223	6	6	243
$^4I_{13/2}$	3919, 3974, 4038, 4077, 4119, 4213, 4276	7	7	357
$^4I_{15/2}$	5817, 5876, 5989, 5142, 6173, 6320, 6448, 6551	8	8	734
$^4F_{3/2}$	11595, 11637	2	2	42
$^4F_{5/2}$	12596, 12613, 12621	3	3	25
$^2H_{9/2}$	12675, 12693, 12755, 12842, 12904	5	5	229
$^4F_{7/2} + {}^4S_{3/2}$	13515, 13591, 13671, 13677, 13710, 13714	6	6	(199)
$^4F_{9/2}$	14835, 14860, 14891, 14927, 14958	5	5	81
$^2H_{11/2}$	15998, 16033, 16045, 16059, 16103	6	5	105*

[a] The levels lying above 12000 cm^{-1} were investigated at 4.2 K.

(*Continued*)

Table 4.3   (*continued*)

[SL]J manifold	Stark-level positions [cm^{-1}]	Number of components		ΔE [cm^{-1}]
		Theory	Exper.	
$^4G_{5/2}+{}^2G_{7/2}$	17304, 17315, 17364, 17512, 17520, 17570, 17601	7	7	(297*)
$^4G_{7/2}$	19147, 19235, 19251, 19325	4	4	176
$^4G_{9/2}$	19568, 19617, 19651, 19685, 19702, 19739, 19801, 19839	5	8	271*
$^2K_{15/2}+{}^2G_{9/2}$	21158, 21176, 21201, 21234, 21254, 21302, 21339, 21351	13	8	(193*)
$^2P_{1/2}$	23468	1	1	—
$^2D_{5/2}$	23991	3	1	—*
	CeF$_3$ crystal, 77 K [265]			
$^4I_{9/2}$	0, 45, 144, 303, 510	5	5	510
$^4I_{11/2}$	1986, 2043, 2077, 2101, 2199, 2237	6	6	251
$^4F_{3/2}$	11598, 11643	2	2	45
$^2P_{1/2}$	23480	1	1	—
	BaF$_2$–LaF$_3$ crystal, 77 K [92]			
$^4I_{9/2}$	0, 108, 260, 337, 410	5	5	410
$^4I_{11/2}$	1900, 1992, 2035, 2070, 2155, 2260	6	6	360
$^4I_{13/2}$	3890, 3956, 3995, 4040*, 4095, 4180*, 4270	7	7	380
$^4I_{15/2}$	5770, 5870, 5925, 6014, 6125*, 6203*, 6290, 6340*	8	8	570*
$^4F_{3/2}$	11480, 11628	2	2	148
	LiYO$_2$ crystal, 77 K [754]			
$^4I_{9/2}$	0, 28, 265, 445, 639	5	5	639
$^4I_{11/2}$	1888, 1915, 2135, 2259, 2329, 2355	6	6	467
$^4I_{13/2}$	4060, 4092, 4202, 4274, 4299, 4323	7	6	263*
$^4I_{15/2}$	5714, 5774, 6031, 6173, 6313, 6398, 6435	8	7	721*
$^4F_{3/2}$	11216, 11410	2	2	194
$^4F_{5/2}+{}^2H_{9/2}$	12186, 12315, 12401, 12436, 12491, 12547, 12638, 12694	8	8	(508)
$^4F_{7/2}+{}^4S_{3/2}$	13151, 13182, 13308, 13387, 13452, 13506	6	6	(355)
$^4F_{9/2}$	14327, 14420, 14493, 14514, 14575	5	5	248
$^2H_{11/2}$	15957, 16142, 16260, 16327	6	4	370*
$^4G_{5/2}+{}^2G_{7/2}$	16656, 16762, 17036, 17123, 17212, 17301	7	6	(745*)
$^4G_{7/2}$	18464, 18594, 18692, 18762	4	4	298
$^2G_{9/2}$	19166, 19275, 19316, 19399, 19608	5	5	440
$^4G_{9/2}+{}^2D_{3/2}+$ $+{}^4G_{11/2}$	20483, 20606, 20799, 20929, 21026, 21124	13	6	(641*)
$^2P_{1/2}$	22915	1	1	—
$^2D_{5/2}$	23337, 23348	3	3	11*
$^2D_{3/2}$	27285, 27473	2	2	188
$^4D_{1/2}$	27933	1	1	—
	LiNbO$_3$ crystal, 77 K [266]			
$^4I_{9/2}$	0, 156, 170, 440, 486	5	5	486
$^4I_{11/2}$	1987, 2034, 2107, 2190, 2228, 2263	6	6	276
$^4I_{13/2}$	3918, 3973, 4035, 4118, 4140, 4184, 4211	7	7	293
$^4I_{15/2}$	5777, 5916, 6005, 6087, 6105, 6217, 6290, 6449	8	8	672
$^4F_{3/2}$	11250, 11409	2	2	159
$^4F_{5/2}+{}^2H_{9/2}$	12133*, 12291, 12396, 12421, 12449, 12463, 12574, 12692	8	8	(559*)

(*Continued*)

Table 4.3 (*continued*)

[SL]J manifold	Stark-level positions [cm^{-1}]	Number of components		ΔE [cm^{-1}]
		Theory	Exper.	
$^4F_{7/2}+{}^4S_{3/2}$	13199, 13291, 13398*, 13437	6	4	(238*)
$^4F_{9/2}$	14498, 14567, 14620, 14664, 14685	5	5	187
$^2H_{11/2}$	15748, 15803, 15873, 15888, 15911, 15926	6	6	178
$^4G_{5/2}$	16753, 16852, 16909	3	3	156
$(^4G,{}^2G)_{7/2}$	16949*, 17071, 17135, 17176	4	4	227*
$(^4G,{}^2G)_{7/2}$	18719, 18786, 18836	4	3	117*
$^4G_{9/2}$	19135, 19260, 19331	5	3	196*
$^2G_{9/2}$	20682, 20764, 20799, 20868, 20938	5	5	256
$^4G_{11/2}+{}^2K_{15/2}+$ $+{}^2(P,D)_{3/2}$	20973, 21022, 21039, 21213, 21372, 21631, 21696	14	7	(723*)
$^2P_{1/2}$	22914	1	1	—
$^2D_{5/2}$	23348, 23554, 23608	3	3	260
$^2(P,D)_{3/2}$	25745, 25882	2	2	137
$^4D_{3/2}$	27365, 27469	2	2	104
$^2D_{1/2}$	28031	1	1	—
	LiNdP$_4$O$_{12}$ crystal, 300 K [609, 764]			
$^4I_{9/2}$	0, 106, 197, 268, 326	5	5	326
$^4I_{11/2}$	1939, 2003, 2053, 2078, 2108, 2136	6	6	197
$^4I_{13/2}$	3893, 3964, 4009, 4053, 4073, 4120, 4132	7	7	293
$^4F_{3/2}$	11484, 11539	2	2	55
	KY(MoO$_4$)$_2$ crystal, 77 K [85]			
$^4I_{9/2}$	0, 94, 162, 253, 430	5	5	430
$^4I_{11/2}$	1960, 1985, 2016, 2079, 2120, 2205	6	6	245
$^4I_{13/2}$	3810, 3865, 3904*, 3940, 4030, 4110, 4195	7	7	385
$^4I_{15/2}$	5750, 5820, 5890*, 5930, 6000, 6180*, 6230, 6290*	8	8	540*
$^4F_{3/2}$	11356, 11438	2	2	82
$^4F_{5/2}$	12369, 12423, 12459	3	3	90
$^2H_{9/2}$	12492, 12545, 12590, 12642, 12670	5	5	178
$^4F_{7/2}$	13215, 13310, 13387, 13435	4	4	220
$^4S_{3/2}$	13479, 13488	2	2	9
$^4F_{9/2}$	14576, 14629, 14698, 14727, 14760	5	5	184
$^2H_{11/2}$	15798, 15835, 15853, 15923, 15970, 15995	6	6	197
$^4G_{5/2}+$ $+(^2G,{}^4G)_{7/2}$	16866, 16909, 16966, 17065, 17185, 17238, 17364	7	7	(498)
$(^2G,{}^4G)_{7/2}$	18702*, 18780, 18797, 18832	4	4	130*
$^2G_{9/2}$	18875, 18928, 18965, 19037*, 19048	5	5	173*
$^2K_{15/2}$	20390, 20400*, 20425, 20483*, 20505*, 20535*, 20590*	8	7	200*
$^4G_{9/2}$	20705, 20730, 20750, 20772, 20825	5	5	120
$^2(P,D)_{3/2}$	20842, 20860	2	2	18
$^4G_{11/2}$	20910, 20995, 21026, 21093, 21128, 21210	6	6	300
$^2P_{1/2}$	23105	1	1	—
$^2D_{5/2}$	23580, 23685*, 23752*	3	3	172*
$^2(P,D)_{3/2}$	25915*, 25975	2	2	60*
	KY(WO$_4$)$_2$ crystal, 77 K [94, 809]			
$^4I_{9/2}$	0, 104, 155, 306, 355	5	5	355
$^4I_{11/2}$	1943, 1959, 2020, 2112, 2142, 2180	6	6	237

(*Continued*)

Table 4.3   (*continued*)

[SL]J manifold	Stark-level positions [cm^{-1}]	Number of components		ΔE [cm^{-1}]
		Theory	Exper.	
$^4I_{13/2}$	3902, 3916, 3975, 4077, 4100, 4145, 4175	7	7	273
$^4I_{15/2}$	5860, 5874, 5952, 5963*, 6074, 6192, 6240, 6287	8	8	427
$^4F_{3/2}$	11300, 11412	2	2	112
$^4F_{5/2}+{}^2H_{9/2}$	12324, 12398, 12425, 12464, 12527, 12560, 12601, 12636	8	8	(312)
$^4F_{7/2}+{}^4S_{3/2}$	13277, 13348*, 13367, 13403, 13441, 13454	6	6	(177)
$^4F_{9/2}$	14545, 14567, 14636, 14743, 14785	5	5	240
$^2H_{11/2}$	15756, 15780, 15838, 15896, 15950, 15968	6	6	212
$^4G_{5/2}+$ $+({}^2G,{}^4G)_{7/2}$	16792, 16903*, 16929, 17047, 17112*, 17224, 17301	7	7	(491)
$({}^2G,{}^4G)_{7/2}$	18755, 18793, 18847, 18936	4	4	181
$^2G_{9/2}$	19138, 19290*, 19324, 19354, 19387	5	5	248
$^2K_{13/2}$	19474, 19535, 19558, 19600, 19662, 19818*	7	6	344*
$^4G_{9/2}+{}^4G_{11/2}$	20768, 20815, 20844, 20859, 20920, 20982, 21042, 21090, 21133	11	9	(365*)
$^2K_{15/2}$	21249, 21418, 21455, 21608, 21678, 21834	8	6	585*
$^2P_{1/2}$	23041	1	1	—
$^2D_{5/2}$	23504, 23652, 23717	3	3	213
$^2P_{3/2}$	25917, 26025	2	2	108
$^4D_{3/2}$	27581, 27660	2	2	79
$^4D_{1/2}$	27757	1	1	—
KGd(WO$_4$) crystal, 77 K				
$^4I_{9/2}$	0, 102, 150, 295, 340	5	5	340
$^4I_{11/2}$	1943, 1960, 2020, 2105, 2130, 2172	6	6	229
$^4I_{13/2}$	3900, 3914, 3970, 4060, 4087, 4135, 4165	7	7	272
$^4F_{3/2}$	11313, 11418	2	2	105
300 K				
$^4I_{11/2}$	1945, 1956, 2020, 2104, 2134, 2172	6	6	227
$^4F_{3/2}$	11315, 11417	2	2	102
CaMg$_2$Y$_2$Ge$_3$O$_{12}$ crystal, 300 K [630, 828]				
$^4I_{9/2}$	0, 148, 187, 347, 807	5	5	807
$^4I_{11/2}$	2010, 2071, 2113, 2153, 2423, 2502	6	6	492
$^4I_{13/2}$	3940, 3972, 4036, 4080, 4393, 4428, 4483	7	7	543
$^4F_{3/2}$	11449, 11570	2	2	121
CaAl$_4$O$_7$ crystal, 77 K [481]				
Stark levels of type I center				
$^4I_{9/2}$	0, 178, 220, 271, 437	5	5	437
$^4I_{11/2}$	1919, 2082, 2088, 2116, 2158, 2226	6	6	307
$^4I_{13/2}$	3862, 3866, 4033, 4049, 4065, 4122, 4220	7	7	358
$^4I_{15/2}$	6030, 6039, 6061, 6110, 6247, 6321	8	6	291*
$^4F_{3/2}$	11372, 11517	2	2	145
$^4F_{5/2}+{}^2H_{9/2}$	12413, 12467, 12514, 12525, 12544, 13579	7	6	(166*)
$^4F_{7/2}+{}^4S_{3/2}$	13366, 13403, 13460, 13530, 13547, 13619	6	6	(253)
$^4F_{9/2}$	14595, 14609, 14633, 14681, 14702	5	5	107
$^2H_{11/2}$	15875, 16033	6	2	158*
$^4G_{5/2}+{}^2G_{7/2}$	16957, 17077, 17093, 17195, 17269, 17333, 17411	7	7	(454)

(Continued)

Table 4.3 (*continued*)

[SL]J manifold	Stark-level positions [cm^{-1}]	Number of components		ΔE [cm^{-1}]
		Theory	Exper.	
$^4G_{7/2}$	18890, 18940, 19001, 19052	4	4	162
$^2G_{9/2}$	19389, 19427, 19480, 19525, 19618	5	5	229
	Stark levels of type II center			
$^4I_{9/2}$	0, 51, 295, 474, 585	5	5	585
$^4I_{11/2}$	1942, 1966, 2215, 2269, 1196	6	5	354*
$^4I_{13/2}$	3881, 3887, 4166, 4226, 4259, 4300, 4339	7	7	458
$^4I_{15/2}$	5920, 5969, 5996, 6012, 6024, 6132, 6348, 6404	8	8	484
$^4F_{3/2}$	11349, 11620	2	2	271
$^4F_{5/2}+{}^2H_{9/2}$	12362, 12537, 12638, 12684, 12737, 12815, 12858	7	7	(496)
$^4F_{7/2}+{}^4S_{3/2}$	13314, 13496, 13584, 13595, 13603, 13703	6	6	(389)
$^4F_{9/2}$	14600, 14750, 14796, 14807	5	4	207*
$^2H_{11/2}$	15940, 15972	6	2	32*
$^4G_{5/2}+G_{7/2}$	17012, 17069, 17106, 17154, 17421, 17491, 17539	7	7	(527)
$^4G_{7/2}$	18870, 18959, 19042, 19116	4	4	246
	Ca(NbO$_3$)$_2$ crystal, 77 K [93, 172]			
$^4I_{9/2}$	0, 100, 158, 237, 522	5	5	522
$^4I_{11/2}$	1952, 1981, 2017, 2093, 2218, 2246	6	6	234
$^4I_{13/2}$	3896, 3919, 3969, 4027, 4180, 4220, 4230	7	7	336
$^4I_{15/2}$	5785*, 5820, 5835, 5935, 6000, 6260, 6315, 6370	8	8	585*
$^4F_{3/2}$	11375, 11423	2	2	48
$^4F_{5/2}$	12306, 12361, 12407	3	3	101
$^2H_{9/2}$	12467, 12502, 12541, 12586, 12692	5	5	225
$^4F_{7/2}$	13277, 13337, 13374, 13448	4	4	171*
$^4S_{3/2}$	13479, 13484	2	2	5
$^4F_{9/2}$	14581, 14612, 14690, 14765, 14812	5	5	231
$^2H_{11/2}$	15741, 15775*, 15818, 15843, 15918, 15987	6	6	246
$^4G_{5/2}$	16798, 16846, 17036	3	3	243
$(^4G,{}^2G)_{7/2}$	17135, 17173, 17212, 17406	4	4	271
$(^4G,{}^2G)_{7/2}$	18748, 18818, 18900*, 18914	4	4	166
$^2K_{13/2}$	19050*, 19157, 19357, 19354, 19391, 19440, 19470*, 19497*	7	7	477*
$^2G_{9/2}$	19619, 19666, 19743, 19814	5	4	195*
$^4G_{9/2}$	20717, 20773, 20790, 20816, 20869*	5	5	152*
$^4G_{11/2}$	20981, 21022, 21053, 21088, 21146, 21218	6	6	237
$^2P_{1/2}$	23095	1	1	—
$^2D_{5/2}$	23624, 23669, 23725	3	3	101
$^2(P,D)_{3/2}$	25979, 26031	2	2	52
	CaMoO$_4$ crystal, 77 K [268]			
$^4I_{9/2}$	0, 109, 162, 223, 458	5	5	458
$^4I_{11/2}$	1982, 2019, 2025, 2062, 2193, 2226	6	6	245
$^4I_{13/2}$	3922, 3950, 3967, 4001, 4055, 4155, 4190	7	7	267
$^4I_{15/2}$	5844, 5895, 5927, 6307, 6348	8	5	504*
$^4F_{3/2}$	11387, 11455	2	2	68
	CaWO$_4$ crystal, 77 K [268, 269]			
$^4I_{9/2}$	0, 95, 165, 230, 472	5	5	472
$^4I_{11/2}$	1976, 2016, 2056, 2189, 2226	6	5	250*

(*Continued*)

Table 4.3 (*continued*)

[SL]J manifold	Stark-level positions [cm^{-1}]	Number of components		ΔE [cm^{-1}]
		Theory	Exper.	
$^4I_{13/2}$	3928, 3955, 3975, 4002, 4085, 4160, 4202	7	7	276
$^4I_{15/2}$	5848, 5903, 5935, 6277, 6305, 6377	8	6	529*
$^4F_{3/2}$	11406, 11469	2	2	63
	SrMoO$_4$ crystal, 77 K [97, 268]			
$^4I_{9/2}$	0, 95, 158, 218, 377	5	5	377
$^4I_{11/2}$	1956, 1999, 2036, 2140	6	4	184*
$^4I_{13/2}$	3915, 3948, 3962, 3996*, 4039, 4100, 4150	7	7	235
$^4F_{3/2}$	11402, 11467	2	2	65
	SrWO$_4$ crystal, 77 K [268]			
$^4I_{9/2}$	0, 97, 163, 222, 395	5	5	395
$^4I_{11/2}$	1968, 2010, 2045, 2152, 2178	6	5	210*
$^4I_{13/2}$	3918, 3950, 3968, 3998, 4122, 4163	7	6	245*
$^4I_{15/2}$	5869, 5913, 5948, 6244	8	4	375*
$^4F_{3/2}$	11420, 11488	2	2	68
	Y$_2$O$_3$ crystal, 4.2 K [144, 770]			
$^4I_{9/2}$	0, 29, 267, 447, 643	5	5	643
$^4I_{11/2}$	1897, 1935, 2147, 2271, 2331, 2359	6	6	462
$^4I_{13/2}$	3814, 3840, 4093, 4200, 4280, 4305, 4329	7	7	515
$^4I_{15/2}$	5709, 5726, 6060, 6162, 6315, 6401, 6443, 6479	8	8	770
$^4F_{3/2}$	11208, 11404	2	2	196
$^4F_{5/2}$	12138, 12321, 12436	3	3	298
$^2H_{9/2}$	12396, 12492, 12554, 12642, 12716	5	5	320
$^4F_{7/2}$	13181, 13310, 13421, 13501	4	4	320
$^4S_{3/2}$	13392, 13415	2	2	23
$^4F_{9/2}$	14426, 14534, 14573, 14723	5	4	297*
$^2H_{11/2}$	15709, 15727, 15756, 15870, 15944, 16024	6	6	315
$^4G_{5/2}$	16590, 17030, 17309	3	3	719
$^2G_{7/2}$	16767, 16903, 17114, 17215	4	4	448
$^4G_{7/2}$	18464, 18592, 18718, 18794	4	4	330
$^4G_{9/2}$	19177, 19193, 19211, 19271, 19286	5	5	109
$^2K_{15/2} + {}^2D_{3/2} +$ $+ {}^2G_{9/2} + {}^4G_{11/2}$	20477, 20506, 20599, 20609, 20718, 20752, 20808, 20927, 20935, 20955, 21034, 21438, 21598, 21697	21	14	(1220*)
$^2P_{1/2}$	22912	1	1	—
$^2D_{5/2}$	23331, 23527	3	2	196*
$^2P_{3/2}$	25697, 25913	2	2	216
$^2I_{11/2} + {}^4D_{3/2} +$ $+ {}^4D_{5/2}$	27212, 27306, 27472, 27754, 27980, 28076, 28197, 28351	11	8	(1139*)
	YAlO$_3$ crystal, 77 K^b [271, 272, 591]			
$^4I_{9/2}$	0, 118, 212, 500, 671	5	5	671
$^4I_{11/2}$	2023, 2097, 2158, 2264, 2323, 2378	6	6	355
$^4I_{13/2}$	3953, 4021, 4092, 4200, 4291, 4328, 4446	7	7	495
$^4I_{15/2}$	5757, 5893, 6011, 6240, 6307, 6402, 6687, 6743	8	8	990
$^4F_{3/2}$	11421, 11550	2	2	129
$^4F_{5/2}$	12411, 12447, 12511	3	3	100
$^2H_{9/2}$	12561, 12593, 12713, 12742, 12883	5	5	322
$^4F_{7/2} + {}^4S_{3/2}$	13323, 13452, 13565, 13589, 13607, 13651	6	6	(328)
$^4F_{9/2}$	14665, 14723, 14740, 14793, 14928	5	5	263

^{b} The positions of some levels of Nd^{3+} ions in YAlO$_3$ crystals at 300 K are shown in Fig. 6.19.

(*Continued*)

Table 4.3   (*continued*)

[SL]J manifold	Stark-level positions [cm^{-1}]	Number of components		ΔE [cm^{-1}]
		Theory	Exper.	
$^2H_{11/2}$	15858, 15893, 15903, 15995, 16095	6	5	237*
$^4G_{5/2}$	16963, 17023, 17116	3	3	153
$^2G_{7/2}$	17295, 17313, 17364, 17456	4	4	161
$^4G_{7/2}$	18846, 18893, 18975, 19077	4	4	231
$^2G_{9/2}$	19245, 19309, 19350, 19425, 19546	5	5	301
$^2K_{13/2}$	19637*, 19652*, 19747, 19806, 19873, 19924	7	6	247*
$^4G_{9/2}$	20865, 20894, 20955, 21041, 21110	5	5	245
$^4G_{11/2} + {}^2K_{15/2}$ $+ {}^2D_{3/2}$	21231, 21276, 21294*, 21367, 21464, 21536, 21580, 21630*, 21654, 21718*, 21748, 21834, 21906*, 21930	16	14	(699*)
$^2P_{1/2}$	23164	1	1	—
$^2D_{5/2}$	23463, 23635, 23759	3	3	296
$^2P_{3/2}$	25981, 26123	2	2	142

Y$_3$Al$_5$O$_{12}$ crystal, 77 K^c [271, 273– 275, 654]

[SL]J manifold	Stark-level positions [cm^{-1}]	Theory	Exper.	ΔE [cm^{-1}]
$^4I_{9/2}$	0, 130, 199, 308, 857	5	5	857
$^4I_{11/2}$	2002, 2029, 2110, 2147, 2468, 2521	6	6	519
$^4I_{13/2}$	3922, 3930, 4032, 4047, 4435, 4442*, 4498	7	7	576
$^4I_{15/2}$	5758, 5814, 5936, 5970, 6570, 6583, 6639, 6734	8	8	976
$^4F_{3/2}$	11427, 11512	2	2	85
$^4F_{5/2}$	12370, 12432, 12519	3	3	149
$^2H_{9/2}$	12575, 12607, 12623, 12819, 12840	5	5	265
$^4F_{7/2} + {}^4S_{3/2}$	13363, 13433, 13563, 13572, 13596, 13633	6	6	(270)
$^4F_{9/2}$	14626, 14678, 14793, 14819, 14916	5	5	290
$^2H_{11/2}$	15743, 15838, 15870, 15957, 16103, 16119	6	6	376
$^4G_{5/2}$	16849, 16992, 17047	3	3	193
$^2G_{7/2}$	17241, 17268, 17322, 17575	4	4	334
$^4G_{7/2}$	18723, 18822, 18843, 18986	4	4	263
$^2K_{13/2} + {}^2G_{9/2}$	19154, 19194, 19294, 19470, 19543, 19596, 19620, 19651, 19814, 20048	12	10	(894*)
$^4G_{9/2}$	20730, 20773, 20790, 20803	5	4	73*
$^4G_{11/2}$	20962, 21029, 21080, 21110, 21159, 21162	6	6	200
$^2K_{15/2}$	21522, 21593, 21661, 21697, 21767, 21791, 21872, 21906	8	8	384
$^2D_{3/2}$	22036	2	1	—*
$^2P_{1/2}$	23155	1	1	—
$^2D_{5/2}$	23674, 23764, 23849	3	3	175
$^2P_{3/2}$	25994	2	1	—*
$^4D_{3/2}$	27571, 27670	2	2	99
$^4D_{1/2}$	27809	1	1	—

YP$_5$O$_{14}$ crystal, 77 K [635]
C$_{2h}^5$ – P2$_1$/c for Y$_{0.83}$Nd$_{0.17}$P$_5$O$_{14}$ phase

[SL]J manifold	Stark-level positions [cm^{-1}]	Theory	Exper.	ΔE [cm^{-1}]
$^4I_{9/2}$	0, 89, 202, 255, 310	5	5	310
$^4I_{11/2}$	1962, 1985, 2049, 2060, 2095, 2174	6	6	212
$^4I_{13/2}$	3924, 3951, 3998, 4046, 4078, 4104, 4169	7	7	245
$^4F_{3/2}$	11468, 11562	2	2	94

C$_{2h}^6$–C2/c for Y$_{0.98}$Nd$_{0.02}$P$_5$O$_{14}$ phase
Stark levels of "1" type center

[SL]J manifold	Stark-level positions [cm^{-1}]	Theory	Exper.	ΔE [cm^{-1}]
$^4I_{9/2}$	0, 104, 181, 390, 446	5	5	446

^{c} The positions of some levels of Nd^{3+} ions in Y$_3$Al$_5$O$_{12}$ crystals at 300 K are shown in Fig. 6.6.

(Continued)

Table 4.3   (*continued*)

[SL]J manifold	Stark-level positions [cm^{-1}]	Number of components		ΔE [cm^{-1}]
		Theory	Exper.	
$^4I_{11/2}$	1966, 1994, 2131, 2141, 2158, 2207	6	6	241
$^4F_{3/2}$	11492, 11523	2	2	31
	Stark levels of "2" type center			
$^4I_{9/2}$	0, 104, 167, 205, 227	5	5	227
$^4I_{11/2}$	1977, 1990, 1993, 2021, 2032, 2048	6	6	71
$^4F_{3/2}$	11491, 11547	2	2	55
	D_{2h}^7–Pnma for $Y_{0.85}Nd_{0.15}P_5O_{14}$ phase			
$^4I_{9/2}$	0, 91, 186, 279, 299	5	5	299
$^4I_{11/2}$	1959, 1981, 2045, 2058, 2086, 2163	6	6	204
$^4I_{13/2}$	3922, 3948, 3996, 4044, 4064, 4099, 4163	7	7	241
$^4F_{3/2}$	11465, 11560	2	2	95
	$Y_3Sc_2Al_3O_{12}$ crystal, 77 K [271, 524]			
$^4I_{9/2}$	0, 111, 179, 301, 821	5	5	821
$^4I_{11/2}$	1983, 2014, 2096, 2132, 2436, 2489	6	6	506
$^4I_{13/2}$	3914, 3937, 4044, 4052, 4424, 4482	7	6	568*
$^4I_{15/2}$	5763, 5795, 5926, 5980, 6538, 6556, 6616, 6706	8	8	943
$^4F_{3/2}$	11428, 11524	2	2	96
	$Y_3Sc_2Ga_3O_{12}$ crystal, 77 K [271, 524]			
$^4I_{9/2}$	0, 97, 168, 252, 778	5	5	778
$^4I_{11/2}$	1984, 2007, 2062, 2100, 2408, 2437	6	6	453
$^4I_{13/2}$	3909, 3921, 3993, 4007, 4391, 4420	7	6	509*
$^4I_{15/2}$	5770, 5804, 5894, 5945, 6506, 6557, 6653	8	7	883*
$^4F_{3/2}$	11439, 11489	2	2	50
	YVO_4 crystal, 77 K [167, 171, 634]			
$^4I_{9/2}$	0, 108, 173, 226, 433	5	5	433
$^4I_{11/2}$	1966, 1988, 2047, 2062, 2154, 2182	6	6	216
$^4I_{13/2}$	3910, 3931, 3980, 4042, 4088, 4158, 4170	7	7	260
$^4I_{15/2}$	5834, 5871, 5917, 6065, 6260, 6264, 6318	8	7	484*
$^4F_{3/2}$	11366, 11384	2	2	18
$^4F_{5/2}$	12366, 12401*, 12405*	3	3	39*
$^2H_{9/2}$	12497, 12542, 12499, 12692	5	4	195*
$^4F_{7/2}$	13319, 13340, 13392, 13461*	4	4	142*
$^4S_{3/2}$	13461*, 13465*	2	2	4*
$^2F_{9/2}$	14579, 14598, 14635, 14719, 14730	5	5	151
$^2H_{11/2}$	15765, 15853	6	2	88*
$^4G_{5/2}$	16824, 16951*, 16970*	3	3	146*
$^2G_{7/2}$	17215, 17244, 17268	4	3	53*
$^4G_{7/2}$	18772, 18832, 18861, 18925	4	4	153
$^2G_{9/2} + ^2K_{13/2}$	19055, 19109, 19172, 19216, 19257, 19298, 19305, 19327, 19361, 19429	12	10	(374*)
$^4G_{9/2} + ^4G_{11/2} + + ^2K_{15/2}$	20794, 20842, 20872, 20912, 21013, 21101, 21204, 21236, 21482, 21561	19	10	(767*)
$^2D_{3/2}$	21935	2	1	—*
$^2P_{1/2}$	23041	1	1	—
$^2D_{5/2}$	23596, 23613, 23641	3	3	45
$^4D_{3/2}$	27632, 27640	2	2	8
$^4D_{1/2}$	27746	1	1	—

(*Continued*)

Table 4.3   (*continued*)

[SL]J manifold	Stark-level positions [cm^{-1}]	Number of components		ΔE [cm^{-1}]
		Theory	Exper.	
	Y$_3$Ga$_5$O$_{12}$^{d} crystal, 77 K [274, 631]			
$^4I_{9/2}$	0, 79, 179, 246, 784	5	5	784
$^4I_{11/2}$	1994, 2007, 2056, 2086, 2420, 2435	6	6	441
$^4I_{13/2}$	3921*, 3985, 3996, 4030*, 4381, 4414	7	6	493*
$^4I_{15/2}$	5765, 5810, 5900, 5920, 6565, 6675	8	7	910*
$^4F_{3/2}$	11443, 11478	2	2	35
$^4F_{5/2}$	12283, 12384, 12418	3	3	135
$^2H_{9/2}$	12514, 12563, 12587, 12811, 12830	5	5	316
$^4F_{7/2}+{}^4S_{3/2}$	13373, 13412, 13556, 13603	6	4	(230*)
$^4F_{9/2}$	14643, 14650, 14773, 14795, 14877	5	5	234
$^2H_{11/2}$	15751, 15833, 15873, 15952, 16082, 16093	6	6	342
$^4G_{5/2}$	16866, 16987, 17047	3	3	181
$^2G_{7/2}$	17250, 17313, 17553	4	3	303*
$^4G_{7/2}$	18755, 18822, 18854, 18975	4	4	220
$^2K_{13/2}+{}^2G_{9/2}$	19246, 19309, 19327, 19436, 19516, 19558, 19577, 19635, 19980, 20000	12	10	(754*)
$^4G_{9/2}$	20713, 20760, 20786, 20867	5	4	154*
$^4G_{11/2}$	20929, 21004, 21044, 21106, 21177, 21182	6	6	253
$^2K_{15/2}+{}^2D_{3/2}$	21612, 21697, 21737*, 21763, 21877, 21949	10	5	(337*)
$^2P_{1/2}$	23166	1	1	—
	300 K			
$^4I_{11/2}$	1988, 2001, 2053, 2084, 2411, 2426	6	6	438
$^4F_{3/2}$	11432, 11469	2	2	37
	Ba$_{0.25}$Mg$_{2.75}$Y$_2$Ge$_3$O$_{12}$ crystal, 300 K [433]			
$^4I_{9/2}$	0, 59, 100, 233, 695	5	5	695
$^4I_{11/2}$	2014, 2084, 2155, 2232, 2429, 2499	6	6	485
$^4F_{3/2}$	11440, 11552	2	2	112
	La$_2$Be$_2$O$_5$ crystal, 77 K [632, 633]			
$^4I_{9/2}$	0, 110, 225, 350, 495	5	5	495
$^4I_{11/2}$	1961, 2040, 2102, 2125, 2213, 2287	6	6	322
$^4I_{13/2}$	3906, 3984, 4046, 4076, 4159, 4237, 4290	7	7	384
$^4I_{15/2}$	5802, 5933, 6003, 6090, 6210, 6298, 6395, 6495	8	8	675
$^4F_{3/2}$	11312, 11525	2	2	213
$^4F_{5/2}+{}^2H_{9/2}$	12347, 12450, 12497, 12538, 12578, 12631, 12697, 12786	8	8	(439)
$^4F_{7/2}+{}^4S_{3/2}$	13265, 13378, 13507, 13518, 13540, 13610	6	6	(345)
$^4F_{9/2}$	14565, 14661, 14715, 14801, 14889	5	5	324
$^2H_{11/2}$	15869, 15878, 15925, 15960, 16003, 16026	6	6	157
$^4G_{5/2}$	16954, 17034, 17108	3	3	154
$(^2G,{}^4G)_{7/2}$	17221, 17261, 17302, 17355	4	4	134
$(^2G,{}^4G)_{7/2}$	18851, 18911, 18938, 19016	4	4	165
$^2K_{13/2}+{}^2G_{9/2}$	19143, 19358, 19373, 19415, 19464, 19512, 19618, 19692, 19800, 20023	12	10	(880*)
$^4G_{9/2}+{}^4G_{11/2}+{}^2K_{15/2}+{}^2D_{3/2}$	20838, 20890, 20912, 20961, 20996, 21024, 21114, 21227, 21278, 21345, 21373*, 21398*, 21475, 21513, 21596, 21618, 21661, 21742, 21862, 22027	21	20	(1189*)
$^2P_{1/2}$	23126	1	1	—
$^2D_{5/2}$	23496, 23745*, 23954*	3	3	458*
$^2P_{3/2}$	25936, 26167*	2	2	231*

^{d} The level positions in this crystal at liquid-helium temperature were studied in [274]. (*Continued*)

Table 4.3 (*continued*)

[SL]J manifold	Stark-level positions [cm^{-1}]	Number of components		ΔE [cm^{-1}]
		Theory	Exper.	
300 K				
$^4I_{9/2}$	0, 110, 225, 350, 495	5	5	495
$^4I_{11/2}$	1962, 2040, 2100, 2120, 2215, 2282	6	6	320
$^4I_{13/2}$	3908, 3984, 4045, 4077, 4160, 4235, 4290	7	7	382
$^4F_{3/2}$	11310, 11518	2	2	208
La$_2$O$_3$ crystal, 10 K [276]				
$^4I_{9/2}$	0, 23, 82, 242, 487	5	5	487
$^4F_{3/2}$	11175, 11322	2	2	147
$^4F_{5/2}$	12195, 12267, 12319	3	3	124
$^2H_{9/2}$	12358, 12391, 12408, 12550	5	4	192*
$^4F_{7/2}$	13141, 13150, 13252, 13272	4	4	131
$^4S_{3/2}$	13336, 13349	2	2	13
$^4F_{9/2}$	14398, 14406, 14472, 14568, 14613	5	5	215
$^2H_{11/2}$	15613, 15665, 15800, 15895	6	4	282*
$^2G_{7/2}$	16666, 16690, 16739, 16778	4	4	112
$^4G_{5/2}$	17021, 17039, 17186	3	3	165
$^2K_{13/2} + {}^4G_{7/2}$	18581, 18603, 18644, 18652, 18695, 18719	11	6	(138*)
$^2G_{9/2}$	19106, 19144, 19178, 19223, 19265	5	5	159
$^4G_{9/2}$	20599, 20645, 20668, 20795	5	5	196*
$^2(P,D)_{3/2}$	20903, 20923	2	2	20
$^4G_{11/2}$	21379	6	1	—*
$^2P_{1/2}$	22884	1	1	—
$^2P_{3/2}$	25720	2	1	—*
$^4D_{3/2}$	27361, 17409	2	2	48
$^2D_{5/2}$	27741, 27760	3	2	29*
$^2I_{11/2}$	27807, 28049	6	2	242*
$^4D_{1/2}$	28409	1	1	—*
$^2I_{13/2}$	29404, 29582, 29636	7	3	232*
LaAlO$_3$ crystal, 300 K [137]				
$^4I_{9/2}$	0, 118, 133, 228*, 604	5	5	604
$^4I_{11/2}$	2072*, 2160*, 2209, 2283*, 2327	6	5	255*
$^4F_{3/2}$	11583, 11606	2	2	23
LaNbO$_4$ crystal, 77 K [277]				
$^4I_{9/2}$	0, 114, 175, 225, 506	5	5	506
$^4I_{11/2}$	1973, 2002, 2042, 2052, 2202, 2255	6	6	282
$^4I_{13/2}$	3920, 3957, 3970, 4001, 4188, 4206, 4242	7	7	322
$^4I_{15/2}$	5840, 5875, 5928, 5985, 6328, 6395	8	6	555*
$^4F_{3/2}$	11389, 11474	2	2	65
NdP$_5$O$_{14}$ crystal, 77 K [635]				
$^4I_{9/2}$	0, 83, 206, 260, 320	5	5	320
$^4I_{11/2}$	1959, 1985, 2044, 2063, 2097, 2179	6	6	220
$^4I_{13/2}$	3919, 3948, 3999, 4041, 4093, 4110, 4174	7	7	255
$^4F_{3/2}$	11476, 11587	2	2	111
300 K [84, 507, 576]				
$^4I_{9/2}$	0, 80, 219, 252, 314	5	5	314
$^4I_{11/2}$	1955, 1978, 2038, 2056, 2092, 2171	6	6	216
$^4I_{13/2}$	3910, 3938, 3990, 4032, 4086, 4106, 4165	7	7	255

(*Continued*)

Table 4.3   (*continued*)

[SL]J manifold	Stark-level positions [cm^{-1}]	Number of components		ΔE [cm^{-1}]
		Theory	Exper.	
$^4I_{15/2}$	5872, 5912, 6011, 6072, 6081, 6210, 6274, 6289	8	8	417
$^4F_{3/2}$	11470, 11582	2	2	112
GdAlO$_3$ crystal, 77 K [471]				
$^4I_{9/2}$	0, 111, 189, 507, 666	5	5	666
$^4I_{11/2}$	2022, 2094, 2155, 2264, 2312, 2369	6	6	347
$^4I_{13/2}$	3948, 4024, 4074, 4184, 4271, 4315, 4440*	7	7	492*
$^4I_{15/2}$	5753*, 5784*, 5915, 6024, 6226*, 6316, 6407*, 6760	8	8	1007*
$^4F_{3/2}$	11449, 11565	2	2	116
$^4F_{5/2}$	12360, 12392, 12438	3	3	78
$^2H_{9/2}$	12533, 12581, 12609, 12761, 12911	5	5	378
$^4F_{7/2} + {}^4S_{3/2}$	13345, 13470, 13580*, 13614, 13631, 13672	6	6	(327)
$^4F_{9/2}$	14724, 14757, 14780, 14816, 14950	5	5	226
$^2H_{11/2}$	15893, 15928, 15944, 16009, 16024, 16116	6	6	223
$^2G_{7/2} + {}^4G_{5/2}$	17016, 17065, 17151, 17340, 17364, 17413, 17483	7	7	(467)
$^4G_{7/2}$	18904, 18954, 19029, 19120	4	4	216
$^2G_{9/2}$	19316, 19385, 19468, 19558, 19706	5	5	390
$^4G_{9/2}$	20948, 20971, 20988, 21027, 21290	5	5	216
$^2P_{1/2}$	23213	1	1	—
Gd$_3$Sc$_2$Al$_3$O$_{12}$ crystal, 77 K [529]				
$^4I_{9/2}$	0, 116, 193, 316, 808	5	5	808
$^4I_{11/2}$	1979, 2023, 2114, 2144, 2422, 2492	6	6	513
$^4I_{13/2}$	3904, 3927, 4048, 4061, 4402, 4477	7	6	573*
$^4I_{15/2}$	5767, 5800, 5942, 5992, 6521, 6545, 6601, 6682	8	8	915
$^4F_{3/2}$	11421, 11537	2	2	116
Gd$_3$Sc$_2$Ga$_3$O$_{12}$ crystal, 77 K [636]				
$^4I_{9/2}$	0, 107, 168, 263, 763	5	5	763
$^4I_{11/2}$	1978, 2010, 2069, 2109, 2393, 2431	6	6	453
$^4I_{13/2}$	3907, 3920, 4000, 4010, 4025, 4380, 4421	7	7	514
$^4I_{15/2}$	5777, 5812, 5914, 5959, 6494*, 6510, 6557, 6647	8	8	870
$^4F_{3/2}$	11434, 11499	2	2	65
300 K				
$^4I_{11/2}$	1975, 2004, 2069, 2108, 2381, 2428	6	6	453
$^4F_{3/2}$	11424, 11492	2	2	68
Gd$_3$Ga$_5$O$_{12}$ crystal, 77 K [504, 594]				
$^4I_{9/2}$	0, 93, 178, 253, 772	5	5	772
$^4I_{11/2}$	1994, 2007, 2063, 2100, 2407, 2433	6	6	439
$^4I_{13/2}$	3925, 3927, 3997, 4010, 4374, 4382, 4415	7	7	490
$^4I_{15/2}$	5770, 5822, 5914, 5932, 6513, 6552, 6653	8	7	883*
$^4F_{3/2}$	11442, 11485	2	2	43
$^4F_{5/2}$	12390, 12427, 12522	3	3	132
$^2H_{9/2}$	12571, 12598, 12604, 12814, 12825	5	5	254
$^4F_{7/2} + {}^4S_{3/2}$	13378, 13421, 13559*, 13563, 13570, 13603	6	6	(225)
$^4F_{9/2}$	14661, 14658, 14771, 14881	5	4	240*
$^2H_{11/2}$	15760, 15840, 15873, 15952, 16085, 16090	6	6	330
$^4G_{5/2}$	16880, 16987, 17053	3	3	73
$^2G_{7/2}$	17256, 17262, 17313, 17544	4	4	288

(*Continued*)

Table 4.3 (*continued*)

[SL]J manifold	Stark-level positions [cm^{-1}]	Number of components		ΔE [cm^{-1}]
		Theory	Exper.	
$^4G_{7/2}$	18761, 18840, 18871, 18984	4	4	223
$^2K_{13/2} + {}^2G_{9/2}$	19238, 19312, 19335, 19448, 19524, 19562, 19581*, 19604, 19822, 19972	12	10	(734*)
$^4G_{9/2}$	20769, 20802, 20811, 20825	5	4	56*
$^4G_{11/2}$	21013, 21060, 21087, 21110, 21174, 21189	6	6	176
$^2K_{15/2}$	21609, 21668, 21690, 21752, 21769	8	6	256*
$^2D_{3/2}$	21956	2	1	—*
$^2P_{1/2}$	23167	1	1	—
$^2D_{5/2}$	23718, 23756, 23790	3	3	72
	LuAlO$_3$ crystal, 77 K [565, 637]			
$^4I_{9/2}$	0, 118, 234, 493, 662	5	5	662
$^4I_{11/2}$	2022, 2099, 2160, 2265, 2320, 2380	6	6	358
$^4I_{13/2}$	3946, 4010, 4085, 4195, 4280, 4330, 4432	7	7	486
$^4I_{15/2}$	5760, 5895, 6010, 6240, 6300, 6405, 6685*, 6735	8	8	975
$^4F_{3/2}$	11393, 11525	2	2	132
$^4F_{5/2}$	12385, 12431, 12487	3	3	102
$^2H_{9/2}$	12540, 12578, 12693, 12737, 12865	5	5	325
$^4F_{7/2} + {}^4S_{3/2}$	13308, 13435, 13540, 13566, 13583, 13638	6	6	(330)
$^4F_{9/2}$	14632, 14697, 14721, 14775, 14908	5	5	276
$^2H_{11/2}$	15837, 15873, 15878, 15979, 15989, 16072	6	6	235
$^4G_{5/2}$	16914, 16986, 17094	3	3	180
$^2G_{7/2}$	17253, 17274, 17331, 17430	4	4	177
$^4G_{7/2}$	18807, 18839, 18932, 19047	4	4	240
$^2G_{9/2}$	19190, 19271, 19323, 19387, 19516	5	5	326
$^2K_{13/2}$	19607*, 19665*, 19719, 19795, 19864, 19896	7	6	289*
$^4G_{9/2}$	20802, 20837, 20907, 20999, 21070	5	5	268
$^4G_{11/2} + {}^2K_{15/2} + {}^2D_{3/2}$	21177, 21231, 21258, 21321, 21436, 21491, 21542, 21598, 21612, 21696, 21724, 21819, 21867, 21896	16	14	(719*)
$^2P_{1/2}$	23111	1	1	—
$^2D_{5/2}$	23582, 23711	3	2	129*
$^2P_{3/2}$	25933, 26082	2	2	149
	300 K [565]			
$^4I_{11/2}$	2026, 2099, 2161, 2265, 2319, 2380	6	6	354
$^4I_{13/2}$	3950, 4015, 4090, 4195, 4282, 4332, 4434	7	7	484
$^4F_{3/2}$	11393, 11521	2	2	128
	Lu$_3$Al$_5$O$_{12}$ crystal, 77 K [191, 278]			
$^4I_{9/2}$	0, 117, 198, 303, 878	5	5	878
$^4I_{11/2}$	2004, 2031, 2101, 2137, 2487, 2529	6	6	525
$^4I_{13/2}$	3926, 3934, 4026, 4040, 4456, 4506	7	6	580*
$^4I_{15/2}$	5759, 5805, 5919, 5954, 6583, 6590, 6653, 6754	8	8	995
$^4F_{3/2}$	11434, 11502	2	2	68
$^4F_{5/2}$	12372, 12421, 12514	3	3	142
$^2H_{9/2}$	12572, 12604, 12626, 12837, 12855	5	5	283
$^4F_{7/2} + {}^4S_{3/2}$	13367, 13426, 13567, 13572, 13596, 13633	6	6	(266)
$^4F_{9/2}$	14633, 14669, 14799, 14819, 14901	5	5	268
$^2H_{11/2}$	15734, 15829, 15864, 15950, 16093	6	5	359*
$^4G_{5/2}$	16841, 16992, 17044	3	3	203

(*Continued*)

Table 4.3   (*continued*)

[SL]J manifold	Stark-level positions [cm^{-1}]	Number of components		ΔE [cm^{-1}]
		Theory	Exper.	
$(^2G,^4G)_{7/2}$	17247, 17262, 17313, 17578	4	4	331
$(^2G,^4G)_{7/2}$	18716, 18804, 18829, 18978	4	4	262
$^2K_{13/2}+$ $+(^2G,^4G)_{9/2}$	19150, 19194, 19290, 19516, 19539, 19589, 19619, 19639, 19837, 20028	12	10	(878*)
$(^2G,^4G)_{9/2}$	20730, 20768, 20790*	5	3	60*
$^4G_{11/2}$	20973, 21035, 21084, 21124, 21168, 21177	6	6	204
$^2K_{15/2}$	21598, 21805, 21820, 21896, 21920, 22036	3	6	438*
$^2P_{1/2}$	23155	1	1	—
$^2D_{5/2}$	23691, 23747, 23810	3	3	160
$^2P_{3/2}$	25994	2	1	—*

Lu$_3$Sc$_2$Al$_3$O$_{12}$ crystal, 77 K [271, 524]

$^4I_{9/2}$	0, 116, 193, 316, 808	5	5	808
$^4I_{11/2}$	1979, 2023, 2114, 2144, 2422, 2492	6	6	513
$^4I_{13/2}$	3904, 3927, 4048, 4061, 4402, 4477	7	6	573*
$^4I_{15/2}$	5767, 5800, 5942, 5992, 6521, 6545, 6601, 6682	8	8	915
$^4F_{3/2}$	11421, 11537	2	2	116

Lu$_3$Ga$_5$O$_{12}$ crystal, 77 K [506]

$^4I_{9/2}$	0, 97, 185, 259, 777	5	5	777
$^4I_{11/2}$	1996, 2010, 2065, 2103, 2411, 2438	6	6	442
$^4I_{13/2}$	3930, 4000, 4016, 4390, 4435	7	5	505*
$^4I_{15/2}$	5765, 5820, 5913, 5928, 5940, 6513, 6556, 6655	8	8	890
$^4F_{3/2}$	11442, 11485	2	2	43

PbMoO$_4$ crystal, 77 K [97, 268, 485]

$^4I_{9/2}$	0, 92, 151, 216, 363	5	5	363
$^4I_{11/2}$	1962, 1992, 2002, 2035, 2134, 2134, 2161	6	6	199
$^4I_{13/2}$	3920, 3947, 3962, 3993, 4032, 4096, 4140	7	7	220
$^4I_{15/2}$	5865, 5910, 5946, 6010, 6175, 6196, 6273	8	7	408*
$^4F_{3/2}$	11397, 11454	2	2	57
$^4F_{5/2}$	12416, 12447, 12496	3	3	80
$^2H_{9/2}$	12510, 12554, 12620, 12653, 12685	5	5	175
$^4F_{7/2}$	13369, 13393, 13466, 13492	4	4	123
$^4S_{3/2}$	13522, 13527	2	2	6
$^4F_{9/2}$	14624, 14646, 14735, 14745, 14789	5	5	165
$^2H_{11/2}$	15834, 15864, 15893, 15935, 15983, 15993	6	6	159
$^2G_{7/2}$	16979, 17048, 17067, 17107	4	4	128
$^4G_{5/2}$	17222, 17286, 17383	3	3	161
$^4G_{7/2}$	18904, 18909, 18980, 19030	4	4	126
$^2G_{9/2}$	19371, 19406, 19463, 19495, 19518	5	5	147
$^2P_{1/2}$	23727	1	1	—
$^2D_{5/2}$	23668, 23712, 13801	3	3	133

Bi$_4$Si$_3$O$_{12}$ crystal, 77 K [566]

$^4I_{9/2}$	0, 47, 60*, 360, 450	5	5	450
$^4I_{11/2}$	1915, 1928, 1954, 2182, 2195, 2209	6	6	294
$^4I_{13/2}$	3860, 3868, 3880, 4152, 4173, 4200	7	6	340*
$^4I_{15/2}$	5823, 5842, 5870, 5925, 6212, 6271, 6302, 6332	8	8	509
$^4F_{3/2}$	11336, 11472	2	2	136

(*Continued*)

Table 4.3   (*continued*)

[SL]J manifold	Stark-level positions [cm^{-1}]	Number of components		ΔE [cm^{-1}]
		Theory	Exper.	
	Bi$_4$Ge$_3$O$_{12}$ crystal, 77 K [567]			
$^4I_{9/2}$	0, 50, 95, 365, 444	5	5	444
$^4I_{11/2}$	1920, 1931, 1956, 2178, 2195, 2207	6	6	287
$^4I_{13/2}$	3870, 3875, 3887, 4153, 4176, 4200	7	6	330*
$^4I_{15/2}$	5790, 5804, 5840, 5885, 6195, 6275, 6303, 6333	8	8	543
$^4F_{3/2}$	11327, 11460	2	2	133
	300 K			
$^4I_{11/2}$	1918, 1930, 1957, 2180*, 2190*, 2200*	6	6	282*
$^4F_{3/2}$	11326, 11454	2	2	128
	LiLa(MoO$_4$)$_2$ crystal, 77 K [97, 424]			
$^4I_{9/2}$	0, 90, 160, 240, 400	5	5	400
$^4I_{11/2}$	1960, 2003, 2102, 2143, 2198*	6	5	238*
$^4I_{13/2}$	3909, 3945, 3973*, 4015, 4065, 4120, 4165	7	7	256
$^4I_{15/2}$	5850, 5895, 5943, 6027, 6240, 6308	8	6	458*
$^4F_{3/2}$	11385, 11467	2	2	82
	LiGd(MoO$_4$)$_2$ crystal, 77 K			
$^4I_{9/2}$	0, 94, 160, 237, 446	5	5	446
$^4I_{11/2}$	1971, 2014, 2132*, 2176	6	4	205*
$^4I_{13/2}$	3928, 3958, 3980*, 4015, 4067, 4140, 4176	7	7	254
$^4I_{15/2}$	5840, 5890, 5935, 6011, 6256, 6366	8	6	526*
$^4F_{3/2}$	11390, 11467	2	2	77
$^4F_{5/2} + {}^2H_{9/2}$	12406, 12447, 12503, 12518, 12539, 12614, 12676, 12714	8	8	(308)
$^4F_{7/2} + {}^4S_{3/2}$	13255, 13290, 13351, 13395, 13487, 13515	6	6	(260)
$^4F_{9/2}$	14470, 14590, 14640, 14720, 14755	5	5	285
$^2H_{11/2}$	15820, 15855, 15880, 15928, 15995*	6	5	175*
$^2G_{7/2}$	16863, 16972, 17038, 17082	4	4	219
$^4G_{5/2}$	17229, 17280, 17394	3	3	165
$^4G_{7/2}$	18786, 18894, 18975, 19011	4	4	225
$^2G_{9/2}$	19275, 19361, 19383, 19474, 19496*	5	5	221*
$^4G_{9/2}$	20790, 20894*, 21050, 21118, 21212	5	5	422
$^2P_{1/2}$	23149	1	1	—
	NaLa(MoO$_4$)$_2$ crystal, 77 K [67, 97, 268]			
$^4I_{9/2}$	0, 92, 156, 229, 403	5	5	403
$^4I_{11/2}$	1953, 1983, 2145, 2173	6	4	220*
$^4I_{13/2}$	3900, 3937, 3962, 4033, 4113, 4158	7	6	258*
$^4I_{15/2}$	5841, 5888, 5933, 6020, 6235, 6307	8	6	446*
$^4F_{3/2}$	11383, 11455	2	2	72
	NaLa(WO$_4$)$_2$ crystal, 77 K [268]			
$^4I_{9/2}$	0, 96, 157, 239, 424	5	5	424
$^4I_{11/2}$	1959, 1996, 2161, 2187	6	4	218*
$^4I_{13/2}$	3913, 3942, 3964, 4132, 4173	7	5	260*
$^4I_{15/2}$	5852, 5898, 5939, 6246, 6334	8	5	482*
$^4F_{3/2}$	11406, 11477	2	2	71
	KLa(MoO$_4$)$_2$ crystal, 77 K [428]			
$^4I_{9/2}$	0, 88, 155, 235, 355	5	5	355

(*Continued*)

Table 4.3 (*continued*)

[SL]J manifold	Stark-level positions [cm⁻¹]	Number of components		ΔE [cm⁻¹]
		Theory	Exper.	
$^4I_{11/2}$	1887*, 1942, 1985, 2010, 2050, 2125	6	6	238*
$^4I_{13/2}$	3900, 3940, 3970, 4020, 4080, 4140	7	6	240*
$^4I_{15/2}$	5855, 5905, 5955, 6025, 6160, 6230	8	6	375*
$^4F_{3/2}$	11389, 11466	2	2	77
	$K_5Nd(MoO_4)_4$ and $K_5Bi(MoO_4)_4$ crystals, 77 K [569]			
$^4I_{9/2}$	0, 50, 230, 315, 474	5	5	474
$^4I_{11/2}$	1955, 1980, 2040, 2150, 2163, 2285	6	6	330
$^4I_{13/2}$	3893, 3942, 3969, 4017, 4157, 4263, 4276	7	7	383
$^4I_{15/2}$	5770, 5810, 5880, 5976, 6130, 6327, 6442, 6655	8	8	885
$^4F_{3/2}$	11363, 11535	2	2	170
	$CaY_4(SiO_4)_3O$ crystal, 77 K [627]			
	Stark levels of C_s center			
$^4I_{9/2}$	0, 139, 214, 314, 446	5	5	446
$^4I_{11/2}$	1975, 1990, 2081, 2165, 2256	6	5	281*
$^4I_{13/2}$	3864, 3937, 4039, 4146	7	4	282*
$^4I_{15/2}$	5899, 5942, 5977, 6163, 6238, 6288, 6404, 6518	8	8	619
$^4F_{3/2}$	11300, 11527	2	2	227
	$CaLa_4(SiO_4)_3O$ crystal, 77 K [627]			
	Stark levels of C_s center			
$^4I_{9/2}$	0, 149, 240, 336, 447	5	5	447
$^4I_{11/2}$	1904, 1975, 2027, 2111, 2273, 2261	6	6	355
$^4I_{13/2}$	3842, 3984, 4051, 4066, 4143	7	5	311*
$^4I_{15/2}$	5811, 5948, 5983, 6155, 6233, 6285, 6406	8	7	595*
$^4F_{3/2}$	11328, 11547	2	2	219
	$YScO_3$ crystal, 77 K [639]			
$^4I_{9/2}$	0, 33, 305, 485, 709	5	5	709
$^4I_{11/2}$	1890, 1952, 2191, 2302, 2388, 2430	6	6	540
$^4I_{13/2}$	3814, 3841, 4140, 4250, 4330, 4354, 4377	7	7	563
$^4F_{3/2}$	11175, 11410	2	2	235
	$Ba_2MgGe_2O_7$ crystal, 300 K [280]			
$^4I_{9/2}$	0, 63, 105, 250, 700	5	5	700
$^4I_{11/2}$	1895, 2040, 2119	6	3	224*
$^4F_{3/2}$	11379, 11514	2	2	135
	$Ba_2ZnGe_2O_7$ crystal, 300 K [280]			
$^4I_{9/2}$	0, 59, 100, 233, 695	5	5	695
$^4I_{11/2}$	1890, 2054, 2127	6	3	237*
$^4F_{3/2}$	11374, 11504	2	2	130
	$Ca_5(PO_4)_3F$ crystal, 77 K [129, 281–283]			
$^4I_{9/2}$	0, 401, 513, 568, 708	5	5	708
$^4I_{11/2}$	1905, 2304, 2365, 2404, 2430, 2514	6	6	606
$^4I_{13/2}$	3818, 4232, 4250, 4302, 4366, 4389, 4402	7	7	584
$^4I_{15/2}$	5723, 6200, 6311, 6353, 6391, 6455, 6579, 6725	8	8	1002
$^4F_{3/2}$	11314, 11676	2	2	362

(*Continued*)

Table 4.3   (*continued*)

[SL]J manifold	Stark-level positions [cm^{-1}]	Number of components		ΔE [cm^{-1}]
		Theory	Exper.	
	Sr$_5$(PO$_4$)$_3$F crystal, 77 K [282]			
$^4I_{9/2}$	0, 573, 596, 635, 740	5	5	740
$^4I_{11/2}$	1896, 2388, 2438, 2490	6	4	594*
$^4F_{3/2}$	11346, 11707	2	2	361
	La$_2$O$_2$S crystal, 77 K [219]			
$^4I_{9/2}$	0, 23, 47, 79, 90	5	5	90
$^4I_{11/2}$	1880, 1889, 1902, 1909, 1959, 2050	6	6	170
$^4F_{3/2}$	11198, 11214	2	2	16
	Pb$_5$(PO$_4$)$_3$F crystal, 77 K [640]			
$^4I_{9/2}$	0, 148, 255, 313, 419	5	5	419
$^4I_{11/2}$	1923, 1955, 2020, 2116, 2142, 2238	6	6	315
$^4I_{13/2}$	3910, 3964, 4084, 4204	7	4	294*
$^4F_{3/2}$	11434, 11622	2	2	188

Table 4.4.   Crystal-field splitting of Sm^{2+} ion manifolds in CaF$_2$ crystal at 20 K [16, 256]

[SL]J manifold	Stark-level positions [cm^{-1}]	Number of components		ΔE [cm^{-1}]
		Theory	Exper.	
$^7F_0$	0	1	1	—
$^7F_1$	263	1	1	—
$^5D_0$	14381	1	1	—
$^5D_1$	14497	1	1	—

Table 4.7.   Crystal-field splitting of Dy^{2+} ion manifolds

[SL]J manifold	Stark-level positions [cm^{-1}]	Number of components		ΔE [cm^{-1}]
		Theory	Exper.	
	CaF$_2$ crystal, 77 K [257]			
$^5I_8$	0, 5, 29, 385, 413, 442, 466	7	7	466
$^5I_7$	4267, 4310, 4326, 4370, 4417, 4441	6	6	174
	SrF$_2$ crystal, 77 K [244]			
$^5I_8$	0, 4.5, 23.4, 322, 348, 374, 394	7	7	394
$^5I_7$	4250, 4283, 4297, 4338, 4377, 4399	6	6	149

Table 4.5. Crystal-field splitting of Eu^{3+} ion manifolds

[SL]J manifold	Stark-level positions [cm^{-1}]	Number of components		ΔE [cm^{-1}]
		Theory	Exper.	
	Y_2O_3 crystal, 77 K [642–644]			
	Stark levels of C_2 center			
$^7F_0$	0	1	1	—
$^7F_1$	199, 359, 543	3	3	543
$^7F_2$	859, 906, 1379	5	3	520*
$^7F_3$	1847, 1867, 1907, 2008, 2021, 2130, 2160	7	7	313
$^7F_4$	2668, 2800, 2846, 3015, 3080, 3119, 3163, 3178, 3190	9	9	522
$^7F_5$	3755, 3825, 3904, 3938, 4019, 4062, 4127 4158, 4227, 4291	11	10	536*
$^7F_6$	4794, 5039	13	2	235*
$^5D_0$	17216	1	1	—
$^5D_1$	18930, 18954	3	2	24*
$^5D_2$	21355, 21357, 21396	5	3	41*
	Stark levels of C_{3i} center			
$^7F_0$	0	1	1	—
$^7F_1$	132, 429	2	2	297
$^5D_0$	17302	1	1	—
$^5D_1$	18991, 19080	2	2	89
	YVO_4 crystal, 4.2 K [284]			
$^7F_0$	0	1	1	—
$^7F_1$	337, 376	2	2	39
$^7F_2$	936, 985, 1038, 1116	4	4	180
$^7F_3$	1854, 1873, 1903, 1904, 1957	5	5	103
$^7F_4$	2700, 2830, 2867, 2879, 2923, 2988, 3063	7	7	363
$^7F_5$	3750, 3800, 3870, 3915, 3928, 3949, 4065	8	7	315*
$^7F_6$	4867, 4916, 4947, 5050, 5053, 5071	10	6	204*
$^5D_0$	17183	1	1	—
$^5D_1$	18932, 18941	2	2	9
$^5D_2$	21359, 21396, 21419, 21455	4	4	96

Table 4.6. Crystal-field splitting of Dy^{3+} ion manifolds in BaY_2F_8 crystal at 77 K [479]

[SL]J manifold	Stark-level positions [cm^{-1}]	Number of components		ΔE [cm^{-1}]
		Theory	Exper.	
$^6H_{15/2}$	0, 8, 40, 48, 68, 109, 216, 590	8	8	590
$^6H_{13/2}$	3520, 3535, 3559, 3581, 3638, 3683, 3836	7	7	316
$^6H_{11/2}$	5858, 5861, 5901, 5958, 6022, 6052	6	6	194
$^6H_{9/2} + {}^6F_{11/2}$	7608, 7647, 7689, 7707, 7778, 7801, 7846, 7861, 7919, 8019, 8074	11	11	(466)
$^6H_{7/2} + {}^6F_{9/2}$	8962, 9005, 9085, 9186, 9224, 9270, 9308, 9430	9	8	(468*)
$^6H_{5/2}$	10169, 10212, 10428	3	3	295
$^6F_{7/2}$	10997, 11089, 11131, 11167	4	4	170
$^6F_{3/2}$	13271	2	1	—*
$^4F_{9/2}$	20950, 21028, 21048, 21119, 21198	5	5	248
$^4I_{15/2}$	22017, 22065, 22082, 22139, 22247, 22282, 22321, 22376	8	8	359

Table 4.8. Crystal-field splitting of Ho^{3+} ion manifolds

[SL]J manifold	Stark-level positions [cm^{-1}]	Number of components		ΔE [cm^{-1}]
		Theory	Exper.	
	LiYF$_4$ crystal, 4 K [645]			
$^5I_8$	0, 7, 24, 48, 56, 72, 217, 270*, 276*, 283*, 290*, 303, 315	13	13	315
$^5I_7$	5153, 5157, 5164, 5185, 5207, 5229, 5233, 5291, 5293, 5293*	11	10	140*
$^5I_6$	8670, 8680, 8686, 8687, 8696, 8701, 8768, 8783, 8796	10	9	126*
$^5I_5$	11240, 11243, 11245, 11252, 11298, 11327, 11333	8	7	93*
$^5I_4$	13184, 13266, 13337, 13405, 13532	7	5	348*
$^5F_5$	15485, 15491, 15508, 15555, 15620, 15627, 15632, 15654	8	8	169
$^5S_2$	18478*, 18481, 18513, 18524	4	4	46*
$^5F_4$	18599, 18605, 18609, 18681, 18679, 18695, 18707	7	7	108
$^5F_3$	20629, 20648, 20700, 20743, 20755	5	5	126
$^5F_2$	21118*, 21122, 21162, 21213	4	4	95*
	BaY$_2$F$_8$ crystal, 20 and 77 K [532]			
$^5I_8$	0, 20, 37, 39, 54, 58, 89, 120, 200, 239, 276, 310, 324, 352, 382, 399	17	16	399*
$^5I_7$	5173, 5177, 5189, 5191, 5197, 5220, 5228, 5256, 5269, 5273, 5276, 5358	15	12	185*
$^5I_5$	11260, 11269, 11279, 11293, 11317, 11390	11	6	130*
$^5F_5$	15493, 15499, 15525, 15578, 15617, 15751, 15695	11	7	202*
	YAlO$_3$ crystal, 2 K [107][a]			
$^5I_8$	0, 6, 37, 48, 58, 71, 100, 126, 137, 193, 211, 222, 289, 327, 425, 474, 499	17	17	499
$^5I_7$	5186, 5187, 5222, 5253, 5255, 5264, 5266, 5268, 5280, 5288, 5318, 5326, 5337, 5346, 5357	15	15	169
	Y$_3$Al$_5$O$_{12}$ crystal, 4.2 K [830]			
	Stark levels of N center			
$^5I_8$	0, 42, 54, 141, 161, 418, 448, 464, 494, 519, 535	17	11	535*
$^5I_7$	5228, 5230, 5244, 5251, 5304, 5312, 5318, 5338, 5350, 5371, 5377, 5406, 5416, 5455	15	14	227
$^5I_6$	8733, 8737, 8742, 8763, 8768, 8771, 8817, 8839, 8851, 8869, 8936, 8940, 8951	13	13	218
$^5I_5$	11320, 11352, 11358, 11387, 11391, 11424, 11428, 11472, 11476	11	9	156*
$^5I_4$	13281, 13339, 13343, 13362, 13366, 13380, 13385, 13405, 13563	9	9	282
$^5F_5$	15457, 15474, 15491, 15513, 15655, 15665, 15670, 15697, 15739, 15744	11	10	287*
$^5S_2 + {}^5F_4$	18451, 18462, 18530, 18539, 18542, 18547, 18582, 18592, 18629, 18664, 18714, 18667, 18728, 18743	14	14	(292)
	Stark levels of A center			
$^5I_8$	0, 27, 41, 63, 83, 101, 233, 310, 354, 387, 431, 462, 486, 502, 528	—	15	528*
$^5I_7$	5209	—	1	—*

[a] The authors of [107] refer to a private communication from *L.A. Riseberg*.   (*Continued*)

Table 4.8   (*continued*)

[SL]J manifold	Stark-level positions [cm^{-1}]	Number of components		ΔE [cm^{-1}]
		Theory	Exper.	
	YVO$_4$ crystal, 4.2 K [646]			
$^5I_8$	0, 12, 21, 40, 48, 105, 124, 226, 264, 288	13	10	288*
$^5I_7$	5117, 5127, 5130, 5137, 5203, 5213, 5217, 5235, 5242	11	9	125*
$^5S_2 + {}^5F_4$	18385, 18393, 18413, 18487, 18537, 18572, 18624	11	7	(239*)
	Y$_3$Fe$_5$O$_{12}$ crystal, 2 K [139]			
	Stark levels of α center			
$^5I_8$	0, 29, 66, 78, 95, 110, 145, 160, 379, 402, 413, 425, 436, 461, 487	17	15	487*
$^5I_7$	5198, 5210, 5234, 5246, 5274, 5294, 5331, 5348, 5369, 5374, 5387, 5407	15	12	209*
	Y$_3$Ga$_5$O$_{12}$ crystal, 2 K [139]			
$^5I_8$	0, 5, 8, 30, 43, 94, 111, 113, 380, 427, 432, 447, 481	17	14	481*
$^5I_7$	5211, 5215, 5218, 5231, 5237, 5269, 5282, 5289, 5324, 5345, 5358, 5365, 5381, 5386, 5391	15	15	180
	LaNbO$_4$ crystal, 4.2 K [406]			
$^5I_8$	0, 24, 30, 37, 52, 62, 83, 212, 248, 260, 273, 295, 316	13	13	316
$^5I_7$	5126, 5161, 5169, 5223, 5236, 5248, 5309	11	7	183*
	GdAlO$_3$ crystal, 77 K [297]			
$^5I_8$	0, 33, 67, 91, 120, 161, 183, 216, 271, 316, 327, 349, 391, 400	17	14	400*
$^5I_7$	5110, 5133, 5195, 5204, 5226, 5247, 5256, 5267, 5288, 5348, 5353, 5358, 5367, 5372, 5377	15	15	267
$^5I_6$	8653, 8671, 8730, 8735, 8741, 8746, 8776, 8796, 8802, 8808, 8815, 8846, 8875	13	13	222
$^5I_5$	11165, 11188, 11212, 11228, 11261, 11306, 11309, 11314, 11337, 11404, 11408	11	11	243
$^5F_5$	15316, 15372, 15386, 15438, 15464, 15501, 15518, 15535, 15560, 15605, 15671	11	11	355
$^5S_2 + {}^5F_4$	18365, 18414, 18426, 18486, 18499, 18509, 18520, 18544, 18579, 18597, 19615, 18657, 18674, 18724	14	14	(359)
$^5F_3$	20576, 20593, 20626, 20666, 20712, 20734, 20746	7	7	170
$^5F_2$	21339, 21390, 21434, 21461*, 21537	5	5	198
	Ho$_3$Al$_5$O$_{12}$ crystal, 4.2 and 77 K [526]^b			
$^5I_8$	0, 16, 41, 74, 89, 104, 123, 162, 270, 316, 335, 378, 413, 452, 488, 520, 551	17	17	551
$^5I_7$	5230, 5242, 5252, 5305, 5313, 5321, 5339, 5349, 5373, 5397, 5405, 5456	15	12	226*
	NaLa(MoO$_4$)$_2$ crystal, 4.2 K [406]			
$^5I_8$	0, 23, 51, 83, 110, 180, 260, 297	13	8	297*
$^5I_7$	5128, 5149, 5263	11	3	135*

^{b} Results require more-accurate definition.

Table 4.9.   Crystal-field splitting of Er^{3+} ion manifolds

[SL]J manifold	Stark-level positions [cm^{-1}]	Number of components		ΔE [cm^{-1}]
		Theory	Exper.	
	LiYF$_4$ crystal, 77 K [287, 647]			
$^4I_{15/2}$	0, 17, 29, 56, 252, 291, 320, 347	8	8	347
$^4I_{13/2}$	6535, 6539, 6579, 6674, 6697, 6724, 6738	7	7	203
$^4I_{11/2}$	10222, 10235, 10283, 10289, 10315	6	5	93*
$^4I_{9/2}$	12360, 12466, 12565, 12660	5	4	300*
$^4F_{9/2}$	15316, 15333, 15349, 15426, 15477	5	5	161
$^4S_{3/2}$	18433, 18494	2	2	61
$^2H_{11/2}$	19154, 19177, 19224, 19307, 19322, 19338	6	6	184
$^4F_{7/2}$	20565, 20657, 20665	4	3	100*
$^4F_{5/2}$	22226, 22259, 22306	3	3	80
$^2H_{9/2}$	24523, 24584, 24640, 24695, 24750	5	5	227
$^4G_{11/2}$	26406, 26464, 26564, 26593, 26636	6	5	230*
	BaY$_2$F$_8$ crystal, 77 K [31]			
$^4I_{15/2}$	0, 26, 47, 103, 284, 329, 370, 410	8	8	410
$^4F_{9/2}$	15312, 15335, 15387, 15420, 15499	5	5	187
	LaF$_3$ crystal, 4.2 K [49, 288, 289]			
$^5I_{15/2}$	0, 52, 122, 200, 220, 320, 401, 444	8	8	444
$^4I_{13/2}$	6604, 6630, 6670, 6701, 6723, 6742, 6825	7	7	211
$^4I_{11/2}$	10302, 10312, 10332, 10346, 10363, 10398	6	6	96
$^4I_{9/2}$	12420, 12520, 12595, 12621, 12692	5	5	272
$^4F_{9/2}$	15390, 15433, 15444, 15475, 15528	5	5	138
$^4S_{3/2}$	18564, 18594	2	2	30
$^2H_{11/2}$	19270, 19310*, 19317*, 19364, 19423	6	5	153*
$^4F_{7/2}$	20658, 20706, 20737, 20791	4	4	133
$^4F_{5/2}$	22373, 22378, 22410	3	3	37
$^4F_{3/2}$	22690, 22755	2	2	65
$^2H_{9/2}$	24608, 24688, 24760, 24843, 24864	5	5	256
$^4G_{11/2}$	26504, 26533, 26562, 26589, 26655, 26717	6	6	213
	CaF$_2$–YF$_3$ crystal^a, 77 K [106]			
$^4I_{15/2}$	0, 107, 185, 230, 295, 374, 440, 530	8	8	530
$^4I_{13/2}$	6506, 6570, 6653, 6711, 6788, 6839, 6930	7	7	424
$^4S_{3/2}$	18535, 18572	2	2	37
	KY(WO$_4$)$_2$ and KEr(WO$_4$)$_2$ crystals, 77 K [792]			
$^4I_{15/2}$	0, 27, 60, 103, 133, 230, 288, 302	8	8	302
$^4I_{13/2}$	6516, 6544, 6570, 6600, 6666, 6718, 6730	7	7	214
$^4I_{11/2}$	10188, 10213, 10229, 10260, 10284, 10290	6	6	102
$^4I_{9/2}$	12392, 12436, 12466, 12497, 12557	5	5	165
$^4F_{9/2}$	15200, 15215, 15285, 15330, 15365	5	5	165
$^4S_{3/2}$	18314, 18378	2	2	64
	KGd(WO$_4$)$_2$ crystal, 77 K [641]			
$^4I_{15/2}$	0, 29, 62, 103, 136, 223, 280, 296	8	8	296
$^4I_{13/2}$	6517, 6546, 6573, 6602, 6660, 6711, 6724	7	7	207
$^4I_{11/2}$	10189, 10214, 10228, 10257, 10280, 10287	6	6	98
$^4I_{9/2}$	12395, 12435, 12468, 12498, 12556	5	5	161
$^4F_{9/2}$	15201, 15219, 15288, 15333, 15368	5	5	167
$^4S_{3/2}$	18327, 18387	2	2	60

^{a} Results require more-accurate definition.

(*Continued*)

Table 4.9 (*continued*)

[SL]J manifold	Stark-level positions [cm^{-1}]	Number of components		ΔE [cm^{-1}]
		Theory	Exper.	
CaWO$_4$ crystal, 77 K [291]				
$^4I_{15/2}$	0, 19, 25, 61, 228, 266, 319	8	7	319*
$^4I_{13/2}$	6376, 6384, 6424, 6505, 6534, 6551, 6569	7	7	193
$^4I_{11/2}$	10046, 10062, 10100, 10119, 10138, 10142	6	6	96
$^4I_{9/2}$	12190, 12323, 12355, 12374, 12451	5	5	261
$^4F_{9/2}$	15074, 15114, 15126, 15190, 15242	5	5	168
$^4S_{3/2}$	18207, 18272	2	2	65
$^2H_{11/2}$	18933, 18948, 18979, 19033, 19038, 19071	6	6	138
$^4F_{7/2}$	20316, 20327, 20420, 20432	4	4	116
$^4F_{5/2}$	22008, 22029, 22053	3	3	45
$^4F_{3/2}$	22325, 22395	2	2	70
$^2H_{9/2}$	24306, 24413, 24430, 24463, 24511	5	5	205
YAlO$_3$ crystal, 4.2 K [109, 467, 675]				
$^4I_{15/2}$	0, 51, 171, 218, 266, 388, 443, 516	8	8	516
$^4I_{13/2}$	6602, 6641, 6669, 6715, 6773, 6814, 6868	7	7	265
$^4I_{11/2}$	10282, 10293, 10322, 10347, 10382, 10402	6	6	120
$^4I_{9/2}$	12393, 12446, 12623, 12648, 12732	5	5	339
$^4F_{9/2}$	15263, 15344, 15374, 15396, 15481	5	5	218
$^4S_{3/2}$	18406, 18487	2	2	81
$^2H_{11/2}$	19119, 19162, 19190, 19240, 19275, 19303	6	6	184
$^4F_{7/2}$	20481, 20554, 20617, 20685	4	4	204
$^4F_{5/2}$	22196, 22227, 22259	3	3	63
$^4F_{3/2}$	22516, 22536	2	2	20
$^2H_{9/2}$	24479, 24526, 24666, 24698, 24765	5	5	286
$^4G_{11/2}$	26307, 26322, 26380, 26458, 26476, 26526	6	6	219
$^4G_{9/2}$	27352, 27381, 27410, 27544, 27670	5	5	318
$^2K_{15/2}$	27399, 27445, 27487, 27732, 27761, 27913, 27997, 28065	8	8	667
$^2G_{7/2}$	27683, 27775, 28044, 28077	4	4	394
$^2P_{3/2}$	31449, 31585	2	2	136
$^2K_{13/2}$	23773, 32823, 32983, 33061, 33162, 33301, 33375	7	7	602
$^4G_{7/2}$	33866, 33962, 34038, 34089	4	4	223
Y$_3$Al$_5$O$_{12}$ crystal, 77 K [292, 293, 590]				
$^4I_{15/2}$	0, 22, 61, 79, 417, 430, 526, 573	8	8	573
$^4I_{13/2}$	6549, 6599, 6606, 6786, 6805, 6823*, 6885	7	7	336
$^4I_{11/2}$	10256, 10287, 10362, 10373, 10414, 10419	6	6	163
$^4I_{9/2}$	12303, 12527, 12577, 12719, 12765	5	5	462
$^4F_{9/2}$	15290, 15315, 15359, 15475, 15520	5	5	230
$^4S_{3/2}$	18397, 18461	2	2	64
$^2H_{11/2}$	19093, 19114, 19151, 19310*, 19347, 19365	6	6	272
$^4F_{7/2}$	20513, 20569, 20648, 20700	4	4	187
$^4F_{5/2}$	22219, 22238, 22285	3	3	66
$^4F_{3/2}$	22585, 22659	2	2	74
$^2H_{9/2}$	24422, 24576, 24591, 24765, 24784	5	5	362
$^4G_{11/2}$	26218, 26280, 26327, 26569, 26577, 26607	6	6	389
300 K [590]				
$^4I_{15/2}$	0, 19, 57, 76, 411, 424, 523, 568	8	8	568
$^4I_{13/2}$	6544, 6596, 6602, 6779, 6800, 6818*, 6879	7	7	335

(*Continued*)

Table 4.9   (*continued*)

[SL]J manifold	Stark-level positions [cm^{-1}]	Number of components		ΔE [cm^{-1}]
		Theory	Exper.	
$^4I_{11/2}$	10252, 10285, 10356, 10367, 10408, 10412	6	6	160
$^4I_{9/2}$	12301, 12524, 12572, 12713, 12760	5	5	459
$^4S_{3/2}$	18392, 18456	2	2	64
	GdAlO$_3$ crystal, 4.2 and 77 K [294, 715]			
$^4I_{15/2}$	0, 50, 183, 224, 269, 394, 448, 533*	8	8	533*
$^4I_{13/2}$	6615, 6656, 6687, 6728, 6794, 6828, 6872	7	7	257
$^4I_{11/2}$	10299, 10310, 10340, 10367, 10401, 10414	6	6	115
$^4I_{9/2}$	12405, 12456, 12649, 12665, 12755	5	5	350
$^4F_{9/2}$	15285, 15373, 15402, 15420, 15489	5	5	204
$^4S_{3/2}$	18440, 18515	2	2	75
$^2H_{11/2}$	10150, 19189, 19211, 19257, 19297, 19324	6	6	174
$^4F_{7/2}$	20515, 20576, 20645, 20716	4	4	201
$^4F_{5/2}$	22229, 22255, 22289	3	3	60
$^4F_{3/2}$	(lines not visible)	—	—	—*
$^2H_{9/2}$	24500, 24501, 24548, 24693, 24736	5	5	236
$^4G_{11/2}$	26344, 26363, 26414, 26479, 26504, 26551	6	6	207
$^2K_{9/2}$	27392, 27417, 27430, 27446, 27478	5	5	86
$^2K_{15/2} + {}^2G_{7/2}$	27526, 27581, 27744, 27780, 27813, 27825, 28016, 28067, 28096, 28210	12	11	(684*)
$^2P_{3/2}$	31485, 31613	2	2	128
	Er$_3$Al$_5$O$_{12}$ crystal, 4.2 K [648, 757]			
$^4I_{15/2}$	0, 27, 58, 79, 423, 436, 530, 574	8	8	574
$^4I_{13/2}$	6551, 6596, 6602, 6790, 6809, 6883, 6885	7	7	334
$^4I_{11/2}$	10258, 10286, 10365, 10376, 10413, 10419	6	6	161
$^4I_{9/2}$	12301, 12529, 12576, 12722, 12766	5	5	465
$^4F_{9/2}$	15291, 15315, 15355, 15474, 15518	5	5	227
$^4S_{3/2}$	18395, 18456	2	2	60
$^2H_{11/2}$	19091, 19113, 19149, 19348, 19366, 19371	6	6	281
$^4F_{7/2}$	20516, 20568, 20649, 20698	4	4	182
$^4F_{5/2}$	22218, 22239, 22284	3	3	66
$^2H_{9/2}$	24417, 24573, 24589, 24764, 24785	5	5	368
$^4G_{11/2}$	26207, 26270, 26315, 26461, 26470, 26501	6	6	294
	77 K			
$^4I_{13/2}$	6550, 6596, 6602, 6789, 6808, 6883*, 6885	7	7	333
$^4I_{11/2}$	10259, 10286, 10365, 10376, 10413, 10419	6	6	160
	300 K			
$^4I_{13/2}$	6549, 6595, 6600, 6783, 6802, 6877*, 6879	7	7	330
$^4I_{11/2}$	10255, 10284, 10357, 10370, 10410, 10413	6	6	158
	Lu$_3$Al$_5$O$_{12}$ crystal, 77 K [590]			
$^4I_{15/2}$	0, 36, 56, 80, 440, 459, 539, 582	8	8	582
$^4I_{13/2}$	6564, 6598, 6606, 6807, 6824, 6851*, 6894	7	7	330
$^4I_{11/2}$	10268, 10289, 10377, 10388, 10420, 10425	6	6	157
$^4I_{9/2}$	12302, 12537, 12585, 12737, 12780	5	5	478
$^4F_{9/2}$	15306, 15327, 15360, 15488, 15530	5	5	224
$^4S_{3/2}$	18408, 18460	2	2	52
$^2H_{11/2}$	19098, 19118, 19154, 19349*, 19360, 19380	6	6	282
$^4F_{7/2}$	20528, 20577, 20662, 20704	4	4	176

(*Continued*)

Table 4.9   (*continued*)

[SL]J manifold	Stark-level positions [cm^{-1}]	Number of components		ΔE [cm^{-1}]
		Theory	Exper.	
$^4F_{5/2}$	22225, 22250, 22290	3	3	65
$^4F_{3/2}$	22601, 22660	2	2	59
$^2H_{9/2}$	24419, 24576, 24595, 24775, 24797	5	5	378
$^4G_{11/2}$	26216, 26277, 26325, 26578, 26586, 26616	6	6	400
	300 K			
$^4I_{15/2}$	0, 32, 53, 76, 434, 450, 534, 578	8	8	578
$^4I_{13/2}$	6559, 6595, 6602, 6798, 6818, 6847*, 6885	7	7	326
$^4I_{11/2}$	10264, 10285, 10370, 10381, 10415, 10420	6	6	156
$^4I_{9/2}$	12300, 12532, 12580, 12731, 12772	5	5	472
$^4S_{3/2}$	18402, 18455	2	2	53

Table 4.10.   **Crystal-field splitting of Tm^{3+} ion manifolds**

[SL]J manifold	Stark-level positions [cm^{-1}]	Number of components		ΔE [cm^{-1}]
		Theory	Exper.	
	CaWO$_4$ crystal, 2 K [649]			
$^3H_6$	0, 26, 53, 245*, 328*, 338*, 370*, 384*	10	8	384*
$^3H_4$	5594, 5733, 5745, 5811, 5926, 5927, 5926	7	7	342
$^3H_5$	8279, 8288, 8299, 8486, 8499, 8531, 8535, 8550	8	8	271
$^3F_4$	12583, 12600, 12628, 12759, 12832, 12835	7	6	252*
$^3F_3$	14481, 14509, 14563, 14572, 14575	5	5	94
$^3F_2$	15039, 15152*, 15156	4	3	113*
$^1G_4$	20986, 21180, 21275, 21292, 21476, 21492, 21520	7	7	534
$^1D_2$	27808, 27855*, 27929, 27940*	4	4	132*
	YAlO$_3$ crystal, 4.2 and 77 K [525, 650–652]			
$^3H_6$	0, 3*, 65*, 144*, 210*, 237, 271, 282, 313, 440, 574, 628*	13	12	628*
$^3H_4$	5622, 5627, 5716, 5722, 5819, 5843, 5935, 5965, 5988	9	9	366
$^3H_5$	8261, 8265, 8322, 8345, 8376, 8459, 8482, 8564, 8589, 8599, 8690	11	11	429
$^3F_4$	12515, 12574, 12667, 12742, 12783, 12872, 12885, 12910, 12950	9	9	435
$^3F_3$	14448, 14478, 14513, 14552, 14593, 14606, 14622	7	7	184
$^3F_2$	15027, 15088, 15177, 15193, 15285	5	5	258
$^1G_4$	21020, 21101, 21193, 21220, 21293, 21321, 21455, 21542, 21620	9	9	600
$^1D_2$	27693, 27708, 27843, 27902, 27926	5	5	233
	Y$_3$Al$_5$O$_{12}$ crystal, 77 K [295]			
$^3H_6$	0, 27, 215, 241, 245, 448, 547, 588, 610	13	9	610*
$^3H_4$	5557, 5737, 5809, 5875, 5907, 6046, 6148, 6177	9	8	560*
	YVO$_4$ crystal, 85 K [296]			
$^3H_6$	0, 54, 119, 138, 158, 192, 208, 332	10	8	332*
$^3H_4$	5550, 5655, 5723, 5774, 5824, 5860, 5879	7	7	329

(*Continued*)

Table 4.10   (*continued*)

[SL]J manifold	Stark-level positions [cm^{-1}]	Number of components		ΔE [cm^{-1}]
		Theory	Exper.	
$^3H_5$	8204, 8232, 8268, 8296, 8338, 8440, 8491	8	7	287*
$^3F_4$	12523, 12563, 12632, 12661, 12704, 12705, 12774	7	7	251
$^3F_3$	14411, 14452, 14453, 14459, 14475	5	5	64
$^3F_2$	15007, 15017, 15069, 15147	4	4	140
$^1G_4$	20938, 21102, 21167, 21234, 21306, 21459	7	6	471*
$^1D_2$	27735, 27736, 27753, 27788	4	4	53
	GdAlO$_3$ crystal, 77 K [286]			
$^3H_6$	0, 64, 110, 140, 226, 303, 339, 425, 526, 551*, 569*, 604*	13	12	604*
$^3H_4$	5623, 5718, 5741, 5827, 5934, 5949, 5976	9	7	353*
$^3H_5$	8262, 8266, 8322, 8343, 8366, 8371, 8457, 8503, 8562, 8667	11	10	405*
$^3F_4$	12494, 12579, 12655, 12743, 12780, 12849, 12886, 12939, 13089	9	9	595
$^3F_3$	14455, 14493, 14597, 14618, 14700, 14735, 14852	7	7	397
$^3F_2$	15052, 15092, 15179, 15289, 15297*	5	5	245*
$^1G_4$	21025, 21126, 21214, 21300, 21451, 21483*, 21547, 21587	9	8	562*
$^1D_2$	27708, 27848, 27918*, 27972*, 28160	5	5	452*

Table 4.11.   Crystal-field splitting of Yb^{3+} ion manifolds

[SL]J manifold	Stark-level positions [cm^{-1}]	Number of components		ΔE [cm^{-1}]
		Theory	Exper.	
	Y$_3$Al$_5$O$_{12}$ crystal, 77 K [298, 483, 484, 654]			
$^2F_{7/2}$	0, 565, 612, 785	4	4	785
$^2F_{5/2}$	10327, 10624*, 10679*, 10902	3	4	575
	Y$_3$Ga$_5$O$_{12}$ crystal, 4.2 and 77 K [298]			
$^2F_{7/2}$	0, 308, 565, 627	4	4	627
$^2F_{5/2}$	10315, 10605, 10765*	3	3	450*
	Lu$_3$Al$_5$O$_{12}$ crystal, 77 K [86, 483, 654]			
$^2F_{7/2}$	0, 600, 621, 761	4	4	761
$^2F_{5/2}$	10335, 10643*, 10685*, 10909	3	4	574

Table 4.12.   Crystal-field splitting of U^{3+} ion manifolds

[SL]J manifold	Stark-level positions [cm^{-1}]	Number of components		ΔE [cm^{-1}]
		Theory	Exper.	
	CaF$_2$ crystal, 77 K [15, 299]			
$^4I_{9/2}$	0, 329, 414, 494, 667	5	5	667
$^4I_{11/2}$	4444, 4505, 4583, 4651, 5050, 5376	6	6	932
	SrF$_2$ crystal, 77 K [15, 299, 300]			
$^4I_{9/2}$	0, 169, 206, 303, 460	5	5	460
$^4I_{11/2}$	4132, 4509, 4613, 4819, 5285, 5494	6	6	1362
	BaF$_2$ crystal, 77 K [15, 299, 300]			
$^4I_{9/2}$	0, 357	5	2	357*
$^4I_{11/2}$	4049, 4175, 4515, 4955, 5187, 5376	6	6	1327

Table 4.13.   Crystal-field splitting of Cr^{3+} ion manifolds in an Al_2O_3 crystal (trigonal symmetry field) at 4.2 K [120, 121, 302]

Manifold	Stark-level positions [cm^{-1}]	Mulliken level designation [303]
$^4A_2\ (t_2^3)$	0	—
	0.38	—
$^2E\ (t_2^3)$	14418 R_1	$\bar{E}$
	14447 R_2	$2\bar{A}$
$^2A_2\ (t_2^3)$	14957	$\bar{E}_b$
$^2E\ (t_2^2)$	15168	$2\bar{A}$
	15190	$\bar{E}_a$
$^4E,\ ^4A_1\ (t_2^2;^4T_2)$	16000–20000	U–band
$^2E\ (t_2^3;^2T_2)$	20993	$2\bar{A}$
	21063	$\bar{E}_a$
$^2A_1\ (t_2^3;^2T_2)$	21357	$\bar{E}_b$
$^4E,^4A_2\ (t_2^2e;^4T_1)$	22500–27000	Y–band
$^4A_2\ (t_2e^2)$	36000–40000	V–band

Table 4.14.   Crystal-field splitting of Co^{2+} ion manifolds in an MgF_2 crystal (distorted octahedral symmetry field) at 77 K [117]

Manifold	Stark-level positions [cm^{-1}]
$^4T_1\ (t_2^5\,e^2)$	0, 152, 798, 1087, 1256, 1398
$^4T_2\ (t_2^4\,e^3)$	6801, 6810, 6880, 6893, 6928

Table 4.15.   Crystal-field splitting of Ni^{2+} ion manifolds in an MgF_2 crystal (distorted octahedral symmetry field) at 20 K [20]

Manifold	Stark-level positions [cm^{-1}]
$^3A_2\ (t_2^6\,e^2)$	0, 1, 6
$^3T_2\ (t_2^5\,e^3)$	6506, 6543, 6630

4.3 Transition Intensities of Ln^{3+} Ions in Laser Crystals[1]

In discussing the problem of the transition intensities of Ln^{3+} ions in crystals, the classic work completed by *Van Vleck* over forty years ago should be mentioned [723]. Using the available experimental results concerning Ln^{3+} ion transition intensities, *Van Vleck* considered three possible mechanisms for interpreting their nature, namely those involving electric quadrupole, magnetic dipole (*md*), and forced electric dipole (*ed*) emission. He showed, in particular, that electric dipole transitions between levels of the same electronic configuration may be attributed to noncentrally symmetric interactions of the activator ion with the crystalline surroundings. A more detailed analysis of the problem of Ln^{3+} ion transition intensity in crystals, performed by *Broer, Gorter,* and *Hoogschagen* in 1945 [724], revealed that the probabilities of forced electric dipole and magnetic dipole transitions may greatly exceed that of electric quadrupole transitions. A variety of experimental and theoretical studies of the spectral properties of crystals (including laser crystals) doped with Ln^{3+} ions in the years that followed conclusively confirmed the results of [723, 724]. Investigations show that in most cases the spectral-emission properties of laser crystals (which are so far available) doped with Ln^{3+} ions are due to forced electric dipole transitions. In some, considerable contribution can also be made by magnetic dipole transitions. The present section takes this fact into account and primary attention is given to electric dipole transitions.

The probabilities for spontaneous electric dipole and magnetic dipole transitions are [726]

$$A^{ed}(i,j) = \frac{64\pi^4 v_{ij}^3}{3hc^3 g_i} \sum_{ij} |\langle i|P|j\rangle|^2,$$

$$A^{md}(i,j) = \frac{64\pi^4 v_{ij}^3}{3hc^3 g_i} \sum_{ij} |\langle i|M|j\rangle|^2,$$

where $\langle i|\ |j\rangle$ are the matrix elements of the operators of the electric and magnetic dipole moments for the $i \rightarrow j$ transition. Thus, the only values in these expressions that depend on crystal properties are the dipole moments P and M. The sum of squared matrix elements in the above equations for spontaneous transition probabilities is called in spectroscopy the line (or transition) strength and is denoted

$$S_{ij} = \sum_{ij} |\langle i|\ |j\rangle|^2. \tag{4.8}$$

The probabilities for transitions between the ith and jth levels, or the Einstein coefficients [see (2.6, 7)] in terms of transition strengths, take form

---

[1] Basic to this section is [725].

$$A_{ij} = \frac{1}{g_i} \frac{64\pi^4 v_{ij}^3}{3hc^3} S_{ij},$$

$$B_{ji} = \frac{1}{g_j} \frac{8\pi^3}{3h^2} S_{ij},$$

$$B_{ij} = \frac{1}{g_i} \frac{8\pi^3}{3h^2} S_{ij}.$$

Thus, among the main problems concerning transition intensity is that of determining the transition strength. As follows from the last equations, this parameter can be found in a number of ways. Experimentally, absorption methods are, certainly, the simplest. We now consider more closely the peculiar features of forced electric dipole or simple electric dipole and magnetic dipole transitions of Ln^{3+} ions in crystals.

4.3.1 Electric Dipole Transitions; Judd–Ofelt Approximation

Electric dipole transitions between the states of the $4f^N$ electron configuration of an isolated Ln^{3+} ion are prohibited by the parity selection rule [726]. This prohibition, as was shown by *Van Vleck* [723] can be more or less avoided due to noncentrally symmetric interactions of the Ln^{3+} ions with the surroundings, which mix states of opposite parity. Examples of these interactions in a crystal are both static (odd terms in the $\mathscr{V}_{cr}^{odd}$ potential expansion for noncentrally symmetric centers that induce pure electron transitions) and dynamic (non-centrally symmetric vibrations of the surroundings that induce electron vibrational transitions) portions of the crystal field. We shall consider now only the static portion that is responsible for the main contribution to the probability for radiative transitions of Ln^{3+} ions in noncentrally symmetric centers.

The $\langle A|$ and $|A'\rangle$ states of Ln^{3+} ions in a crystal can be represented as linear combinations of the wave functions of the ground $4f^N$ configuration $|4f^N\psi JJ_z\rangle$ with the wave functions of excited $|\beta\rangle$ configurations of the opposite parity

$$\langle A| = \langle 4f^N\psi JJ_z| - \sum_\beta \frac{\langle 4f^N\psi JJ_z|\mathscr{V}_{cr}^{odd}|\beta\rangle\langle\beta|}{E(4f^N\psi JJ_z) - E(\beta)}, \tag{4.9}$$

$$|A'\rangle = |4f^N\psi'J'J_z'\rangle - \sum_\beta \frac{|\beta\rangle\langle\beta|\mathscr{V}_{cr}^{odd}|4f^N\psi'J'J_z'\rangle}{E(4f^N\psi'J'J_z') - E(\beta)}. \tag{4.10}$$

The operator of the electric dipole moment P can be expanded in the x, y and z components of $(P_q^{(1)})$ which induce transitions of various polarizations. When $q = 0$, π polarization occurs (z component); when $q = \pm 1$, σ polarization occurs (x, y components). Nonzero matrix elements of the $P_q^{(1)}$ operator relate the states of opposite parity admixed to $\langle A|$ and $|A'\rangle$

$$\langle A|P_q^{(1)}|A'\rangle = -\sum_\beta \frac{\langle \psi JJ_z|\mathcal{V}_{cr}^{odd}|\beta\rangle\langle\beta|P_q^{(1)}|\psi'J'J_z'\rangle}{E(4f^N\psi JJ_z) - E(\beta)}$$

$$-\sum_\beta \frac{\langle \psi JJ_z|P_q^{(1)}|\beta\rangle\langle\beta|\mathcal{V}_{cr}^{odd}|\psi'J'J_z'\rangle}{E(4f^N\psi'J'J_z') - E(\beta)}. \tag{4.11}$$

Here the summation is over all the configuration-state components $|\beta\rangle$ of opposite parity. The terms in the denominators of expressions (4.9–11), $E(4f^N\psi JJ_z)$ and $E(\beta)$, are the energy levels of the ground and excited electron configurations. It is rather difficult to calculate the right-hand side of expression (4.11) because, for this purpose, it is necessary to know not only the $E(\beta)$ energies and the wave functions of the excited $|\beta\rangle$ configuration levels, but also the odd portion of the crystal-field potential which is responsible for mixing opposite parity states.

The next significant step in developing the theory of transition intensities of Ln^{3+} ions in crystals was due to *Judd* [559] and *Ofelt* [560]. Working independently and simultaneously, they made an assumption, which was evident from the physical viewpoint, that permitted simplification of expression (4.11) considerably. They replaced $E(4f^N\psi JJ_z) - E(\beta)$ and $E(4f^N\psi'J'J_z') - E(\beta)$ by the constant ΔE independent of ψ, J, and β. This is equivalent to the assumption that electron configuration splitting is negligible compared with the energy gap between the levels. Then the energy denominator in (4.11) is taken outside the summation sign, and the expression becomes

$$\langle A|P_q^{(1)}|A'\rangle = \sum_{t,m} Y(t, m, q) \cdot \langle 4f^N\psi JJ_z|U_{m+q}^{(t)}|4f^N\psi'J'J_z'\rangle. \tag{4.12}$$

Here t is an even number; $\langle 4f^N\psi JJ_z|U_{m+q}^{(t)}|4f^N\psi'J'J_z'\rangle$ is the matrix element of the $(m + q)$th component of the irreducible tensor operator of rank t; $Y(t, m, q)$ is the constant determined by

$$Y(t, m, q) = \sum_k (-1)^{m+q} (2t + 1)^{1/2} A_{km} \begin{pmatrix} 1 & t & k \\ q - (m + q) & m \end{pmatrix} Z(k, t),$$

where $Z(k, t)$ is the value proportional to the overlap integral of the radial parts of the wave functions for states belonging to the ground and excited electron configurations of the opposite parity and inversely proportional to the energy gap therein; A_{km} is the odd parameter of the crystal-field potential ($k \leqslant 7$). Depending on the presence of odd terms in the crystal-field potential expansion in spherical harmonics and, hence, on the mechanism of forced electric dipole transition between levels of Ln^{3+} ions in the activator center of a given symmetry, the thirty-two crystallographic groups can be divided into two classes. The first class includes all the centrally symmetric groups plus the noncentrally symmetric group O which has odd terms of $k \leqslant 7$. The second class contains all other noncentrally symmetric groups.

For forced electric dipole transitions, the following selection rules characterize the states of an isolated ion in terms of quantum numbers:

$$\Delta l = \pm 1;$$

$$\Delta S = 0;$$

$$|\Delta L| \leqslant 2l;$$

$$|\Delta J| \leqslant 2l.$$

The first rule concerns excited configurations of the opposite parity whose admixture may lead to such transitions. The second and the third rules pertain to the S and L values. It is worth noting at this point that spin–orbit interaction leads to violation of these selection rules. The last rule will hold as long as J remains a good quantum number. This condition usually obtains for Ln^{3+} ions, though there are examples of its violation because of J mixing [558]. If the terminal and the initial states both have $J = 0$, there are additional selection rules for transitions between them, since k is an odd number whereas t may be even. It follows from this that ΔJ must be an even number.

According to the value of q in crystal fields with cubic symmetry, selection rules are established by polarization. It is known that if J is half-integral, the levels in a crystal can be characterized by the quantum number μ alone. If J is integral (as in, say, the groups D_{3h} and C_{3h}) transitions that are equivalent from the viewpoint of the selection rules in μ may differ with regard to selection rules in the irreducible representations Γ_y. In such circumstances, it is evidently more convenient to employ irreducible representations, which provide more-complete selection rules for any point symmetry.

Let Γ_1 and Γ_2 be irreducible representations that transform the components of the operator P associated with transitions of π and σ polarization. Then, by group-theoretic arguments, polarization selection rules can be established for the $\Gamma_i \rightarrow \Gamma_j$ transitions in the symmetry group concerned. If $\Gamma_i \times \Gamma_j = \Sigma_n a_n \Gamma_n$, the $\Gamma_i \rightarrow \Gamma_j$ transition will be

$$
\begin{aligned}
\pi \text{ polarized} \quad &\text{when } a_1 \neq 0,\, a_2 = 0; \\
\sigma \text{ polarized} \quad &\text{when } a_1 = 0,\, a_2 \neq 0; \\
\pi\sigma \text{ polarized} \quad &\text{when } a_1 \neq 0,\, a_2 \neq 0; \\
\text{forbidden} \quad &\text{when } a_1 = a_2 = 0.
\end{aligned}
$$

Tables of irreducible representations according to which the x, y, and z components of the operator P are transformed are given in the work by *Prather* [727] for some symmetry groups. Cubic groups with similar irreducible representations according to which the operator P components are transformed will undergo only nonpolarized transitions.

Summation of expression (4.12) over all J_z values of the initial and terminal states will remove the anisotropy associated with polarization and yield the matrix element determining the probabilities for intermanifold transitions. In this case, the expression for line strength is

$$s^{ed}(\psi J; \psi'J') = \frac{1}{e^2} \sum_{J_z J'_z} |\langle A|P_q^{(1)}|A'\rangle|^2$$

$$= \sum_{t=2,4,6} \Omega_t |\langle \psi J \| U^{(t)} \| \psi'J'\rangle|^2. \tag{4.13}$$

Here $\langle \psi J \| U^{(t)} \| \psi'J'\rangle$ is a reduced matrix element of the irreducible tensor operator of rank t and Ω_t are the intensity parameters[2] determined by the expression

$$\Omega_t = (2t + 1) \sum_{k,m} |A_{km}|^2 Z^2(k, t)/(2k + 1),$$

where A_{km} are the odd parameters of the expansion of the crystal-field operator [558].

The line strength s determined by (4.13) is associated with the spontaneous-emission probability and with the integrated absorption coefficient (2.21) by

$$A(\psi J; \psi'J') = \frac{64\pi^4 e^2}{3h(2J+1)\bar{\lambda}^3} \frac{n(n^2+2)^2}{9} s(\psi J; \psi'J') \tag{4.14}$$

and

$$\int k(\lambda) \, d\lambda = \frac{8\pi^3 N_0 \bar{\lambda} e^2}{3ch(2J+1)} \frac{(n^2+2)^2}{9n} s(\psi J; \psi'J'), \tag{4.15}$$

where $\bar{\lambda}$ is the mean wavelength that corresponds to the $J \to J'$ transition.

Since for electrons of the $4f^N$ configuration an intermediate-coupling scheme applies, where the wave functions are linear combinations of the Russell–Saunders states [558], the effective matrix element should be taken in expression (4.13). Here, it can be represented as a sum of matrix elements over all

---

[2] *Judd* [559] used the intensity parameters T_t. However, most authors subsequently used the more convenient Ω_t parameters related to T_t by

$$\Omega_t = \frac{3h(2J+1)}{8\pi^2 m} T_t \frac{9n}{(n^2+2)^2},$$

where J is the total angular moment of the ground multiplet. To facilitate calculations by (4.13), the authors of [514, 662] adopted another designation for the line strength related to the conventional one by $s = S/e^2$ and measured in cm^2.

the linear combinations of Russell–Saunders states taken with corresponding weights

$$\langle[\psi_1]J_1\|U^{(t)}\|[\psi_2]J_2\rangle = \sum_{\psi_1'\psi_2'} C_{\psi_1 J_1}(\psi_1') \cdot C_{\psi_2 J_2}(\psi_2')\langle\psi_1 J_1\|U^{(t)}\|\psi_2' J_2'\rangle.$$

(4.16)

Since the $\langle\,\|U^{(t)}\|\,\rangle$ matrix elements for a given Ln^{3+} ion vary only slightly from medium to medium, they may be considered unchanged. Table 4.16 lists the values of $|\langle\,\|U^{(t)}\|\,\rangle|^2$ for absorption transitions from the ground state of Ln^{3+} ions (the aquo-ion case); Table 4.17 gives the values for transitions from excited states. To avoid overloading these tables, only information on laser-crystal activators and their initial laser metastable states is included (see Fig. 3.2a). More complete information about the matrix elements can be found in the original works cited in these tables.

Thus, to describe the intensities of transitions observed in absorption and luminescence of crystals doped with Ln^{3+} ions in this approximation, it is sufficient to have three parameters Ω_t ($t = 2, 4$, and 6). As a rule, these parameters are chosen from measured intermultiplet (integrated) absorption coefficients. The line strength s is related to the Ω_t parameters by a system of linear equations like (4.13) which can be presented in vector form as

$$s = \mathscr{R}\Omega.$$

Here s is the q-vector whose components are the calculated line strengths s_{cal}; Ω is the vector whose components are the intensity parameters Ω_t; and $\mathscr{R}$ is the matrix whose elements are given by

$$a_{it} = |\langle\psi J\|U^{(t)}\|\psi_i J_i\rangle|^2,$$

(4.17)

where $\langle\psi J|$ is the initial and $|\psi_i J_i\rangle$ is the terminal state of the given manifold i transition.

If the absorption band is a superposition of lines assigned to several intermanifold transitions, the matrix element (4.17) can be taken to be the sum of the corresponding squared matrix elements because of the additivity of the integrated absorption coefficient of this band. By requiring that the sum of the squared differences between the calculated and measured line strengths be minimized, the following relation can be derived for Ω:

$$\Omega = (\mathscr{R}^\dagger\,\mathscr{R})^{-1}\,\mathscr{R}^\dagger s.$$

Here the components of the Ω vector are the desired intensity parameters that provide minimum root-mean-square deviation between the calculated s_{cal} and the measured s_{exp} values, and $\mathscr{R}^\dagger$ is the transposed matrix. In this case, the root-mean-square error is

Table 4.16. Calculated values of the squares of the reduced-matrix elements of $|\langle [SL]J \| U^{(t)} \| [S'L']J' \rangle|^2$ for Ln^{3+} (aquo) ions for transitions from the ground state $[SL]J$ [a]

$[S'L']J'$	State energy [cm^{-1}]	$t = 2$	$t = 4$	$t = 6$
		Pr^{3+} [728]		
$^3H_5$	2322	0.1095	0.2017	0.6109
$^3H_6$	4496	0.0001	0.0330	0.1395
$^3F_2$	5149	0.5089	0.4032	0.1177
$^3F_3$	6540	0.0654	0.3469	0.6983
$^3F_4$	6973	0.0187	0.0500	0.4849
$^1G_4$	9885	0.0012	0.0072	0.0266
$^1D_2$	16840	0.0026	0.0170	0.0520
$^3P_0$	20706	0	0.1728	0
$^3P_1$	21330	0	0.1707	0
$^1I_6$	21500	0.0093	0.0517	0.0239
$^3P_2$	22535	0	0.0362	0.1355
		Nd^{3+} [728]		
$^4I_{11/2}$	2007	0.0194	0.1073	1.1652
$^4I_{13/2}$	4005	0.0001	0.0136	0.4557
$^4I_{15/2}$	6080	0	0.0001	0.0452
$^4F_{3/2}$	11527	0	0.2293	0.0549
$^4F_{5/2}$	12573	0.0010	0.2371	0.3970
$^2H_{9/2}$	12738	0.0092	0.0080	0.1154
$^4F_{7/2}$	13460	0	0.0027	0.2352
$^4S_{3/2}$	13565	0.0010	0.0422	0.4245
$^4F_{9/2}$	14854	0.0009	0.0092	0.0417
$^2H_{11/2}$	16026	0.0001	0.0027	0.0104
$^4G_{5/2}$	17167	0.8979	0.4093	0.0359
$^2G_{7/2}$	17333	0.0757	0.1848	0.0314
$^2K_{13/2}$	19018	0.0068	0.0002	0.0312
$^4G_{7/2}$	19103	0.0550	0.1570	0.0553
$^4G_{9/2}$	19554	0.0046	0.0608	0.0406
$^2K_{15/2}$	21016	0	0.0052	0.0143
$^2G_{9/2}$	21171	0.0010	0.0148	0.0139
$(^2D, {}^2P)_{3/2}$	21266	0	0.0188	0.0002
$^4G_{11/2}$	21563	≈ 0	0.0053	0.0080
$^2P_{1/2}$	23140	0	0.0367	0
$^2D_{5/2}$	23865	≈ 0	0.0002	0.0021
$(^2P, {}^2D)_{3/2}$	26260	0	0.0014	0.0008
$^4D_{3/2}$	28312	0	0.1960	0.0170
$^4D_{5/2}$	28477	0.0001	0.0567	0.0275
$^2I_{11/2}$	28624	0.0049	0.0146	0.0034
$^4D_{1/2}$	28894	0	0.2584	0
$^2L_{15/2}$	29260	0	0.0248	0.0097
		Eu^{3+} [729][b]		
$^7F_2$	1018	0.1375	0	0
	668*	0.0518*	0	0
$^7F_3$	1880	0	0	0
	1530*	0.2093*	0.1281*	0
$^7F_4$	2866	0	0.1402	0
	2517*	0	0.1741*	0

[a] These data are adduced only for transitions that connect with manifolds lying below 30000 cm^{-1}.

[b] The asterisks indicate data that correspond to transitions from $^7F_1$ manifold, placed above the ground state at 350 cm^{-1}.

(*Continued*)

Table 4.16 (*continued*)

[S'L']J'	State energy [cm^{-1}]	t = 2	t = 4	t = 6
$^7F_5$	3927	0	0	0
	3578*	0	0.1193*	0.0544*
$^7F_6$	5049	0	0	0.1450
	4679*	0	0	0.3773*
$^5D_0$	17286	0	0	0
	16936*	0	0	0
$^5D_1$	19026	0	0	0
	18676*	0.0026*	0	0
$^5D_2$	21499	0.0008	0	0
	21149*	0.0001*	0	0
$^5D_3$	24389	0	0	0
	24039*	0.0004*	0.0012*	0
$^5L_6$	25375	0	0	0.0155
	25025*	0	0	0.0090*
$^5G_2$	26296	0.0006	0	0
	25946*	0.0006*	0	0
$^5L_7$	26469	0	0	0
	26119*	0	0	0.0183*
$^5G_3$	26535	0	0	0
	26185*	0.0003*	0.0014*	0
$^5G_4$	26672	0	0.0007	0
	26332*	0	0.0002*	0
$^5G_5$	26733	0	0	0
	26383*	0	0.0005*	0.0100*
$^5G_6$	26762	0	0	0.0038
	26412*	0	0	0.0051*
$^5L_8$	27435	0	0	0
	27085*	0	0	0
$^5D_4$	27641	0	0.0011	0
	27291*	0	0.0007*	0
$^5L_9$	28244	0	0	0
	27894*	0	0	0
$^5L_{10}$	28813	0	0	0
	28463*	0	0	0
Tb^{3+} [730]				
$^7F_5$	2112	0.5376	0.6418	0.1175
$^7F_4$	3370	0.0889	0.5159	0.2654
$^7F_3$	4344	0	0.2324	0.4126
$^7F_2$	5028	0	0.0482	0.4695
$^7F_1$	5481	0	0	0.3763
$^7F_0$	5703	0	0	0.1442
$^5D_4$	20545	0.0010	0.0008	0.0013
$^5D_3$	26336	0	0.0002	0.0014
$^5G_6$	26425	0.0017	0.0045	0.0118
$^5L_{10}$	27146	0	0.0004	0.0592
$^5G_5$	27795	0.0012	0.0018	0.0135
$^5D_2$	28150	0	≈ 0	0.008
$^5G_4$	28319	0.0001	0.0003	0.0091
$^5L_9$	28503	0	0.0021	0.0466
$^5G_3$	29007	0	0.0001	0.0031

(*Continued*)

Table 4.16   (*continued*)

$[S'L']J'$	State energy [cm^{-1}]	$t = 2$	$t = 4$	$t = 6$
$^5L_8$	29202	0	0.0001	0.0235
$^5L_7$	29406	0.0005	0.0001	0.0119
$^5L_6$	29550	0.0001	0.0001	0.0003
$^5G_2$	29577	0	≈ 0	0.0005
		Dy^{3+} [728]		
$^6H_{13/2}$	3560	0.2457	0.4139	0.6824
$^6H_{11/2}$	5833	0.0923	0.0366	0.6410
$^6H_{9/2}$	7692	0	0.0176	0.1985
$^6F_{11/2}$	7730	0.9387	0.8292	0.2048
$^6F_{9/2}$	9087	0	0.5736	0.7213
$^6H_{7/2}$	9115	0	0.0007	0.0392
$^6H_{5/2}$	10169	0	0	0.0026
$^6F_{7/2}$	11025	0	0.1360	0.7146
$^6F_{5/2}$	12432	0	0	0.3452
$^6F_{3/2}$	13212	0	0	0.0610
$^6F_{1/2}$	13760	0	0	0
$^4F_{9/2}$	21144	0	0.0047	0.0295
$^4I_{15/2}$	22293	0.0073	0.0003	0.0654
$^4G_{11/2}$	23321	0.0004	0.0145	0.0003
$^4F_{7/2}$	25754	0	0.0768	0.0263
$^4I_{13/2}$	25919	0.0041	0.0013	0.0248
$^4M_{21/2}$	26341	0	0.0102	0.0822
$^4K_{17/2}$	26365	0.0109	0.0048	0.0935
$^4M_{19/2}$	27219	0.0004	0.0166	0.1020
$(^4P, {}^4D)_{3/2}$	27254	0	0	0.0448
$^4P_{5/2}$	27503	0	0	0.0697
$^4I_{11/2}$	28152	0.0001	≈ 0	0.0074
$^4P_{7/2}$	28551	0	0.5222	0.0125
$(^4M, {}^4I)_{15/2}$	29244	0.0023	0.0005	0.0009
$(^4F, {}^4D)_{15/2}$	29593	0	0	0.0249
$^4I_{9/2}$	29885	0	0.0003	0.0003
		Ho^{3+} [728]		
$^5I_7$	5116	0.0250	0.1344	1.5216
$^5I_6$	8614	0.0084	0.0386	0.6921
$^5I_5$	11165	0	0.0100	0.0936
$^5I_4$	13219	0	≈ 0	0.0077
$^5F_5$	15519	0	0.4250	0.5687
$^5S_2$	18354	0	0	0.2268
$^5F_4$	18612	0	0.2392	0.7071
$^5F_3$	20673	0	0	0.3460
$^3K_8$	21308	0.0208	0.0334	0.1578
$^5G_6$	22094	1.5201	0.8410	0.1411
$^5F_1$	22375	0	0	0
$(^5G, {}^3G)_5$	23887	0	0.5338	0.0002
$^5G_4$	25826	0	0.0315	0.0359
$^3K_7$	26117	0.0058	0.0046	0.0338
$(^3G, {}^3H)_5$	27653	0	0.0790	0.1610
$^3H_6$	27675	0.2155	0.1179	0.0028
$(^5F, {}^3F, {}^5G)_2$	28301	0	0	0.0041
$^3G_5$	28816	0	0	0.0133
$^3L_9$	29020	0.0185	0.0052	0.1536

(*Continued*)

Table 4.16   (*continued*)

$[S'L']J'$	State energy [cm^{-1}]	$t = 2$	$t = 4$	$t = 6$
		Er^{3+} [728]		
$^4I_{13/2}$	6610	0.0195	0.1173	1.4316
$^4I_{11/2}$	10219	0.0282	0.0003	0.3953
$^4I_{9/2}$	12378	0	0.1733	0.0099
$^4F_{9/2}$	15245	0	0.5354	0.4618
$^4S_{3/2}$	18462	0	0	0.2211
$^2H_{11/2}$	19256	0.7125	0.4125	0.0925
$^4F_{7/2}$	20422	0	0.1469	0.6266
$^4F_{5/2}$	22074	0	0	0.2232
$^4F_{3/2}$	22422	0	0	0.1272
$(^2G, {}^4F, {}^2H)_{9/2}$	24505	0	0.0189	0.2256
$^4G_{11/2}$	26496	0.9183	0.5262	0.1172
$^4G_{9/2}$	27478	0	0.2416	0.1235
$^2K_{15/2}$	27801	0.0219	0.0041	0.0758
$^2G_{7/2}$	28000	0	0.0174	0.1163
		Tm^{3+} [728]		
$(^3F, {}^3H)_4$	5811	0.5375	0.7261	0.2382
$^3H_5$	8390	0.1074	0.2314	0.6383
$(^3H, {}^3F)_4$	12720	0.2373	0.1090	0.5947
$^3F_3$	14510	0	0.3164	0.8411
$^3F_2$	15116	0	≈ 0	0.2581
$^1G_4$	21374	0.0483	0.0748	0.0125
$^1D_2$	28032	0	0.3156	0.0928

Table 4.17.   Calculated values of the squares of the reduced matrix-elements of $|\langle [SL]J \| U^{(t)} \| [S'L']J' \rangle|S$ for Ln^{3+} (aquo) ions for transitions from the $[SL]J$ excited (laser) state[a]

$[SL]J$	$[S'L']J'$	State energy [cm^{-1}]	$t = 2$	$t = 4$	$t = 6$
			Pr^{3+} [731]		
$^1G_4$	$^3F_4$	2910	0.0467	0.1203	0.2844
	$^3F_3$	3340	0.0026	0.0031	0.0452
	$^3F_2$	4740	0.0001	0.0138	0.0032
	$^3H_6$	5390	0.1927	0.1905	0.1865
	$^3H_5$	7560	0.0307	0.0715	0.3344
	$^3H_4$	9880	0.0188	0.0044	0.0119
$^1D_2$	$^1G_4$	6950	0.3865	0.0493	0.0844
	$^3F_4$	9870	0.5143	0.0004	0.0147
	$^3F_3$	10300	0.0300	0.0168	0
	$^3F_2$	11690	0.0131	0.0814	0
	$^3H_6$	12340	0	0.0649	0.0058
	$^3H_5$	14520	0	0.0019	0.0003
	$^3H_4$	16840	0.0020	0.0165	0.0493
$^3P_0$	$^1D_2$	3870	0.0134	0	0
	$^1G_4$	10820	0	0.0425	0
	$^3F_4$	13730	0	0.1213	0
	$^3F_3$	14170	0	0	0
	$^3F_2$	15560	0.2943	0	0

[a] These data are adduced only for transitions that connect with manifolds lying below 30000 cm^{-1}.   (*Continued*)

Table 4.17   (*continued*)

$[SL]J$	$[S'L']J'$	State energy [cm^{-1}]	$t = 2$	$t = 4$	$t = 6$
	$^3H_6$	16210	0	0	0.0726
	$^3H_5$	18380	0	0	0
	$^3H_4$	20700	0	0.1713	0
$^3P_1$	$^3P_0$	620	0	0	0
	$^1D_2$	4490	0.0749	0	0
	$^1G_4$	11440	0	0.0605	0
	$^3F_4$	14360	0	0.2851	0
	$^3F_3$	14790	0.5714	0.1964	0
	$^3F_2$	16180	0.2698	0	0
	$^3H_6$	16830	0	0	0.1246
	$^3H_5$	19000	0	0.2857	0.0893
	$^3H_4$	21330	0	0.1721	0
$^3P_2$	$^1I_6$	1030	0	0.0257	0.1405
	$^3P_1$	1200	0.4231	0	0
	$^3P_0$	1830	0.1929	0	0
	$^1D_2$	5690	0.0011	0.0718	0
	$^1G_4$	12650	0.5640	0.0341	0.0184
	$^3F_4$	15560	0.5233	0.1170	0.0072
	$^3F_3$	15990	0.2584	0.3082	0
	$^3F_2$	17390	0.0323	0.3000	0
	$^3H_6$	18040	0	0.5010	0.0544
	$^3H_5$	20210	0	0.1888	0.1316
	$^3H_4$	22530	0.0001	0.0362	0.1373
		Nd^{3+} [732]			
$^4F_{3/2}$	$^4I_{15/2}$	5450	0	0	0.028
	$^4I_{13/2}$	7520	0	0	0.212
	$^4I_{11/2}$	9520	0	0.142	0.407
	$^4I_{9/2}$	11530	0	0.230	0.056
		Eu^{3+} [735]			
$^5D_0$	$^7F_6$	12260	0	0	0.0043
	$^7F_4$	14420	0	0.0030	0
	$^7F_2$	16270	0.0039	0	0
		Tb^{3+} [733]			
$^5D_4$	$^7F_0$	14840	0	0.0025	0
	$^7F_1$	15060	0	0.0015	0
	$^7F_2$	15520	0.0009	0.0004	0.0001
	$^7F_3$	16200	0.0027	0.0005	0.0007
	$^7F_4$	17180	0.0003	0.0019	0.0019
	$^7F_5$	18430	0.0146	0.0011	0.0036
	$^7F_6$	20470	0.0007	0.0016	0.0019
		Ho^{3+} [734]			
$^5I_7$	$^5I_8$	5150	0.0250	0.1344	1.5218
$^5I_6$	$^5I_7$	3500	0.0319	0.1336	0.9309
	$^5I_8$	8540	0.0083	0.0383	0.6918
$^5F_5$	$^5I_4$	2300	0.0001	0.0059	0.0040
	$^5I_5$	4350	0.0068	0.0271	0.1649
	$^5I_6$	6910	0.0103	0.1213	0.4995
	$^5I_7$	10410	0.0178	0.3299	0.4341
	$^5I_8$	15450	0	0.4278	0.5686

(*Continued*)

Table 4.17 (*continued*)

[SL]J	[S'L']J'	State energy [cm^{-1}]	$t=2$	$t=4$	$t=6$
$^5S_2$	$^5F_5$	2740	0	0.0110	0.0036
	$^5I_4$	5030	0.0014	0.0262	0.2795
	$^5I_5$	7090	0	0.0043	0.1063
	$^5I_6$	9650	0	0.0207	0.1542
	$^5I_7$	13150	0	0	0.4097
	$^5I_8$	18190	0	0	0.2270
		Er^{3+} [49]			
$^4I_{13/2}$	$^5I_{15/2}$	6480	0.0188	0.1176	1.4617
$^4I_{11/2}$	$^4I_{13/2}$	3640	0.021	0.11	1.04
	$^4I_{15/2}$	10120	0.0259	0.0001	0.3994
$^4F_{9/2}$	$^4I_{9/2}$	2880	0.096	0.0061	0.012
	$^4I_{11/2}$	5110	0.0671	0.0088	1.2611
	$^4I_{13/2}$	8750	0.0096	0.1576	0.0870
	$^4I_{15/2}$	15240	0	0.5655	0.4651
$^4S_{3/2}$	$^4I_{9/2}$	6080	0	0.0729	0.2285
	$^4I_{11/2}$	8240	0	0.0037	0.3481
	$^4I_{13/2}$	11870	0	0	0.0789
	$^4I_{15/2}$	18350	0	0	0.2560
$^4H_{9/2}$	$^4F_{9/2}$	9290	0.010	0.030	0.17
	$^4I_{9/2}$	12180	0.0076	0.0050	0.41
	$^4I_{11/2}$	14400	0.077	0.11	0.096
	$^4I_{13/2}$	18050	0.073	0.12	0.0028
	$^4I_{15/2}$	24530	0	0.078	0.056
		Tm^{3+} [563]			
$^3H_4$	$^3H_6$	5810	0.527	0.718	0.228
$^3F_4$	$^3H_5$	4320	0.011	0.480	0.004
	$^3H_4$	6910	0.129	0.133	0.213
	$^3H_6$	12720	0.249	0.118	0.608

$$\delta = \left[\sum_{i=1}^{n} (s_{cal} - s_{exp})^2 \frac{1}{q-p} \right]^{1/2},$$

where q is the number of analyzed line groups (intermanifold transitions) and p is the number of parameters sought, which in our case is three.

Numerous calculations of spectral intensities of Ln^{3+} ions in a variety of media indicate fairly good agreement of the calculated and measured line strengths (see, for instance, [49, 50, 174, 448, 561, 736–738]). Several workers have analyzed in detail the transition probabilities between Stark components (see, for example, [739, 740]). Such calculations are justified, however, only when the crystal splitting has been analyzed and the wave functions of the levels under study have been established. Unfortunately, at present it is extremely difficult to obtain such reliable information, particularly in the case of low-symmetry activator centers.

An analysis of the transition intensities for Ln^{3+} ions in crystals has revealed systematic variations of the Ω_t parameters for adjacent Ln ions in the lanthanide group for one and the same matrix. This fact has been used by some scientists to estimate the intensity parameters of Ln^{3+} ions when no experimental intensity data are available (e.g., for the Pm^{3+} ion [738]) or when the number of observed intermanifold transitions does not suffice to permit empirical determination of Ω_t (e.g., for the Yb^{3+} ion [737]). Table 4.18 contains all the known results for the Ω_t parameters for laser crystals doped with Ln^{3+} ions.

Table 4.18. Intensity parameters of Ln^{3+} ions in laser crystals

Crystal	Ln^{3+} ion site symmetry	Ion	$\Omega_t[10^{-20}\ cm^2]$			Reference
			$t = 2$	$t = 4$	$t = 6$	
$LiYF_4$	S_4	Nd^{3+}	1.9	2.7	5.0	[267]
LaF_3	C_{2v}	Pr^{3+}	0.12	1.77	4.78	[561]
			0.13	0.70	10.0	[174][a]
		Nd^{3+}	0.35	2.57	2.50	[561]
		Er^{3+}	1.07	0.28	0.63	[49]
BaF_2–CeF_3	—	Nd^{3+}	0.43	2.30	4.50	[736]
BaF_2–LuF_3	—	Nd^{3+}	0.67	2.46	4.58	[736]
Y_2O_3	C_2	Nd^{3+}	8.55	5.25	2.89	[561]
		Eu^{3+}	9.86	2.23	$\gtrsim$0.32	[50]
			6.3	0.7	0.5	[267]
$YAlO_3$	C_s	Nd^{3+}	1.24	4.68	5.85	[741]
		Ho^{3+}	1.82	2.38	1.53	[734]
		Er^{3+}	1.06	2.63	0.78	[741]
		Tm^{3+}	0.67	2.30	0.74	[741]
$Y_3Al_5O_{12}$	D_2	Nd^{3+}	0.37	2.29	5.97	[736]
			0.2	2.7	5.0	[448]
		Tm^{3+}	0.7	1.2	0.5	[267]
$Y_3Ga_5O_{12}$[b]	D_2	Er^{3+}	0.33	0.26	0.23	[743]
$Gd_3Ga_5O_{12}$	D_2	Nd^{3+}	0.0	3.3	3.7	[744]
$Lu_3Sc_2Al_3O_{12}$	D_2	Nd^{3+}	0.22	3.07	5.27	[736]
ZrO_2–Y_2O_3	—	Nd^{3+}	0.23	1.20	1.36	[736]

[a] Calculated data.
[b] Stimulated emission by Er^{3+} ions in $Y_3Ga_5O_{12}$ crystals has not yet been obtained.

4.3.2 Magnetic-Dipole Transitions

Magnetic-dipole transitions obey the selection rules:

$$\Delta l = 0,$$

$$\Delta S = 0,$$

$$\Delta L = 0,$$

$$|\Delta J| \leq 1, \text{ (but not } 0 \leftrightarrow 0),$$

from which it is evident that such transitions are allowed between states of the same parity. The line strength of a magnetic-dipole transition is determined by

$$s^{md}(\psi J, \psi J') = \left(\frac{eh}{4m\pi c}\right)^2 |\langle [\psi]J \| L + 2S \| [\psi']J' \rangle|^2.$$

To calculate the matrix elements of the operator of the $L + 2S$ magnetic-dipole moment between the states of the $4f^N$ configuration of Ln^{3+} ions, *Wybourne* [558] gives corresponding formulas.

Oscillator strengths for magnetic-dipole transitions between multiplets of Ln^{3+} ions have been calculated by *Carnall*, *Fields*, and *Rajnak* [745]. Some of these results that pertain to laser ions are listed in Table 4.19. As was pointed out above, the probabilities for magnetic-dipole transitions are lower than those for forced electric-dipole transitions. In some cases, however, they may contribute

Table 4.19. Calculated magnetic-dipole oscillator strengths for the laser Ln^{3+} ions for some transitions from the $[SL]J$ ground state [745]

Ion	$[S'L']J'$	State energy [cm^{-1}]	$f \times 10^8$
Pr^{3+}	$^2H_5$	2322	9.76
	$^3F_3$	6540	0.02
	$^3F_4$	6973	0.49
	$^1G_4$	9885	0.25
Nd^{3+}	$^4I_{11/2}$	2007	14.11
	$^2H_{9/2}$	12738	1.12
	$^2F_{9/2}$	14854	0.20
	$^2G_{7/2}$	17333	0.02
	$^2I_{11/2}$	28524	0.05
Eu^{3+}	$^7F_1$	350	17.73
	$^5D_1$	19026	1.62
	$^5F_1$	33429	2.16
Tb^{3+}	$^7F_5$	2112	12.11
	$^5G_8$	26425	5.03
	$^5G_5$	27795	0.36
	$^5L_6$	29550	0.14
Dy^{3+}	$^6H_{13/2}$	3506	22.68
	$^4I_{15/2}$	22293	5.95
	$^4I_{13/2}$	25919	0.41
	$^4K_{17/2}$	26365	0.09
	$(^4M, {}^4I)_{15/2}$	29244	0.69
Ho^{3+}	$^5I_7$	5116	29.47
	$^3K_8$	21308	6.39
	$^3K_7$	26117	0.28
	$^3L_9$	29020	0.12
Er^{3+}	$^4I_{13/2}$	6610	30.82
	$^2K_{15/2}$	27801	3.69
Tm^{3+}	$^3H_5$	8390	27.25
	$^1I_6$	34886	1.40
Yb^{3+}	$^2F_{5/2}$	10400	17.76

considerably to the total radiative-transition probability. This is the case, for example, for the $^5D_1 \to {}^7F_{1,2}$ laser transitions of Eu^{3+} ions [644], the $^5I_6 \to {}^5I_7$ and $^5I_7 \to {}^5I_8$ laser transitions of Ho^{3+} ions [562], and the $^4I_{11/2} \to {}^4I_{13/2}$ and $^4I_{13/2} \to {}^4I_{15/2}$ laser transitions of Er^{3+} ions [49] (see also [50]).

4.4 Spectroscopic-Quality Parameter of Laser Condensed Media Doped with Nd^{3+} Ions

Currently, Nd^{3+} ions are the most widely used activators in laser condensed media. Due to their spectroscopic properties, they readily exhibit stimulated emission at room temperature on the principal $^4F_{3/2} \to {}^4I_{11/2}$ and the additional $^4F_{3/2} \to {}^4I_{13/2}$ transitions. Extensive spectroscopic information accumulated by numerous research groups engaged in studying neodymium-doped laser crystals has permitted elucidation of the most important regularities of their properties, which have proved very useful in the search for new materials.

The parameter that characterizes the possibility of exciting stimulated emission in a given channel is the intermanifold luminescence branching ratio

$$\beta_{JJ'} = \frac{A(J, J')}{\sum_{J'} A(J, J')}. \tag{4.18}$$

Analysis of expression (4.14) reveals that the probability of spontaneous transitions $^4F_{3/2} \to {}^4I_{J'}$ depends primarily on the intensity parameters Ω_t of fourth and sixth order, because the matrix element (4.16) of rank 2 for transitions between these states is equal to zero ($|\Delta J| > 2$) [514, 662]. The luminescence branching ratios can then be represented as dependent on only one parameter which can be taken as

$$X = \Omega_4/\Omega_6.$$

By substituting (4.14) into (4.18), the analytical dependences can be obtained

$$\beta_{JJ'}(X) = \frac{(a_{J'}X + b_{J'})/\bar{\lambda}_{J'}^3}{\sum_{J'} (a_{J'}X + b_{J'})/\bar{\lambda}_{J'}^3}, \tag{4.19}$$

where the $a_{J'}$ and the $b_{J'}$ constants are equal to the squared matrix elements of the irreducible tensor operators of ranks 4 and 6

$$a_{J'} = |\langle {}^4F_{3/2} \| U^{(4)} \| {}^4I_{J'} \rangle|^2$$

and

$$b_{J'} = |\langle {}^4F_{3/2} \| U^{(6)} \| {}^4I_{J'} \rangle|^2$$

respectively.

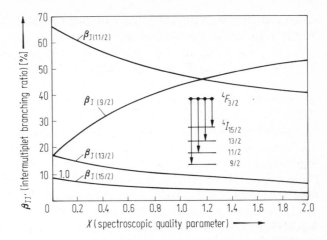

Fig. 4.2. Dependences of $\beta_{JJ'}(X)$ for $^4F_{3/2} \to {}^4I_{J'}$ transitions of Nd³⁺ ions on the spectroscopic-quality parameter X. Note that $\beta_{J(15/2)}$ is plotted on a reduced scale, given on the right of the vertical axis

Figure 4.2 is a plot of the analytical dependence (4.19). In the calculation, the $a_{J'}$ and $b_{J'}$ values were taken from *Krupke* [732][3]. It is evident that each crystal has its own X parameter which, in turn, determines a set of four $\beta_{JJ'}$ coefficients. As a glance at Fig. 4.2 shows, the maximum possible value of the $\beta_{JJ'}$ coefficient for the principal emission $^4F_{3/2} \to {}^4I_{11/2}$ channel is about 0.66 (for $X = 0$), and that for the additional $^4F_{3/2} \to {}^4I_{13/2}$ channel is about 0.17. Maximum luminescence intensity, however, can be observed on another additional $^4F_{3/2} \to {}^4I_{9/2}$ transition (maximum value about 0.75) for very high values of X.

Since determination of the parameter X requires only the ratio Ω_4/Ω_6, it is not necessary to calculate completely all the observed absorption bands. To this end, those multiplets may be chosen whose levels are rather isolated and transitions to which depend on the Ω_4 and Ω_6 parameters. The authors of [514, 662] chose two such states, namely, $^2P_{1/2}$ and $^4I_{15/2}$ (other variants are also possible [746]). For them, the line strength is expressed by the following simple equations in terms of the parameters Ω_t:

$$s(^4I_{9/2}, {}^2P_{1/2}) = 0.0367 \, \Omega_4$$

and

$$s(^4I_{9/2}, {}^4I_{15/2}) = 0,0001 \, \Omega_4 + 0.0452 \, \Omega_6.$$

---

[3]Due to the latest results on the accurate definition of wave-functions for Ln³⁺ ions by *H. M. Crosswhite, H. Crosswhite, Kaseta,* and *Sarup* of the Argonne National Laboratory in the USA (see p. 65 of [722], and [768]), there are some small changes in the values of $a_{J'}$ and $b_{J'}$. This same remark also applies to the data of Tables 4.16, 17.

If the first term in $s(^4I_{9/2}, {}^4I_{15/2})$ is neglected (its effect is significant only when $\Omega_4/\Omega_6 \cong 100$), X can be written as

$$X = 1.23\, s(^4I_{9/2}, {}^2P_{1/2})/s(^4I_{9/2}, {}^4I_{15/2}). \tag{4.20}$$

Equation (4.20) is convenient also because only relative values of s for the $^4I_{9/2} \rightarrow {}^4I_{15/2}$ and the $^4I_{9/2} \rightarrow {}^2P_{1/2}$ transitions, rather than their absolute values, are needed in determining X. Hence, it is not necessary to determine the Nd^{3+} ion concentrations in the crystals under study; this greatly simplifies the experiment.

For several laser crystals, the values of the X parameter thus obtained have been used to find the $\beta_{JJ'}$ coefficients and to compare them with values measured by independent experiments. The agreement is very good. This method of calculating the spectroscopic-quality parameter was later applied successfully to investigations of some types of neodymium-doped laser glasses [746, 747]. Information on the spectroscopic-quality parameter for several neodymium-doped laser crystals is summarized in Table 4.20.

To conclude this section, we note that the above method for determining the parameter X may be employed for studying multicenter crystalline disordered media, whose spectroscopic properties satisfy the "quasi-center" concept [92, 106] (for more details, see Sect. 2.3). This method may also be applied successfully to study anisotropic crystals doped with Nd^{3+} ions [272, 565, 741,

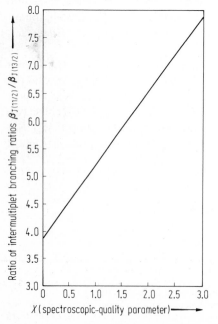

Fig. 4.3. Dependence of $\beta_{J(11/2)}/\beta_{J(13/2)}$ on the spectroscopic-quality parameter X for Nd^{3+} ions

749, 751]. In some cases, a more accurate determination of the parameter X requires high-temperature absorption measurements, in which the populations of the Stark levels of the ground state will be about the same. As our investigations have shown, the parameter X may also be defined by the ratio of the luminescence intensities of Nd^{3+} ions in the different $^4F_{3/2} \rightarrow {}^4I_{J'}$ emission channels. In this case, the definition of the ratio of luminescence intensities in the $^4F_{3/2} \rightarrow {}^4I_{11/2}$ and $^4F_{3/2} \rightarrow {}^4I_{13/2}$ channels is more convenient. By comparison of this ratio with the ratio $\beta_{J(11/2)}/\beta_{J(13/2)}$ (see Fig. 4.3), the value of X can be determined using the known dependences $X(\beta_{JJ'})$ (by direct recalculation of the data in Fig. 4.2). Use of absorption and luminescence methods for the definition of the X parameter in $Y_3Al_5O_{12}$, $LuAlO_3$, $YAlO_3$, and $K_5Nd(MoO_4)_4$ crystals doped with Nd^{3+} ions has given results in good agreement (not worse than 20% in accuracy)[4]. A number of interesting results on transition intensities in laser crystals have been obtained in Refs. [658, 748]. The luminescence method was also discussed in [769].

Table 4.20. Spectroscopic-quality parameter X of Nd^{3+} ion doped laser crystals[a]

X	Crystal	Symmetry type	Reference
0.22	α-NaCaYF$_6$	$O_h^5 - Fm3m$	[514]
0.22	$CaF_2 - CdF_3$	$O_h^5 - Fm3m$	[514]
0.24	$SrF_2 - LaF_3$	$O_h^5 - Fm3m$	[514]
0.265	$LuAlO_3$	$D_{2h}^{16} - Pbnm$	[565]
0.27	$SrF_2 - YF_3$	$O_h^5 - Fm3m$	[514]
0.27	$CaF_2 - LaF_3$	$O_h^5 - Fm3m$	[514]
0.27	$BaF_2 - LaF_3$	$O_h^5 - Fm3m$	[514]
0.29	$BaF_2 - LuF_3$	$O_h^5 - Fm3m$	[514]
0.29	$Lu_3Al_5O_{12}$	$O_h^{10} - Ia3d$	[514]
0.3	$YAlO_3$	$D_{2h}^{16} - Pbnm$	[751]
0.3	$Y_3Al_5O_{12}$	$O_h^{10} - Ia3d$	[514]
0.32	$Y_3Sc_2Al_3O_{12}$	$O_h^{10} - Ia3d$	[514]
0.32	$Gd_3Sc_2Al_3O_{12}$	$O_h^{10} - Ia3d$	[514]
0.32	$BaF_2 - CeF_3$	$O_h^5 - Fm3m$	[514]
0.32	$Y_3Sc_2Ga_3O_{12}$	$O_h^{10} - Ia3d$	[514]
0.35	$Lu_3Ga_5O_{12}$	$O_h^{10} - Ia3d$	[514]
0.38	$Y_3Ga_5O_{12}$	$O_h^{10} - Ia3d$	[514]
0.41	$Gd_3Ga_5O_{12}$	$O_h^{10} - Ia3d$	[514]
0.51	$Bi_4Ge_3O_{12}$	$T_d^6 - I\bar{4}3d$	[567]
0.74	$HfO_2 - Y_2O_3$	$O_h^5 - Fm3m$	[514]
0.84	$ZrO_2 - Y_2O_3$	$O_h^5 - Fm3m$	[514]
0.87	$K_5Nd(MoO_4)_4$	$D_{3d}^5 - R\bar{3}m$	[569]
0.87	$K_5Bi(MoO_4)_4$	$D_{3d}^5 - R\bar{3}m$	[569]
1.32	Y_2O_3	$T_h^7 - Ia3$	[514]
1.5	$YScO_3$	$T_h^7 - Ia3$	[639]

[a] The values of the parameter X for some crystals listed here, determined by method [662], and those obtained on the basis of full calculation by $^4I_{9/2} \rightarrow [S'L']J'$ absorption, are in satisfactory agreement with each other, with an accuracy of 5–20% [514]. X has also been measured for a number of crystals in another recent study [769].

---

[4] *S. E. Sarkisov*: Dissertation (The A. V. Shubnikov Institute of Crystallography).

5. Summary of the Properties of Activated Laser Crystals

In Table 5.1, laser crystals are divided into five groups. The classification is based on the crystal-chemical principle first suggested and applied by the author of the present book and *Osiko* in their reviews on inorganic laser materials with ionic structure [6–8]. The experience of many years of work by various research groups in both the Soviet Union and other countries indicates that such an approach is very useful [438]. On the one hand, active media can be classified according to their physical and chemical properties, which is technologically important. On the other hand, it shows well the relationships between the crystalline structure of substances, and the formation within this structure of varied types of activator centers, upon the properties of which all spectroscopic and emission parameters of laser crystals depend.

The overwhelming majority of known laser crystals are either fluorides or oxygen-containing compounds; these, in turn, are subdivided according to the manner of formation of the activator centers, into simple systems (mainly, single-center ordered crystals) and mixed multicentered systems that differ from the former by their disordered crystal structure (see Sect. 2.3). A small series of compounds whose anions are represented simultaneously by fluorine and oxygen or by sulphur and oxygen, as well as chloride and bromide crystals, are tentatively assigned to the fifth group. In order to avoid overloading the table with laser-crystal subclasses, the oxygen-containing compounds with complex anions have been combined with oxides into a single group. This should be acceptable, since we are largely concerned with the physical properties of laser crystals in the table; moreover, the number of activated oxide lasers developed so far is not so great as to require us to separate these oxides in an independent group.

The first column of the table contains the chemical formulas of the crystals (in the form conventional in spectroscopic and laser studies) and also indicates the crystal space group, which characterizes the type and symmetry of the structure, as well as the symmetry of the position of the cation(s) replaced by the activator or sensitizing ions. The symbol "?" in parentheses indicates that the crystal structure is not yet determined or requires further investigation. In certain cases, the symmetry of the activator center is indicated. For convenience, in each group of the table, crystals are described in the order of increasing atomic number of their first cation, the chemical formulas being written in the conventional form. For example, calcium compounds precede those containing strontium and barium. Some laser crystals contain, besides the emitting activator ions, sensitizing or deactivating impurities. In cases where their concentration is approximately that of the laser activator, the chemical formula is given, for instance, in the form $LiYF_4: Er^{3+}-Ho^{3+}$. Here, the Er^{3+} ion serves as a sensitizer. Sometimes, the sensitizing influence is exerted by ions that enter into the structure of the matrix itself. For example, the formula for the $LiYF_4$ crystal, which contains yttrium and erbium in about equal proportions, has been written in the table as $Li(Y, Er)F_4$. A more complicated case is also possible, for example, $Li(Y, Er)F_4: Tm^{3+}-Ho^{3+}$. Here, the concentration of Tm^{3+} ions (which act as a second sensitizer) is small, being about equal to the concentration of the emitting Ho^{3+} ions.

In the second column of the table, the type and valence of the activator ion, and also the concentration in the units in which it was given in the original publication, is indicated. Usually, it is given in terms of the atomic or weight percentage. For fluoride crystals with Ln^{2+} ions, the impurity content is indicated for the Ln^{3+} ions in the initial substance. For anisotropic crystals, the mutual orientation of the geometrical axis F of the active elements and its crystallographic axes (a, b, c) (for biaxial crystals) or optical axis (c) (for unaxial crystals) is indicated, because it determines the spectral composition of the stimulated emission.

The third column gives the emission wavelengths λ_g (not recalculated to vacuum), the polarization of the radiation and the letter notation of lines used in the original works. (Russian letters have been replaced here by capital Roman letters in alphabetical order.) The range of temperature tuning of λ_g for a given induced transition is also indicated in the third column.

The fourth column contains information about the operating temperature of the emitting crystal.

The mode of laser operation and the threshold energies E_{thr} (or excitation powers P_{thr}) are shown in the fifth and sixth columns. The generation modes are indicated by the following notation:

P—pulsed mode,
CW—continuous-wave mode,
q-CW—quasi-continuous mode,
Q—Q-switched mode.

In the seventh column, the spectral range of excitation of the active medium is indicated. In some cases, the duration of the exciting pulse is also given.

The next two columns (eight and nine) give the luminescence lifetime of the metastable state τ_{lum}, and the luminescence linewidth $\Delta\nu_{lum}$, associated with a given induced transition. The luminescence spectra for mixed-type crystals (so-called solid solutions) are characterized by broad bands that represent the superposition of individual lines belonging to the various activator centers. For such media, the table gives the values for the luminescence bandwidths at 0.5 of their maximum intensity.

The laser-emission transitions and the positions of the terminal laser levels E_{term} are indicated in the tenth and eleventh columns.

In view of the fact that the values for E_{thr} and P_{thr} are different for the same crystals in different excitation systems, the table gives (columns 12 and 13), for convenience of comparison between known lasers, the characteristics of illuminating chambers, types of pumping sources, cooling systems and filters, and also the geometrical dimensions (length, diameter) of the laser crystals and the parameters of the optical resonator (types of mirrors and their optical transmittance). The following notation is used:

SE—stimulated emission,
ER—elliptical illuminating reflector,
W—tungsten-iodine incandescent lamp,
DM—dielectric mirrors (multilayer),
LP—laser pumping,
PS—pyrochemical source of excitation.

The liquefied gases employed to cool the emitting crystals are indicated by the conventional symbols:

O_2—oxygen,
N_2—nitrogen,
He—helium,
Ne—neon.

In addition, the following abbreviations have been used:

EMDM—end multilayer dielectric mirrors,
EDAC—end dielectric antireflection coating (wavelengths and, in some cases, coating material are indicated),
DR—dispersion resonator,
LPG and TPG—longitudinal (end)- and transverse (side)-pumping geometry,
QLP—quartz light pipes,
TQC—tubular quartz cryostat.

The properties of the widely used liquid filters (solutions of $NaNO_3$ and $K_2Cr_2O_7$) are described in [304]. The symbol $\parallel$ indicates that the ends of the active element are plane parallel. The letter R indicates that the surface of the resonator mirrors is spherical, their radii of curvature being given sometimes. The distance between mirrors (resonator base) is designated as d_r. For the confocal resonator, $R_1 = R_2 = R = d_r$.

In the next to last column of the table are presented, as a supplement, such spectroscopic and generation characteristics as the emitted power P_g and energy

E_g, duration of emitted pulse τ_g, peak cross sections for emission σ_e and absorption σ_a transitions (in cm^2), and rates of change with temperature of the wavelength $\delta\lambda/\delta T$ and of the frequency $\delta\nu/\delta T$ of the laser lines (in 10^{-2}Å K^{-1} and 10^{-2} cm^{-1} K^{-1}, respectively). In many cases, the emitted linewidths $\Delta\nu_g$ are given. This parameter has most commonly been measured for an excitation energy two or three times higher than the threshold energy in the absence of special selectors in the optical resonator. The same column indicates, for some lasers, the emission efficiency η_g (in % units), whose value has been determined in most cases from the slope of the linear part of the dependences $P_g(P_{exc})$ or $E_g(E_{exc})$. One of the most important thermophysical characteristics of a laser crystal is the thermal conductivity coefficient K, which is given in mW · cm^{-1} · K^{-1} units at 300 K. Refractive indices are also indicated (for the ordinary and extraordinary waves, the mean value is sometimes given); and the repetition frequency f of the laser pulses is indicated. In addition, the column contains data designated as follows:

ECC—excess charge compensation,
EET—excitation energy transfer (for sensitized crystals),
RED—residual excitation deactivation (for crystals with deactivator ions),
PAT—phonon-assisted transition.

The remaining symbols follow usual conventions.

In compiling the table, the first communications about the development of individual crystal lasers (type, generation mode, induced transition, specificity of excitation conditions) and also publications describing the measurements or refinement of the basic spectroscopic parameters responsible for the stimulated emission were used. The cited references always indicate the pioneering work that led to the discovery of the stimulated-emission effect for a given crystal, even though the characteristics described in that work are sometimes omitted in the summary table. Only data that have been refined and checked in different references have been included, as far as possible. In most cases, the spectroscopic and emission characteristics are those measured in several studies; accordingly, all references on each crystal are grouped together.

The table was compiled from papers published mostly before the beginning of 1980.

For reference, Table 5.2 (p. 312) presents the site symmetries of all possible atomic positions in the various crystal structures. The table also gives the International and the Schönflies symbols for the 230 crystal space groups.

(*Chapter* 6: *page* 319)

Table 5.1.   **Spectral and stimulated emission characteristics of activated laser crystals**

Crystal host, space group, cation-site symmetry	Activator [%]	Emission wavelength [μm]	Temperature [K]	Mode of operation	Threshold energy [J] or power [W]	Excitation spectral range [μm]	Luminescence lifetime [ms]
							Simple fluoride laser crystals
LiYF$_4$ C_{4h}^6–$I4_1/a$ $S_4(Y^{3+})$	Pr^{3+} 0.2	(π) 0.479[a]	300	P	0.002[b]	LP; 0.444 τ_{exc} = 300 ns	0.05
		$\approx$0.64[a,c]	300	P	–	LP; 0.444 τ_{exc} = 300 ns	0.05
LiYF$_4$ C_{4h}^6–$I4_1/a$ $S_4(Y^{3+})$	Nd^{3+} $\approx$2	(π) 1.0471 (σ) 1.0530	300 300	P P	– 2	0.4–0.9	0.57
LiYF$_4$ C_{4h}^6–$I4_1/a$ $S_4(Y^{3+})$	Ho^{3+} $\approx$2	0.7498	$\approx$90	P	13.5	0.3–0.53	0.05
	$\approx$2	0.9794	$\approx$90	P	18	0.3–0.63	0.1
	$\approx$2	1.0143	$\approx$90	P	18	0.3–0.53	0.05
	$\approx$2	1.3960	300	P	78	0.3–0.53	0.05
	$\approx$2	2.0672	$\approx$90	P	75	0.3–1.94	$\approx$10

[a] λ_g value requires more-accurate definition.
[b] The absorbed pump power (energy) at which stimulated emission begins is indicated.
[c] The detailed spectroscopic investigations were not carried out.

Lumines- cence linewidth [cm^{-1}]	Laser transition	Energy terminal level [cm^{-1}]	Excitation system, pump source, filter, cooling	Rod length, rod diameter [mm]; resonator cavity parameters, mirrors (coating, transmittance)	Remarks	Reference
with ordered structure						
22	$^3P_0 \rightarrow {}^3H_4$	0	P coumarin dye laser, LPG	$8 \times 6 \times 2$ $\parallel$, plane and spherical mirrors $(R = 1 \text{ m})$, DM, 1%	$\sigma_e = 4.4 \times 10^{-19}$ $\Delta\nu_g = 0.1 \text{ cm}^{-1}$	[606]
–	$^3P_0 \rightarrow {}^3F_2$	≈ 5200	P coumarin dye laser, LPG			[123]
12 12.5	$^4F_{3/2} \rightarrow {}^4I_{11/2}$	2042 2042	–	50; 5	$\sigma_e = 3 \times 10^{-19}$ $n_o \approx 1.46$ $n_e \approx 1.48$ $n_2 \approx 0.6 \times 10^{-13}$	[261, 305, 306]
–	$^5S_2 \rightarrow {}^5I_7$	≈ 5300	ER with Xe flashtube, N$_2$ vapor	30; 6 $\parallel$, $R = 500$ mm, DM, $\approx 2\%$		[568]
–	$^5F_5 \rightarrow {}^5I_7$	≈ 5300	ER with Xe flashtube, N$_2$ vapor	30; 6 $\parallel$, $R = 500$ mm DM, $\approx 2\%$		[568]
–	$^5S_2 \rightarrow {}^5I_6$	≈ 8800	ER with Xe flashtube, N$_2$ vapor	30; 6 $\parallel$, $R = 500$ mm, DM, $\approx 2\%$		[568]
–	$^5S_2 \rightarrow {}^5I_5$	≈ 11300	ER with Xe flashtube	30; 6 $\parallel$, $R = 500$ mm, DM, $\approx 2\%$		[568]
–	$^5I_7 \rightarrow {}^5I_8$	≈ 350	ER with Xe flashtube, N$_2$ vapor	30; 6 $\parallel$, $R = 500$ mm, DM, $\approx 2\%$		[568]

(Continued)

Table 5.1 (*continued*)

Crystal host, space group, cation-site symmetry	Activator [%]	Emission wavelength [μm]	Temperature [K]	Mode of operation	Threshold energy [J] or power [W]	Excitation spectral range [μm]	Luminescence lifetime [ms]
$LiYF_4$ $C_{4h}^6-I4_1/a$ $S_4(Y^{3+})$	Er^{3+} 2	(π) 0.8500	300	P	10	0.3–0.55	0.2
$LiYF_4: Gd^{3+}$ $C_{4h}^6-I4_1/a$ $S_4(Y^{3+})$	Tb^{3+} ≈ 10	0.5445	300	P	75	0.3–0.5 $\tau_{exc} = 100\ \mu s$	5
$LiYF_4: Er^{3+}$ $C_{4h}^6-I4_1/a$ $S_4(Y^{3+})$	Ho^{3+} ≈ 2 $a \parallel F$	2.066	77	P	40	0.4–2 Excitation into Ho^{3+} and Er^{3+} bands	20
$Li(Y,Er)F_4:$ $:Tm^{3+}$ $C_{4h}^6-I4_1/a$ $S_4(Y^{3+}, Er^{3+})$	Ho^{3+} 1.7	2.0654	300	P	20	0.4–2 Excitation into Ho^{3+}, Er^{3+}, and Tm^{3+} bands	12
$LiHoF_4$ $C_{4h}^6-I4_1/a$ $S_4(Ho^{3+})$	Ho^{3+} 100	0.979[a]	≈ 90	P	6–7	0.3–0.7	0.04
	100	1.486[a]	≈ 90	P	6–7	0.3–0.7	0.04
	100	2.352[a]	≈ 90	P	6–7	0.3–0.7	0.04
$LiErF_4$ $C_{4h}^6-I4_1/a$ $S_4(Er^{3+})$	Er^{3+} 100	1.732[a]	≈ 90	P	7.5	0.3–0.55	–

[a] λ_g value requires more-accurate definition.

Luminescence linewidth [cm^{-1}]	Laser transition	Energy terminal level [cm^{-1}]	Excitation system, pump source, filter, cooling	Rod length, rod diameter [mm]; resonator cavity parameters, mirrors (coating, transmittance)	Remarks	Reference
10	$^4S_{3/2} \rightarrow ^4I_{13/2}$	6714	Cylindrical illuminator with Xe flashtube	30; 3–5 $\parallel$, DM		[104]
–	$^5D_4 \rightarrow ^7F_5$	≈ 2000	Illuminator with Xe flashtube	40; 4	$Gd^{3+} \approx 10$ at.% $E_g = 75$ mJ EET to Tb^{3+}	[83]
–	$^5I_7 \rightarrow ^5I_8$	≈ 300	ER with Xe flashtube FX-85-3, N$_2$	24; 4.5 $\parallel$, Al, 5 and 15%	$Er^{3+} = 5$ wt.% EET to Ho^{3+}	[131]
–	$^5I_7 \rightarrow ^5I_8$	≈ 300	Illuminator with Xe flashtube FX-38A-3	32; 3.2 $\parallel$, DM, 4%	$Er^{3+} = 50$ at.%, $Tm^{3+} = 6.7$ at.% $\eta_g = 1.3$ EET to Ho^{3+}	[132]
2–5	$^5F_5 \rightarrow ^5I_7$	≈ 5300	ER with Xe flashtube ISP-250, N$_2$ vapor	15; 3 $\parallel$, $R = 500$ mm DM, 0.5%	$n_o = 1.464$ $n_e = 1.498$ ($\lambda = 0.589$ μm)	[545]
2–5	$^5F_5 \rightarrow ^5I_6$	≈ 8000	ER with Xe flashtube ISP-250, N$_2$ vapor	15; 3 $\parallel$, $R = 500$ mm DM, 0.5%		[545]
2–5	$^5F_5 \rightarrow ^5I_5$	≈ 11500	ER with Xe flashtube ISP-250, N$_2$ vapor	15; 3 $\parallel$, $R = 500$ mm, DM, 0.5%		[545]
2–5	$^4S_{3/2} \rightarrow ^4I_{9/2}$	≈ 12200	ER with Xe flashtube ISP-250, N$_2$	15; 3 $\parallel$, $R = 500$ mm, DM, 0.5%	$n_o = 1.464$ $n_e = 1.497$ ($\lambda = 0.589$ μm)	[545]

(*Continued*)

Table 5.1 (*continued*)

Crystal host, space group, cation-site symmetry	Activator [%]	Emission wavelength [μm]	Temperature [K]	Mode of operation	Threshold energy [J] or power [W]	Excitation spectral range [μm]	Luminescence lifetime [ms]
MgF_2 $D_{4h}^{14}-P4_2/mnm$ $D_{2h}(Mg^{2+})$	V^{2+} 0.2	1.1217	77	P	1070	0.4–1.1	2.3
MgF_2 $D_{4h}^{14}-P4_2/mnm$ $D_{2h}(Mg^{2+})$	Co^{2+} 1	1.750[a] 1.8035 ≈ 1.99[a] ≈ 2.05[a]	77 77 77 77	P P P P	690 730 666 700	0.4–1.4	1.3
MgF_2 $D_{4h}^{14}-P4_2/mnm$ $D_{2h}(Mg^{2+})$	Ni^{2+} 0.5–1.5	1.623 1.636 1.674→1.676 1.731→1.756 1.785→1.797	77 77–82 82–100 100–192 198–240	P P CW CW P	150 160 240 65 1650	0.4–1.4	11.5 3.7 (295 K)
$KMgF_3$ O_h^1-Pm3m	Co^{2+} –	1.821	77	P	530	0.4–1.4	3.1
CaF_2 O_h^5-Fm3m $C_{4v}(Ca^{2+} + + F^-)$	Nd^{3+} 0.4–0.6	1.0448 C 1.0457 A 1.0467 B 1.0448 C 1.0456 A 1.0466 B 1.0480 F 1.0507 D 1.0648 E	77 77 77 50 50 50 50 50 50	P P P P P P P P P	400 30 160 240 70 160 1350 870 1100	0.5–0.9	1.1

[a] λ_g value requires more-accurate definition.

Lumines-cence linewidth [cm^{-1}]	Laser transition	Energy terminal level [cm^{-1}]	Excitation system, pump source, filter, cooling	Rod length, rod diameter [mm]; resonator cavity parameters, mirrors (coating, transmittance)	Remarks	Reference
–	$^4T_2 \rightarrow {}^4A_2$	1166	Illuminator with Xe flashtube FT-524, N$_2$	–	$\sigma_e = 8 \times 10^{-21}$ SE in PAT $n = 1.38$ $n_2 \approx 0.3 \times 10^{-13}$	[113, 114]
– – – –	$^4T_2 \rightarrow {}^4T_1$	1087 1256 1780 1930	Illuminator with Xe flashtube FT-524, N$_2$	–	SE at 1.99– –2.05 μm due to PAT	[113, 117]
– – – – –	$^3T_2 \rightarrow {}^3A_2$	340 390 ≈ 530 ≈ 800 ≈ 935	Under P mode, illuminator with Xe flash-tube FT-524 and under CW mode, ER with W lamp (500 W), N$_2$ and its vapor	30–50 $R = 500$ mm, DM, 0.1%, DR	SE in PAT For $P_{exc} = 500$ W $P_g = 1$ W $K = 31$	[20, 113]
–	$^4T_2 \rightarrow {}^4T_1$	1420	Illuminator with Xe flashtube FT-524, N$_2$	–	SE in PAT	[113]
≈ 2 ≈ 2 ≈ 2 ≈ 2 ≈ 2 ≈ 2 ≈ 2 ≈ 3 ≈ 5	$^4F_{3/2} \rightarrow {}^4I_{11/2}$	2031 2032 2028 2031 2032 2028 2053 2061 2190	Low-temp. ER with Xe IFP-800, TQC, ZhS-17, N$_2$ and He and their vapors	75; 6.35 $\parallel$, 10″, DM, 0.6%	$n = 1.43$ $n_2 \approx 0.65 \times 10^{-13}$ $K = 97$	[24, 32, 45, 243, 308–316]

(*Continued*)

Table 5.1 (*continued*)

Crystal host, space group, cation-site symmetry	Activator [%]	Emission wavelength [μm]	Temperature [K]	Mode of operation	Threshold energy [J] or power [W]	Excitation spectral range [μm]	Luminescence lifetime [ms]
		1.0370 K	460–550	P	–		
		1.0461→	300→530	P	–		1.2
		→1.0468 A					
		1.0628→	300→	P	–		
		→1.0623 G					
		1.0448 C	120	P	78	0.3–0.9	
		1.0661 H	120	P	–		
		1.0461 A	300	q-CW	–		
CaF_2 O_h^5–Fm3m $C_{4v}(Ca^{2+} + F^-)$	Ho^{3+} 0.4–0.8	0.5512	77	P	1200	0.4–0.55	0.7
		2.092^a	77	P	260	0.4–2	–
CaF_2 O_h^5–Fm3m	Er^{3+} 4–8	0.8456	77	P	100	0.3–0.85	1
		0.8548	77	P	–		
$C_{4v}(Ca^{2+} + F^-)$	0.1	1.617^a	77	P	1000	0.3–1.6	20
CaF_2: Nd^{3+} O_h^5–Fm3m	Yb^{3+} ≈5	1.0336	120	P	48	0.3–1 Excitation into Nd^{3+} and Yb^{3+} bands	
CaF_2 O_h^5–Fm3m $O_h(Ca^{2+})$	Sm^{2+} 0.05	0.7083	20	P	0.01	0.4–0.7	0.002
			15	P	0.3		
			35	P	0.8		
			45	P	1.5		

^{a} λ_g value requires more-accurate definition.

Luminescence linewidth [cm⁻¹]	Laser transition	Energy terminal level [cm⁻¹]	Excitation system, pump source, filter, cooling	Rod length, rod diameter [mm]; resonator cavity parameters, mirrors (coating, transmittance)	Remarks	Reference
–		2030	High-temp.	75; 6.5		
$20\rightarrow$		2032	ER with Xe IFP-800, ZhS-17	$\parallel$, 10″, $R = 576$ mm, DM, 0.7%	$\sigma_e = 3 \times 10^{-21}$ $\delta\lambda/\delta T = 3.2$	
$30\rightarrow$		–			$\delta\lambda/\delta T = 2$	
9		2031	ER with Xe flashtube, N₂ and its vapor	25, 3		
–				$\parallel$, Ag		
20		2032	Round illuminator with PS	45; 6.5 $\parallel$, $R = 576$ mm, DM, 0.7%		
≈ 6	$^5S_2\rightarrow{}^5I_8$	≈ 370	ER with Xe flashtube IFP-800, TQC, N₂	75; 6.5 $\parallel$, 30″, DM, 0.1 and 0.7%	$\Delta\nu_g = 0.6$ cm⁻¹	[102, 319]
–	$^5I_7\rightarrow{}^5I_8$	–	Illuminator with Xe flash-tube FT-524, N₂	52; 3 R, Ag, 1%		[183]
6–8	$^4S_{3/2}\rightarrow{}^4I_{13/2}$	≈ 7000	ER with Xe flashtube IFP-1000, N₂	53; 5 $\parallel$, Ag, 5%		[105]
–		≈ 7000				
–	$^4I_{13/2}\rightarrow{}^4I_{15/2}$	≈ 400	Illuminator with Xe flash-tube FT-524, N₂	51; 9.6 $\parallel$, Al		[320]
–	$^2F_{5/2}\rightarrow{}^2F_{7/2}$	–	ER with Xe flashtube PEK-1-3, N₂ vapor	25; 3 $\parallel$, Ag	$NdF_3 = 2$ wt.% EET to Yb^{3+}	[152]
1.5	$5d(A_{1u})\rightarrow{}^7F_1$	≈ 263	Four-mirror ER with Xe flashtube, H₂	20; 3 $\parallel$, 30″, Ag		[14, 118, 256, 322–325]
–			ER with Xe flashtube ISP-2000, He vapor	40; 8 $\parallel$, DM, 40%	$\bar{P}_g = 10$ W $(f = 50$ Hz)	
–					$\eta_g = 0.5$	

(Continued)

Table 5.1 (*continued*)

Crystal host, space group, cation-site symmetry	Activator [%]	Emission wavelength [μm]	Temperature [K]	Mode of operation	Threshold energy [J] or power [W]	Excitation spectral range [μm]	Luminescence lifetime [ms]
	$3 \times 10^{18} \text{cm}^{-3}$	0.7083	65–90	P	–	LP; 0.6943	
		0.7207	85–90	P	–		
		0.7287	110–130	P	–		
		0.7310	155	P	–		
		0.745	210	P			
CaF_2 O_h^5–Fm3m $O_h(Ca^{2+})$	Dy^{2+} 0.03	2.35867	77	CW	15–100	0.3–1	40–50
			27	CW			
			15	CW			
			4.2	CW			
		2.35867	4.2→120	P	–		
		2.35867	77	CW	–	Solar emission	
		2.35867	77	CW		0.3–1	
		2.35867	77	q-CW		0.4–1	
	10^{17} cm^{-3}	2.35867	77	Q	–	0.3–1	
CaF_2 O_h^5–Fm3m $O_h(Ca^{2+})$	Tm^{2+} 0.05	1.160	4.2	P	50	0.4–0.7	4
			4.2	CW	600		
			27	CW	1000		

Luminescence linewidth [cm^{-1}]	Laser transition	Energy terminal level [cm^{-1}]	Excitation system, pump source, filter, cooling	Rod length rod diameter [mm]; resonator cavity parameters, mirrors (coating, transmittance)	Remarks	Reference
–		≈ 263	Ruby laser emission with $E_g = 0.5$ J	20; 8 $\parallel$, 5″, DM, 1 and 7%	SE at 0.72– –0.74 μm due to PAT	
–		≈ 506				
–		≈ 658				
–		≈ 700				
–		≈ 958				
	$^5I_7 \rightarrow {}^5I_8$					
0.3		29	Illuminator with Hg lamp GEAH6, supercooled O_2 and Ne, and He vapor	25 × 6 × 3		[39–42, 226, 227, 244, 257, 327–331]
0.1						
–						
0.04						
0.04→ 3		29				
0.3		29	Spherical mirror with 450 mm diameter and conical condenser with crystal, N_2	26 × 3 × 4	γ irradiation (10^6 rad)	
0.3		29	Illuminator with CW lamps, N_2	–	$P_g = 150$ W	
0.3		29	Illuminator with PS, N_2	120; 8 $\parallel$, 15″, DM, 5%	γ irradiation (10^7 rad) $E_g = 4$ J, $\tau_g = 50$ ms	
0.3		29	Closed illuminator with Xe flashtubes, supercooled N_2	70; 7 $\parallel$, DM, rotatable 90° prism (50–500 Hz)	For $f = 200$ Hz $\bar{P}_g = 0.05$ W	
	$^2F_{5/2} \rightarrow {}^2F_{7/2}$					
0.03 (77 K)		0	Spherical mirror with Xe flashtube FX-100, He	25.4 × 6 × 6		[221, 332]
			Two spherical mirrors with diameter 450 mm, Hg lamp GEAH6, solution $NaNO_3$ filter, He and Ne	25; 6.35 R, Ag	γ irradiation	

(*Continued*)

Table 5.1 (*continued*)

Crystal host, space group, cation-site symmetry	Activator [%]	Emission wavelength [μm]	Temperature [K]	Mode of operation	Threshold energy [J] or power [W]	Excitation spectral range [μm]	Luminescence lifetime [ms]
CaF_2 O_h^5–Fm3m $C_{4v}(Ca^{2+} +$ $+ F^-)$ $C_{3v}(Ca^{2+} +$ $+ O^{2-})$	U^{3+} 0.05	$\approx 2.6^a$ $2.234^{(2)}$ 2.439 2.511 2.571 2.613	4.2 77 77 77 77 77 300 90 77	P P P P P P P P CW	– 1200 4.35 –	0.5–1.3	0.14 0.015 0.095 0.13
CaF_2 $^{(3)}$ O_h^5–Fm3m $C_{3v}(Ca^{2+} +$ $+ O^{2-})$	Nd^{3+} 0.2–0.5	$1.0885 \rightarrow$ $\rightarrow 1.0889$	$300 \rightarrow 420$	P	170	0.5–0.9	1.25
CaF_2 $^{(3)}$ O_h^5–Fm3m $C_{3v}(Ca^{2+} +$ $+ O^{2-})$	Er^{3+} 0.1	$\approx 1.26^a$	77	P	2000	0.3–0.55	0.3
		1.696^a 1.715^a 1.726^a	77 77 77	P P P	2000 1000 1000	0.3–0.55	0.3
	0.05	1.5298 1.5308	77 77	P P	20 8	0.3–1.53 $\tau_{exc} = 35\ \mu s$	0.8–1.3
CaF_2 $^{(3)}$ O_h^5–Fm3m $C_{3v}(Ca^{2+} +$ $+ O^{2-})$	Tm^{3+} –	$\approx 1.9^{(1)}$	77	P	–	0.3–1.9	–
CaF_2 $^{(3)}$ O_h^5–Fm3m $O_h(Ca^{2+})$	Sm^{2+} 7×10^{17} cm^{-3}	0.7085	15–20 4.2	P P	– 1	0.4–0.7	

^{a} λ_g value requires more-accurate definition.
$^{(1)}$ The detailed data are not available.

Lumines- cence linewidth [cm^{-1}]	Laser transition	Energy terminal level [cm^{-1}]	Excitation system, pump source, filter, cooling	Rod length, rod diameter [mm]; resonator cavity parameters, mirrors (coating, transmittance)	Remarks	Reference
	$^4I_{11/2} \rightarrow {}^4I_{9/2}$					
–		505	–	37 ; 9		[13, 15,
–		0		$\parallel$, 15″, Ag,		18, 333–
–		398		1.5%		335]
–		470				
–		609				
–		609				
–						
–			ER with Hg lamp GEAH6, solution K$_2$Cr$_2$O$_7$ filter, O$_2$ and N$_2$	R, Ag, 1%		
	$^4F_{3/2} \rightarrow {}^4I_{11/2}$					
25		≈ 2000	High-temp. ER with Xe IFP-800 flashtube, ZhS-17	75 ; 6 $\parallel$, 15″, $R =$ $= 600$ mm, DM, 2%	$\delta\lambda/\delta T = 3.3$	[32, 318]
	$^4S_{3/2} \rightarrow {}^4I_{11/2}$					
–		11000	Illuminator with Xe flashtube, N$_2$	60 ; 8 $\parallel$, Ag		[108]
	$^4S_{3/2} \rightarrow {}^4I_{9/2}$					
–		13000	Illuminator with Xe flashtube, N$_2$	60 ; 8 $\parallel$, Ag		[108]
–						
	$^4I_{13/2} \rightarrow {}^4I_{15/2}$					
–		–	ER with Xe flashtube, N$_2$	38 ; 5 $\parallel$, 2″, Ag, 1.5% at 1.5 μm	γ irradiation EET from color center to Er^{3+}	[146]
–		–				
	$^3H_4 \rightarrow {}^3H_6$					
–		–	–	–		[146]
	$5d(A_{1u}) \rightarrow {}^7F_1$					
–		≈ 260	–	Plane and spherical resonators	$E_g = 10$ mJ	[326]

(2) According to [299], this line may be attributed to U^{4+} ions located on the trigonal sites of the CaF$_2$ crystal.
(3) The crystals contained oxygen.

(*Continued*)

Table 5.1 (*continued*)

Crystal host, space group, cation-site symmetry	Activator [%]	Emission wavelength [μm]		Temperature [K]	Mode of operation	Threshold energy [J] or power [W]	Excitation spectral range [μm]	Luminescence lifetime [ms]
MnF_2 $D_{4h}^{14}-P4_2/mnm$ $D_{2h}(Mn^{2+})$	Ni^{2+} 1	$\approx 1.6^a$		77	P	–	0.4–1.4	≈ 11
		1.865		20	P	740		
		1.915		77	P	740		
		1.922		77	P	210		
		1.929		85	CW	270		
		1.939		85	CW	240		
ZnF_2 $D_{4h}^{14}-P4_2/mnm$ $D_{2h}(Zn^{2+})$	Co^{2+} 1	2.165		77	P	430	0.4–1.4	0.4
SrF_2 O_h^5-Fm3m $C_{4v}(Sr^{2+} + + F^-)$	Nd^{3+} 0.1–0.8	1.0370 1.0437	A B	295 77	P P	480 150	0.5–0.9	1.25 (0.1%)
		1.0370 1.0445	A B	300 300	P P	40 95		1.1 (0.8%)
		1.0446 1.0370→ →1.0395A 1.0445	B	500–550 300→530 300–320	P P P	– – –		
		1.0370	A	300	q-CW	–	0.4–0.9	
SrF_2 O_h^5-Fm3m $C_{4v}(Sr^{2+} + + F^-)$	Tm^{3+} –	1.972^a		77	P	1600	0.3–1.97	–
SrF_2 O_h^5-Fm3m $O_h(Sr^{2+})$	Sm^{2+} 0.1	0.6969		4.2	P	–	0.4–0.69	14–15

^{a} λ_g value requires more-accurate defination.

Lumines-cence linewidth [cm^{-1}]	Laser transition	Energy terminal level [cm^{-1}]	Excitation system, pump source, filter, cooling	Rod length, rod diameter [mm]; resonator cavity parameters, mirrors (coating, transmittance)	Remarks	Reference
	$^3T_2 \rightarrow {}^3A_2$					
–		–	Under P mode,	DR	EET from Mn^{2+}	[113]
–		560	illuminator		to Ni^{2+}	
–		580	with Xe flash-		SE in PAT	
–		600	tube FT-524			
–		620	and under CW			
–		650	mode, ER with			
			W lamp(500 W),			
			O$_2$ super-			
			cooled with N$_2$			
	$^4T_2 \rightarrow {}^4T_1$					
–		1895	Illuminator	–	SE in PAT	[113, 117]
			with Xe flash-			
			tube FT-524,			
			N$_2$			
	$^4F_{3/2} \rightarrow {}^4I_{11/2}$					
20		2008	Illuminator	52; 3,	$\sigma_e = 8 \times 10^{-22}$	[36, 183,
3		2008	with Xe flash-	R, Ag, 1%	$n = 1.44$	243, 315,
			tube FT-524,		$n_2 \approx 0.6 \times 10^{-13}$	336–338]
			N$_2$			
20		2008	ER with Xe	30; 5.8		
25		2008	flashtube	$\parallel$, 20″, R =		
			IFP-400,	= 576 mm,		
			ZhS-17	DM, 0.7%		
–		2045	High-temp.			
20→		2008	ER with Xe		$\delta\lambda/\delta T \approx 11$	
			flashtube			
25		2008	IFP-800,			
			ZhS-17			
20		2008	Illuminator		$\tau_g = 10$ ms	
			with PS			
	$^3H_4 \rightarrow {}^3H_6$					
–		–	Illuminator	52; 3		[183]
			with Xe flash-	R, Ag, 1%		
			tube FT-524,			
			N$_2$			
	$^5D_0 \rightarrow {}^7F_1$					
1		≈ 270	Illuminator	$\parallel$, 15″, Ag		[322]
			with Xe flash-	1%		
			tube FT-524,			
			He			

(*Continued*)

Table 5.1 (*continued*)

Crystal host, space group, cation-site symmetry	Activator [%]	Emission wavelength [μm]	Temperature [K]	Mode of oper-ation	Threshold energy [J] or power [W]	Excitation spectral range [μm]	Lumines-cence lifetime [ms]
SrF_2 O_h^5–Fm3m $O_h(Sr^{2+})$	Dy^{2+} 0.03	2.3659	77	P	100	0.3–1.2	≈ 50
SrF_2 O_h^5–Fm3m $C_{4v}(Sr^{2+} + + F^-)$	U^{3+} 0.1	2.407	90 20	P P	38 8	0.4–1.3	0.06 0.11
BaF_2 O_h^5–Fm3m $C_{4v}(Ba^{2+} + + F^-)$	Nd^{3+} –	1.060[a]	77	P	1600	0.4–0.9	–
BaF_2 O_h^5–Fm3m $C_{4v}(Ba^{2+} + + F^-)$	U^{3+} –	2.556	20	P	12	1–1.5	0.15
BaY_2F_8 C_{2h}^3–C2/m	Ho^{3+} 10	2.362 [a] 2.375 [a] 2.363 [a] 2.377 [a]	77 77 20 20	P P P P	555 755 302 295	0.3–0.65	0.07 $\geqslant 0.07$
BaY_2F_8 C_{2h}^3–C2/m	Er^{3+} 0.005– –0.2	0.5540	77	P	–	0.4–0.55	0.83
	0.12– –0.2	0.5617	77	P	–	0.4 μm band	0.025

[a] λ_g value requires more-accurate definition.
[4] In 1964, the authors of Ref. [705] observed at 77 K pulsed emission of a SrF_2–Dy^{2+} polycrystal, synthesized by the hot pressure method.

Luminescence linewidth [cm^{-1}]	Laser transition	Energy terminal level [cm^{-1}]	Excitation system, pump source, filter, cooling	Rod length, rod diameter [mm]: resonator cavity parameters, mirrors (coating, transmittance)	Remarks	Reference
≈ 0.5	$^5I_7 \to {}^5I_8$	23.4	ER with Xe flashtube	55$^{(4)}$ DM, 5%	γ irradiation (10^7 rad)	[244, 339]
– –	$^4I_{11/2} \to {}^4I_{9/2}$	334	Illuminator with Xe flashtube FT-524, He and N$_2$ vapor	Ag		[341]
–	$^4F_{3/2} \to {}^4I_{11/2}$	≈ 2000	Illuminator with Xe flashtube FT-524, N$_2$	52; 3 R, Ag, 1%	$n = 1.47$ $n_2 \approx 10^{-13}$	[183]
–	$^4I_{11/2} \to {}^4I_{9/2}$	107	Illuminator with Xe flashtube, FT-524, Ne	–		[342]
– – – –	$^5F_5 \to {}^5I_5$	11260 11317 11260 11293	Illuminator with Xe flashtube FT–524, N$_2$ and Ne(?)	$R = 52$ mm, DM, loss 5–6%		[532]
–	$^4S_{3/2} \to {}^4I_{15/2}$	406	Illuminator with Xe flashtube FT–524, solution NaNO$_2$ filter, N$_2$	EMDM		[103]
–	$^2H_{9/2} \to {}^4I_{13/2}$	6740	Illuminator with Xe flashtube FT–524, solution NaNO$_2$ filter, N$_2$	EMDM	$E_{thr}(0.5617\ \mu m)$ is 1.5 times as much as E_{thr} of 0.6709 μm line	[103]

(Continued)

Table 5.1 (*continued*)

Crystal host, space group, cation-site symmetry	Activator [%]	Emission wavelength [μm]	Temperature [K]	Mode of operation	Threshold energy [J] or power [W]	Excitation spectral range [μm]	Luminescence lifetime [ms]
	0.2	0.6709	77	P	–	0.4–0.67	–
	0.2	0.7037	77	P	–	0.4 μm band	0.025
BaY$_2$F$_8$: :Er^{3+},Tm^{3+} C$^3_{2h}$–C2/m	Ho^{3+} 1.5	2.171 [a] 2.065 [a] 2.0746	295 77 77	P P P CW	450 46 30 50	0.3–2 Excitation into Ho^{3+}, Er^{3+}, and Tm^{3+} bands	≈ 16 ≈ 9
		2.0866 2.0555 2.074 [a]	77 20 20	P CW P P	30 50 9 55		
Ba(Y,Er)$_2$F$_8$ C$^3_{2h}$–C2/m	Dy^{3+} 2	3.022 [5]	77	P	510	0.5–2.8 Excitaion into Dy^{3+} and Er^{3+} bands	7
Ba(Y,Yb)$_2$F$_8$ C$^3_{2h}$–C2/m	Ho^{3+} 0.5	0.5515	77	P	635	IR excitation	–
				P	355	0.4–0.9	

[a] λ_g value requires more-accurate definition.
[5] The luminescence spectra of Dy^{3+} ions in Ba(Y,Er)$_2$F$_8$ crystals at 77 K is characterized by overlapping lines connected with both electron and electron-vibrational transitions. Therefore, the value of λ_g is shifted about 5 cm^{-1} from the pure electron-transition frequency [479]. The possibility of increasing the emission efficiency of BaY$_2$F$_8$–Dy^{3+} crystals by coactivation with Ho^{3+}, Er^{3+}, Tm^{3+} and Yb^{3+} ions is also considered in [479].

Lumines- cence linewidth [cm^{-1}]	Laser transition	Energy terminal level [cm^{-1}]	Excitation system, pump source, filter, cooling	Rod length, rod diameter [mm]; resonator cavity parameters, mirrors (coating, transmittance)	Remarks	Reference
—	$^4F_{9/2} \rightarrow {}^4I_{15/2}$	≈ 400	Illuminator with Xe flash-tube FT–524, solution NaNO$_2$ filter, N$_2$	EMDM		[103]
—	$^2H_{9/2} \rightarrow {}^4I_{11/2}$	10336	Illuminator with Xe flash-tube FT–524, solution NaNO$_2$ filter, N$_2$	EMDM	$E_{thr}(0.7037 \ \mu m)$ is 1.5 times more than E_{thr} of 0.6709 μm line	[103]
— — — — — —	$^5I_7 \rightarrow {}^5I_8$	≈ 400 352 352 381 310 352	Illuminators with Xe flash-tube FT–524 and W lamp 500–WT3Q, liquid filter (dye 9–acet-ylanthracene in CCl$_4$), N$_2$ and Ne(?)	$R \approx 52$ mm, DM, 0.4%	Er^{3+} = 15%, Tm^{3+} = 1.5% SE in PAT For P_{exc} = = 225 W $P_g \approx 560$ mW EET to Ho^{3+}	[532]
≈ 60	$^6H_{13/2} \rightarrow {}^6H_{15/2}$	216	Illuminator with Xe flash-tube FT–524, N$_2$	DM, 0.4%	Er^{3+} = 35% EET to Dy^{3+}	[479]
—	$^5S_2 \rightarrow {}^5I_8$	385	Illuminator with Xe flash-tube FT-524, N$_2$	EMDM	Yb^{3+} = 30% Energy cumula-tion on Yb^{3+} and transfer to Ho^{3+}	[31]

(*Continued*)

Table 5.1 (*continued*)

Crystal host, space group cation-site symmetry	Activator [%]	Emission wavelength [μm]	Temperature [K]	Mode of operation	Threshold energy [J] or power [W]	Excitation spectral range [μm]	Lumines-cence lifetime [ms]
Ba(Y,Yb)$_2$F$_8$ C_{2h}^3–C2/m	Er^{3+} 2–3	0.6700	77	P	170	IR excitation, LP	–
		0.6709	77	P	195		
		0.6700	77	P	165	0.4–0.9	
LaF$_3$ D_{3d}^4–P$\bar{3}$c1 C_2(La^{3+})	Pr^{3+} 1	0.5985^d	76	P	60	0.4–0.6	–
LaF$_3$ D_{3d}^4–P$\bar{3}$c1 C_2(La^{3+})	Nd^{3+} 1–2	1.04065 A	300	P	2.8	0.5–0.9	0.6–0.7
		1.06335 B	300	P	8		
		1.0400 A	77	P	7.5		
		1.0451 D	77	P	15		
		1.06305 B	77	P	5		
	∠cF = 20°	1.04065 A	300	CW	4000		
		1.0523 G	77	P	–		
		1.0583 C	77	P	–		
		1.0670 E	77	P	–		
		1.04065→ →1.0410A	300→430	P	–		
		1.06335→ →1.0638B	300→650	P	–		
		1.0595→ →1.0613C	380→	P	–		
		1.0632→ →1.0642F	400→	P	–		
		1.0465 A	300	q-CW	–		
	4 ∠cF = 20°	1.3310* 1.3675	300 300	P P	— 40	0.5–0.9	0.6–0.7
	2 ∠cF = 73°	1.3310 1.3595*	300 300	P P	7 –		
	4 ∠cF = 20°	1.3125* 1.3235 1.3305* 1.3670	77 77 77 77	P P P P	– 60 – 40		
	2 ∠cF = 73°	1.3125 1.3235* 1.3305 1.3670*	77 77 77 77	P P P P	20 – 5 –		

^{d} The spectroscopic data confirming this result are not published.

Luminescence linewidth [cm⁻¹]	Laser transition	Energy terminal level [cm⁻¹]	Excitation system, pump source, filter, cooling	Rod length, rod diameter [mm]; resonator cavity parameters, mirrors (coating, transmittance)	Remarks	Reference
	$^4F_{9/2} \to {}^4I_{15/2}$				$Yb^{3+} = 38\%$	
–		410	Illuminator	EMDM	Energy cumulation on Yb^{3+}	[31]
–		410	with Xe flashtube FT-524 and laser diode GaAs: Si (0.9 μm), N₂		and transfer to Ho^{3+}	
–		410				
	$^3P_0 \to {}^3H_6$					
–		≈4200	ER with Xe flashtube, N₂	32; 0.6 R, 1%	$n \approx 1.58$ $K = 51$	[75]
	$^4F_{3/2} \to {}^4I_{11/2}$					
25		1983	ER with Xe	25–40; 5–6	$\Delta v_g = 2.3$ cm⁻¹	[24, 32,
35		2188	flashtube	∥, 30″, $R = 500$	$\Delta v_g = 6$ cm⁻¹	37, 183,
6.5		1980	IFP-400,	mm, DM, 0.7%	$\Delta v_g \leqslant 1$ cm⁻¹	207, 243,
6.5		2069	ZhS-17, N₂		$\Delta v_g \leqslant 1$ cm⁻¹	264, 343,
15		2188			$\Delta v_g \approx 5$ cm⁻¹	344]
25		1983	ER with Xe flashtube, water	38; 5 ∥, 10″, $R =$ $= 576$ mm, DM, 0.7%	$n \approx 1.6$ $n_2 \approx 1.5 \times 10^{-13}$	
6.5		2092			In CAM laser	
6.5		2188				
18		2223				
25→		1983	High-temp. ER with Xe		$\delta\lambda/\delta T \approx 3$	
35→		2190	flashtube IFP-800,		$\delta\lambda/\delta T \approx 1.5$	
–		2190	ZhS-17		$\delta\lambda/\delta T \approx 4$	
–		2220			$\delta\lambda/\delta T \approx 3.1$	
25		1983	ER with PS			
	$^4F_{3/2} \to {}^4I_{13/2}$					
21		≈4070	ER with Xe	25; 5	*In CAM	[87, 88]
25		≈4270	flashtube IFP-400,	∥, $R = 600$ mm, DM, 1%	laser	
21		≈4070	ZhS-17			
25		≈4270				
6		3974	Low-temp.			
8		4038	ER with Xe			
10		4077	flashtube			
13		4276	IFP-400,			
6		3974	ZhS-17, N₂			
8		4038				
10		4077				
13		4276				

(Continued)

Table 5.1 (*continued*)

Crystal host, space group cation-site symmetry	Activator [%]	Emission wavelength [μm]	Temperature [K]	Mode of operation	Threshold energy [J] or power [W]	Excitation spectral range [μm]	Lumines-cence lifetime [ms]
LaF$_3$ D_{3d}^4–P$\bar{3}$c1 C_2(La^{3+})	Er^{3+} 0:05	1.6113	7	P	500	0.4–1.6	–
CeF$_3$ D_{3d}^4–P$\bar{3}$c1 C_2(Ce^{3+})	Nd^{3+} $\approx$4 $c \perp F$	1.0410 A 1.0638 B	300 300	P P	25 8.5	0.4–0.9	$\approx$0.27
		1.0404 A 1.0639 B	77 77	P P	3.5 6		$\approx$0.25
	$\angle cF = 10°$	1.0410 A 1.0638 B 1.0404 A 1.0410 A 1.0638 B	300 300 77 300 300	P P P q-CW q-CW	8.5 7.5 2.5 – –		
	2 $c \perp F$	1.3170* 1.3320 1.3690* 1.3130* 1.3240 1.3310 1.3675	300 300 300 77 77 77 77	P P P P P P P	– 20 – – 20 10 20	0.5–0.9	$\approx$0.27
HoF$_3$ D_{2h}^{16}–Pnma C_s(Ho^{3+})	Ho^{3+} 100	2.090	77	P	$\approx$1	0.3–2	2.6
						Mixed fluoride laser crystals	
5NaF·9YF$_3$ O_h^5–Fm3m	Nd^{3+} 1	1.0506B 1.0595C	300 300	P P	45 83	0.5–0.9	0.96
	1	1.3070	300	P	250	0.5–0.9	0.96

[6] The authors of [345] found electron-excitation-energy transfer from Ce^{3+} to Nd^{3+} ions.

Luminescence linewidth [cm⁻¹]	Laser transition	Energy terminal level [cm⁻¹]	Excitation system, pump source, filter, cooling	Rod length, rod diameter [mm]; resonator cavity parameters, mirrors (coating, transmittance)	Remarks	Reference
	$^4I_{13/2} \rightarrow {}^4I_{15/2}$					
–		≈ 400	ER with Xe flashtube PEK, N_2 vapor	25.4; 5 $\parallel$, 5″, Ag, 1%		[289]
	$^4F_{3/2} \rightarrow {}^4I_{11/2}$					
≈ 25		1983	ER with Xe	25; 5	$E \perp c$ axis	[243, 265,
≈ 35		2189	flashtube IFP-400	$\parallel$, 10″, $R = 576$ mm, DM, 0.7%	$E \parallel c$ axis $n \approx 1.62$ $n_2 \approx 1.6 \times 10^{-13}$	345]
6.5		1996	Low-temp.			
15		2199	ER with Xe flashtube IFP-400, N_2			
≈ 25		1983			$\Delta v_g = 2.5$ cm⁻¹	
≈ 35		2189			$\Delta v_g = 9$ cm⁻¹	
6.5		1996			$\Delta v_g \approx 1$ cm⁻¹	
≈ 25		1983	ER with PS			
≈ 35		2189			EET[6] to Nd^{3+}	
	$^4F_{3/2} \rightarrow {}^4I_{13/2}$					
28		≈ 4040	ER with Xe	25; 5	*In CAM laser	[89]
30		≈ 4070	flashtube	$\parallel$, $R = 600$ mm,		
40		≈ 4280	IFP-400	DM, 1%		
6		≈ 3970	Low-temperat.			
8		≈ 4040	ER with Xe			
10		≈ 4080	flashtube			
15		4280	IFP-400, N_2			
	$^5I_7 \rightarrow {}^5I_8$					
≈ 1.5		≈ 380	ER with Xe flashtube, TQC, N_2	15; 3 R, DM, 0.1%	$E \perp (100)$ $n \approx 1.58$	[346]
with disordered structure						
	$^4F_{3/2} \rightarrow {}^4I_{11/2}$					
100		≈ 2000	ER with Xe	15; 6	$\Delta v_g \approx 7.5$ cm⁻¹	[66]
–		≈ 2100	flashtube IFP-400, ZhS-17	$\parallel$, 30″, $R = 576$ mm, DM, 2%	$\Delta v_g \approx 13$ cm⁻¹	
	$^4F_{3/2} \rightarrow {}^4I_{13/2}$					
200		≈ 4000	ER with Xe flashtube IFP-400, ZhS-17	15; 6 $\parallel$, 30″, $R = 600$ mm, DM, 1%		[88]

(Continued)

Table 5.1 (*continued*)

Crystal host, space group, cation-site symmetry	Activator [%]	Emission wavelength [μm]		Temperature [K]	Mode of operation	Threshold energy [J] or power [W]	Excitation spectral range [μm]	Luminescence lifetime [ms]
CaF_2–SrF_2 O_h^5–Fm3m	Nd^{3+} 0.5	1.0369		300	P	140	0.5–0.9	1.4
CaF_2–YF_3 O_h^5–Fm3m	Nd^{3+} 0.2–12							
		1.0461	A	300	P	180	0.5–0.9	0.48 (0.2%)
		1.0540	B	300	P	6		
		1.0632	C	300	P	7		
		1.0540	B	300–430	P	–		
		1.0632→ →1.0603C		300→950	P	–		
		1.0540	B	300	q-CW	–		
		1.0632	C	300	q-CW	–		
	1–2	1.3255		77	P	40	0.5–0.9	0.48 (0.2%)
		1.3380		77	P	70		
		1.3600		77	P	200		
		1.3270		300	P	12		
		1.3370		300	P	10		
		1.3585		300	P	20		
CaF_2–YF_3 O_h^5–Fm3m	Ho^{3+} 1–3	2.0318		90	P	280	0.3–2	–
				77	P	145		
				77	q-CW	–		
CaF_2–YF_3 O_h^5–Fm3m	Er^{3+} 0.5–5	0.8430	B	77	P	200	0.3–0.55	0.5 (0.5%)
		0.8456	A	77	P	40		
	10	0.8456	A	77	P	65		

Luminescence linewidth [cm⁻¹]	Laser transition	Energy terminal level [cm⁻¹]	Excitation system, pump source, filter, cooling	Rod length, rod diameter [mm]; resonator cavity parameters, mirrors (coating, transmittance)	Remarks	Reference
20	$^4F_{3/2} \to {}^4I_{11/2}$	≈ 2000	ER with Xe flashtube IFP-400, ZhS-17	20; 5 $\parallel$, 15″, R, 1%	$CaF_2/SrF_2 = 1$ $\Delta v_g = 2.5$ cm⁻¹	[347]
	$^4F_{3/2} \to {}^4I_{11/2}$				$YF_3 = (2-16)$ wt.%	
30		2032	ER with Xe	40–75; 5	$\Delta v_g = 0.8$ cm⁻¹	[23, 28,
65		–	flashtube	$\parallel$, 30″, $R =$	$\Delta v_g = 9.5$ cm⁻¹	32, 243,
90		2170	IFP type, ZhS-17	$= 500$ mm, DM, 1%	$\Delta v_g = 25$ cm⁻¹	313, 348– 351]
65		–	High-temp.	45; 5.5		
90→		2170	ER with Xe flashtube IFP-800, ZhS-17	$\parallel$, 30″, $R =$ $= 576$ mm, DM, 0.7%	$\delta\lambda/\delta T \approx 4.9$	
65		–	Illuminator	110, 10	$\tau_g = 10$ ms	
90		2170	with PS	$\parallel$, 30″, $R = 576$ mm, DM, 2%	$E_g = 0.35$ J	
	$^4F_{3/2} \to {}^4I_{13/2}$				$YF_3 = 5$ wt.%	
45		≈ 4040	Low-temp.	39; 5		[87, 90]
55		≈ 4090	ER with Xe	$\parallel$, 10″, $R =$		
60		≈ 4200	flashtube IFP-400, ZhS-17, N₂	$= 600$ mm, DM, 1%		
70		≈ 4040	ER with Xe			
80		≈ 4090	flashtube			
120		≈ 4200	IFP-400, ZhS-17			
	$^5I_7 \to {}^5I_8$					
≈ 90		≈ 80	Low-temp.	45; 5.5		[243, 352,
≈ 90			ER with Xe flashtube IFP-400, N₂ and its vapor	$\parallel$, 25″, $R =$ $= 576$ mm, DM, 2%		353]
≈ 90			Illuminator with PS, N₂		$\tau_g = 15$ ms	
	$^4S_{3/2} \to {}^4I_{13/2}$				$YF_3 = (2-3)$ wt.%	
9.5		6711	Low-temp.	45; 5.5	$\sigma_e = 4 \times 10^{-20}$	[25, 106,
8.5		6711	ER with Xe flashtube	$\parallel$, 20″, $R =$ $= 576$ mm,	$\sigma_e = 8 \times 10^{-20}$ $\Delta v_g \approx 1$ cm⁻¹	546]
8.5		6711	IFP-400, QLP, N₂	DM, 2%		

(Continued)

Table 5.1 (*continued*)

Crystal host, space group, cation-site symmetry	Activator [%]	Emission wavelength [μm]	Temperature [K]	Mode of operation	Threshold energy [J] or power [W]	Excitation spectral range [μm]	Luminescence lifetime [ms]
	0.5–2	1.547[a]	77	P	250	0.3–1.5	≈ 10
			77	q-CW	–		
$Ca_2Y_5F_{19}$ D_{3d}^4–P$\bar{3}$c1	Nd^{3+} 1	1.0498[7]	300–600	P	15	0.5–0.9	0.36
	1	1.3200	77	P	250	0.5–0.9	0.36
		1.3190	300	P	150		0.36
		1.3525	300	P	180		
CaF_2–LaF_3 O_h^5–Fm3m	Nd^{3+} ≈ 2	1.0645	300	P	25	0.5–0.9	–
	≈ 2	1.3190	300	P	70	0.5–0.9	–
CaF_2–CeO_2 O_h^5–Fm3m	Nd^{3+} 1	1.0885[a]	300	P	850	0.5–0.9	≈ 1.2
CaF_2–CeF_3 O_h^5–Fm3m	Nd^{3+} 0.5–1	1.0657$\rightarrow$ $\rightarrow$1.0640	300$\rightarrow$700	P	86	0.5–0.9	≈ 0.7

[a] λ_g value requires more-accurate definition.
[7] *Osiko* et al., observed the laser effect too.

Lumines-cence linewidth [cm^{-1}]	Laser transition	Energy terminal level [cm^{-1}]	Excitation system, pump source, filter, cooling	Rod length, rod diameter [mm]; resonator cavity parameters, mirrors (coating, transmittance)	Remarks	Reference
	$^4I_{13/2} \rightarrow \, ^4I_{15/2}$					
≈ 12		107	Low-temp. ER with Xe flashtube IFP-400, QLP, N$_2$	45; 5.5 $\parallel$, 20″, $R = 600$ mm, DM, 1%	$\sigma_e = 10^{-19}$ In CAM laser	[25, 106, 243, 353]
≈ 12		107	Illuminator with PS, N$_2$		$\tau_g = 15$ ms	
	$^4F_{3/2} \rightarrow \, ^4I_{11/2}$					
≈ 100		≈ 2000	High-temp. ER with Xe flashtube IFP-800, ZhS-17	34; 5 $\parallel$, $R = 576$ mm, DM, 0.7%		[32, 90]
	$^4F_{3/2} \rightarrow \, ^4I_{13/2}$					
25 ≈ 80 ≈ 100		≈ 4100 ≈ 4100 ≈ 4200	Low- and room-temp. ER with Xe flash-tube IFP-400, ZhS-17, N$_2$	21; 5 $\parallel$, $R = 600$ mm, DM, 1%		[90]
	$^4F_{3/2} \rightarrow \, ^4I_{11/2}$				LaF$_3$ = 10 wt.%	
≈ 100		≈ 2100	ER with Xe flashtube IFP-400, ZhS-17	21; 5 $\parallel$, $R = 576$ mm, DM, 1%		[536]
	$^4F_{3/2} \rightarrow \, ^4I_{13/2}$					
≈ 120		≈ 4100	ER with Xe flashtube IFP-400, ZhS-17	21; 5 $\parallel$, $R = 600$ mm, DM, 1%		[536]
	$^4F_{3/2} \rightarrow \, ^4I_{11/2}$					
≈ 25		≈ 2100	ER with Xe flashtube IFP-800, ZhS-17	45; 6 $\parallel$, 30″, $R = 576$ mm, DM, 0.7%		[349]
	$^4F_{3/2} \rightarrow \, ^4I_{11/2}$				CeF$_3$ = 30 wt.%	
≈ 100		≈ 2000	High-temp. ER with Xe flashtube IFP-800, ZhS-17	45; 5 $\parallel$, 20″, $R = 576$ mm, DM, 0.7%	$\Delta v_g = 3$ cm^{-1} $\delta\lambda/\delta T \approx 4.2$	[22, 32, 354, 355]

(*Continued*)

Table 5.1 *(continued)*

Crystal host, space group, cation-site symmetry	Activator [%]	Emission wavelength [μm]	Temperature [K]	Mode of operation	Threshold energy [J] or power [W]	Excitation spectral range [μm]	Luminescence lifetime [ms]
	0.5	1.3190	300	P	300	0.5–0.9	≈ 0.7
CaF_2–GdF_3 O_h^5–Fm3m	Nd^{3+} ≈ 2	1.0654	300	P	30	0.5–0.9	–
	≈ 2	1.3185	300	P	90	0.5–0.9	–
CaF_2–ErF_3 O_h^5–Fm3m	Ho^{3+} 0.5	2.030	77	P	140	0.3–1.9 Excitation into Ho^{3+} and Er^{3+} bands	≈ 10
	Er^{3+} 2	0.8456	77	P	400	0.3–0.55	0.5
CaF_2–ErF_3 O_h^5–Fm3m	Er^{3+} 10	2.7307	300 300	P Q	400 –	0.3–1.1	–
CaF_2–ErF_3 O_h^5–Fm3m	Tm^{3+} 0.5	1.860	77	P	13	0.3–1.8 Excitation into Er^{3+} and Tm^{3+} bands	11.5
		1.860	77	q-CW	–		
		1.860	77	CW	3000		6

Luminescence linewidth [cm^{-1}]	Laser transition	Energy terminal level [cm^{-1}]	Excitation system, pump source, filter, cooling	Rod length, rod diameter [mm]; resonator cavity parameters, mirrors (coating, transmittance)	Remarks	Reference
≈ 100	$^4F_{3/2} \rightarrow {}^4I_{13/2}$	≈ 4000	ER with Xe flashtube IFP-400, ZhS-17	45; 5.5 $\parallel$, $R = 600$ mm, DM, 1%		[88]
≈ 110	$^4F_{3/2} \rightarrow {}^4I_{11/2}$	≈ 2100	ER with Xe flashtube IFP-400, ZhS-17	21; 5 $\parallel$, $R = 576$ mm, DM, 1%	$GdF_3 = 10$ wt.%	[536]
≈ 100	$^4F_{3/2} \rightarrow {}^4I_{13/2}$	≈ 4000	ER with Xe flashtube IFP-400, ZhS-17	21; 5 $\parallel$, $R = 600$ mm, DM, 1%		[536]
≈ 70	$^5I_7 \rightarrow {}^5I_8$	≈ 200	Low-temp. ER with Xe flashtube IFP-400, QLP, N_2	45; 6 $\parallel$, 15″, $R = 100$ m, DM, 2%	$ErF_3 = 2$ wt.% $\sigma_e \approx 10^{-19}$ EET to Ho^{3+}	[133]
8.5	$^4S_{3/2} \rightarrow {}^4I_{13/2}$	6711	Low-temp. ER with Xe flashtube IFP-400, QLP, N_2	45; 6 $\parallel$, 15″, $R = 576$ mm, DM, 1%	In CAM laser	[25]
–	$^4I_{11/2} \rightarrow {}^4I_{13/2}$	≈ 6500	Illuminator with two Xe flashtubes IFP-1200, water	120; 9 $\parallel$, plane and spherical DM, 30%	For $E_{exc} = 3000$ J $E_g = 4$ J	[469]
≈ 100	$^3H_4 \rightarrow {}^3H_6$	≈ 400	Low-temp. ER with Xe flashtube IFP-400, QLP, N_2	45; 5.5 $\parallel$, 20″, $R = 100$ m, DM, 2%	$ErF_3 = 2$ wt.% $\sigma_e \approx 10^{-19}$ EET to Tm^{3+} In CAM laser	[145, 148, 243]
≈ 100		≈ 400	Illuminator with PS, N_2 ER with Xe flashtube			
–			ER with Xe flashtubes, supercooled N_2	80; 5 $\parallel$, 10″, Ag,	$ErF_3 = 10$ wt.%	

(*Continued*)

Table 5.1 (*continued*)

Crystal host, space group, cation-site symmetry	Activator [%]	Emission wavelength [μm]	Temperature [K]	Mode of oper-ation	Threshold energy [J] or power [W]	Excitation spectral range [μm]	Lumines-cence lifetime [ms]
	Er^{3+} 2	0.8456	77	P	360	0.3–0.55	0.5
SrF_2-YF_3 O_h^5-Fm3m	Nd^{3+} 2	1.0567	300	P	15	0.5–0.9	0.36
	2	1.3225	300	P	140	0.5–0.9	0.36
		1.3300	300	P	160		
		1.3320	77	P	200		–
$Sr_2Y_5F_{19}$ $D_{3d}^4-P\bar{3}c1$	Nd^{3+} 2	1.0493	300	P	40	0.5–0.9	0.4
	2	1.3190	300	P	75	0.5–0.9	0.4
SrF_2-LaF_3 O_h^5-Fm3m	Nd^{3+} 0.5–1	1.0597→ →1.0583	300→800	P	18	0.5–0.9	≈0.8
	1	1.3250	300	P	45	0.5–0.9	≈0.8
		1.3160	77	P	150		–
		1.3235	77	P	180		
		1.3355	77	P	250		

Lumines-cence linewidth [cm^{-1}]	Laser transition	Energy terminal level [cm^{-1}]	Excitation system, pump source, filter, cooling	Rod length, rod diameter [mm]; resonator cavity parameters, mirrors (coating, transmittance)	Remarks	Reference
8.5	$^4S_{3/2} \rightarrow {}^4I_{13/2}$	6711	Low-temp. ER with Xe flashtube IFP-400, QLP, N$_2$	45; 5.5 $\parallel$, 20″, $R = 576$ mm, DM, 1%	In CAM laser	[25]
≈ 65	$^4F_{3/2} \rightarrow {}^4I_{11/2}$	≈ 2000	ER with Xe flashtube ISP-1000, ZhS-17	20; 6 $\parallel$, 15″, $R = 576$ mm, DM, 0.7%	YF$_3$ = 10 wt.% $\Delta v_g = 24$ cm^{-1}	[356]
≈ 20 ≈ 75 25	$^4F_{3/2} \rightarrow {}^4I_{13/2}$	≈ 4000 ≈ 4000 ≈ 4000	Low- and room-temp. ER with Xe flash-tube IFP-400, ZhS-17, N$_2$	20; 6 $\parallel$, 15″, $R = 600$ mm, DM, 1%		[88]
≈ 110	$^4F_{3/2} \rightarrow {}^4I_{11/2}$	≈ 2000	ER with Xe flashtube IFP-400, ZhS-17	16; 5 $\parallel$, 5″, $R = 576$ mm, DM, 1%	$\Delta v_g = 15$ cm^{-1}	[317]
≈ 130	$^4F_{3/2} \rightarrow {}^4I_{13/2}$	≈ 4000	ER with Xe flashtube IFP-400, ZhS-17	16; 5 $\parallel$, 5″, $R = 600$ mm, DM, 1%		[317]
≈ 80	$^4F_{3/2} \rightarrow {}^4I_{11/2}$	≈ 2000	High-temp. ER with Xe flashtube IFP-800, ZhS-17	45; 5 $\parallel$, 15″, $R = = 576$ mm, DM, 0.7%	LaF$_3$ = 30 wt.% $\Delta v_g = 20$ cm^{-1} $\delta\lambda/\delta T \approx 2.8$	[22, 32, 355, 357]
≈ 90 30 30 40	$^4F_{3/2} \rightarrow {}^4I_{13/2}$	≈ 4000 ≈ 4000 ≈ 4000 ≈ 4100	Low- and room-temp. ER with Xe flash-tube IFP-400, QLP, ZhS-17, N$_2$	45; 5 $\parallel$, 15″, $R = 600$ mm, DM, 1%		[88]

(*Continued*)

Table 5.1 (*continued*)

Crystal host, space group, cation-site symmetry	Activator [%]	Emission wavelength [μm]	Temperature [K]	Mode of oper- ation	Threshold energy [J] or power [W]	Excitation spectral range [μm]	Lumines- cence lifetime [ms]
SrF_2-CeF_3 O_h^5-Fm3m	Nd^{3+} ≈ 2	1.0590^c	300	P	14	0.5-0.9	–
	≈ 2	1.3255^c	300	P	35	0.5-0.9	–
SrF_2-GdF_3 O_h^5-Fm3m	Nd^{3+} 2	1.0528	300	P	7	0.5-0.9	0.4
	2	1.3260 1.3250	300 77	P P	10 15	0.5-0.9	0.4 –
SrF_2-LuF_3 O_h^5-Fm3m	Nd^{3+} ≈ 2	1.0560	300	P	11	0.5-0.9	–
	2	1.3200	300	P	35	0.5-0.9	–
CdF_2-YF_3 O_h^5-Fm3m	Nd^{3+} 1-2	$1.0656\rightarrow$ $\rightarrow 1.0629$	$300\rightarrow 600$	P	12	0.5-0.9	≈ 0.3

^{c} The detailed spectroscopic investigations were not carried out.

Luminescence linewidth [cm^{-1}]	Laser transition	Energy terminal level [cm^{-1}]	Excitation system, pump source, filter, cooling	Rod length, rod diameter [mm]; resonator cavity parameters, mirrors (coating, transmittance)	Remarks	Reference
≈ 110	$^4F_{3/2} \rightarrow {}^4I_{11/2}$	≈ 2000	ER with Xe flashtube IFP-400, ZhS-17	21; 5 $\parallel$, 5″, $R = 576$ mm, DM, 1%	$CeF_3 = 35$ wt.%	[271]
≈ 130	$^4F_{3/2} \rightarrow {}^4I_{13/2}$	≈ 4000	ER with Xe flashtube IFP-400, ZhS-17	21; 5 $\parallel$, 5″, $R = 600$ mm, DM, 1%		[271]
≈ 100	$^4F_{3/2} \rightarrow {}^4I_{11/2}$	≈ 2000	ER with Xe flashtube ISP-1000, ZhS-17	30; 5 $\parallel$, 5″, $R = 576$ mm, DM, 1%	$GdF_3 = 10$ wt.%	[91, 358]
≈ 75 25	$^4F_{3/2} \rightarrow {}^4I_{13/2}$	≈ 4000 ≈ 4000	Low- and room-temp. ER with Xe flashtube ISP-1000, QLP, ZhS-17, N$_2$	30; 5 $\parallel$, 5″, $R = 600$ mm, DM, 1%		[91]
≈ 70	$^4F_{3/2} \rightarrow {}^4I_{11/2}$	≈ 2000	ER with Xe flashtube IFP-400, ZhS-17	20–25; 5–6 $\parallel$, 5″, $R = 576$ mm, DM, 1%	$LuF_3 = 10$ wt.% $\Delta v_g = 15$ cm^{-1}	[487, 536]
≈ 75	$^4F_{3/2} \rightarrow {}^4I_{13/2}$	≈ 4000	ER with Xe flashtube IFP-400, ZhŠ-17	20; 6 $\parallel$, 5″, $R = 600$ mm, DM, 1%		[536]
≈ 100	$^4F_{3/2} \rightarrow {}^4I_{11/2}$	≈ 2000	High-temp. ER with Xe flashtube ISP-1000, ZhS-17	23; 6 $\parallel$, 20″, $R = 576$ mm, DM, 0.7%	$YF_3 = 15$ wt.% $\Delta v_g = 19$ cm^{-1} $\delta\lambda/\delta T \approx 9$	[34, 359]

(*Continued*)

Table 5.1 (*continued*)

Crystal host, space group, cation-site symmetry	Activator [%]	Emission wavelength [μm]	Temperature [K]	Mode of operation	Threshold energy [J] or power [W]	Excitation spectral range [μm]	Luminescence lifetime [ms]
	2	1.3245	300	P	40	0.5–0.9	≈ 0.3
		1.3165	77	P	150		
CdF_2–LaF_3 O_h^5–Fm3m	Nd^{3+} ≈ 1	1.0665^a	300	P	500	0.5–0.9	≈ 0.3
BaF_2–YF_3 O_h^5–Fm3m	Nd^{3+} 2	1.0521	300	P	13	0.5–0.9	0.3
	2	1.3200	300	P	55	0.5–0.9	0.3
BaF_2–LaF_3 O_h^5–Fm3m	Nd^{3+} 0.5– –4.5	1.0534 A	300	P	2.7	0.5–0.9	≈ 0.43 (0.5%)
		1.0538 A	77	P	15		
		1.0580 B	77	P	50		
		1.0534→ →1.0563A	300→920	P	–		
		1.0534 A	300	q-CW	–	0.4–0.9	
	1	1.3185	300	P	10	0.5–0.9	≈ 0.43
		1.3280	300	P	7		
		1.3290	77	P	30		

^{a} λ_g value requires more-accurate definition.

Luminescence linewidth [cm⁻¹]	Laser transition	Energy terminal level [cm⁻¹]	Excitation system, pump source, filter, cooling	Rod length, rod diameter [mm]; resonator cavity parameters, mirrors (coating, transmittance)	Remarks	Reference
≈ 120 55	$^4F_{3/2} \rightarrow {}^4I_{13/2}$	≈ 4000 ≈ 4000	Low- and room-temp. ER with Xe flashtube IFP-400, QLP, ZhS-17, N_2	23; 6 $\parallel$, 20″, $R = 600$ mm, DM, 1%		[89]
≈ 100	$^4F_{3/2} \rightarrow {}^4I_{11/2}$	≈ 2000	ER with Xe flashtube IFP-400, ZhS-17	20; 4.5 $\parallel$, 15″, $R = 576$ mm, DM, 0.7%		[359]
≈ 60	$^4F_{3/2} \rightarrow {}^4I_{11/2}$	≈ 2000	ER with Xe flashtube IFP-400, ZhS-17	15; 5 $\parallel$, $R = 576$ mm, DM, 1%	$YF_3 = 5$ wt.% $\Delta v_g \approx 7$ cm⁻¹	[487]
≈ 100	$^4F_{3/2} \rightarrow {}^4I_{13/2}$	≈ 4000	ER with Xe flashtube IFP-400, ZhS-17	15; 5 $\parallel$, $R = 600$ mm, DM, 1%		[487]
≈ 75 ≈ 30 ≈ 35 $75\rightarrow$ ≈ 75	$^4F_{3/2} \rightarrow {}^4I_{11/2}$	1992 1992 2035 1992 1992	Low- and room-temp. ER with Xe flash-tubes IFP type, QLP, ZhS-17, N_2 High-temp. ER with Xe flashtube IFP-800, ZhS-17 ER with PS	13–80; 5–6 $\parallel$, 15″, $R = 576$ mm, DM, 1%	$LaF_3 = 30$ wt.% $\sigma_e = 2 \times 10^{-20}$ $\Delta v_g = 17$ cm⁻¹ $\Delta v_g = 30$ cm⁻¹ $\delta\lambda/\delta T \approx 5.2$ $\tau_g = 10$ ms	[22, 92, 243, 355, 360]
≈ 55 ≈ 80 30	$^4F_{3/2} \rightarrow {}^4I_{13/2}$	≈ 4000 ≈ 4000 3956	Low- and room-temp. ER with Xe flash-tube IFP-400, QLP, ZhS-17, N_2	40; 6 $\parallel$, $R = 600$ mm, DM, 1%		[89, 92]

(*Continued*)

Table 5.1   (*continued*)

Crystal host, space group, cation-site symmetry	Activator [%]	Emission wavelength [μm]		Temperature [K]	Mode of oper-ation	Threshold energy [J] or power [W]	Excitation spectral range [μm]	Lumines-cence lifetime [ms]
BaF_2–CeF_3 O_h^5–Fm3m	Nd^{3+} 2	1.0537		300	P	28	0.5–0.9	0.31
BaF_2–GdF_3 O_h^5–Fm3m	Nd^{3+} 2	1.0526		300	P	25	0.5–0.9	0.3
BaF_2–LuF_3 O_h^5–Fm3m	Nd^{3+} 2	1.0523[c]		300	P	100	0.5–0.9	–
LaF_3–SrF_2 D_{3d}^4–P$\bar{3}$c1	Nd^{3+} 2 $c \perp F$	1.0486 1.0635		300 300	P P	5.5 11	0.5–0.9	0.22
				300	q-CW	–	0.4–0.9	
	2 $c \perp F$	1.3315 1.3170 1.3275 1.3325		300 77 77 77	P P P P	13 60 40 50	0.5–0.9	0.22
α-NaCaYF$_6$ O_h^5–Fm3m	Nd^{3+} 0.5–4	1.0539 1.0629	B C	300 300	P P	11 17	0.5–0.9	0.36 (0.5%)
				300	q-CW	–	0.4–0.9	
		1.0539→ →1.0549B		300→550	P	–	0.5–0.9	
		1.0629→ →1.0597C		300→ →≈1000	P	–		

[c] The detailed spectroscopic investigations were not carried out.

Luminescence linewidth $[cm^{-1}]$	Laser transition	Energy terminal level $[cm^{-1}]$	Excitation system, pump source, filter, cooling	Rod length, rod diameter [mm]; resonator cavity parameters, mirrors (coating, transmittance)	Remarks	Reference
≈ 105	$^4F_{3/2} \rightarrow {}^4I_{11/2}$	≈ 2000	ER with Xe flashtube IFP-400, ZhS-17	15; 6 $\parallel$, $R = 576$ mm, DM, 1%	$CeF_3 = 5$ wt.% $\Delta v_g = 7$ cm^{-1}	[487]
≈ 75	$^4F_{3/2} \rightarrow {}^4I_{11/2}$	≈ 2000	ER with Xe flashtube IFP-400, ZhS-17	20; 6 $\parallel$, $R = 576$ mm, DM, 1%	$GdF_3 = 5$ wt.% $\Delta v_g = 8$ cm^{-1}	[487]
≈ 50	$^4F_{3/2} \rightarrow {}^4I_{11/2}$	≈ 2000	ER with Xe flashtube IFP-400, ZhS-17	15; 5 $\parallel$, 10″, $R = 576$ mm, DM, 1%	$LuF_3 = 5$ wt.%	[271]
≈ 65 ≈ 65	$^4F_{3/2} \rightarrow {}^4I_{11/2}$	≈ 2000 ≈ 2000	ER with Xe flashtube IFP-400, ZhS-17 ER with PS	26; 5 $\parallel$, 10″, $R = 576$ mm, DM, 1%	$\Delta v_g = 9$ cm^{-1} $\Delta v_g = 6$ cm^{-1}	[243, 361]
≈ 55 30 30 30	$^4F_{3/2} \rightarrow {}^4I_{13/2}$	≈ 4000 ≈ 4000 ≈ 4000 ≈ 4100	Low- and room-temp. ER with Xe flashtube IFP-400, QLP, ZhS-17, N$_2$	26; 5 $\parallel$, 10″, $R = 600$ mm, DM, 1%		[93]
≈ 70 ≈ 90 $70 \rightarrow$ $90 \rightarrow$	$^4F_{3/2} \rightarrow {}^4I_{11/2}$	≈ 2000 ≈ 2000 ≈ 2000 ≈ 2000	ER with Xe flashtube IFP-400, ZhS-17 ER with PS High-temp. ER with Xe flashtube IFP-800, ZhS-17	29; 6 $\parallel$, 15″, $R = 576$ mm, DM, 0.7%	$\tau_g = 10$ ms $\delta\lambda/\delta T \approx 4$ $\delta\lambda/\delta T \approx 4.9$	[32, 202, 243]

(*Continued*)

Table 5.1   (*continued*)

Crystal host, space group, cation-site symmetry	Activator [%]	Emission wavelength [μm]	Temperature [K]	Mode of oper- ation	Threshold energy [J] or power [W]	Excitation spectral range [μm]	Lumines- cence lifetime [ms]
	1–2	1.3285	300	P	20	0.5–0.9	0.36
		1.3375	300	P	18		
		1.3600	300	P	35		
		1.3260	77	P	45		
		1.3390	77	P	70		
α-NaCaCeF$_6$ O_h^5–Fm3m	Nd^{3+} 1	1.0653	300	P	7	0.5–0.9	0.44
		1.0653→ →1.0633	300→920	P	–		
			300	q-CW	–	0.4–0.9	
		1.3190	300	P	40	0.5–0.9	0.44
		1.3165	77	P	140		–
α-NaCaErF$_6$ O_h^5–Fm3m	Ho^{3+} ≈1	2.0345	150	P	300	0.3–2	–
		2.0312	77	P	180	Excitation into Ho^{3+} and Er^{3+} bands	–
		2.0377	77	P	210		
α-NaCaErF$_6$ O_h^5–Fm3m	Tm^{3+} ≈1	1.8580	150	P	95	0.3–1.8	–
		1.8885	150	P	120	Excitation into Er^{3+} and Tm^{3+} bands	
		1.8580	77	P	65		
		1.8885	77	P	100		
CaF$_2$–YF$_3$– –NdF$_3$ O_h^5–Fm3m	Nd^{3+} ≈16	1.0632	300	P	250	0.5–0.9	0.48 (0.2%)

Luminescence linewidth [cm⁻¹]	Laser transition	Energy terminal level [cm⁻¹]	Excitation system, pump source, filter, cooling	Rod length, rod diameter [mm]; resonator cavity parameters, mirrors (coating, transmittance)	Remarks	Reference
	$^4F_{3/2} \rightarrow {}^4I_{13/2}$					
≈ 70		≈ 4000	Low- and room-	29; 5		[89]
≈ 80		≈ 4100	temp. ER	‖, 15″, $R = 600$		
≈ 120		≈ 4100	with Xe flash-	mm, DM, 1%		
30		≈ 4000	tube IFP-400,			
45		≈ 4100	QLP, ZhS-17, N_2			
	$^4F_{3/2} \rightarrow {}^4I_{11/2}$					
≈ 100		≈ 2000	ER with Xe flashtube IFP-800, ZhS-17	20; 6 ‖, 20″, $R = 576$ mm, DM, 0.7%		[32, 65, 243]
$100 \rightarrow$		≈ 2000	High-temp. ER with Xe flashtube IFP-800 ZhS-17		$\delta\lambda/\delta T \approx 4.3$	
≈ 100		≈ 2000	ER with PS		$\tau_g = 10$ ms	
	$^4F_{3/2} \rightarrow {}^4I_{13/2}$					
≈ 95		≈ 4000	Low- and room-	20; 6		[89]
40		≈ 4000	temp. ER with Xe flash- tube IFP-400, QLP, ZhS-17, N_2	‖, 20″, $R = 600$ mm, DM, 1%		
	$^5I_7 \rightarrow {}^5I_8$					
≈ 120		≈ 120	Low-temp.	20; 6	EET to Ho^{3+}	[134, 352]
≈ 100		≈ 120	ER with Xe	‖, 30″, $R = 576$		
≈ 100		≈ 120	flashtube IFP-400, QLP, N_2 and its vapor	mm, DM, 2%		
	$^3H_4 \rightarrow {}^3H_6$					
≈ 75		≈ 470	Low-temp.	20; 6	EET to Tm^{3+}	[134, 352]
≈ 50		≈ 540	ER with Xe	‖, 30″, $R = 576$		
≈ 70		≈ 470	flashtube	mm, DM, 2%		
≈ 50		≈ 540	IFP-400, QLP, N_2 and its vapor			
	$^4F_{3/2} \rightarrow {}^4I_{11/2}$				$YF_3 = 15$ wt.%	
≈ 100		2170	ER with Xe flashtube IFP-400, ZhS-17	40; 6 ‖, 10″, $R = 576$ mm, DM, 1%		[546]

(*Continued*)

Table 5.1   (*continued*)

Crystal host, space group, cation-site symmetry	Activator [%]	Emission wavelength [μm]	Temperature [K]	Mode of operation	Threshold energy [J] or power [W]	Excitation spectral range [μm]	Luminescence lifetime [ms]
CaF_2–ErF_3– –TmF_3 O_h^5–Fm3m	Er^{3+} 12.5	$\approx 2.69^a$	298	P	$10^{(8)}$	0.3–0.8	4.5
	Tm^{3+} 0.5	1.860^a	100	P	$5^{(8)}$	0.3–1.8 Excitation into Er^{3+} and Tm^{3+} bands	2.9
SrF_2–CeF_3– –GdF_3 O_h^5–Fm3m	Nd^{3+} 2	1.0589	300	P	54	0.5–0.9	0.46
CdF_2–YF_3– –LaF_3 O_h^5–Fm3m	Nd^{3+} ≈ 1	1.065^a	300	P	550	0.5–0.9	≈ 0.4
CaF_2–ErF_3– –TmF_3–YbF_3 O_h^5–Fm3m	Ho^{3+} 0.5–1	$\approx 2.05^a$ $\approx 2.06^a$ $\approx 2.1^a$	100 298 77	P P P	16 100 90	0.2–2 Excitation into Ho^{3+}, Er^{3+}, Tm^{3+}, and Tb^{3+} bands	–
CaF_2–SrF_2– –BaF_2–YF_3– –LaF_3 O_h^5–Fm3m	Nd^{3+} 1	1.0535 A 1.0623 C	300 300	P P	30 150	0.5–0.9	0.4

---

[a] λ_g value requires more-accurate definition.
[8] The E_{thr} value depends ErF_3 and TmF_3 concentration.
[9] Weight ratio of ternary fluorides to binary fluorides.

Luminescence linewidth [cm^{-1}]	Laser transition	Energy terminal level [cm^{-1}]	Excitation system, pump source, filter, cooling	Rod length, rod diameter [mm]: resonator cavity parameters, mirrors (coating, transmittance)	Remarks	Reference
–	$^4I_{11/2} \to {}^4I_{13/2}$	≈ 7000	ER with Xe flashtube PEK 1–3	25; 3 $R = 2$ m, Ag, 2%	$Tm^{3+} = 0.5\%$	[111]
–	$^3H_4 \to {}^3H_6$	≈ 300	ER with Xe flashtube PEK 1–3, N_2 vapor	25; 3 $R = 2$ m, Ag, 2%	$ErF_3 = 12.5\%$ EET to Tm^{3+}	[111]
≈ 100	$^4F_{3/2} \to I_{11/2}$	≈ 2000	ER with Xe flashtube ISP–1000, ZhS–17	45; 5 ‖, 5″, $R = 576$ mm, DM, 1%	$CeF_3 = GdF_3 = 3$ wt.%	[91, 358]
≈ 120	$^4F_{3/2} \to {}^4I_{11/2}$	≈ 2000	ER with Xe flashtube IFP–400, ZhS–17	20; 5 ‖, 15″, $R = 576$ mm, DM, 0.7%		[359]
≈ 6 – –	$^5I_7 \to {}^5I_8$	≈ 250	ER with Xe flashtube PEK 1–3, N_2 vapor ER with Xe flashtubes, supercooled N_2	$R = 2$ m, Ag, $\approx 2\%$ 80; 5 ‖, 8–10″, Ag, 3%	$ErF_3 = TmF_3 = YbF_3 = 3$ wt.% EET to Ho^{3+}	[111, 145]
≈ 100 ≈ 120	$^4F_{3/2} \to {}^4I_{11/2}$	≈ 2000 ≈ 2000	ER with Xe flashtube IFP–400, ZhS–17	30; 5 ‖, 30″, $R = 500$ mm, DM, 0.7%	$\Sigma MeF_3/\Sigma MeF_2 = 0.43^{(9)}$ $\Delta v_g = 20$ cm^{-1} $\Delta v_g = 25$ cm^{-1}	[32, 64, 243, 349]

(Continued)

Table 5.1 *(continued)*

Crystal host, space group, cation-site symmetry	Activator [%]	Emission wavelength [µm]	Temperature [K]	Mode of operation	Threshold energy [J] or power [W]	Excitation spectral range [µm]	Luminescence lifetime [ms]
		1.0535 A	300	q–CW	–	0.4–0.9	
		1.0535→ →1.0547 A	300→550	P	–		
		1.0623→ →1.0585 C	300→700	P	–		
							Simple oxide laser crystals
LiYO$_2$ D_{4h}^{19}–I4$_1$/amd	Nd^{3+} 2(?)	1.0515 1.0751 1.0748	300 300 77	P P P	32 26 16	–	0.15
LiNbO$_3$ C_{3v}^6–R3c	Nd^{3+} 0.2–2 $c \perp F$	(π) 1.0846A	300	P	3.5	0.5–0.9	0.085
	$c \parallel F$	(σ) 1.0933B	300	P	3		
		(σ) 1.0933→ →1.0922 B	300→620	P	–		
		1.0787→ →1.0782 C	450→590	P	–		
	$c \perp F$	(π) 1.0846A (π) 1.084^a A	295 77	P P	6 10		
		(π) 1.0846A	200–600	P	–		
	0.5 $c \perp F$	(π) 1.0846A	300	q–CW	0.28	LP; 0.7525 $\tau_{\text{exc}} = 0.5$ ms (f = 60 Hz)	0.1

^{a} λ_g value requires more-accurate definition.
^{c} The detailed data are not available. The results are doubtful and require confirmation or correction.

Luminescence linewidth [cm^{-1}]	Laser transition	Energy terminal level [cm^{-1}]	Excitation system, pump source, filter, cooling	Rod length, rod diameter[mm]; resonator cavity parameters, mirrors (coating, transmittance)	Remarks	Reference
≈ 100		≈ 2000	Illuminator with PS			
$100 \rightarrow$		≈ 2000	High-temp. ER with Xe		$\delta\lambda/\delta T \approx 4.8$	
$120 \rightarrow$		≈ 2000	flashtube IFP–800, ZhS–17		$\delta\lambda/\delta T \approx 9.5$	

with ordered structure

	$^4F_{3/2} \rightarrow {}^4I_{11/2}$					
–		≈ 2000	ER with Xe	20; 3	c	[754]
–		≈ 2000	flashtube	$\parallel$, DM, 20%		
		≈ 2000	IFP–400, N$_2$			

	$^4F_{3/2} \rightarrow {}^4I_{11/2}$					
20		2034	ER with Xe flashtube IFP–800	19; 5 $\parallel$, $R = 500$ mm, DM, 1%		[34, 156, 162–165, 266, 584]
30		2107	ER with Xe flashtube IFP–400, ZhS–17	12; 3.5 $\parallel$, $R = 576$ mm, DM, 0.7%	$\Delta v_g = 2$ cm^{-1}	
$30 \rightarrow$		2107	High-temp. ER with Xe		$\delta\lambda/\delta T \approx 3.3$	
–		1990	flashtube IFP–800, ZhS–17		$\delta\lambda/\delta T \approx 3.5$	
20		2034	Illuminator with Xe flashtube FT–524, N$_2$	EMDM	$n_0 = 2.286$ $n_e = 2.2$ $K = 56$	
–		–				
20		2034	High-temp. ER with Xe flashtube IFP–800, ZhS–17	15; 5 $\parallel$, $R = 576$ mm, DM, 0.7%		
20		2034	CW Kr ion laser	$10 \times 3 \times 3$ DM, 0.3%	Active element in immersion liquid CCl$_4$ ($n = 1.46$ at $\lambda = 1.06$ μm)	

(Continued)

Table 5.1 (*continued*)

Crystal host, space group, cation-site symmetry	Activator [%]	Emission wavelength [μm]	Temperature [K]	Mode of operation	Threshold energy [J] or power [W]	Excitation spectral range [μm]	Luminescence lifetime [ms]
	0.3	1.3550*	300	P	–	0.5–0.9	0.085
	$c \parallel F$	1.3870	300	P	4.5		
	0.3	1.3745	300	P	6		
	$c \perp F$	1.3870*	300	P	–		
$LiNbO_3$ C_{3v}^6–R3c	Ho^{3+} 0.25–1	(σ) 2.0786	77	P^f	425	–	–
$LiNbO_3$ C_{3v}^6–R3c	Tm^{3+} 0.25–1	(σ) 1.8532	77	P^f	220	–	–
$LiNbO_3:Mg^{2+}$ C_{3v}^6–R3c	Nd^{3+} 1 $c \perp F$	(π) 1.0846A	300	P	1.5	0.5–0.9	0.085
	$c \parallel F$	(σ) 1.0933B	300	P	2		
	1 $c \parallel F$	1.3870	300	P^f	3.5	0.5–0.9	0.085
$LiNdP_4O_{12}$ C_{2h}^6–C2/c $C_2(Nd^{3+})$	Nd^{3+} 100 $a \parallel F$	1.0477	300	P q–CW	0.013 b	LP 0.4579, 0.4765, 0.5145, 0.5965	0.325* ($\approx 1\%$)
	$c \parallel F$		300	CW	2×10^{-4} b	LP; 0.5145	

[b] The absorbed pump power (energy) at which stimulated emission begins is indicated.

[f] The composition of the stimulated-emission spectrum and the threshold energies depend on the mutual orientation of the geometric axis of the rod F (axis of the laser element) and of the crystallographic directions of the crystal.

Luminescence linewidth [cm^{-1}]	Laser transition	Energy terminal level [cm^{-1}]	Excitation system, pump source, filter, cooling	Rod length, rod diameter [mm]; resonator cavity parameters, mirrors (coating, transmittance)	Remarks	Reference
	$^4F_{3/2} \rightarrow {}^4I_{13/2}$					
18		≈ 4035	ER with Xe	20–35; 4–6	*In CAM laser	[89]
20		≈ 4035	flashtube	$\parallel$, $R = 600$ mm,		
20		≈ 3970	IFP–400,	DM, 1%		
20		≈ 4035	ZhS–17	15; 5		
				$\parallel$, $R = 600$ mm,		
				DM, 1%		
–	$^5I_7 \rightarrow {}^5I_8$	≈ 310	Illuminator with Xe flash-tube FT–524, N$_2$	EMDM		[156]
–	$^3H_4 \rightarrow {}^3H_6$	271	Illuminator with Xe flash-tube FT-524, N$_2$	EMDM	Laser with frequency self-doubling	[156]
	$^4F_{3/2} \rightarrow {}^4I_{11/2}$					
20		2034	ER with Xe	20; 4	Mg^{2+} ions	[530]
			flashtube	$\parallel$, 10″, $R = 576$	increase Nd^{3+}	
30		2107	IFP–400,	mm, DM, 1%	ion concentra-	
			ZhS–17		tion in LiNbO$_3$	
					crystal	
	$^4F_{3/2} \rightarrow {}^4I_{13/2}$					
20		4035	ER with Xe flashtube IFP–400, ZhS–17	20; 4 $\parallel$, 10″, $R = 576$ mm, DM, 1%		[530]
	$^4F_{3/2} \rightarrow {}^4I_{11/2}$					
		1939	P and CW Ar lasers and rhodamine 6G dye laser, LPG and TPG	to 2 mm $\parallel$, R, DM	$\sigma_e^a = 1.5 \times 10^{-19}$ $\sigma_e^b = 1.3 \times 10^{-19}$ $\sigma_e^c = 3.2 \times 10^{-19}$ $n \approx 1.58$ at $\lambda = 0.6328$ μm	[580, 608–611, 621, 764]
				0.3 $\parallel$, $R_1 = 90$ mm, $d_r = 90$ mm, DM, 0.3–3%	$P_g = 7$ mW $\eta_g = 43$ *In isostructural LiLaP$_4$O$_{12}$ crystal	

(*Continued*)

Table 5.1 (*continued*)

Crystal host, space group, cation-site symmetry	Activator [%]	Emission wavelength [μm]	Temperature [K]	Mode of operation	Threshold energy [J] or power [W]	Excitation spectral range [μm]	Luminescence lifetime [ms]
			300	CW	1.4×10^{-4} [b]	LP; 0.87	
	$a \parallel F$		260–315	P	–	0.8	
	100 $c \parallel F$	1.317 [a]	300	CW	1.7×10^{-2} [b]	LP; 0.5145	0.325*
$BeAl_2O_4$ D_6^6–$P6_322$	Cr^{3+} 0.05	0.6799 [c]	77	P	–[11]	0.35–0.68	2.8 0.3 (300 K)
MgO O_h^5–Fm3m	Ni^{2+} –	1.3144	77	P	230	0.4–1.3	–
$NaNdP_4O_{12}$ C_{2h}^5–$P2_1/c$ $C_2(Nd^{3+})$	Nd^{3+} 100 $c \parallel F$	1.051[a]	300	P	–	LP; 0.5145	≈ 0.3 ($\approx 1\%$)
	100 $b \parallel F$	≈ 1.32 [a]	300	CW	0.048 [b]	LP; 0.5145	≈ 0.3
Al_2O_3 D_{3d}^6–$R\bar{3}c$ $C_3(Al^{3+})$	Cr^{3+} 0.05	0.6929 R_2	300	P	100	0.35–0.69	3
		0.6943 R_1	300	P	13		
		0.6934 R_1	77	P	4		

[a] λ_g value requires more-accurate definition.
[b] The absorbed pump power (energy) at which stimulated emission begins is indicated.
[c] The detailed spectroscopic investigations were not carried out.
[10] The calculated value is given.
[11] A copper pipe was used for crystal cooling.

Luminescence linewidth [cm^{-1}]	Laser transition	Energy terminal level [cm^{-1}]	Excitation system, pump source, filter, cooling	Rod length, rod diameter [mm]; resonator cavity parameters, mirrors (coating, transmittance)	Remarks	References
			Semiconductor laser $(Ga_xAl_{1-x}As)$ LED based on $Al_xGa_{1-x}As$, LPG and TPG	1.33; 0.3 $\parallel$, DM 4.85 $\times$ 4 $\times$ 1.9 and 0.3	$P_g = 38$ mW $\eta_g = 15$	
$-$	$^4F_{3/2} \to {}^4I_{13/2}$	≈ 4000	CW Ar ion laser, LPG	0.3 $\parallel$, $R_1 = 90$ mm, $d_r = 90$ mm, DM, to 3%	$\sigma_e \approx 7 \times 10^{-20}$	[752, 764]
$-$	$^2E(\bar{E}, 2\bar{A}) \to {}^4A_2$	0	Low-temp. illuminator with Xe flash-tubes, N_2	64 $\times$ 3 $\times$ 4.6 $\parallel$, DM	$E_g = 1$ J	[753]
$-$	$^3T_2 \to {}^3A_2$	398	Illuminator with Xe flash-tube FT-524, N_2	$-$	$K = 585^{(10)}$	[113]
$-$	$^4F_{3/2} \to {}^4I_{11/2}$	≈ 2000	Ar ion laser	5 $\times$ 5 $\times$ 5	$\sigma_e^c = 2.1 \times 10^{-19}$	[578]
$-$	$^4F_{3/2} \to {}^4I_{13/2}$	≈ 4000	CW Ar ion laser, LPG	0.3 $\parallel$, $R_1 = 90$ mm, $d_r = 90$ mm, DM, to 3.2%		[752]
7.5	$^2E \to {}^4A_2$ $^2E(2\bar{A}) \to {}^4A_2$	0	Illuminator with Xe flash-tube FT-524	28 Filter for selection R_2 line	$n_o = 1.763$ $n_e = 1.755$ $n_2 \approx 1.4 \times 10^{-13}$	[12, 32, 122,190, 203–206, 243, 246, 301, 362–371, 373–375]
≈ 10	$^2E(\bar{E}) \to {}^4A_2$	0	ER with Xe flashtube	20–75; 3 $\parallel$, DM, 1%	$\sigma_e = 2.5 \times 10^{-20}$ $\sigma_a^{exc} = 3 \times 10^{-21}$	
0.1		0	ER with Xe flashtube, N_2	30; 3 $\parallel$, 30″, R, DM, 0.5%		

(Continued)

Table 5.1 *(continued)*

Crystal host, space group, cation-site symmetry	Activator [%]	Emission wavelength [μm]	Temperature [K]	Mode of operation	Threshold energy [J] or power [W]	Excitation spectral range [μm]	Luminescence lifetime [ms]
	0.5– –0.7	0.7009 N_2 0.7041 N_1	77 77	P P	3000		1.1 1.2
	0.05	0.06943 R_1	300	CW	1400		
		0.6943 → →0.6952 R_1	300→550	P	–		
		0.6943 R_1	77	q–CW	–		
	Isotope 50 $c \perp F$	0.6934382	70	Q	≈ 600		4.3
	Isotope 52	0.6934089	70	Q	≈ 400		–
		0.6934255	70	Q	≈ 400		–
	0.05	0.7670	300	P	0.2	LP; 0.6943	3
		(12)					
$KY(MoO_4)_2$ D_{2h}^{14}–Pbna (Pbcn) $C_2(Y^{3+})$	Nd^{3+} 2 $a \parallel F$	1.0669A	300	P^f	10	0.3–0.9	0.13 (0.3%)
	2 $a \parallel F$	1.3485	300	P^f	20	0.3–0.9	0.13
$KY(WO_4)_2$ C_{2h}^6–C2/c $C_2(Y^{3+})$	Nd^{3+} 0.3– –2.5 $b \parallel F$	0.9137 C	77	P^f	5.5	0.5–0.9	0.11 (0.3%)

f The composition of the stimulated-emission spectrum and the threshold energies depend on the mutual orientation of the geometric axis of the rod F (axis of the laser element) and of the crystallographic directions of the crystal.

(12) In [376], simultaneous emission of Cr^{3+} (0.6943 μm) and Eu^{3+} (0.6130 μm) ions in Al_2O_3 crystal is mentioned. The results require confirmation.

(13) Crystallographic faces (natural-growth surfaces) or cleavage planes of these crystals were used as the laser-rod end faces.

Luminescence linewidth $[\text{cm}^{-1}]$	Laser transition	Energy terminal level $[\text{cm}^{-1}]$	Excitation system, pump source, filter, cooling	Rod length, rod diameter [mm]; resonator cavity parameters, mirrors (coating, transmittance)	Remarks	Reference
≈ 2	$14299\ \text{cm}^{-1}\rightarrow$	35	Illuminator with Xe flash-tube FT–524, N_2	40; 3 $\parallel$, Ag, 1%	Pair centers responses for SE	
≈ 2	$14233\ \text{cm}^{-1}\rightarrow$	33				
≈ 10	$^2E(\bar{E})\rightarrow{}^4A_2$	0	ER with Hg lamp PEK–A, water	50; 3 $\parallel$, DM, 1 and 25%	$P_g = 2.37$ W $\eta_g = 0.1$	
$10\rightarrow$		0	High-temp. ER with Xe IFP–800	28; 6 $\parallel$, 15″, $R = 576$ mm, DM, 1%	$\delta\lambda/\delta T = 6$	
0.1		0	Illuminator with PS, N_2		$\tau_g = 10$ ms	
0.1	$^2E(\bar{E})\rightarrow{}^4A_2$	0	ER with Xe flashtube, N_2	75; 6.5 Rotating DM, 55%		
0.1		0				
0.1		0				
11	$^2E\rightarrow{}^4A_2$	–	Q-switched ruby laser	–	SE in PAT	
12	$^4F_{3/2}\rightarrow{}^4I_{11/2}$	1983	ER with Xe flashtube IFP–400	6.5 $\parallel^{(13)}$, $R = 576$ mm, DM, 1%	$\Delta\nu_g = 4\ \text{cm}^{-1}$ $\sigma_e = 2\times10^{-19}$ $n \approx 1.94$	[377]
35	$^4F_{3/2}\rightarrow{}^4I_{13/2}$	≈ 3940	ER with Xe flashtube IFP–400	5 $\parallel^{(13)}$, $R = 600$ mm, DM, 1%		[93]
25	$^4F_{3/2}\rightarrow{}^4I_{9/2}$	355	Low-temp. ER with Xe flashtube IFP–400, QLP, N_2	14; 4 $\parallel$, 20″, EDAC at $\lambda = 1.06\ \mu$m, $R = 576$ mm, DM, 1%	$n \approx 1.9$	[77, 94, 379, 380]

(Continued)

Table 5.1 (*continued*)

Crystal host, space group, cation-site symmetry	Activator [%]	Emission wavelength [μm]	Temperature [K]	Mode of oper- ation	Threshold energy [J] or power [W]	Excitation spectral range [μm]	Lumines- cence lifetime [ms]
	2.5 $b \parallel F$	1.0688 A	300	P	0.3	0.5–0.9	0.11
			300	CW	400		
		1.0687 A	77	P	0.5		0.11
		1.0688 A	300	q-CW	–	0.4–0.9	
		1.0687→ →1.0690 A	77→600	P	–	0.5–0.9	
	2.5 $b \parallel F$	1.3525 1.3545	300 300	P P	0.9 3	0.5–0.9	0.11
		1.3525	300	CW	1100		
		1.3515 1.3545	77 77	P P	1 1		
K(Y, Er) (WO₄)₂: Tm³⁺ C_{2h}^6–C2/c C_2(Y³⁺, Er³⁺)	Ho³⁺ ≈3 $b \parallel F$	2.0720ᶠ	≈110 ≈220	P P	55 220	0.4–2 Excitation into Ho³⁺, Er³⁺, and Tm³⁺ bands	–

ᶠ The composition of the stimulated-emission spectrum and the threshold energies depend on the mutual orientation of the geometric axis of the rod *F* (axis of the laser element) and of the crystallographic directions of the crystal.

Lumines- cence linewidth [cm^{-1}]	Laser transition	Energy terminal level [cm^{-1}]	Excitation system, pump source, filter, cooling	Rod length, rod diameter [mm]; resonator cavity parameters, mirrors (coating, transmittance)	Remarks	Reference
20		1944	ER with Xe flashtube IFP–400, ZhS–17	14–25; 4–5 $\parallel$, 10–20″, $R =$ = 576 mm, DM, 1%		[77, 94, 210, 379, 380]
			ER with Xe lamp DKsTV– –3000, water	25; 5 $\parallel$, 10″, $R = 576$ mm, DM, 1%	$P_g = 0.55$ W	
3.3		1943	Low-temp. ER with Xe flashtube IFP–400, QLP, ZhS–17, N$_2$			
20		1944	Closed PS– –illuminator		$\tau_g = 7$ ms	
3.3→		1943	Low-and high- temp. ER with Xe flash- tube IFP–400, ZhS–17, N$_2$		$\delta\lambda/\delta T = 0.85$	
	$^4F_{3/2} \rightarrow {}^4I_{13/2}$					
13		≈3900	ER with Xe flashtube IFP–400, ZhS–17	25; 5 $\parallel$, $R = 600$ mm, DM, 1%		[77, 93, 94, 222]
14		≈3916				
13		≈3900	ER with Xe lamp DKsTV– –3000, ZhS–17, water			
6.5		3902	Low-temp. ER with Xe flashtube IFP–400, QLP, ZhS–17, N$_2$			
7		3916				
	$^5I_7 \rightarrow {}^5I_8$				Er^{3+} ≈ 30 at.%, Tm^{3+} ≈ 3 at.%	
–		≈400	Low-temp. ER with Xe flashtube ISP–250, N$_2$ vapor	55; 5.5 $\parallel$, $R = 600$ mm, DM, 1%	EET to Ho^{3+}	[595]
–						

(*Continued*)

Table 5.1 *(continued)*

Crystal host, space group, cation-site symmetry	Activator [%]	Emission wavelength [μm]	Temperature [K]	Mode of oper-ation	Threshold energy [J] or power [W]	Excitation spectral range [μm]	Lumines-cence lifetime [ms]
K(Y,Er) (WO$_4$)$_2$:Ho^{3+}, Tm^{3+} C^6_{2h}–C2/c C_2(Y^{3+}, Er^{3+})	Er^{3+} ≈ 30 $b \parallel F$	2.6887	300–350	P	10^9	0.3–1.1 $\tau_{exc} = 40 - 60$ μs	–
KNdP$_4$O$_{12}$ C^2_2–P2$_1$	Nd^{3+} 100	1.052 [a]	300	CW q–CW [b]	4.5 $\times$ 10^{-4}	LP; 0.58	≈ 0.275* (≈ 1%)
	100 $b \parallel F$	≈ 1.32 [a]	300	CW	4.3 $\times$ 10^{-2} [b]	LP; 0.5145	≈ 0.275*
KGd(WO$_4$)$_2$ C^6_{2h}–C2/c C_2(Gd^{3+})	Nd^{3+} 2.5 $b \parallel F$	1.0672	300 300	P CW	0.4 [f] 400	0.5–0.9	0.11 (≈ 0.3%)
	2.5 $b \parallel F$	1.3510	300	P	0,9 [f]	0.5–0.9	0.11
KGd(WO$_4$)$_2$ C^6_{2h}–C2/c C_2(Gd^{3+})	Ho^{3+} 3 5 $b \parallel F$	2.9342	300 300	P P	25 [f,g] 28	0.5–1.1 $\tau_{exc} = 40 - 60$ μs	–
KGd(WO$_4$)$_2$ C^6_{2h}–C2/c C_2(Gd^{3+})	Er^{3+} 3 $c \parallel F$ 10 $b \parallel F$	0.8468 0.8468	300 300	P P	34[f,g] 70	0.25–0.55 $\tau_{exc} = 40 - 60$ μs	–

[a] λ_g value requires more-accurate definition.

[b] The absorbed pump power (energy) at which stimulated emission begins is indicated.

[f] The composition of the stimulated-emission spectrum and the threshold energies depend on the mutual orientation of the geometric axis of the rod F (axis of the laser element) and of the crystallographic directions of the crystal.

[g] The E_{thr} value depends on τ_{exc}.

Luminescence linewidth [cm^{-1}]	Laser transition	Energy terminal level [cm^{-1}]	Excitation system, pump source, filter, cooling	Rod length, rod diameter [mm]; resonator cavity parameters, mirrors (coating, transmittance)	Remarks	Reference
–	$^4I_{11/2} \rightarrow {}^4I_{13/2}$	≈ 6500	ER with Xe flashtube ISP–250	55; 5.5 $\parallel$, $R = 600$ mm, DM, 1%	$Ho^{3+} = Tm^{3+} \approx$ ≈ 3 at.% RED from Er^{3+}	[595]
≈ 47	$^4F_{3/2} \rightarrow {}^4I_{11/2}$	≈ 2000	Dye laser ($f = 150$ Hz)	≈ 0.05 $\parallel$, DM, 0.3%	$\sigma_3 \approx 1.5 \times 10^{-19}$ $P_g = 6.6$ mW $\eta_g = 35$ *In isostructural $KGdP_4O_{12}$ crystal	[579, 581, 613]
–	$^4F_{3/2} \rightarrow {}^4I_{13/2}$	≈ 4000	CW Ar ion laser, LPG	0.3 $\parallel$, $R_1 = 90$ mm, $d_r = 90$ mm, DM, to 3.2%		[752]
12.5	$^4F_{3/2} \rightarrow {}^4I_{11/2}$	1945	ER with Xe IFP–400 and DKsTV–3000 lamps, ZhS–17, water	27; 2.5 $\parallel$, $R = 600$ mm, DM, 1%		[595]
12.5	$^4F_{3/2} \rightarrow {}^4I_{13/2}$	3906	ER with Xe flashtube IFP–400, ZhS–17	27; 2.5 $\parallel$, $R = 600$ mm, DM, 1%		[595]
–	$^5I_6 \rightarrow {}^5I_7$	≈ 5400	ER with Xe flashtube ISP–250, ZhS–17	37–48; 5.5 $\parallel$, 5″, $R = 600$ DM, 1%		[641]
≈ 18	$^4S_{3/2} \rightarrow {}^4I_{13/2}$	6517	ER with Xe flashtube ISP–250	25–27; 6 $\parallel$, 5″, $R = 600$ mm, DM, 1%		[641]

(Continued)

Table 5.1 (*continued*)

Crystal host, space group, cation-site symmetry	Activator [%]	Emission wavelength [μm]	Temperature [K]	Mode of operation	Threshold energy [J] or power [W]	Excitation spectral range [μm]	Luminescence lifetime [ms]
	3 $c \parallel F$	1.7155	300	P	$30^{f,g}$	0.25–0.55 $\tau_{exc} = 40 - 60$ μs	–
	3 $b \parallel F$	1.7325	300	P	10		
	5 $b \parallel F$	1.7325	300	P	13		
	10 $b \parallel F$	1.7325	300	P	22		
	3 $c \parallel F$	2.7222	300	P	$50^{f,g}$	0.5–1.1 $\tau_{exc} = 40 - 60$ μs	–
		2.7990	300	P	95		
	3 $b \parallel F$	2.7222	300	P	65		
	5 $b \parallel F$	2.7990	300	P	40		
	10 $b \parallel F$	2.7990	300	P	36		
	25 $b \parallel F$	2.7990	300	P	15		
$CaMg_2Y_2$ Ge_3O_{12} O_h^{10}–Ia3d	Nd^{3+} ≈ 6	1.05896	300	P	65	0.4–0.9	0.305 (0.5%)
$CaAl_4O_7$ C_{2h}^6–C2/c $C_{2v}(Ca^{2+})$, $C_1(Al^{3+})$	Nd^{3+} 0.4	1.0786	300	P	8^f	0.5– 0.9	0.28
		1.059 *	300	P	–		
		1.0636 *	300	P	–		
		1.0786 *	300	P	–		
		1.05895	77	P	70	0.4–0.9	0.28
		1.06585	77	P	10		
		1.07655	77	P	35		
		1.0772	77	P	40		

^{f} The composition of the stimulated-emission spectrum and the threshold energies depend on the mutual orientation of the geometric axis of the rod F (axis of the laser element) and of the crystallographic directions of the crystal.

^{g} The E_{thr} value depends on τ_{exc}.

Luminescence linewidth [cm^{-1}]	Laser transition	Energy terminal level [cm^{-1}]	Excitation system, pump source, filter, cooling	Rod length, rod diameter [mm]; resonator cavity parameters, mirrors (coating, transmittance)	Remarks	Reference
	$^4S_{3/2} \rightarrow {}^5I_{9/2}$					
–		12498	ER with Xe flashtube	25–42; 3.5–6 $\parallel$,5″,$R = 600$		[641]
–		12556	ISP–250	mm, DM, 1%		
–		12556				
–		12556				
	$^4I_{11/2} \rightarrow {}^4I_{13/2}$					
–		≈6520	ER with Xe	24–45; 3.5–6		[641]
–		≈6710	flashtube	$\parallel$, 5″, $R = 600$		
–		≈6520	ISP–250, ZhS–17	mm, DM, 1%		
–		≈6710				
–		≈6710				
–		≈6710				
	$^4F_{3/2} \rightarrow {}^4I_{11/2}$					
≈37		2008	Illuminator with two Xe flashtubes ILC–L–268, filter (Corning 3060)	26; 5 $\parallel$, $d_r = 200$ mm, DM, 2.5%	$\sigma_e = 6 \times 10^{-20}$ $n \approx 1.83$	[630]
	$^4F_{3/2} \rightarrow {}^4I_{11/2}$					
28		≈2100	ER with Xe flashtube	25; 6 $\parallel$, 5″$R = 576$	$\Delta\nu_g = 3$ cm^{-1} $n \approx 1.64$	[95]
30		≈2100	IFP–400,	mm, DM, 1%	*In CAM laser	
11		≈2100	ZhS–17			
28		≈2100				
3.5		≈2100	Low-temp.			
1.7		≈2100	ER with XE			
4.8		≈2100	flashtube			
6.2		≈2100	IFP–400, LQP, N$_2$			

(*Continued*)

Table 5.1 (*continued*)

Crystal host, space group, cation-site symmetry	Activator [%]	Emission wavelength [μm]	Temperature [K]	Mode of operation	Threshold energy [J] or power [W]	Excitation spectral range [μm]	Luminescence lifetime [ms]
	0.4	1.3420	300	P	21 [f]	0.5–0.9	0.28
		1.3710	300	P	15		
		1.3400	77	P	65		0.28
		1.3675	77	P	50		
$CaAl_4O_7$ C_{2h}^6–C2/c $C_{2v}(Ca^{2+})$, $C_1(Al^{3+})$	Er^{3+} 0.7–1	1.5500	77	P	180 [f]	0.4–1.5	–
		1.5815	77	P	250		
$CaAl_{12}O_{19}$ (?)	Nd^{3+} 1–2	1.0497	300	P	50 [f]	0.5–0.9	0.34
$CaSc_2O_4$ D_{2h}^{16}–Pnam	Nd^{3+} 1	1.0720	300	P	2.3 [f]	0.5–0.9	0.105
		1.0755	300	P	2.7		
		1.0868	300	P	1		
		1.0730	77	P	0.7		0.105
		1.0867	77	P	0.4		
	1	1.3565	300	P	1.5 [f]	0.5–0.9	0.105
		1.3870	77	P	6		0.105
$Ca_3(VO_4)_2$ C_{2h}^6–B2/b (C2/c)	Nd^{3+} ≈ 2	1.067 [a]	300	P	3000	0.4–0.9	0.15

[a] λ_g value requires more-accurate definition.
[f] The composition of the stimulated-emission spectrum and the threshold energies depend on the mutual orientation of the geometric axis of the rod F (axis of the laser element) and of the crystallographic directions of the crystal.

Luminescence linewidth [cm⁻¹]	Laser transition	Energy terminal level [cm⁻¹]	Excitation system, pump source, filter, cooling	Rod length, rod diameter [mm]; resonator cavity parameters, mirrors (coating, transmittance)	Remarks	Reference
	$^4F_{3/2} \rightarrow {}^4I_{13/2}$					
16		≈ 4100	Low- and room-	25; 6		[88, 95]
20		≈ 4300	temp. ER	$\parallel$, 5″, $R = 600$		
4		≈ 4100	with Xe flash-	mm, DM, 1%		
6		≈ 4100	tube IFP–400, QLP, ZhS–17, N_2			
	$^4I_{13/2} \rightarrow {}^4I_{15/2}$					
6.8		≈ 400	Low-temp.	32; 6		[95]
8		≈ 500	ER with Xe flashtube IFP–400, QLP, N_2	$\parallel$, 5″, $R = 600$ mm, DM, 1%		
	$^4F_{3/2} \rightarrow {}^4I_{11/2}$					
≈ 20		≈ 2000	ER with Xe flashtube IFP–400, ZhS–17	10 $\parallel$, 10″, $R = 576$ mm, DM, 1%	$n \approx 1.7$	[96]
	$^4F_{3/2} \rightarrow {}^4I_{11/2}$					
≈ 22		≈ 2100	Low- and room-	25; 5	$\Delta v_g = 5$ cm⁻¹	[488, 489]
≈ 22		≈ 2100	temp. ER	$\parallel$, 2″, $R = 576$	$\Delta v_g = 1.5$ cm⁻¹	
≈ 25		≈ 2250	with Xe flash-	mm, DM, 1%	$\Delta v_g \approx 1.5$ cm⁻¹	
3.5		≈ 2100	tube IFP–400,		$\Delta v_g < 0.5$ cm⁻¹	
4.5		≈ 2250	QLP, ZhS–17, N_2		$\Delta v_g < 0.5$ cm⁻¹	
	$^4F_{3/2} \rightarrow {}^4I_{13/2}$					
≈ 30		≈ 4050	Low- and room-	25; 5		[488, 489]
≈ 12		≈ 4200	temp. ER with Xe flash- tube IFP–400, QLP, ZhS–17, N_2	$\parallel$, 2″, $R = 600$ mm, DM, 1%		
	$^4F_{3/2} \rightarrow {}^4I_{11/2}$					
≈ 160		≈ 2000	Illuminator LS–3 with Xe flashtube FT–100–B	52; 3.2–6.3 $\parallel$, Ag, 5%	$\Delta v_g = 20$ cm⁻¹ ECC–Na^+ $n = 1.89$ $K = 14$	[382]

(*Continued*)

Table 5.1 (*continued*)

Crystal host, space group, cation-site symmetry	Activator [%]	Emission wavelength [μm]	Temperature [K]	Mode of operation	Threshold energy [J] or power [W]	Excitation spectral range [μm]	Luminescence lifetime [ms]
$Ca(NbO_3)_2$ D_{2h}^{14}–Pcan (Pbcn) $C_2(Ca^{2+})$	Pr^{3+} –	≈ 1.04 [e]	77	P	20–25	0.4–1	–
$Ca(NbO_3)_2$ D_{2h}^{14}–Pcan (Pbcn) $C_2(Ca^{2+})$	Nd^{3+} 1–2	1.0615 A 1.0612 A	300 77	P P	1[f] 0.8	0.5–0.9	0.12 0.12
		1.0615 A 1.0612 A	300 77	CW CW	350 400		
		1.0615→ →1.0625 A	300→650	P	–		
		1.0588*N_3 1.0612*A 1.0614*N_1 1.0626*N_2	77 77 77 77	P P P P	45 5.5 18 35		
	1	1.3380 1.3425 1.3370 1.3415	300 300 77 77	P P P P	4.5 6 8 20	0.5–0.9	0.12 0.12
$Ca(NbO_3)_2$ D_{2h}^{14}–Pcan (Pbcn) $C_2(Ca^{2+})$	Ho^{3+} 0.5	2.047	77	P	90[f]	–	2.2
$Ca(NbO_3)_2$ D_{2h}^{14}–Pcan (Pbcn) $C_2(Ca^{2+})$	Er^{3+} –	≈ 1.61 [a]	77	P	800 [f]	–	–

[a] λ_g value requires more-accurate definition.
[e] The detailed data are not available. The results are doubtful and require confirmation or correction.

Luminescence linewidth [cm^{-1}]	Laser transition	Energy terminal level [cm^{-1}]	Excitation system, pump source, filter, cooling	Rod length, rod diameter [mm]; resonator cavity parameters, mirrors (coating, transmittance)	Remarks	Reference
–	$^1G_4 \rightarrow {}^3H_4$	–	Illuminator with Xe flash-tube FT–524, N$_2$	30; 3 R, Ag, 2%	ECC–Ti^{4+}	[383]
10 4.5	$^4F_{3/2} \rightarrow {}^4I_{11/2}$	1949 1952	Low- and room-temp. ER with Xe flash-tube IFP–400, QLP, ZhS–17, N$_2$	11; 3.5 $\parallel$, $R = 576$ mm, DM, 0.7%	Without ECC	[34, 172, 173, 212, 213, 383]
10 4.5		1949 1959	Low- and room-temp. ER with Xe lamp DKsTV–3000, QLP, ZhS–17, water and N$_2$	30; 4 $\parallel$, 15″, $R = 576$ mm, DM, 0.7%	$P_g = 1$ W	
10→		1949	High-temp. ER with Xe flashtube IFP–800, ZhS–17	20; 5 $\parallel$,15″,$R = 576$ mm, DM, 0.7%	$\delta\lambda/\delta T = 2.9$	
– 4.5 – –		– 1952 – –	ER with Xe flashtube IFP–400, QLP, ZhS–17, N$_2$	–	*In CAM laser using Ca(NbO$_3$)$_2$–Nd^{3+} and BaF$_2$–LaF$_3$––Nd^{3+} or neodymium glass LGS–2 (300 K)	
13 16 7 10	$^4F_{3/2} \rightarrow {}^4I_{13/2}$	3898 ≈3970 3896 3969	Low- and room-temp. ER with Xe flash-tube IFP–400, QLP, ZhS–17, N$_2$	15; 6 $\parallel$, $R = 600$ mm, DM, 1%		[93]
–	$^5I_7 \rightarrow {}^5I_8$	–	Illuminator with Xe flash-tube FT–524, N$_2$	30; 3 R, Ag, 2%	ECC–Ti^{4+}	[383]
–	$^4I_{13/2} \rightarrow {}^4I_{15/2}$	–	Illuminator with Xe flash-tube FT–524,N$_2$	30; 3 R, Ag, 2%	ECC–Ti^{4+}	[383]

ᶠ The composition of the stimulated-emission spectrum and the threshold energies depend on the mutual orientation of the geometric axis of the rod F (axis of the laser element) and of the crystallographic directions of the crystal.

(Continued)

Table 5.1   (*continued*)

Crystal host, space group, cation-site symmetry	Activator [%]	Emission wavelength [μm]	Temperature [K]	Mode of operation	Threshold energy [J] or power [W]	Excitation spectral range [μm]	Luminescence lifetime [ms]
$Ca(NbO_3)_2$ D_{2h}^{14}–Pcan (Pbcn) $C_2(Ca^{2+})$	Tm^{3+} –	≈ 1.91 [a]	77	P	125 [f]	–	–
$CaMoO_4$ C_{4h}^6–$I4_1/a$ $S_4(Ca^{2+})$	Nd^{3+} 1.8	1.061	300	CW P	1200 [f] 1	0.4–0.9	0.12
		1.0673	77 300	CW P	1200 7		–
	1.3	1.0573	295	P	10		
$CaMoO_4$:Er^{3+} C_{4h}^6–$I4_1/a$ $S_4(Ca^{2+})$	Ho^{3+} 0.5	2.0556 2.0707 2.074 2.074	77 77 77 20	P P P P	310 [f] 170 107 45	0.4–2 Excitation into Ho^{3+} and Er^{3+} bands	1.3
$CaMoO_4$:Er^{3+} C_{4h}^6–$I4_1/a$ $S_4(Ca^{2+})$	Tm^{3+} 0.5	1.9060 1.9115	77 77	P P	20 [f] 19	0.4–1.9 Excitation into Ho^{3+} and Tm^{3+} bands	0.9
$CaWO_4$ C_{4h}^6–$I4_1/a$ $S_4(Ca^{2+})$	Pr^{3+} 0.5	1.0468 [14]	20–90	P	20	0.4–0.6	0.023
$CaWO_4$ C_{4h}^6–$I4_1/a$ $S_4(Ca^{2+})$	Nd^{3+} 0.5–3	(σ) 0.9145	77	P	4.6	0.4–0.9	0.18 (0.5%)

[a] λ_g value requires more-accurate difinition.

[f] The composition of the stimulated-emission spectrum and the threshold energies depend on the mutual orientation of the geometric axis of the rod F (axis of the laser element) and of the crystallographic directions of the crystal.

[14] In [385, 452] this line is attributed to the $^1G_4 \rightarrow {}^3H_4$ transition. In [452] the E_{thr} value for the 1.047 μm line at 77 K is equal to 12.8 J. A rod 40 mm in length and 5 mm in diameter with plane-parallel silvered end faces was used.

Luminescence linewidth [cm^{-1}]	Laser transition	Energy terminal level [cm^{-1}]	Excitation system, pump source, filter, cooling	Rod length, rod diameter [mm]; resonator cavity parameters, mirrors (coating, transmittance)	Remarks	Reference
–	$^3H_4 \rightarrow {}^3H_6$	–	Illuminator with Xe flash-tube FT–524, N$_2$	30; 3 R, Ag, 2%	ECC–Ti^{4+}	[383]
–	$^4F_{3/2} \rightarrow {}^4I_{11/2}$	≈ 2000	ER with Hg lamp GEAH6, solution	25; 2.5 $R = 52$ mm, DM	ECC–Nb^{5+} $n_o \approx 1.957$	[183, 211, 384]
–		≈ 2000	NaNO$_3$ filter		$n_e \approx 1.963$ $K \approx 39$	
12		≈ 2000	Illuminator LS–3 with Xe flashtube ET–100 B, N$_2$ vapor	56; 6.5 $\parallel$, Ag, 5%	$\Delta v_g = 0.8$ cm^{-1}	
–	$^5I_7 \rightarrow {}^5I_8$	≈ 250	Illuminator with Xe flashtube FT–524, N$_2$ and Ne(?)	–	Er^{3+} = 0.75% EET to Ho^{3+} $n_o \approx 1.939$ $n_e \approx 1.945$	[21]
–	$^3H_4 \rightarrow {}^3H_6$	≈ 325	Illuminator with Xe flashtube FT– 524, N$_2$	–	Er^{3+} = 0.75% EET to Tm^{3+}	[21]
–	$^1D_2 \rightarrow {}^3F_4$	≈ 6700	Illuminator with Xe flashtube FT–524, Ne(?) and N$_2$	36; 3 R, Ag, 2%	$n_o \approx 1.89$ $n_e \approx 1.90$ $K \approx 40$	[385, 386]
15	$^4F_{3/2} \rightarrow {}^4I_{9/2}$	471	Illuminator with Xe flashtube FT–524, N$_2$	–	ECC–Na$^+$ $n_o \approx 1.898$ $n_e \approx 1.912$	[78]

(*Continued*)

Table 5.1 (*continued*)

Crystal host, space group, cation-site symmetry	Activator [%]	Emission wavelength [µm]	Temperature [K]	Mode of oper-ation	Threshold energy [J] or power [W]	Excitation spectral range [µm]	Lumines-cence lifetime [ms]
	0.5–3	(π) 1.0582 D	300	P	0.5	0.5–0.9	0.18
		(σ) 1.0652 A	300	P	1.3		
		(σ) 1.0649 A	77	P	0.5		0.18
		1.0587 N₁	77	P	1		–
		1.0601 N₂	77	P	1		–
		(σ) 1.0649 A	77	P	0.5		
		1.0634 M₁	77	P	80		–
		(σ) 1.0650 A	77	P	0.5		
		(π) 1.0582 D	300	CW	1200		
		(π) 1.0582 D	300	q–CW	–	0.4–0.9	
		(σ) 1.0652 A	300	q–CW	–		
		(π) 1.0582→ →1.0597 D	300→700	P	–	0.5–0.9	
	1 c ∥ F	(σ) 1.3340	300	P	4	0.5–0.9	0.18
		(σ) 1.3475	300	P	10		
		(σ) 1.3310	77	P	40		0.18
		(σ) 1.3459	77	P	15		
	1 c ⊥ F	(π) 1.3370	300	P	1		
		(π) 1.3390	300	P	1		
		(π) 1.3885*	300	P	–		
		(π) 1.3370	300	CW	1500	0.3–0.9	
		(π) 1.3345	77	P	1	0.5–0.9	
		(π) 1.3372	77	P	2		
		1.3459	77	P	2		
		(π) 1.3880	77	P	5		

Luminescence linewidth [cm^{-1}]	Laser transition	Energy terminal level [cm^{-1}]	Excitation system, pump source, filter, cooling	Rod length, rod diameter [mm]; resonator cavity parameters, mirrors (coating, transmittance)	Remarks	Reference
	$^4F_{3/2} \rightarrow {}^4I_{11/2}$					
≈ 20		2016	Low- and room-	30; 5	ECC–Na$^+$	[17, 32,
≈ 20		2016	temp. ER	$\parallel$, 20″, $R = 576$	$\sigma_e = 4 \times 10^{-19}$	78,178,
7		2016	with Xe flash-	mm, DM, 1%		184, 185,
7		–	tube IFP–800		ECC–Nb^{5+}	208, 209,
–		–	and IFP–400,			243, 387,
7		2016	QLP, ZhS–17,		Without ECC	389]
–		–	N$_2$			
–		2016				
≈ 20		2016	ER with Hg	50; 2	$P_g = 10$ mW	
			lamp GEAH6,	R, Ag, 1%	ECC–Na$^+$	
			solution			
			NaNO$_3$ filter,			
			water			
≈ 20		2016	ER with PS	42; 3	$\tau_g = 10$ ms	
≈ 20		2016		$R = 42$ mm,		
				DM, 1%		
20$\rightarrow$		2016	High-temp.	45; 4.5	$\delta\lambda/\delta T \approx 3.9$	
			ER with Xe	$\parallel$, 20″, $R = 576$		
			flashtube	mm, DM, 1%		
			IFP–800,			
			ZhS–17			
	$^4F_{3/2} \rightarrow {}^4I_{13/2}$					
20		≈ 3970	Low- and room-	45; 5	ECC–Na$^+$	[78, 93,
25		≈ 3970	temp. ER	$\parallel$, $R = 600$ mm,	$n_o \approx 1.87$	222]
5		3925	with Xe flash-	DM, 1%	$n_e \approx 1.90$	
7		3975	tube IFP–400,			
			QLP, ZhS–17, N$_2$			
23		≈ 3925	ER with Xe		ECC–Na$^+$	
25		≈ 4000	flashtube			
28		≈ 4200	IFP–400,		*In CAM laser	
			ZhS–17 and			
			illuminator			
			with Xe flash-			
			tube FT–524			
23		≈ 3925	ER with Xe	30; 5	ECC–Na$^+$	
			lamp DKsTV–			
			–3000, water			
8		3975	Low-temp.			
–		3928	ER with Xe			
7		3975	flashtube			
10		4202	IFP–400, QLP,			
			ZhS–17,N$_2$ and			
			illuminator			
			with Xe flash-			
			tube FT–524,			
			N$_2$			

(*Continued*)

Table 5.1 (*continued*)

Crystal host, space group, cation-site symmetry	Activator [%]	Emission wavelength [μm]	Temperature [K]	Mode of oper- ation	Threshold energy [J] or power [W]	Excitation spectral range [μm]	Lumines- cence lifetime [ms]
$CaWO_4$ $C^6_{4h}-I4_1/a$ $S_4(Ca^{2+})$	Ho^{3+} 0.5	2.046 2.059	77 77	P P	80 [f] 250	0.4–2	–
$CaWO_4$ $C^6_{4h}-I4_1/a$ $S_4(Ca^{2+})$	Er^{3+} 1	1.612[a]	77	P	800 [f]	0.4–1.6	–
$CaWO_4$ $C^6_{4h}-I4_1/a$ $S_4(Ca^{2+})$	Tm^{3+} 0.3 0.5	1.911 1.916	77 77	P P	60 [f] 73	0.4–1.9	–
$SrAl_4O_7$ $C^6_{2h}-C2/c$ $C_{2v}(Sr^{2+})$, $C_1(Al^{3+})$	Nd^{3+} 0.7	1.0576 1.0828 1.0566 1.0568 1.0627	300 300 77 77 77	P P P P P	28 25 4.5 15 18	0.2–0.9	0.34 0.38
	0.7	1.3345 1.3665 1.3320 1.3530 1.3680*	300 300 77 77 77	P P P P P	80 [f] 120 25 30 –	0.2–0.9	0.34 0.38
$SrAl_{12}O_{19}$ (?)	Nd^{3+} 1.5	1.0491 1.0621*	300 300	P P	14 [f] –	0.5–0.9	≈0.4
	1.5	1.3065	300	P	30 [f]	0.5–0.9	≈0.4
$SrMoO_4$ $C^6_{4h}-I4_1/a$ $S_4(Sr^{2+})$	Pr^{3+} –	≈1.04 [d]	–	P	– [(15)]	–	–

[a] λ_g value requires more-accurate definition.

[d] The spectroscopic data confirming this result are not published.

[f] The composition of the stimulated-emission spectrum and the threshold energies depend on the mutual orientation of the geometric axis of the rod F (axis of the laser element) and of the crystallographic directions of the crystal.

[(15)] The emission channel of this crystal, according to the data of [386], requires more-accurate definition.

Luminescence linewidth [cm⁻¹]	Laser transition	Energy terminal level [cm⁻¹]	Excitation system, pump source, filter, cooling	Rod length, rod diameter [mm]; resonator cavity parameters, mirrors (coating, transmittance)	Remarks	Reference
– –	$^5I_7 \rightarrow \,^5I_8$	250 –	Illuminator with XE flash-tube FT–524, N_2	52; 3 R, Ag, 1%	$n_o \approx 1.879$ $n_e \approx 1.892$	[183, 390]
–	$^4I_{13/2} \rightarrow \,^4I_{15/2}$	375	Illuminator with Xe flash-tube FT–524, N_2	25.4 ×6 ×6	ECC–Li$^+$	[391]
– –	$^3H_4 \rightarrow \,^3H_6$	325 –	Illuminator with Xe flash-tube FT–524, N_2	52; 3 R, Ag, 1%		[183, 392]
≈20 ≈30 1.5 1.6 1.8	$^4F_{3/2} \rightarrow \,^4I_{11/2}$	≈2100 ≈2100	Low- and room-temp. ER with Xe flash-tube IFP–400, QLP, N_2	20; 6 ‖, 5″, $R = 576$ mm, DM, 1%	$\Delta v_g = 2$ cm⁻¹ $\Delta v_g = 4$ cm⁻¹ $\Delta v_g = 0.5$ cm⁻¹ $\Delta v_g = 0.5$ cm⁻¹ $\Delta v_g = 0.5$ cm⁻¹ $n \approx 1.62$	[88, 96]
≈20 ≈33 1.7 2 2.3	$^4F_{3/2} \rightarrow \,^4I_{13/2}$	≈4100 ≈4100 ≈4000	Low- and room-temp. ER with Xe flash-tube IFP–400, QLP, N_2	20; 6 ‖, 5″, $R = 600$ mm, DM, 1%	*In CAM laser	[88, 96]
≈13 ≈13	$^4F_{3/2} \rightarrow \,^4I_{11/2}$	≈2000 ≈2100	ER with Xe flashtube IFP–400, ZhS–17	13 ‖, 10″, $R = 576$ mm, DM, 1%	$\Delta v_g = 2$ cm⁻¹ *In CAM laser	[96, 489]
≈15	$^4F_{3/2} \rightarrow \,^4I_{13/2}$	≈4000	ER with Xe flashtube IFP–400, ZhS–17	13 ‖, 10″, $R = 600$ mm, DM, 1%	$n \approx 1.7$	[489]
–	$^1G_4 \rightarrow \,^3H_4$	–	–	–	$n_o \approx 1.878$ $n_e \approx 1.880$	[69]

(Continued)

Table 5.1 (*continued*)

Crystal host, space group, cation-site symmetry	Activator [%]	Emission wavelength [μm]	Temperature [K]	Mode of operation	Threshold energy [J] or power [W]	Excitation spectral range [μm]	Lumines-cence lifetime [ms]
SrMoO₄ C_{4h}^6–I4₁/a $S_4(Sr^{2+})$	Nd³⁺ 1–3	1.0576 F	295	P	10	0.4–0.9	≈0.17
		1.0643 A	295	P	125		
		1.059ᵃ C	77	P	150		≈0.17
		1.0611 E	77	P	500		
		1.0627 D	77	P	170		
		1.0640 A	77	P	17		
		1.0652 B	77	P	70		
	≈2	1.0645 A	300	q–CW	–		
	1 $c \parallel F$	1.3325	300	P	35	0.4–0.9	≈0.17
		1.3300*	77	P	–		≈0.17
		1.3440	77	P	50		
		1.3790*	77	P	–		
SrWO₄ C_{4h}^6–I4₁/a $S_4(Sr^{2+})$	Nd³⁺ –	1.063ᵃ B	295	P	180 ᶠ	0.4–0.9	0.2
		1.0574 A	77	P	4.7		–
		1.0607 C	77	P	7.6		
		1.0627 B	77	P	5.1		
Y₂O₃ T_h^7–Ia3 $C_2(Y^{3+})$, $C_{3i}(Y^{3+})$	Nd³⁺ 1	1.073 ᵃ	77	P	260	0.4–0.9	0.26
		1.078 ᵃ	77	P	350		
	1	1.073ᵃ	300	P	–		
Y₂O₃ T_h^7–Ia3 $C_2(Y^{3+})$, $C_{3i}(Y^{3+})$	Eu³⁺ 5	0.6113	220	P	85	0.3–0.5	0.87

ᵃ λ_g value requires more-accurate definition.

ᶠ The composition of the stimulated-emission spectrum and the threshold energies depend on the mutual orientation of the geometric axis of the rod *F* (axis of the laser element) and of the crystallographic directions of the crystal.

[16] CW generation of single-crystal fibers (approximately 25 mm long and of various diameters from 60 to 200 μm) of Y₂O₃–Nd³⁺ (≈1.5 wt.%) was obtained at 300 K, due to the two $^4F_{3/2} \rightarrow {}^4I_{11/2}$ and $^4F_{3/2} \rightarrow {}^4I_{13/2}$ channels, excited by a krypton ion laser operating at 0.7525 μm.

Luminescence linewidth [cm⁻¹]	Laser transition	Energy terminal level [cm⁻¹]	Excitation system, pump source, filter, cooling	Rod length, rod diameter [mm]; resonator cavity parameters, mirrors (coating, transmittance)	Remarks	Reference
≈ 35	$^4F_{3/2} \rightarrow {}^4I_{11/2}$	≈ 2000	Illuminators	52; 3	ECC–Li$^+$	[183, 243, 384, 393]
5			with Xe flash-	R, Ag, 1%		
–		≈ 2000	tubes FT–524			
–			and FT–100 B,			
–			N$_2$ vapor and			
–			N$_2$			
15		≈ 2000	ER with PS	34; 6.4 $\parallel$, 20″, $R = 576$ mm, DM, 0.7%	$E_{\text{thr}} = 2$J in ER with Xe flashtube IFP–400	
20	$^4F_{3/2} \rightarrow {}^4I_{13/2}$	≈ 3960	Low- and room-	21; 5	ECC–Na$^+$	[97]
5		3948	temp. ER	$\parallel$, $R = 600$ mm,	*In CAM laser	
6		≈ 3962	with Xe flash-	DM, 1%	$n_o \approx 1.872$	
12		4150	tube IFP–400, QLP, N$_2$		$n_e \approx 1.875$	
–	$^4F_{3/2} \rightarrow {}^4I_{11/2}$	≈ 2000	Illuminator	52; 3		[183]
3		≈ 2000	with Xe flash-	R, Ag, 1%		
–			tube FT–524,			
–			N$_2$			
2	$^4F_{3/2} \rightarrow {}^4I_{11/2}$	1892	Cryostat in	2; 2	$n \approx 1.91$	[144, 394, 653]
2			helical Xe	$\parallel$, DM, 1%	$K = 270$ (16)	
12		1892	flashtube, N$_2$	–	SE excited up to 370 K	
			–			
28	$^5D_0 \rightarrow {}^7F_2$	859	Illuminator with four Xe flashtubes GE–100, N$_2$ vapor	12; 1 $\parallel$, 30″, 3%		[395]

(Continued)

Table 5.1 *(continued)*

Crystal host, space group, cation-site symmetry	Activator [%]	Emission wavelength [μm]	Temperature [K]	Mode of operation	Threshold energy [J] or power [W]	Excitation spectral range [μm]	Lumines-cence lifetime [ms]
YAlO$_3$ D_{2h}^{16}–Pbnm (Pnma) C_s(Y^{3+})	Nd^{3+} 1 $c \parallel$ F	0.930	300	P	0.15	LP	0.18
	1–3	1.0796 A	300	P	<0.5 ^{f}	0.5–0.9	0.18
		1.0644 C	300	P	<1		(0.3%)
		1.06405 C	77	P	<1		0.18
		1.07255 D	77	P	<1		
		1.0585* B	300	P	25		
		1.0644* C	300	P	7		
		1.0726* B	300	P	7		
		1.0796* A	300	P	4		
	$b \parallel$ F	1.0644 C	300	CW	1800		
		1.0796 A	300	CW	1200		
		1.06405→ →1.0654 C	77→500	P	–		
		1.0652→ →1.0659 E	310→ →500	P	–		
		1.07255→ →1.0730 D	77→490	P	–		
		1.0796→ →1.0803 I	600→ →700	P	–		
		1.07955→ →1.0802 A	77→700	P	–		
		1.0842 G	300	P	–		
		1.0847 G	530	P	–		
		1.0913 H	530	P	–		
		1.0990 F	300	P	–		
		1.0991 F	500	P	–		
	0.9 $b \parallel$ F	1.3393[17]	300	P	10	0.5–0.9	0.18
				CW	–		
		1.3413[17]	300	P	13		
				CW	–		

^{f} The composition of the stimulated-emission spectrum and the threshold energies depend on the mutual orientation of the geometric axis of the rod F (axis of the laser element) and of the crystallographic directions of the crystal.

[17] In [98], λ_g values of 1.3391 and 1.3411 μm are given for these lines.

Luminescence linewidth [cm⁻¹]	Laser transition	Energy terminal level [cm⁻¹]	Excitation system, pump source, filter, cooling	Rod length, rod diameter [mm]; resonator cavity parameters, mirrors (coating, transmittance)	Remarks	Reference
	$^4F_{3/2} \rightarrow {}^4I_{9/2}$					
30		670	Xe ion laser	$5 \times 8.3 \times 16.6$ ‖, 10″, EDAC at $\lambda = 1.06\ \mu m$, DM, 6%	$n \approx 1.96$ $K = 110 - 140$	[79, 272]
	$^4F_{3/2} \rightarrow {}^4I_{11/2}$					
9.8		2157	Low- and room-	15–43; 5–6		[10, 34,
6.5		2026	temp. ER	‖, 10″, $R = 576$		98, 176,
1.6		2023	with Xe flash-	mm, DM, 1%	$\Delta v_g = 1\ cm^{-1}$	220, 272,
2		2097	tube IFP–400,			396–399,
			QLP, ZhS–17,			492, 700,
10		2097	N₂		*In CAM laser	751]
6.5		2026				
8.5		2097				
9.8		2157				
6.5		2026	Double ER with	75; 6	$P_g = 35\ W$	
			Kr lamps, water	‖, EDAC at	$\eta_g = 0.8$	
9.8		2157		$\lambda = 1.07\ \mu m$,	$P_g = 100\ W$	
				$d_r = 400\ mm$,	$\eta_g = 2.0$	
				DM, 3%		
1.6→		2030	Low- and high-	14; 4	$\delta v/\delta T = 4.5$	
			temp. ER	‖, 3″, $R = 576$		
–		2156	with Xe flash-	mm, DM, 1%	$\delta v/\delta T = 3.6$	
			tubes IFP–400	Crystal con-		
2→19		2097	and IFP–800,	sisted of	$\delta v/\delta T = 2.6$	
			QLP, ZhS–17,	two disorien-		
–		2270	N₂ and its	ted blocks	$\delta v/\delta T = 6.4$	
			vapor			
2.5→		2156			$\delta v/\delta T = 1.7$	
14.2		2322				
–		2318				
–		2373				
14		2322				
18		2318				
	$^4F_{3/2} \rightarrow {}^4I_{13/2}$					
7.5		3955	Illuminator	76; 6.3		[93, 98,
			with Kr lamp,	‖, EDAC at		222,
			water	$\lambda = 1.07\ \mu m$,		591, 751]
10.5		3955		DM, 2%	$P_g = 14\ W$	
					$\eta_g = 0.4$	

(Continued)

Table 5.1 (*continued*)

Crystal host, space group, cation-site symmetry	Activator [%]	Emission wavelength [μm]	Temperature [K]	Mode of oper-ation	Threshold energy [J] or power [W]	Excitation spectral range [μm]	Lumines-cence lifetime [ms]
	$b \parallel$ F	$1.3413^{(17)}$	300	P	1		
				CW	1500		
		1.3512*	300	P	–		
		1.3391	77	P	0.5		0.18
		1.3514	77	P	1		
		1.3644	77	P	5		
		1.3849*	77	P	–		
		1.4026*	77	P	–		
$YAlO_3$ D_{2h}^{16}–Pbnm (Pnma) $C_s(Y^{3+})$	Ho^{3+} 2 $[112] \parallel$ F 10 $c \approx \parallel$ F	2.9180	300	P	7.5 ^{g}	0.5–1.2 $\tau_{exc} = 40 - 60$ μs	0.8–1.0
		3.0312	300	P	50 ^{g}		
$YAlO_3$ D_{2h}^{16}–Pbnm (Pnma) $C_s(Y^{3+})$	Er^{3+} 0.5–2 $[112] \parallel$ F	0.84975	300	P	100 ^{g}	0.35–0.55 $\tau_{exc} = 40 - 60$ μs	≈ 0.13
		0.8594	300	P	65 ^{g}		
		0.84965	77	P	≈ 20		
		0.85165	77	P	≈ 35		
	2 $[112] \parallel$ F	1.6632	300	P	20 ^{g}	0.35–0.55	≈ 0.13
	2 $[112 \parallel$ F	2.7309	300	P	40 ^{g}	0.5–1.0 $\tau_{exc} = 40 \mu$s	≈ 0.9
$YAlO_3$:Cr^{3+} D_{2h}^{16}–Pbnm (Pnma) $C_s(Y^{3+})$, $C_i(Al^{3+})$	Nd^{3+} 0.9 $a \parallel$ F $c \parallel$ F	1.0644	300	Q	9	0.4–0.9 Excitation into Nd^{3+} and Cr^{3+} bands	0.18
			300	CW	–		

^{g} The E_{thr} value depends on τ_{exc}.
$^{(17)}$ In [98], λ_g values of 1.3391 and 1.3411 μm are given for these lines.

Lumines-cence linewidth [cm^{-1}]	Laser transition	Energy terminal level [cm^{-1}]	Excitation system, pump source, filter, cooling	Rod length, rod diameter [mm]; resonator cavity parameters, mirrors (coating, transmittance)	Remarks	Reference
10.5		3955	ER with Xe lamps IFP–400 and DKsTV––5000, ZhS–17, water	25; 5 $\parallel$, $R = 600$ mm, DM, 1%		
9		4020	ER with Xe flashtube IFP–400, ZhS–17		*In CAM laser	
1.6		3953	Low-temp.			
1.8		4021	ER with Xe			
2.5		4092	flashtube			
4.0		4200	IFP–400, QLP,			
4.5		4291	ZhS–17, N$_2$			
≈ 7	$^5I_6 \rightarrow {}^5I_7$	≈ 5300	ER with Xe flashtube	15–47; 5–5.5 $\parallel$, 10″, $R = 600$		[547, 565, 655]
≈ 8		≈ 5600	ISP–250, ZhS–17	mm, DM, 1%		
6.5	$^4S_{3/2} \rightarrow {}^4I_{13/2}$	6641	ER with Xe	50; 5	$\sigma_e \approx 2 \times 10^{-20}$	[107, 547, 565]
13		6774	flashtube ISP–250	$\parallel$, 5″, $R = 600$ mm, DM, 1%		
1.5		6643	ER with Xe			
1.5		6671	flashtube ISP–250, TQC and QLP, N$_2$			
≈ 8	$^4S_{3/2} \rightarrow {}^4I_{9/2}$	12397	ER with Xe flashtube L–213 B or ISP–250	50; 5–6 $\parallel$, 5″, $R = 600$ mm, DM 0.1–1%	In [109, 110] EDAC at $\lambda = 1.4$–1.7 μm were used	[109, 110, 547, 565]
≈ 5.5	$^4I_{11/2} \rightarrow {}^4I_{13/2}$	6642	ER with Xe flashtube ISP–250, ZhS–17	50; 5 $\parallel$, 5″, $R = 600$ mm, DM, 1%		[547, 565]
6.5	$^4F_{3/2} \rightarrow {}^4I_{11/2}$	2026	Illuminator with Xe flash-tube L–213 B	50; 5 $\parallel$, EDAC at $\lambda = 1.06$ μm	Cr^{3+} = 0.3 at.% EET to Nd^{3+} $P_g = 6.5$ W $\eta_g = 0.3$	[127]

(Continued)

Table 5.1 (continued)

Crystal host, space group, cation-site symmetry	Activator [%]	Emission wavelength [μm]	Temperature [K]	Mode of operation	Threshold energy [J] or power [W]	Excitation spectral range [μm]	Luminescence lifetime [ms]
$YAlO_3$:Cr^{3+} D_{2h}^{16}–Pbnm (Pnma) $C_s(Y^{3+})$, $C_i(Al^{3+})$	Tm^{3+} 1.5 [112]∥F	1.856 [a]	≈90	P	–	0.3–1.7 Excitation into Tm^{3+} and Cr^{3+} bands	≈10
		1.883 [a]	≈90	P	8		
		1.933 [a]	≈90	P	–		
	0.8 [111]∥F	2.348 [a]	300	P	–	0.3–0.8 Excitation into Tm^{3+} and Cr^{3+} bands	0.63
		2.349 [a]	300	P	–		≈0.77*
	1.0 [112]∥F Cr^{3+}	(π) 2.274 [a]	300	P	31		
		(σ) 2.318 [a]	300	P	132		
		(π,σ) 2.353 [a]	300	P	110		
	0.25%	(π) 2.354 [a]	300	P	144		
		(σ) 2.355 [a]	300	P	140		
	1.0 [111]∥F Cr^{3+} 0.75%	(π) 2.274 [a]	300	P	94		
		(σ) 2.318 [a]	300	P	188		
		(π,σ) 2.353 [a]	300	P	325		
	1.0 [112]∥F Cr^{3+}	≈2.34 [a]	300	P	12.5	$\tau_{exc} = 100$ μs	≈0.2
			≈90	P	5 [g]		
$Y_3Al_5O_{12}$ O_h^{10}–Ia3d $D_2(Y^{3+})$	Nd^{3+} 1	0.8910	300	P	0.1	LP; 0.49–-0.54 $\tau_{exc} = 290$ μs 0.4–0.9	0.255 (0.3%)
		0.8999	300	P	0.1		
		0.9385	300	P	0.1		
		0.9460	300	P	38		
	0.3–-1.3	1.0615 B	300	P	100	0.5–0.9	0.255 (0.3%)
		1.06415A	300	P	0.1		
		1.0610 B	77	P	<0.5		0.255

[a] λ_g value requires more-accurate definition.

[g] The E_{thr} value depends on τ_{exc}.

[18] Here an effective (peak) cross section due to partial resonance of two A (1.06415 μm) and A' (1.0644 μm) transitions is given (see Chapter 6).

Luminescence linewidth [cm⁻¹]	Laser transition	Energy terminal level [cm⁻¹]	Excitation system, pump source, filter, cooling	Rod length, rod diameter [mm]; resonator cavity parameters, mirrors (coating, transmittance)	Remarks	Reference
	$^3H_4 \to {}^3H_6$				$Cr^{3+} = 0.15$ at.% EET to Tm^{3+}	
–		≈ 237	ER with Xe flashtube ISP–250, N_2 vapor	20; 5 $\parallel$, $R = 500$ mm, DM, 0.5%		[525]
–		≈ 313				
–		≈ 440				
	$^3F_4 \to {}^3H_5$				$Cr^{3+} = 0.1$ at.% EET to Tm^{3+}	[112, 525, 563]
–		≈ 8250	–	55; 5		
–		≈ 8250		$\parallel$, DM, 1%	$E_g = 0.01$ J	
–		≈ 8250	ER with Xe and Kr flash-	50–75, 3–5	for $E_{exc} = 550$ J	
–		≈ 8250	tubes	$\parallel$, $d_r = 250$	*Calculated	
–		≈ 8250	(ILC7L3,	mm, EDAC at	value [563]	
–		≈ 8250	ILC7L4 and	$\lambda = 2.3$ μm,		
–		≈ 8250	ILC L–3F1G)	DM, 1%		
–		≈ 8250				
–		≈ 8250				
–		≈ 8250	ER with Xe flashtube ISP–250, N_2 vapor	20; 5 $\parallel$, $R = 500$ mm, DM, 0.5% at $\lambda = 1.8$–2.4 μm	EET to Tm^{3+} In cascade operation scheme the emission was simultaneously observed at 1.9334 μm ($^3H_4 \to {}^3H_6$ transition)	
	$^4F_{3/2} \to {}^4I_{9/2}$					
–		200	Ar ion laser	13		[80, 81]
–		311		$\parallel$, DM, DR		
–		852				
9		852	Er with Xe flashtube	75; 3	$n = 1.822$	
	$^4F_{3/2} \to {}^4I_{11/2}$					
3.6		2002	Low- and room-	40; 5	$^{(18)} \sigma_e =$	[9, 33,
≈ 5		2110	temp. ER	$\parallel$, 5″, $R = 576$	$= 3.3 \times 10^{-19}$	38, 99,
≈ 1		2006	with Xe flash-	mm, DM, 1%	$n = 1.816$	167–170,
			tube IFP–400,		$n_2 = 4.08 \times 10^{-13}$	187, 216,
			QLP, ZhS–17, N_2			217, 228,

(Continued)

Table 5.1   (*continued*)

Crystal host, space group, cation-site symmetry	Activator [%]	Emission wavelength [μm]	Temperature [K]	Mode of oper- ation	Threshold energy [J] or power [W]	Excitation spectral range [μm]	Lumines- cence lifetime [ms]
		1.0521 C	300	P	25		
		1.0615 B	300	P	12		
		1.06415 A	300	P	9		
		1.0682 I	300	P	50		
		1.0737 D	300	P	20 [19]		
		1.0779 H	300	P	38		
	1 [111]$\parallel$F	1.06415 A	300	CW	0.007*	0.8	
		1.0521 C	300	CW	594	0.4–0.9	
		1.0615 B	300	CW	330		
		1.06415 A	300	CW	100		
		1.0737 D	300	CW	348		
		1.1119 C	300	CW	623		
		1.1158 E	300	CW	646		
		1.1225 G	300	CW	676		
	1	1.06415 A	300	CW	–		
		1.06415 A	300	CW	–	Solar emis- sion	
		1.06415 A	300	q–CW	–	0.4–0.9	
		1.0637→ →1.0670 A	170→ →900	P	–	0.5–0.9	
		1.0610→ →1.0627 B	77→600	P			

[19] In [601] the λ_g value given for the D line is 1.0738 μm.

[20] In [629] *Marling* discusses the room temperature CW laser action of a $Y_3Al_5O_{12}$–Nd^{3+} crystal at 10 lines in the principal $^4F_{3/2} \rightarrow {}^4I_{11/2}$ channel, and at 9 lines in the additional $^4F_{3/2} \rightarrow {}^4I_{13/2}$ channel; moreover significant output powers were obtained at many lines. In this work, a dispersion resonator with single-tuned etalons was used. Unfortunately, we were unable to include any of these results either in this or any other tables of this book.

Luminescence linewidth [cm⁻¹]	Laser transition	Energy terminal level [cm⁻¹]	Excitation system, pump source, filter, cooling	Rod length, rod diameter [mm]; resonator cavity parameters, mirrors (coating, transmittance)	Remarks	Reference
4.5		2002	Illuminator with Xe flash-tube	50; 5 $\parallel$, EDAC at $\lambda = 1.06\ \mu m$, DR, $d_r = 600$ mm	Resonator dispersion 0.002 μm (angular minute)$^{-1}$	[242, 271, 401–405, 407, 601, 623, 629[20]]
3.6		2002				
≈5		2110				
6.5		2146				
4.6		2110				
7		2146				
≈5		2110	Heterostructure LED (AlGaAs), LPG	5; 0.08 (microelement), $\parallel$, DM	Laser threshold current ≈ 45 mA	
4.5		2002	ER with W lamp, water	28; 2.5 $\parallel$, two types of DR		
3.6		2002				
≈5		2110				
4.6		2110				
10.2		2461				
10.6		2514				
9.9		2514				
≈5		2110	Special illuminator with Kr lamps, water	Active element consisted of five crystals	$P_g \approx 750$ W $\eta_g = 1.7$	
≈5		2110	Parabolic Al mirror with diameter 610 mm and conical condenser, water	24; 3 $R = 50$ mm, DM, 0.5%	$P_g = 1-1.5$ W	
≈5		2110	Special PS-illuminator	80; 4.5 $\parallel$, 10″, $R = 576$ mm, DM, 6%	$E_g = 1$ J $\tau_g = 7$ ms	
3.5→		2110	Low- and high-temp. ER with Xe flash-tube IFP–800, QLP, ZhS–17,	40; 5 $\parallel$, 5″, $R = 576$ mm, DM, 1%	$\delta\lambda/\delta T = 4.85$ For ≈225 K E_{thr} (A) = $= E_{thr}$ (B)	
1→16		2002	N$_2$ and its vapor		$\delta\lambda/\delta T \approx 4$ In CAM laser	

(*Continued*)

Table 5.1   *(continued)*

Crystal host, space group, cation-site symmetry	Activator [%]	Emission wavelength [μm]	Temperature [K]	Mode of operation	Threshold energy [J] or power [W]	Excitation spectral range [μm]	Luminescence lifetime [ms]
	1	1.3187	300 [21]	P	1.2	0.5–0.9	0.255
		1.3335	300 [21]	P	–		(0.3%)
		1.3351	300 [21]	P	–		
		1.3381	300 [21]	P	1.5		
		1.3533	300 [21]	P	–		
		1.3572	300 [21]	P	1.9		
		1.3187	300	CW	457		
		1.3550	77	P	1		
	1	1.833 [a]	230	P	7.5	0.5–0.9	0.255
			293	P	9		(0.3%)
$Y_3Al_5O_{12}$ O_h^{10}–Ia3d $D_2(Y^{3+})$	Gd^{3+} ≈ 2	0.3146 [e]	300	P	2400	–	0.008
$Y_3Al_5O_{12}$ O_h^{10}–Ia3d $D_2(Y^{3+})$	Ho^{3+} 2	2.0914 B	77	P	1760	0.4–2	4–5
		2.0975 A	77	P	44		
		2.1223 C	77	P	400		
	≈ 10	2.9403	300	P	25 [g]	0.4–1.2 $\tau_{exc} = 100$ μs	0.06

[a] λ_g value requires more-accurate definition.

[e] The detailed data are not available. The results are doubtful and require confirmation or correction.

[g] The E_{thr} value depends on τ_{exc}.

[20] In [629] *Marling* discusses the CW laser action of a $Y_3Al_5O_{12}$–Nd^{3+} crystal laser at 300 K in the principal $^4F_{3/2} \rightarrow {}^4I_{11/2}$ channel (10 lines), in the additional $^4F_{3/2} \rightarrow {}^4I_{13/2}$ channel (9 lines); moreover significant output powers were obtained at many lines. In this work, a dispersion resonator with single-tuned etalons was used. Unfortunately, we were unable to include any of these results either in this or any other tables of this book.

[21] The λ_g values for the $^4F_{3/2} \rightarrow {}^4I_{13/2}$ transition, at 300 K are given in [99] as: 1.318; 1.330; 1.335; 1.338; 1.353 and 1.358 μm.

Luminescence linewidth [cm⁻¹]	Laser transition	Energy terminal level [cm⁻¹]	Excitation system, pump source, filter, cooling	Rod length, rod diameter [mm]; resonator cavity parameters, mirrors (coating, transmittance)	Remarks	Reference
	$^4F_{3/2} \to {}^4I_{13/2}$					
≈4		3924	Illuminator	52; 6	$n \approx 1.81$	[99, 100,
≈3		3924	with Xe flash-	$\parallel$, DM, 5%,		271, 567,
3,3		3933	tube, filter	DR		591,
4		4034				629 [20]]
≈4		4034				
≈4		4055				
≈4		3924	Ex-focal ellip-soid with Xe lamp	DR	$P_g = 0.03$ W	
2.5		4047	Low-temp. ER with Xe flashtube IFP– 400, QLP, ZhS–17, N₂	40; 5 $\parallel$, $R = 600$ mm, DM, 1%		
	$^4F_{3/2} \to {}^4I_{15/2}$					
–		5965	ER with Kr	50; 3	$n \approx 1.80$	[101, 629]
≈8			flashtube, filter	$\parallel$, DM, 2%		
	$^6P_{7/2} \to {}^8S_{7/2}$					
≈20		0	Illuminator with two XE flashtubes IFP–2000	20; 3 $\parallel$, Ag		[285]
	$^5I_7 \to {}^5I_8$					
–		532	Illuminator	≈25	$n \approx 1.80$	[128, 135]
–		462	with Xe flash-			
–		518	tube FT–524, N₂			
	$^5I_6 \to {}^5I_7$					
≈15		≈5400	ER with Xe flashtube ISP–250	25–40; 5 $\parallel$, 10″, $R = 600$ mm, DM, 1%	$n \approx 1.79$	[547, 565]

(*Continued*)

Table 5.1 (*continued*)

Crystal host, space group, cation-site symmetry	Activator [%]	Emission wavelength [μm]	Temperature [K]	Mode of operation	Threshold energy [J] or power [W]	Excitation spectral range [μm]	Luminescence lifetime [ms]
$Y_3Al_5O_{12}$ O_h^{10}–Ia3d $D_2(Y^{3+})$	Er^{3+} ≈ 20 2.5	0.86275 0.86275	77 300	P P	80 ^{g} 45	0.3–0.55 $\tau_{exc} = 40$–60 μs	0.12 0.12 (0.3%)
	1	1.6452 B 1.6602 A	77 77	P P	470 80	0.4–1.6	≈ 8
	0.4	1.6449 [22]	295	P	75–200		9.1
	1.2	1.632 ^{a}	300	P	320		6.5
	≈ 1.2	1.7757	300 300	P Q	22 ^{g} 250	0.3–0.55	0.12
	2.5– –10	1.7757	300	P	300	0.3–0.55 $\tau_{exc} = 40$ μs	0.12
	10–20	2.8302 2.9364	300 300	P P	≈ 15 ^{g} ≈ 15	0.35–1.2 $\tau_{exc} = 40$ μs	0.11 ($\approx 1\%$)
	2.5	2.9363	300	P	≈ 90		
$Y_3Al_5O_{12}$ O_h^{10}–Ia3d $D_2(Y^{3+})$	Tm^{3+} –	1.8834 B 2.0132 C 2.0132 C	77 77 85	P P CW	590 208 315	0.4–2	15

^{a} λ_g value requires more-accurate definition.
^{g} The E_{thr} value depends on τ_{exc}.

Luminescence linewidth [cm^{-1}]	Laser transition	Energy terminal level [cm^{-1}]	Excitation system, pump source, filter, cooling	Rod length, rod diameter [mm]; resonator cavity parameters, mirrors (coating, transmittance)	Remarks	Reference
	$^4S_{3/2} \to {}^4I_{13/2}$					
6.5		6805	Low- and room-	25–46; 4–6		[546, 547,
≈ 10		6800	temp. ER with Xe flash-tubes ISP–250 or IFP–400, QLP, N$_2$	$\parallel$, 10″, $R = 600$ mm, DM, 1%		565]
	$^4I_{13/2} \to {}^4I_{15/2}$					
–		525	Illuminator with Xe flash-tube FT–524, N$_2$	≈ 25		[128, 135, 147, 537]
≈ 20		459	Illuminator with Xe flash-tube	75; 6 $\parallel$, DM, 20%	$\Delta\nu_g = 2.2$ cm^{-1} $E_g = 0.03$ J Yb^{3+} = 5 wt.%	
–		426	ER with Xe flashtube IFP–400	57; 5 $\parallel$, DM		
	$^4S_{3/2} \to {}^4I_{9/2}$					
–		12766	ER with Xe flashtube	57; 5 $\parallel$, DM, electro-optical element (LiNbO$_3$)	For $E_{exc} = 280$ J $E_g = 0.036$ J	[537, 546, 547, 565]
		12766	ER with Xe flashtube ISP–250	26–46, 4–6 $\parallel$, 10″, $R = 600$ mm, DM, 1%		
	$^4I_{11/2} \to {}^4I_{13/2}$					
≈ 8.5		≈ 6880	ER with Xe	40–80, 5–6	$n \approx 1.79$	[490, 546,
≈ 9		≈ 6880	flashtubes ISP and IFP types	$\parallel$, 10″, DM, 1%		572, 655]
≈ 9		6880		46; 6 $\parallel$, 10″, $R = 600$ mm, DM, 1%		
	$^3H_4 \to {}^3H_6$					
–		240	Under P condi-	≈ 25		[128, 135]
–		582	tions, illumi-			
–		582	nator with Xe flashtube FT–524; under CW conditions, with W lamp, O$_2$ supercooled with N$_2$			

[22] According to measurements by *Thornton* et al. (see [437]) the Y$_3$Al$_5$O$_{12}$: Yb^{3+}–Er^{3+} crystal has $\lambda_g = 1.6459$ μm at 300 K.

(*Continued*)

Table 5.1 (*continued*)

Crystal host, space group, cation-site symmetry	Activator [%]	Emission wavelength [μm]	Temperature [K]	Mode of operation	threshold energy [J] or power [W]	Excitation spectral range [μm]	Luminescence lifetime [ms]
$Y_3Al_5O_{12}$ O_h^{10}–Ia3d $D_2(Y^{3+})$	Yb^{3+} 0.7	1.0293 [23]	77	P	9	0.9–1.1	1.12
$Y_3Al_5O_{12}$ O_h^{10}–Ia3d $C_{3i}(Al^{3+})$	Cr^{3+} 0.5	0.6874 R_1	≈ 77	P	300	–	9
$Y_3Al_5O_{12}$: Cr^{3+} Nd^{3+} O_h^{10}–Ia3d $D_2(Y^{3+})$, $C_{3i}(Al^{3+})$	1.3	1.0641 A	300	P	1	0.4–0.9 Excitation into Nd^{3+} and Cr^{3+} bands	0.23 (77 K) 0.21 (300 K)
		1.0615 B	300 300	CW CW	800 * 750 **		
		1.0610 B	77 77	CW CW	440 * 180 **		
		1.06415 A	300	CW	–	Solar emission	
$Y_3Al_5O_{12}$: Cr^{3+} Ho^{3+} O_h^{10}–Ia3d $D_2(Y^{3+})$, $C_{3i}(Al^{3+})$	–	2.0975 A	77 85	P CW	25 210	0.4–2 Excitation into Ho^{3+} and Cr^{3+} bands	–
		2.1223 C	77 85	P CW	25 250		
$Y_3Al_5O_{12}$: Cr^{3+} Tm^{3+} O_h^{10}–Ia3d $D_2(Y^{3+})$, $C_{3i}(Al^{3+})$	–	2.0132 C	77 77	P CW	30 160	0.4–2 Excitation into Tm^{3+} and Cr^{3+} bands	–
		2.019 D	295	P	640		

[11] A copper pipe was used for crystal cooling.
[23] According to [128] $\lambda_g = 1.0296$ μm.

Lumines-cence linewidth [cm^{-1}]	Laser transition	Energy terminal level [cm^{-1}]	Excitation system, pump source, filter, cooling	Rod length, rod diameter [mm]; resonator cavity parameters, mirrors (coating, transmittance)	Remarks	Reference
≈ 7	$^2F_{5/2} \rightarrow {}^2F_{7/2}$	612	Low-temp. ER with Xe flashtube IFP–400, QLP, N$_2$	24; 6 $\parallel$, 5″, $R = 576$ mm, DM, 2%	$\sigma_e = 1.4 \times 10^{-19}$ $\Delta v_g = 1$ cm^{-1}	[86, 128, 483]
1.4	$^2E(\bar{E}) \rightarrow {}^4A_2$	0	Illuminator with Xe flash-tubes [11]	83; 10 $\parallel$	$n \approx 1.829$	[400, 701, 702]
—	$^4F_{3/2} \rightarrow {}^4I_{11/2}$	2110	Illuminator with Xe flash-tube FT–524	30; 3 $R = 52$ mm, DM	Cr^{3+} = 1% EET to Nd^{3+}	[126, 381]
—		2002	Illuminator with CW lamps, solution NaNO$_3$ filter, water and N$_2$		*With GEAH6 lamp **With W lamp	
—		2006				
—		2110	Mirror with diameter 624 mm and conical condenser	—	For multimode condition $P_g = 4.8$ W and for TEM$_{00}$ mode condition $P_g = 0.8$ W	
—	$^5I_7 \rightarrow {}^5I_8$	≈ 460	Under P condi-tions, illumina-tor with Xe flashtube FT––524; under CW conditions, with W lamp, O$_2$ supercooled with N$_2$	≈ 25	Cr^{3+} = 0.5% EET to Ho^{3+}	[128]
—		518				
—						
—	$^3H_4 \rightarrow {}^3H_6$	582	Under P condi-tions, illumina-tor with Xe flashtube FT––524; under CW conditions, with W lamp, N$_2$	≈ 25	Cr^{3+} = 0.5% EET to Tm^{3+}	[128]
—		600				

(*Continued*)

Table 5.1   (*continued*)

Crystal host, space group, cation-site symmetry	Activator [%]	Emission wavelength [μm]	Temperature [K]	Mode of operation	Threshold energy [J] or power [W]	Excitation spectral range [μm]	Luminescence lifetime [ms]
	1	2.324 [a]	300	P	200	0.35–0.8 Excitation into Tm^{3+} and Cr^{3+} bands	0.79 [10]
$Y_3Al_5O_{12}$: Nd^{3+} Yb^{3+} O_h^{10}–Ia3d $D_2(Y^{3+})$	≈ 2	1.0293 1.0297	77 200	P P	4.5 145	0.5–1.1 Excitation into Nd^{3+} and Yb^{3+} bands	1.12 1.2
	Nd^{3+} 0.8	1.0612 B	200	P	160	0.5–0.9	–
$Y_3Al_5O_{12}$: : Nd^{3+},Cr^{3+} O_h^{10}–Ia3d $D_2(Y^{3+})$, $C_{3i}(Al^{3+})$	Yb^{3+} ≈ 2	1.0293 1.0298	77 210	P P	2 115	0.35–1.1 Excitation into Nd^{3+}, Yb^{3+}, and Cr^{3+} bands	1.12 1.2
	Nd^{3+} 0.8	1.0612 B 1.0638 A	210 210	P P	100 120	0.35– 0.9 Excitation into Nd^{3+} and Cr^{3+} bands	–

[a] λ_g value requires more-accurate definition.
[10] The calculated value is given.

Luminescence linewidth [cm^{-1}]	Laser transition	Energy terminal level [cm^{-1}]	Excitation system, pump source, filter, cooling	Rod length, rod diameter [mm]; resonator cavity parameters, mirrors (coating, transmittance)	Remarks	Reference
—	$^3F_4 \rightarrow {}^3H_5$	≈ 8300	ER with Xe flashtube ILC7L3	50; 3 $\parallel$, DM	$Cr^{3+} = 0.1$ at.% EET to Tm^{3+}	[563]
≈ 7 ≈ 25	$^2F_{5/2} \rightarrow {}^2F_{7/2}$	612 ≈ 612	Low-temp. ER with Xe flashtube IFP–400, QLP, ZhS–17, N$_2$ and its vapor	24; 6 $\parallel$, 5″, $R = 576$ mm, DM, 2%	$Nd^{3+} = 0.8$ at.% $\sigma_e = 1.4 \times 10^{-19}$ $\Delta v_g = 4$ cm^{-1} EET to Yb^{3+}	[86, 483]
≈ 9	$^4F_{3/2} \rightarrow {}^4I_{11/2}$	2005	Low-temp. ER with Xe flashtube IFP–400, QLP, ZhS–17, N$_2$ vapor	24; 6 $\parallel$, 5″, $R = 576$ mm, DM, 1%	$\Delta v_g = 1$ cm^{-1} At 200 K, simultaneous emission from Yb^{3+} and Nd^{3+} ions was observed	[483]
≈ 7 ≈ 25	$^2F_{5/2} \rightarrow {}^2F_{7/2}$	612 ≈ 612	Low-temp. ER with Xe flashtube IFP–400, QLP, N$_2$ and its vapor	35; 6 $\parallel$, 10″, $R = 576$ mm, DM, 2%	$Cr^{3+} = 0.5\%$, $Nd^{3+} = 0.8$ at.% $\sigma_e = 1.4 \times 10^{-19}$ $\Delta v_g \approx 1$ cm^{-1} $\Delta v_g = 4$ cm^{-1} EET to Yb^{3+}	[86, 483]
≈ 3 ≈ 4	$^4F_{3/2} \rightarrow {}^4I_{11/2}$	2005 2110	Low-temp. ER with Xe flashtube IFP–400, QLP, N$_2$ vapor	35; 6 $\parallel$, 10″, $R = 576$ mm, DM, 1%	$Cr^{3+} = 0.5\%$ $\Delta v_g \approx 1$ cm^{-1} $\Delta v_g \approx 1$ cm^{-1} EET to Nd^{3+} At 210 K simultaneous emission from Yb^{3+} and Nd^{3+} ions was ovserved	[86, 483]

(*Continued*)

Table 5.1   (*continued*)

Crystal host, space group, cation-site symmetry	Activator [%]	Emission wavelength [μm]	Temperature [K]	Mode of operation	Threshold energy [J] or power [W]	Excitation spectral range [μm]	Luminescence lifetime [ms]
Y_2SiO_5 C_{2h}^6–C2/c $C_1'(Y_I^{3+})$, $C_1''(Y_{II}^{3+})$	Nd^{3+} 1–2	1.0715	300	P	4 [f]	0.5–0.9	0.24
		1.0742	300	P	5		
		1.0782	300	P	2		
		1.0710	77	P	1.5		0.24
		1.0781	77	P	0.5		
	≈ 5	1.0644	77	P	≈ 6		
		1.0710	77	P	≈ 5		
		1.0740	77	P	≈ 2		
		1.0742	300	P	8		
	1–2	1.3585	300	P	4 [f]	0.5–0.9	0.24
		1.3580	77	P	2		0.24
$Y_2SiO_5:Er^{3+}$ C_{2h}^6–C2/c $C_1'(Y_I^{3+})$, $C_1''(Y_{II}^{3+})$	Ho^{3+} 0.5–1	2.085 [a]	≈ 110	P	8.3–18	0.3–2 Excitation into Ho^{3+} and Er^{3+} bands	≈ 10 (0.5%)
		2.092 [a]	≈ 110	P	7 [f]		
		2.105 [a]	≈ 110	P	3.4–23		
	0.5	2.105 [a]	≈ 220	P	60		
$Y_2SiO_5: Er^{3+}$: Tm^{3+} C_{2h}^6–C2/c $C_1'(Y_I^{3+})$, $C_1''(Y_{II}^{3+})$	Ho^{3+} 0.8	2.085 [a]	≈ 110	P	4.5 [f]	0.3–2 Excitation into Ho^{3+}, Er^{3+}, and Tm^{3+} bands	≈ 4.5
		2.085	300	P	60		
YP_5O_{14} C_{2h}^6–C2/c (Y_I^{3+}), (Y_{II}^{3+})	Nd^{3+} 16	1.0525	300	CW	–	LP	0.22

---

[a] λ_g value requires more-accurate definition.

[f] The composition of the stimulated-emission spectrum and the threshold energies depend on the mutual orientation of the geometric axis of the rod F (axis of the laser element) and of the crystallographic directions of the crystal.

(24) Emission from a Y_2SiO_5–Nd^{3+} crystal on the two $^4F_{3/2} \to {}^4I_{9/2}$ and $^4F_{3/2} \to {}^4I_{11/2}$ transitions was observed in [509]. The data of [509] are doubtful and criticized in [82].

(25) The length and diameter of the laser rod are not indicated. The excitation thresholds are in $J\ cm^{-1}$.

Luminescence linewidth [cm⁻¹]	Laser transition	Energy terminal level [cm⁻¹]	Excitation system, pump source, filter, cooling	Rod length, rod diameter [mm]; resonator cavity parameters, mirrors (coating, transmittance)	Remarks	Reference
	$^4F_{3/2} \rightarrow {}^4I_{11/2}$					
13		≈ 2100	Low- and room-	20; 5	$n \approx 1.8$	[82, 546]
17		≈ 2100	temp. ER	$\parallel$, 5″, $R = 576$	[24]	
20		≈ 2100	with Xe flash-	mm, DM, 1%	$\Delta\nu_g = 3$ cm⁻¹	
2		≈ 2100	tube IFP–400,		$\Delta\nu_g < 1$ cm⁻¹	
4		≈ 2100	QLP, ZhS–17,		$\Delta\nu_g < 1$ cm⁻¹	
–		≈ 2000	N₂	12; 4		
2		≈ 2100		$\parallel$, 5″, $R = 576$		
–		≈ 2100		mm, DM, 1%		
≈ 20		≈ 2100				
	$^4F_{3/2} \rightarrow {}^4I_{13/2}$					
20		≈ 4100	Low- and room-	20; 5		[82]
5		≈ 4100	temp. ER with	$\parallel$, 5″, $R = 600$		
			Xe flashtube	mm, DM, 1%		
			IFP–400, QLP,			
			ZhS–17, N₂			
	$^5I_7 \rightarrow {}^5I_8$				Er^{3+} upon 10%	
–		≈ 500	Illuminator	–[25]	EET to Ho^{3+}	[638]
–		≈ 500	with Xe flash-	$\parallel$, $R = 500$ mm,		
–		≈ 500	tube, N₂	DM, 1%		
–		≈ 500	vapor		$Er^{3+} = 2.9$ at.%	
	$^5I_7 \rightarrow {}^5I_8$				$Er^{3+} = 8$ at.%,	
					$Tm^{3+} = 0.8$ at.%	
–		≈ 500	Illuminator	–[25]	EET to Ho^{3+}	[638]
–		≈ 500	with Xe flash-	$\parallel$, $R = 500$ mm,		
			tube, N₂	DM, 1%		
			vapor			
	$^4F_{3/2} \rightarrow {}^4I_{11/2}$				In $Y_x Nd_{1-x} P_5 O_{14}$	[249, 615]
–		≈ 2000	CW laser	–	This phase	
					existed at	
					$0.83 \leqslant x \leqslant 1$	

(Continued)

Table 5.1 (*continued*)

Crystal host, space group, cation-site symmetry	Activator [%]	Emission wavelength [μm]	Temperature [K]	Mode of operation	Threshold energy [J] or power [W]	Excitation spectral range [μm]	Luminescence lifetime [ms]
YP_5O_{14} C_{2h}^5–$P2_1/c$ $C_1(Y^{3+})$	Nd^{3+} >17	1.051 [a]	300	CW	–	LP	0.32
	99	1.0515	300	P	–	LP; 0.581	
	90	1.0515	300	P	–	LP; 0.4765, 0.5017, 0.5145	
$Y_3Sc_2Al_3O_{12}$ O_h^{10}–Ia3d $D_2(Y^{3+})$	Nd^{3+} ≈1	1.0595 1.0622 1.0587	300 300 77	P P P	5 ≈7 3	0.5–0.9	≈0.28 ≈0.28
	≈1	1.3360	300	P	10	0.5–0.9	≈0.28
$Y_3Sc_2Ga_3O_{12}$ O_h^{10}–Ia3d $D_2(Y^{3+})$	Nd^{3+} ≈1	1.0583 1.0615 1.0575	300 300 77	P P P	1.5 3 <1	0.5–0.9	≈0.28 ≈0.28
	≈1	1.3310	300	P	5	0.5–0.9	≈0.28
YVO_4 D_{4h}^{19}–I4/amd $D_{2d}(Y^{3+})$	Nd^{3+} ≈1 $a\|F$	(π) 1.0641 A	300	CW	0.1 [b]	LP; 0.5145	0.09
	≈1 $c\|F$	(π) 1.0641 A (σ) 1.0664 B	300 300	P P	60 1.3	0.5–0.9	

[a] λ_g value requires more-accurate definition.
[b] The absorbed pump power (energy) at which stimulated emission begins is indicated.

Luminescence linewidth [cm⁻¹]	Laser transition	Energy terminal level [cm⁻¹]	Excitation system, pump source, filter, cooling	Rod length, rod diameter [mm]; resonator cavity parameters, mirrors (coating, transmittance)	Remarks	Reference
	$^4F_{3/2} \to {}^4I_{11/2}$					
–		≈ 2000	CW laser	–	In $Y_x Nd_{1-x} P_5 O_{14}$ This phase	[249, 615]
–		≈ 2000	P dye laser (Zeiss)		existed at $0.83 \geqslant x \geqslant 0$	
–			P multimode Ar ion laser	$4.5 \times 3 \times 2$ $R = 100$ mm, DM, 2%	$P_g = 240$ mW $\eta_g = 24$ $E \parallel b$ axis	
≈ 6	$^4F_{32} \to {}^4I_{11/2}$	1976	Low- and room-	20; 5		[488, 529,
≈ 8		2100	temp. ER with	$\parallel$, 10″, $R = 576$		536, 654]
≈ 2		1983	Xe flashtube IFP–400, QLP, ZhS–17, N_2	mm, DM, 1%		
≈ 13	$^4F_{3/2} \to {}^4I_{13/2}$	≈ 3930	ER with Xe flashtube IFP–400, ZHS–17	20; 5 $\parallel$, 10″, $R = 600$ mm, DM, 1%		[488, 529, 654]
≈ 8	$^4F_{3/2} \to {}^4I_{11/2}$	1979	Low- and room-	10; 5		[536]
–		2060	temp. ER	$\parallel$, 10″, $R = 576$		
≈ 2		1983	with Xe flash-tube IFP–400, QLP, ZhS–17, N_2	mm, DM, 1%		
≈ 14	$^4F_{3/2} \to {}^4I_{13/2}$	3915	ER with Xe flashtube IFP–400, ZhS–17	10; 5 $\parallel$, 10″, $R = 600$ mm, DM, 1%		[536]
6.9	$^4F_{3/2} \to {}^4I_{11/2}$	1964	CW Ar ion laser, LPG	30; 3 $\parallel$, DM, 1–20%	$\sigma_e = 3 \times 10^{-18}$ $P_g = 1$ W for $P_{exc} = 4.5$ W $\eta_g = 25$	[34, 171, 243, 409– 412, 605, 614]
6.9		1964	ER with Xe	20; 5	$\sigma_e = 1.7 \times 10^{-18}$	
7.7		1985	flashtube IFP–400,	$\parallel$, 15″, $R = 576$ mm, DM, 0.7%	$\sigma_e = 1.3 \times 10^{-18}$	

(Continued)

Table 5.1 (*continued*)

Crystal host, space group, cation-site symmetry	Activator [%]	Emission wavelength [μm]	Temperature [K]	Mode of oper- ation	Threshold energy [J] or power [W]	Excitation spectral range [μm]	Lumines- cence lifetime [ms]
		(σ) 1.0625* C	300	P	2.6		
		(σ) 1.0648* D	300	P	50		
		(σ) 1.0664 B	300	q–CW	–	0.4–0.9	
		(σ) 1.0664→	300→	P	–	0.5–0.9	
		→1.0672 B	→690				
	≈ 1	1.3425	300	P	9	0.5–0.9	0.09
		1.3415	77	P	15		
	2	≈ 1.34 [a]	300	CW	0.5 [b]	LP; 0.5145	
YVO$_4$ D_{4h}^{19}–I4/amd $D_{2d}(Y^{3+})$	Eu^{3+} –	0.6193 [d]	90	P	60	0.4–0.6	–
YVO$_4$ D_{4h}^{19}–I4/amd $D_{2d}(Y^{3+})$	Tm^{3+} –	≈ 2.0 [d]	77	P	– [(26)]	–	–
YVO$_4$: : Er^{3+}, Tm^{3+} D_{4h}^{19}–I4/amd $D_{2d}(Y^{3+})$	Ho^{3+} 1 $c \perp F$	2.0412	77	P	–	0.4–2 Excitation into Ho^{3+}, Er^{3+}, and Tm^{3+} bands	–
			77	CW	–	LP; 0.4880 and 0.5145	
Y$_3$Fe$_5$O$_{12}$ O_h^{10}–Ia3d $D_2(Y^{3+})$	Ho^{3+} –	≈ 2.1 [a]	77	P	≈ 0.02	LP; 1.06 Excitation into Fe^{3+} band ($^4T_{1g}$)	3–4

[a] λ_g value requires more-accurate definition.
[b] The absorbed pump power (energy) at which stimulated emission begins is indicated.
[d] The spectroscopic data confirming this result are not published.

Luminescence linewidth [cm^{-1}]	Laser transition	Energy terminal level [cm^{-1}]	Excitation system, pump source, filter, cooling	Rod length, rod diameter [mm]; resonator cavity parameters, mirrors (coating, transmittance)	Remarks	Reference
7.8		1964	ZhS−17		*In CAM laser	
6.5		1985				
7.7		1985	ER with PS		$\tau_g = 7$ ms	
7.7→		1985	High-temp. ER with Xe flashtube IFP−400, ZhS−17		$\delta\lambda/\delta T \approx 2.1$ $n_0 = 1.958$ $n_e = 2.168$ $K \approx 51$	
	$^4F_{3/2} \to {}^4I_{13/2}$					
8		3913	Low- and room-temp. ER with Xe flash-tube IFP−400, QLP, ZhS−17, N$_2$	20; 5 $\parallel$, $R = 600$ mm, DM, 1%		[88, 614]
≈3		3912				
−		≈3900	CW Ar ion laser, LPG	30; 3 $\parallel$, DM	$\sigma_e = 3.6 \times 10^{-19}$ $P_g = 35$ mW for $P_{exc} = 5$ W $\eta_g = 7$	
	$^5D_0 \to {}^7F_2$					
−		≈1000	Illuminator with Xe flash-tubes FX−51	≈16; 6.5		[409]
	$^3H_4 \to {}^3H_6$					
−		−	−	−	$n_0 = 1.93$ $n_e = 2.13$	
	$^5I_7 \to {}^5I_8$				Er^{3+} = 25% and Tm^{3+} = 7%	
−		229	ER with Xe flashtube, N$_2$	7; 5 $\parallel$	E$\parallel c$ axis EET to Ho^{3+}	[538, 660]
			CW Ar ion laser, LPG			
	$^5I_7 \to {}^5I_8$					
−		≈400	P neodymium glass laser, TPG, N$_2$	18 × 0.8 × 0.8 $R = 18$ mm, DM, 0.04%	EET to Ho^{3+}	[573]

[26] The authors of [413] refer to a private communication of *Johnson* and *Thomas*.

(*Continued*)

Table 5.1 (*continued*)

Crystal host, space group, cation-site symmetry	Activator [%]	Emission wavelength [μm]	Temperature [K]	Mode of operation	Threshold energy [J] or power [W]	Excitation spectral range [μm]	Luminescence lifetime [ms]
		≈ 2.1 [a]	77	P	0.012	LP; 0.6943 Excitation into Fe^{3+} band ($^4T_{2g}$)	3–4
$Y_3Fe_5O_{12}$: : Er^{3+}, Tm^{3+} O_h^{10}–Ia3d $D_2(Y^{3+})$	Ho^{3+} 2	2.086 A	77	P	30	1.1–2 Excitation into Ho^{3+}, Er^{3+}, and Tm^{3+} bands	3.6
		2.089 B	77	P	30		
		2.107 C	77	P	50		
$Y_3Ga_5O_{12}$ O_h^{10}–Ia3d $D_2(Y^{3+})$	Nd^{3+} ≈ 1	1.0589 C	300	P	5	0.5–0.9	≈ 0.27
		1.0603 B	300	P	4		
		1.06205 A	300	P	2		
		1.0583 C	77	P	1		≈ 0.27
		1.05975 B	77	P	7		
		1.0614 A	77	P	13		
	≈ 1	1.3305	300	P	12.5	0.5–0.9	≈ 0.27
$Y_3Ga_5O_{12}$ O_h^{10}–Ia3d $D_2(Y^{3+})$	Ho^{3+} –	2.086	77	P	70	0.4–2	–
		2.114	77	P	250		
$Y_3Ga_5O_{12}$: : Nd^{3+} O_h^{10}–Ia3d $D_2(Y^{3+})$	Yb^{3+} ≈ 2	1.0233	77	P	2	0.5–1.1 Excitation into Nd^{3+} and Yb^{3+} bands	≈ 1.3

[a] λ_g value requires more-accurate definition.

Luminescence linewidth [cm^{-1}]	Laser transition	Energy terminal level [cm^{-1}]	Excitation system, pump source, filter, cooling	Rod length, rod diameter [mm]; resonator cavity parameters, mirrors (coating, transmittance)	Remarks	Reference
–		≈ 400	P ruby laser	$25.4 \times 0.8 \times 0.8$ DM	EET to Ho^{3+}	
≈ 16 ≈ 17 ≈ 20	$^5I_7 \rightarrow {}^5I_8$	402 413 413 ≈ 425 461 487	Illuminator with Xe flashtube FT–524, N$_2$, Si-concentrator	5 $\parallel$, DM	Er^{3+} = Tm^{3+} = 5% EET to Ho^{3+} Laser lines constitute superposition of two components close in frequency. Magnetic retuning 0.007 μm (16 cm^{-1}) wavelength of line A from 2.0842 μm to 2.0912 μm	[138, 139, 414]
7.2 7.5 7.8 2.5 3.0 3.0	$^4F_{3/2} \rightarrow {}^4I_{11/2}$	1988 2001 2053 1994 2007 2.56	Low- and room-temp. ER with Xe flashtube IFP–400, QLP, ZhS–17, N$_2$	20; 5 $\parallel$, 5″, $R = 576$ mm, DM, 1%	$\Delta v_g < 1$ cm^{-1} $\Delta v_g < 1$ cm^{-1} $\Delta v_g < 1$ cm^{-1} $n = 1.915$	[9, 504, 631]
8	$^4F_{3/2} \rightarrow {}^4I_{13/2}$	≈ 3920	ER with Xe flashtube IFP–400, ZhS–17	20; 5 $\parallel$, 5″, $R = 600$ mm, DM, 1%	$n = 1.9$ $K = 90$	[504, 631]
– –	$^5I_7 \rightarrow {}^5I_8$	418 481	Illuminator with Xe flashtube FT–524, N$_2$		Metastable level $^5I_7$ – –5211 cm^{-1}	[138, 139]
≈ 8	$^2F_{5/2} \rightarrow {}^2F_{7/2}$	≈ 600	Low-temp. ER with Xe flashtube IFP–400, QLP, ZhS–17, N$_2$	15; 5 $\parallel$, 7″, $R = 576$ mm, DM, 1%	Nd^{3+} = 1.5% $\Delta v_g \approx 1$ cm^{-1} EET to Yb^{3+}	[483, 486]

(*Continued*)

Table 5.1 (*continued*)

Crystal host, space group, cation-site symmetry	Activator [%]	Emission wavelength [μm]	Temperature [K]	Mode of oper- ation	Threshold energy [J] or power [W]	Excitation spectral range [μm]	Lumines- cence lifetime [ms]
(Y,Er)AlO$_3$ D_{2h}^{16}–Pbnm (Pnma) C_s(Y^{3+}, Er^{3+})	Tm^{3+} 1 [001]$\|$F	1.861	77	P	25	0.3–1.8 Excitation into Er^{3+} and Tm^{3+} bands	4
(Y,Er)AlO$_3$: : Tm^{3+} D_{2h}^{16}–Pbnm (Pnma) C_s(Y^{3+}, Er^{3+})	Ho^{3+} $\approx$2 [001]$|$F	2.123 [27]	300 77	P P	$\approx$300 $\approx$1	0.3–2 Excitation into Ho^{3+}, Er^{3+}, and Tm^{3+}, bands	6.6
(Y,Er)$_3$Al$_5$O$_{12}$ O_h^{10}–Ia3d D_2(Y^{3+}, Er^{3+})	Ho^{3+} $\approx$0.7	2.0917 B 2.0979 A 2.123 C	77 77 77 85	P P P CW	390 11 3800 47	0.4–2 Excitation into Ho^{3+} and Er^{3+} bands	4
(Y,Er)$_3$Al$_5$O$_{12}$ O_h^{10}–Ia3d D_2(Y^{3+}, Er^{3+})	Er^{3+} $\approx$30 [100]$|$F 30–50	2.9364 2.8302 2.9365	300 300 300 300	P Q P P	190 – 11 12	0.5–1.1 0.5–1.1 $\tau_{exc} = 40$–60 μs	0.1 0.1– –0.08
(Y,Er)$_3$Al$_5$O$_{12}$ O_h^{10}–Ia3d D_2(Y^{3+}, Er^{3+})	Tm^{3+} $\approx$0.7	1.880 A 1.884 2.014 C	77 77 77 85	P P P CW	264 180 170 520	0.4–1.8 Excitation into Er^{3+} and Tm^{3+} bands	15

[27] According to [141], $\lambda_g = 2.119$ μm.

[28] Crystals with different orientations were used; one face of each was covered by a 100% dielectric mirror and the other by a broad-band antireflection coating. External output dielectric mirrors with different transmittance were used in the measurements.

Luminescence linewidth [cm⁻¹]	Laser transition	Energy terminal level [cm⁻¹]	Excitation system, pump source, filter, cooling	Rod length, rod diameter [mm]; resonator cavity parameters, mirrors (coating, transmittance)	Remarks	Reference
5.5	$^3H_4 \rightarrow {}^3H_6$	≈ 240	ER with Xe flashtube, N_2	50; 5 $\parallel$, DM, 40%	$Er^{3+} = 30\%$ EET to Tm^{3+} $\eta_g = 0.13$	[107]
17 9.8	$^5I_7 \rightarrow {}^5I_8$	474	ER with Xe flashtube L–213 B, Pyrex cryostat, N_2	50; 2–6 $\parallel$, DM (28)	$Er^{3+} = 30–50\%$ $Tm^{3+} = 3– 7\%$ $\sigma_e = 2 \times 10^{-19}$ $E_g = 1.2$ J $\eta_g \approx 0.7$ EET to Ho^{3+}	[107, 141]
– – – –	$^5I_7 \rightarrow {}^5I_8$	532 462 518	Under P conditions, illuminator with Xe flashtube FT–524; and under CW conditions, W and Hg (GEAH6) lamps, O_2 supercooled with N_2	≈ 25	$Er^{3+} = 47\%$ EET to Ho^{3+}	[128, 135]
– 8,5 ≈ 9	$^4I_{11/2} \rightarrow {}^4I_{13/2}$	≈ 6880 ≈ 6880 (6818) ≈ 6880	ER with Xe flashtube IFP type, filter ER with Xe flashtube ISP--250, ZhS–17	60; 3 $\parallel$, DM 40; 4–5 $\parallel$, 10″, $R = 600$ mm, DM, 1%	$E_g = 0.3$ J $E_g = 0.005$ J $n \approx 1.79$	[490, 572]
– – – –	$^3H_4 \rightarrow {}^3H_6$	228 240 582	Under P conditions, illuminator with Xe flashtube FT- –524; and under CW conditions, W lamp, O_2 supercooled with N_2	≈ 25	$Er^{3+} = 47$ at.% EET to Tm^{3+}	[128, 135]

(*Continued*)

Table 5.1 (*continued*)

Crystal host, space group, cation-site symmetry	Activator [%]	Emission wavelength [μm]	Temperature [K]	Mode of operation	Threshold energy [J] or power [W]	Excitation spectral range [μm]	Luminescence lifetime [ms]
$(Y,Er)_3Al_5O_{12}$: Ho^{3+} : Tm^{3+}							
O_h^{10}–Ia3d $D_2(Y^{3+}, Er^{3+})$	1.7–5	2.0982	77	P	–	0.4–2 Excitation into Ho^{3+}, Er^{3+}, and Tm^{3+} bands	≈5
		2.1227	85	P	–		
		2.1288	295	P	1240		
		2.1227	85	CW	30		
	2	2.0990	77	P	3		≈5
		2.1285	77	P	14		
		2.0990	77	CW	45		
	1.65	≈2.13 [a]	90	P	4		–
			300	P	60		–
		≈2.13 [a]	77	P	0.6		
				CW	5 [(29)]		
		2.098 [a]	77	CW	45		
	≈1.7	2.0977	77	CW	≈40	0.4–2	≈7
		2.0990	77	q–CW	–	0.4–2	
$(Y,Er)_3Al_5O_{12}$: Ho^{3+} :Tm^{3+},Yb^{3+}							
O_h^{10}–Ia3d $D_2(Y^{3+}, Er^{3+})$	–	2.1227	85	CW	–	0.4–2 Excitation into Ho^{3+}, Er^{3+}, Tm^{3+}, and Yb^{3+} bands	–
$(Y,Yb)_3Al_5O_{12}$ O_h^{10}–Ia3d $D_2(Y^{3+}, Yb^{3+})$	Yb^{3+} ≈35 ≈50	1.0293	77 77	P P	2 >50	0.9–1.1	≈0.35 ≈1.12 (≈1%)

[a] λ_g value requires more-accurate definition.
[(29)] Results obtained by *Hopkins* et al. (see [437]).

Luminescence linewidth [cm^{-1}]	Laser transition	Energy terminal level [cm^{-1}]	Excitation system, pump source, filter, cooling	Rod length, rod diameter [mm]; resonator cavity parameters, mirrors (coating, transmittance)	Remarks	Reference
	$^5I_7 \rightarrow {}^5I_8$				$Tm^{3+} = 6.7$ at.%	
–		462	Illuminator	32	EET to Ho^{3+}	[140, 223,
–		518	with Xe flash-	DM, 0.5%		224, 243,
–		532	tube FT–524, N_2			415, 597]
		518	Illuminator with W lamp, N_2	–	$P_g = 15$ W $\eta_g = 5$	
–		≈ 462	Illuminator	50	$Er^{3+} = 50$ at.%,	
–		538	with Xe flash-	DM	$Tm^{3+} = 6$ at.%	
–		≈ 462	tube IFP–800,N_2			
–		532	Illuminator	50; 3		
–			with Xe flash- tube FT–524, N_2 vapor	DM, 1%		
–		532	Illuminator with W lamp,N_2	30; 3		
–		≈ 462	ER (with Au coating) with W (1000 W) lamp,N_2	50; 3 $\parallel$, $R = 1000$ mm, DM	$\Delta \nu_g = 1$ cm^{-1} $P_g = 5$ W for TEM$_{00}$ mode $P_g = 20.5$ W for multimode conditions, $\eta_g = 1$	
–		≈ 462	ER with W (1000 W) lamp, O_2 supercooled with N_2	70; 4 $\parallel$, DM, 10%	$P_g = 50$ W $\eta_g = 6.5$ $Er^{3+} = 50$ at.%, $Tm^{3+} = 6.6$ at.%	
–		≈ 462	Illuminator with PS, N_2	20; 2.5	$\tau_g = 15$ ms	
	$^5I_7 \rightarrow {}^5I_8$					
–		518	Illuminator with W lamp, N_2	–	EET to Ho^{3+}	[140, 491]
≈ 10	$^2F_{5/2} \rightarrow {}^2F_{7/2}$	≈ 612	Low-temp. ER with Xe flash- tube IFP–400, N_2	23; 5 $\parallel$, 10″, $R = 576$ mm, DM, 1%		[546]

(Continued)

Table 5.1   (*continued*)

Crystal host, space group, cation-site symmetry	Activator [%]	Emission wavelength [μm]	Temperature [K]	Mode of operation	Threshold energy [J] or power [W]	Excitation spectral range [μm]	Lumines- cence lifetime [ms]
$(Y,Lu)_3Al_5O_{12}$ $O_h^{10}-Ia3d$ $D_2(Y^{3+},$ $Lu^{3+})$	Nd^{3+} ≈ 0.3	1.0642 1.0608 1.0636 1.0726	295 77 77 77	P P P P	– – – –	0.3–0.9	0.2 0.2
$Ba_{0.25}Mg_{2.75}Y_2$ Ge_3O_{12} $O_h^{10}-Ia3d$	Nd^{3+} ≈ 2	1.0615	300	P	22.5	0.4–0.8	≈ 0.3
$La_2Be_2O_5$ C_{2h}^6-C2/c $C_1(La^{3+})$	Nd^{3+} 1 $b(Y)\|F$	1.0698	300 300	P Q	1.8 8–9	0.5–0.9	0.155 (0.3%)
	$X\|F$	1.079 [a]	300 300	P Q	10–11 14		
	1 $b(Y)\|F$	1.0698	300	CW	–		
	≈ 1	1.3510	300	P	2.5 [f]	0.4–0.9	0.155 (0.3%)
La_2O_3 $D_{3d}^3-P\bar{3}m1$ $C_{3v}(La^{3+})$	Nd^{3+} 1–2	1.079 [a]	77	P	230 [f]	0.4–0.9	0.12
$LaAlO_3$ $D_{3d}^5-R\bar{3}m$ $C_{3v}(La^{3+})$	Nd^{3+} ≈ 2	1.0804	300	P	120 [f]	0.5–0.9	0.14

[a] λ_g value requires more-accurate definition.
[f] The composition of the stimulated-emission spectrum and the threshold energies depend on the mutual orientation of the geometric axis of the rod F (axis of the laser element) and of the crystallographic directions of the crystal.

Luminescence linewidth [cm^{-1}]	Laser transition	Energy terminal level [cm^{-1}]	Excitation system, pump source, filter, cooling	Rod length, rod diameter [mm]; resonator cavity parameters, mirrors (coating, transmittance)	Remarks	Reference
	$^4F_{3/2} \rightarrow {}^4I_{11/2}$				$Lu^{3+} = 28.4\%$	
–		≈ 2100	ER with Xe	$25 \times 4 \times 4$		[416]
–		≈ 2000	flashtube	$\parallel$, Au		
–		≈ 2100	IFP–800, N_2			
–		≈ 2100				
	$^4F_{3/2} \rightarrow {}^4I_{11/2}$					
30		2014	Illuminator with two Xe flashtube L–268, filter	$\approx 19; 5$ $\parallel$, DM, 2.5%	$\sigma_e = 6 \times 10^{-20}$ $n \approx 1.8$	[433]
	$^4F_{3/2} \rightarrow {}^4I_{11/2}$					
≈ 27		1962	Under P conditions, Xe flashtubes FX–42–C–3 and IFP–400; and CW conditions, with two W lamps	$25–75; 5–6.5$ $\parallel$, $d_r = 750$ mm, DM, upon 65%	$\sigma_e^X = 4 \times 10^{-20}$ $\eta_q = 2.38$ $\mathbf{E}\parallel$X direction	[596, 633, 657]
≈ 40		2041			$\sigma_e^Y = 6 \times 10^{-20}$ $\eta_q = 2.75$ $\mathbf{E}\parallel$Y direction	
≈ 27		1962	Q1500T4/4CL, water	$50; 5$ $\parallel$, $d_r = 320$ mm, DM	$P_g = 9$ W for $P_{exc} = 3000$ W $K \approx 47$	
	$^4F_{3/2} \rightarrow {}^4I_{13/2}$					
≈ 32		3908	ER with Xe flashtube IFP–400, ZhS–17	$40; 8$ $\parallel$, $R = 600$ mm, DM, 1%	$n_2 = 2.1 \times 10^{-13}$	[633]
	$^4F_{3/2} \rightarrow {}^4I_{11/2}$					
11		≈ 2000	Cryostat in helical Xe (10 cm) flashtube, N_2	$8; 4$ $\parallel$, DM		[417]
	$^4F_{3/2} \rightarrow {}^4I_{11/2}$					
20		2327	ER with Xe flashtube IFP–400, ZhS–17	$8 \times 4 \times 3$ $\parallel$, 20″, $R = 576$ mm, DM, 0.5%		[137]

(*Continued*)

Table 5.1   (*continued*)

Crystal host, space group, cation-site symmetry	Activator [%]	Emission wavelength [μm]	Temperature [K]	Mode of operation	Threshold energy [J] or power [W]	Excitation spectral range [μm]	Luminescence lifetime [ms]
LaP_5O_{14} $C_{2h}^5-P2_1/c$ $C_1(La^{3+})$	Nd^{3+} ≈ 25	1.051 [a]	300	CW	$-$ [30]	LP	0.31 ($\approx 1\%$)
$LaNbO_4$ C_{2h}^6-C2/c $C_2(La^{3+})$	Nd^{3+} 1–2	1.0618 [31]	300	P	10 [f]	0.5–0.9	0.12
$LaNbO_4$ C_{2h}^6-C2/c $C_2(La^{3+})$	Ho^{3+} ≈ 2	2.8510 [c]	300	P	50 [f]	0.4–1.2 $\tau_{exc} = 60-100$ μs	–
$LaNbO_4:Er^{3+}$ C_{2h}^6-C2/c $C_2(La^{3+})$	Ho^{3+} 0.9	≈ 2.07 [a]	≈ 90	P	20 [f]	0.5–2 Excitation into Ho^{3+} and Er^{3+} bands	20
CeP_5O_{14} $C_{2h}^5-P2_1/c$ $C_1(Ce^{3+})$	Nd^{3+} ≈ 10	1.051 [a]	300	P	–	LP; 0.4765, 0.5017 and 0.5145	≈ 0.28 ($\approx 1\%$)
$NdAl_3(BO_3)_4$ D_3^7-R32 $D_3(Nd^{3+})$	Nd^{3+} 100 $c \parallel F$	1.0635 [a]	300	q-CW	≈ 0.001 [b]	LP; 0.58	0.019 $\approx 0.05^*$ (1%)
			300	q-CW	≈ 0.017 [b]	LP; 0.82	

---

[a] λ_g value requires more-accurate definition.
[b] The absorbed pump power (energy) at which stimulated emission begins is indicated.
[c] The detailed spectroscopic investigations were not carried out.
[f] The composition of the stimulated-emission spectrum and the threshold energies depend on the mutual orientation of the geometric axis of the rod F (axis of the laser element) and of the crystallographic directions of the crystal.
[30] For the case of excitation of stimulated emission by light from an LED, only the theoretical estimate of the P_{thr} value is given.
[31] The λ_g value corresponds to more-accurate data (unpublished) by the authors of [418].

Lumines-cence linewidth [cm^{-1}]	Laser transition	Energy terminal level [cm^{1-}]	Excitation system, pump source, filter, cooling	Rod length, rod diameter [mm]; resonator cavity parameters, mirrors (coating, transmittance)	Remarks	Reference
35–40	$^4F_{3/2} \rightarrow {}^4I_{11/2}$	1964	LED	–	$\sigma_e^b = 1.8 \times 10^{-19}$	[576]
15	$^4F_{3/2} \rightarrow {}^4I_{11/2}$	1973	ER with Xe flashtube IFP–800	30; 5 $\parallel$, 10″	$n \approx 2$	[277, 418]
–	$^5I_6 \rightarrow {}^5I_7$	≈ 5500	ER with Xe flashtube ISP--250,ZhS–17	22; 6 $\parallel$, 5″, $R = 600$ mm, DM, 1%		[607]
≈ 8	$^5I_7 \rightarrow {}^5I_8$	≈ 295	ER with Xe flashtube ISP–250, filter, N$_2$ vapor	20; 5–6 $\parallel$, $R = 500$ mm, mm, DM, 0.5%	$Er^{3+} \approx 0.5\%$ EET to Ho^{3+}	[406]
–	$^4F_{3/2} \rightarrow {}^4I_{11/2}$	≈ 2000	P multimode Ar ion laser	Several mm $R = 100$ mm, 2–2.5 %	$\mathbf{E} \parallel b$ axis	[615]
≈ 28	$^4F_{3/2} \rightarrow {}^4I_{11/2}$	≈ 2000	CW dye laser Semiconductor heterostructure laser (Ga$_{1-x}$Al$_x$As)	0.136–0.34 $R = 50$ mm, d$_r = 100$ mm, DM, 0.3% 0.173 $\parallel$, $R = 50$ mm, DM, 0.3%	$\sigma_e = 8 \times 10^{-19}$ $P_g = 7$ mW $\eta_g \approx 20$ *In isostructural GdAl$_3$(BO$_3$)$_4$ crystal	[579, 589, 624]

(Continued)

Table 5.1 (*continued*)

Crystal host, space group, cation-site symmetry	Activator [%]	Emission wavelength [μm]	Temperature [K]	Mode of operation	Threshold energy [J] or power [W]	Excitation spectral range [μm]	Lumines-cence lifetime [ms]
NdP_5O_{14} C_{2h}^5–$P2_1/c$ $C_1(Nd^{3+})$	Nd^{3+} 100 $a\,\|\,F$	1.0512	300	P	4×10^{-5} [b]	LP; 0.58	≈ 0.31* ($\approx 1\%$)
			300	CW	0.016 [b]	LP; 0.6764, 0.7525 and 0.7993	
			300	CW	0.004 [b]	LP; 0.58	
			300	q-CW	0.007 [b]	LP; 0.82	
	$c\,\|\,F$	1.0513	300	CW	–	LP; 0.514	
		1.0521	300	CW	–		
		1.0529	300	CW	–		
		1.0629– –1.0638	300	CW			
	$c\,\|\,F$	1.0512	300	P	0.11	0.3–0.9	
			300	Q	–		
Gd_2O_3 C_{2h}^3–C2/m $C_s(Gd^{3+})$	Nd^{3+} –	1.0741	300	P	36 [f]	0.4–0.9	0.12
		1.0789	300	P	9		
		1.0776	77	P	12		
		1.0789	77	P	3		
$GdAlO_3$ D_{2h}^{16}–Pbnm (Pnma) $C_s(Gd^{3+})$	Nd^{3+} 1	1.0690 [a]	300	P	38	–	–
		1.0760	300	P	210 [f]	0.5–0.9	0.1
		1.0689 [a]	77	P	24	–	–
		1.0759 [a]	77	P	41		

[a] λ_g value requires more-accurate definition.

[b] The absorbed pump power (energy) at which stimulated emission begins is indicated.

[f] The composition of the stimulated-emission spectrum and the threshold energies depend on the mutual orientation of the geometric axis of the rod F (axis of the laser element) and of the crystallographic directions of the crystal.

[13] Crystallographic faces (natural-growth surfaces) or cleavage planes of these crystals were used as the laser-rod end faces.

[32] In [471], without any description of the experimental conditions stimulated emission at two 0.8880 and 0.9273 μm lines at 77 K in $GdAlO_3$–Nd^{3+} crystal is reported. The results require confirmation.

Lumines-cence linewidth [cm^{-1}]	Laser transition	Energy terminal level [cm^{-1}]	Excitation system, pump source, filter, cooling	Rod length, rod diameter [mm]; resonator cavity parameters, mirrors (coating, transmittance)	Remarks	Reference
	$^4F_{3/2} \rightarrow {}^4I_{11/2}$					
≈ 40		1964	P rhodamine 6G dye laser	0.035 ‖ [13], $R = 73.5$ mm, DM, 1.5%	$\sigma_e = 2 \times 10^{-19}$ $\Delta v_q \approx 12.5$ cm^{-1} $n \approx 1.6$ $\eta_g = 63$	[84, 574, 575, 612, 615–617, 624, 742]
			CW Kr ion laser, LPG	– ‖, $R = 25$ mm, DM, 0.02%		
			CW rhodamine 6G dye laser, LPG	0.76 ‖, $R = 50$ mm, DM, 0.03%	$\eta_g = 3.5$	
			Semiconductor heterostructure laser	0.565 ‖, $R = 50$ mm, DM, 0.3%	$P_g = 3.4$ mW $\eta_g = 7.5$ *In isostructural LaP$_5$O$_{14}$ crystal	
–		≈ 1964	CW Ar ion laser, LPG	0.46–2.96 ‖, DM, 2%	$\mathbf{E} \parallel a, b$ axes $n_a = 1.607$	
–		–			$n_b = 1.602$	
–		≈ 2068 (2186)			$\mathbf{E} \parallel a$ axis $\eta_g = 5-7$	
≈ 40		≈ 1964	Closed illumi-nator (l = 13 mm) with Xe flashtube	7 × 1.2 $R = 100$ mm, DM, 0.15%	$E_g = 1$ mJ for $E_{exc} = 1$ J and $f_g = 2.1$ kHz	
	$^4F_{3/2} \rightarrow {}^4I_{11/2}$					
–		≈ 2000	ER with Xe flashtube	6; 2 ‖, DM	$n \approx 2.1$	[419]
20–30						
–		≈ 2000	FX–33, N$_2$			
	$^4F_{3/2} \rightarrow {}^4I_{11/2}$					
–		2095	ER with Xe flashtube IFP–400	20; 2.5	[32]	[137, 471]
15		≈ 2166	ER with Xe flashtube ISP–1000, ZhS–17	5 × 3 × 3 ‖, 15″, $R = 576$ mm, DM, 0.5%		
–		2094	–	20; 2.5	[32]	
–		2155				

(*Continued*)

Table 5.1   (*continued*)

Crystal host, space group, cation-site symmetry	Activator [%]	Emission wavelength [μm]	Temperature [K]	Mode of operation	Threshold energy [J] or power [W]	Excitation spectral range [μm]	Luminescence lifetime [ms]
GdAlO$_3$ D_{2h}^{16}–Pbnm (Pnma) C_s(Gd^{3+})	Ho^{3+} 3	1.9925 ^{e}	77	P	240	–	–
GdAlO$_3$ D_{2h}^{16}–Pbnm (Pnma) C_s(Gd^{3+})	Er^{3+} 3	1.5646 ^{e}	77	P	18	0.4–1.5	–
GdAlO$_3$ D_{2h}^{16}–Pbnm (Pnma) C_s(Gd^{3+})	Tm^{3+} 3	1.8529 ^{e}	77	P	216	–	–
GdP$_5$O$_{14}$ C_{2h}^5–P2$_1$/c C_1(Gd^{3+})	Nd^{3+} $\approx$10	1.051 ^{a}	300	P	–	LP; 0.4765, 0.5017 and 0.5145	$\approx$0.28 0.32 ($\approx$1%)
GdScO$_3$ (?)	Nd^{3+} 1.5	1.08515	300	P	3.5 ^{f}	0.5–0.9	0.135
Gd$_3$Sc$_2$Al$_3$O$_{12}$ O_h^{10}–Ia3d D_2(Gd^{3+})	Nd^{3+} $\approx$1	1.05995 1.0620* 1.05915	300 300 77	P P P	3 – 7	0.5–0.9	$\approx$0.28
		1.060 ^{a}	300	CW	700	0.4–0.9	
	$\approx$1	1.3360	300	P	7	0.5–0.9	$\approx$0.28

^{a} λ_g value requires more-accurate definition.

^{e} The detailed data are not available. The results are doubtful and require confirmation or correction.

^{f} The composition of the stimulated-emission spectrum and the threshold energies depend on the mutual orientation of the geometric axis of the rod F (axis of the laser element) and of the crystallographic directions of the crystal.

$^{(33)}$ In [297] the description of experimental conditions is absent. Moreover, the data of [473] for the values of $\lambda_g =$ = 1.9908 μm, for the GdAlO$_3$–Ho^{3+} crystal, are different. The results are doubtful and require confirmation.

$^{(34)}$ In [286] there is no description of the experiment. Moreover, according to [473], in which there is also no information about the measurement conditions, the value of λ_g = 1.8515 μm is different. In [473], values of λ_g for YAlO$_3$ crystals doped with Ho^{3+} ions are given at 77 K (2.116 μm), Er^{3+} ions (1.5763 μm) and Tm^{3+} ions (1.8560 μm). The results are doubtful and require confirmation. In view of this, the results of [766] also seem doubtful.

Luminescence linewidth [cm⁻¹]	Laser transition	Energy terminal level [cm⁻¹]	Excitation system, pump source, filter, cooling	Rod length, rod diameter [mm]; resonator cavity parameters, mirrors (coating, transmittance)	Remarks	Reference
–	$^5I_7 \rightarrow \, ^5I_8$	91	ER with Xe flashtube IFP–400, N_2	20; 2	(33)	[297]
–	$^4I_{13/2} \rightarrow \, ^4I_{15/2}$	224	ER with Xe flashtube IFP–400, N_2	20; 2.5 ‖, Au		[294]
–	$^3H_4 \rightarrow \, ^3H_6$	226	ER with Xe flashtube IFP–400, N_2	20; 2	(34)	[286]
–	$^4F_{3/2} \rightarrow \, ^4I_{11/2}$	≈ 2000	P multimode Ar ion laser	Several mm $R = 100$ mm, DM, 1–2.5%	$\mathbf{E} \Vert b$ axis	[615]
≈ 18	$^4F_{3/2} \rightarrow \, ^4I_{11/2}$	2065	ER with Xe flashtube IFP–400, ZhS–17	12; ≈ 4 ‖, 10″, $R = 576$ mm, DM, 1%	$\Delta\nu_g = 2$ cm⁻¹ (35)	[639]
≈ 10 ≈ 15 ≈ 5	$^4F_{3/2} \rightarrow \, ^4I_{11/2}$	1978 2111 1979	Low- and room-temp. ER with Xe flashtube IFP–400, QLP, ZhS–17, N_2	25; 6 ‖, 10″, $R = 576$ mm, DM, 1%	$n = 1,884$ *In CAM laser	[86, 488, 529, 531]
11.5		1978	Illuminator with W lamp, water	29; 2.5 ‖, $R \approx 1$ m, $d_r =$ $= 500$ mm, 0.4%	$P_g = 0.55$ W $\sigma_e \approx 3.2 \times 10^{-19}$	
12	$^4F_{3/2} \rightarrow \, ^4I_{13/2}$	3931	ER with Xe flashtube IFP–400, ZhS–17	25; 6 ‖, 10″, $R = 600$ mm, DM, 1%		[86, 488, 529]

(35) [472] reports stimulated emission from a $GdScO_3$–Nd^{3+} crystal at 77 and 300 K at $\lambda_g = 0.9054$ μm, without any description of experimental conditions. The result requires confirmation.

(*Continued*)

Table 5.1　(*continued*)

Crystal host, space group, cation-site symmetry	Activator [%]	Emission wavelength [μm]	Temperature [K]	Mode of operation	Threshold energy [J] or power [W]	Excitation spectral range [μm]	Lumines-cence lifetime [ms]
$Gd_3Sc_2Al_3O_{12}$: Nd^{3+} O_h^{10}–Ia3d $D_2(Gd^{3+})$	Yb^{3+} ≈ 2	1.0299	77	P	8	0.5–1.1 Excitation into Nd^{3+} and Yb^{3+} bands	≈ 1.2
$Gd_3Sc_2Ga_3O_{12}$ O_h^{10}–Ia3d $D_2(Gd^{3+})$	Nd^{3+} ≈ 1	1.05755 C 1.0580* 1.06045 A 1.0612　A	77 77 77 300	P P P P	3 12 10 7	0.5–0.9	≈ 0.26 (0.3%) 0.26
$Gd_3Ga_5O_{12}$ O_h^{10}–Ia3d $D_2(Gd^{3+})$	Nd^{3+} 1–1.5	1.0580　C 1.0599　B 1.0600　B 1.0615　A 1.0621　A 1.0591　C 1.0606* B 1.0621　A	77 77 ≈ 120 ≈ 120 300 300 300 300	P P P P P P P CW	1 2 1.5 1.5 3 – – 1000	0.5–0.9	≈ 0.27 (0.3%) ≈ 0.27
	1.5	1.3307 1.3315	77 300	P P	7 5.3	0.5–0.9	≈ 0.27 (0.3%)
$Gd_3Ga_5O_{12}$: Nd^{3+} O_h^{10}–Ia3d $D_2(Gd^{3+})$	Yb^{3+} 2	1.0232	77	P	2,5	0.5–1.1 Excitation into Nd^{3+} and Yb^{3+} bands	≈ 1.3
$Gd_2(MoO_4)_3$ C_{2v}^8–Pba2	Nd^{3+} 3 $c \parallel F$	1.0701 1.0606*	300 300	P P	2.5 –	0.5–0.9	0.15

Luminescence linewidth [cm⁻¹]	Laser transition	Energy terminal level [cm⁻¹]	Excitation system, pump source, filter, cooling	Rod length, rod diameter [mm]; resonator cavity parameters, mirrors (coating, transmittance)	Remarks	Reference
	$^2F_{5/2} \rightarrow {}^2F_{7/2}$				$Nd^{3+} = 1.5\%$	
≈13		≈600	Low-temp. ER with Xe flashtube IFP–400, QLP, ZhS–17, N_2	20; 5.5 ‖, 7″, $R = 576$ mm, DM, 1%	$\Delta v_g = 1.5$ cm⁻¹ EET to Yb^{3+}	[483, 488, 529]
	$^4F_{3/2} \rightarrow {}^4I_{11/2}$					
6.5		1978	Low- and room-temp. ER with Xe flash-tube IFP–400, QLP, ZhS–17, N_2	22; 5 ‖, 7″, $R = 576$ mm, DM, 1%	$\Delta v_g = 1$ cm⁻¹ $\Delta v_g = 0.5$ cm⁻¹ $\Delta v_g = 0.5$ cm⁻¹ *Does not belong to main center.	[636]
–		–				
8.5		2069				
≈14		2070				
	$^4F_{3/2} \rightarrow {}^4I_{11/2}$					
≈2.3		1994	Low- and room-temp. ER with Xe flash-tube IFP–400, QLP, ZhS–17, N_2 and its vapor	25; 5 ‖, 5″, $R = 576$ mm, DM, 1%	$\Delta v_g < 1$ cm⁻¹ $\Delta v_g < 1$ cm⁻¹ At 120 K E_{thr} (A) ≈ ≈ E_{thr}(B) *In CAM laser	[9, 504, 594]
≈2.3		2007				
–		≈2007				
–		≈2063				
7.2		2064				
6.7		1992				
7.0		2005				
7.2		2064	ER with Xe lamp DKsTV– –3000, ZhS–17, water	36; 5 ‖, $R = 600$ mm, DM, 1%	$K ≈ 90$	
	$^4F_{3/2} \rightarrow {}^4I_{13/2}$					
≈2.5		3927	Low- and room-temp. ER with Xe flash-tube IFP–400, QLP, ZhS–17, N_2	18–25; 5 ‖, $R = 600$ mm, DM, 1%		[504, 594]
≈7		3925				
	$^2F_{5/2} \rightarrow {}^2F_{7/2}$				$Nd^{3+} = 2$ at.%	
≈9		≈600	Low-temp. ER with Xe flashtube IFP–400, QLP, ZhS–17, N_2	15; 5 ‖, 7″, $R = 576$ mm, DM, 1%	$\Delta v_g = 1$ cm⁻¹ EET to Yb^{3+}	[483, 486]
	$^4F_{3/2} \rightarrow {}^4I_{11/2}$					
≈40		2010	ER with Xe flashtube IFP––400, ZhS–17	5, 10 ‖, 10″, $R = 576$ mm, DM, 1%	$\Delta v_g = 7$ cm⁻¹ $n ≈ 1.848$ *In CAM laser	[164–166, 420]
≈35		≈1926				

(Continued)

Table 5.1 (*continued*)

Crystal host, space group, cation-site symmetry	Activator [%]	Emission wavelength [μm]	Temperature [K]	Mode of operation	Threshold energy [J] or power [W]	Excitation spectral range [μm]	Luminescence lifetime [ms]
$Ho_3Al_5O_{12}$ O_h^{10}–Ia3d $D_2(Ho^{3+})$	Ho^{3+} 100	2.1224 2.1294	≈ 90 ≈ 90	P P	42 27	0.4–2	≈ 1
		2.1227 2.1297 2.097 [a]	77 77 77	P P P	50 50 ≈ 400		
$Ho_3Sc_2Al_3O_{12}$ O_h^{10}–Ia3d $D_2(Ho^{3+})$	Ho^{3+} 100	2.1170 2.1285	77 77	P P	≈ 150 ≈ 150	0.4–2	–
$Ho_3Ga_5O_{12}$ O_h^{10}–Ia3d $D_2(Ho^{3+})$	Ho^{3+} 100	2.1135 2.086 [a]	77 77	P P	65 ≈ 400	0.4–2	–
Er_2O_3 T_h^7–Ia3 $C_2(Er^{3+})$, $C_{3i}(Er^{3+})$	Ho^{3+} 1	2.121	77 145	P CW P	5 200 20	0.4–2 Excitation into Ho^{3+} and Er^{3+} bands	10
Er_2O_3 T_h^7–Ia3 $C_2(Er^{3+})$, $C_{3i}(Er^{3+})$	Tm^{3+} 0.5	1.934	77	P CW	3 500	0.4–1.9 Excitation into Er^{3+} and Tm^{3+} bands	2.9

[a] λ_g value requires more-accurate definition.

Luminescence linewidth [cm^{-1}]	Laser transition	Energy terminal level [cm^{-1}]	Excitation system, pump source, filter, cooling	Rod length, rod diameter [mm]; resonator cavity parameters, mirrors (coating, transmittance)	Remarks	Reference
	$^5I_7 \rightarrow {}^5I_8$					
–		≈ 530	ER with Xe	30; 3		[526, 546]
–		≈ 550	flashtube	$\parallel$, $R = 500$ mm,		
			ISP–250, N$_2$	DM, 0.5%		
			vapor			
≈ 8		≈ 530	Low-temp.	23; 4.5		
≈ 8		≈ 550	ER with Xe	$\parallel$, 10″, $R = 600$		
≈ 12		≈ 500	flashtube	mm, DM, 1%		
			ISP–1000, QLP,			
			N$_2$			
	$^5I_7 \rightarrow {}^5I_8$					
≈ 16		≈ 530	Low-temp.	20; 5		[271, 546]
≈ 18		≈ 550	ER with Xe	$\parallel$, 10″, $R = 600$		
			flashtube	mm, DM, 1%		
			ISP–1000, QLP,			
			N$_2$			
	$^5I_7 \rightarrow {}^5I_8$					
≈ 12		≈ 550	Low-temp.	37; 6.3		[546]
≈ 25		≈ 500	ER with Xe	$\parallel$, 10″, $R = 600$		
			flashtube	mm, DM, 1%		
			ISP–1000, QLP,			
			N$_2$			
	$^5I_7 \rightarrow {}^5I_8$					
–		–	Under P conditions, illuminator with Xe flashtube; and under CW conditions, with W lamp (l = 65 mm), N$_2$	12 $\parallel$, DM	EET to Ho^{3+}	[136]
	$^3H_4 \rightarrow {}^3H_6$					
–		–	– Round illuminator with 12 W lamps, N$_2$	10 $\parallel$, Ag	EET to Tm^{3+}	[149]

(Continued)

Table 5.1 (*continued*)

Crystal host, space group, cation-site symmetry	Activator [%]	Emission wavelength [μm]	Temperature [K]	Mode of operation	Threshold energy [J] or power [W]	Excitation spectral range [μm]	Luminescence lifetime [ms]
$ErAlO_3$ D_{2h}^{16}–Pbnm (Pnma) $C_s(Er^{3+})$	Ho^{3+} ≈2	2.1205 [36]	77	P	80 [f]	0.4–2 Excitation into Ho^{3+} and Er^{3+} bands	–
$ErAlO_3$ D_{2h}^{16}–Pbnm (Pnma) $C_s(Er^{3+})$	Tm^{3+} –	1.872 [a]	77 150	P P	15 [f] 300	0.4–1.9 Excitation into Er^{3+} and Tm^{3+} bands	5.5 3.5
$Er_3Al_5O_{12}$ O_h^{10}–Ia3d $D_2(Er^{3+})$	Er^{3+} 100	2.9367 [37]	300	P	12 [g]	0.5–1.1 $\tau_{exc} = 40$ μs	0.07
$Er_3Al_5O_{12}$ O_h^{10}–Ia3d $D_2(Er^{3+})$	Tm^{3+} –	1.9–2.0	77	P	– [e]	0.4–2 Excitation into Er^{3+} and Tm^{3+} bands	–
Er_2SiO_5 C_{2h}^6–C2/c $C_I'(Er_I^{3+})$, $C_I''(Er_{II}^{3+})$	Ho^{3+} 0.01	2.085 [a]	≈110	P	10 [f]	0.3–2 Excitation into Ho^{3+} and Er^{3+} bands	0.08
$Er_3Sc_2Al_3O_{12}$: :Tm^{3+} O_h^{10}–Ia3d $D_2(Er^{3+})$	Ho^{3+} ≈2	2.0985	77	P	10	1.4–2 Excitation into Ho^{3+}, Er^{3+}, and Tm^{3+} bands	–

[a] λ_g value requires more-accurate definition.
[e] The spectroscopic data confirming this result are not published.
[f] The composition of the stimulated-emission spectrum and the threshold energies depend on the mutual orientation of the geometric axis of the rod F (axis of the laser element) and of the crystallographic directions of the crystal.
[g] The E_{thr} value depends on τ_{exc}.
[25] The length and diameter of the laser rod are not indicated. The excitation thresholds are in J cm^{-1}.
[36] A misprint of the value of λ_g appears in [137].

Lumines-cence linewidth $[cm^{-1}]$	Laser transition	Energy terminal level $[cm^{-1}]$	Excitation system, pump source, filter, cooling	Rod length, rod diameter [mm]; resonator cavity parameters, mirrors (coating, transmittance)	Remarks	Reference
≈ 10	$^5I_7 \rightarrow {}^5I_8$	≈ 570	Low-temp. ER with Xe flashtube IFP–400, QLP, N_2	15; 4 $\parallel$, 10″, $R = 600$ mm, DM, 0.5%	EET to Ho^{3+}	[137]
–	$^3H_4 \rightarrow {}^3H_6$	≈ 200	ER with Xe flashtube IFP–2000, N_2	12; 4 $\parallel$, Ag, 2%	EET to Tm^{3+}	[225]
≈ 10	$^4I_{11/2} \rightarrow {}^4I_{13/2}$	≈ 6880	ER with Xe flashtube ISP-–250, ZhS–17	32; 4 $\parallel$, $R = 600$ mm, DM, 1%		[572]
–	$^3H_4 \rightarrow {}^3H_6$	–	–	≈ 18 [38]	EET to Tm^{3+}	[150]
–	$^5I_7 \rightarrow {}^5I_8$	≈ 500	Illuminator with Xe flash-tube, N_2 vapor	– [25] $\parallel$, $R = 500$ mm, DM, 1%	EER to Ho^{3+}	[271, 638]
≈ 22	$^5I_7 \rightarrow {}^5I_8$	≈ 470	Low-temp. ER with Xe flashtube, QLP, N_2	20; 5.5 $\parallel$, 10″, $R = 600$ mm, DM, 1%	$Tm^{3+} = 5\%$ EET to Ho^{3+}	[86, 488, 529]

[37] Later, more-accurate measurements show the value of λ_g for the $Er_3Al_5O_{12}$–Er^{3+} crystal at 300 K must be taken as 2.9367 ± 0.0005 μm, and not 2.9370 ± 0.0005 μm, as in [572].

[38] The authors of [150] refer to a private communication of *Johnson* and *Geusic*.

(*Continued*)

Table 5.1 (*continued*)

Crystal host, space group, cation-site symmetry	Activator [%]	Emission wavelength [μm]	Temperature [K]	Mode of operation	Threshold energy [J] or power [W]	Excitation spectral range [μm]	Luminescence lifetime [ms]
ErVO$_4$:Tm^{3+} D_{4h}^{19}–14/amd D_{2d}(Er^{3+})	Ho^{3+} 1 $c \perp$ F	2.0416	≈ 77	P	–	0.4–2 Excitation into Ho^{3+} and Tm^{3+} bands	–
			≈ 77	CW	–	LP; 0.4880 and 0.5145	
(Er,Tm,Yb)$_3$ Al$_5$O$_{12}$ O_h^{10}–Ia3d D_2(Er^{3+},Tm^{3+}, Yb^{3+})	Ho^{3+} ≈ 6	2.1010	77	P	10	0.4–2 Excitation into Ho^{3+}, Er^{3+},Tm^{3+}, and Yb^{3+} bands	–
(Er,Yb)$_3$Al$_5$O$_{12}$ O_h^{10}–Ia3d D_2(Er^{3+}, Yb^{3+})	Tm^{3+} ≈ 3	1.8850 2.0195	77 77	P P	48 25	0.4–2 Excitation into Er^{3+}, Tm^{3+}, and Yb^{3+} bands	–
(Er,Lu)AlO$_3$ D_{2h}^{16}–Pbnm (Pnma) C_s(Er^{3+}, Lu^{3+})	Ho^{3+} 2–5	2.0010 2.1205	77 77	P P	75 [f] 30	0.4–2 Excitation into Ho^{3+} and Er^{3+} bands	–
(Er,Lu)AlO$_3$ D_{2h}^{16}–Pbnm (Pnma) C_s(Er^{3+}, Lu^{3+})	Tm^{3+} 2–5	1.8845	77	P	45 [f]	0.4–1.9 Excitation into Er^{3+} and Tm^{3+} bands	–
(Er,Lu)$_3$Al$_5$O$_{12}$ O_h^{10}–Ia3d D_2(Er^{3+}, Lu^{3+})	Er^{3+} ≈ 35	2.8298 2.9395	300 300	P P	12 [g] 12 [g]	0.4–1.1 $\tau_{exc} = 40$–60 μs	≈ 0.11 ($\approx 1\%$)

[f] The composition of the stimulated-emission spectrum and the threshold energies depend on the mutual orientation of the geometric axis of the rod F (axis of the laser element) and of the crystallographic directions of the crystal.

[g] The E_{thr} value depends on τ_{exc}.

(39) Fresnel reflection from the crystal end faces was used.

Lumines-cence linewidth [cm^{-1}]	Laser transition	Energy terminal level [cm^{-1}]	Excitation system, pump source, filter, cooling	Rod length, rod diameter [mm]; resonator cavity parameters, mirrors (coating, transmittance)	Remarks	Reference
–	$^5I_7 \rightarrow {}^5I_8$	≈ 230	ER with Xe flashtube, N$_2$	7; 5 $\parallel$ [(39)]	Tm^{3+} = 1% $\mathbf{E} \parallel c$ axis EET to Ho^{3+}	[660]
			CW Ar ion laser, LPG			
≈ 12	$^5I_7 \rightarrow {}^5I_8$	≈ 500	Low-temp. ER with Xe flashtube, QLP, N$_2$	31; 6 $\parallel$, $R = 600$ mm, DM, 2%	Er^{3+} = Tm^{3+} = = Yb^{3+} = 31.3% EET to Ho^{3+}	[86, 486]
≈ 7 ≈ 12	$^3H_4 \rightarrow {}^3H_6$	≈ 240 ≈ 580	Low-temp. ER with Xe flashtube IFP–400, QLP, N$_2$	20; 5 $\parallel$, $R = 600$ mm, DM, 2%	Er^{3+} = Yb^{3+} $\approx 48\%$ EET to Tm^{3+}	[86, 486]
6 10	$^5I_7 \rightarrow {}^5I_8$	≈ 200 ≈ 570	Low-temp. ER with Xe flashtube IFP–400, N$_2$	25; 5 $\parallel$, $R = 600$ mm, DM, 1%	EET to Ho^{3+}	[421]
5.5	$^3H_4 \rightarrow {}^3H_6$	≈ 400	Low-temp. ER with Xe flashtube IFP–400, N$_2$	25; 5 $\parallel$, $R = 600$ mm, DM, 1%	EET to Tm^{3+}	[421]
≈ 8.5 ≈ 9	$^4I_{11/2} \rightarrow {}^4I_{13/2}$	≈ 6885 (6847) ≈ 6885	ER with Xe flashtube ISP–250	31; 5 $\parallel$, 10″, $R = 600$ mm, DM, 1%		[546, 565, 569, 572]

(*Continued*)

Table 5.1   (*continued*)

Crystal host, space group, cation-site symmetry	Activator [%]	Emission wavelength [μm]	Temperature [K]	Mode of operation	Threshold energy [J] or power [W]	Excitation spectral range [μm]	Luminescence lifetime [ms]
$(Er,Lu)_3Al_5O_{12}$: Er^{3+} :Ho^{3+}, Tm^{3+} O_h^{10}–Ia3d $D_2(Er^{3+}, Lu^{3+})$	≈ 33	2.6990	300	P	20 ^{g}	0.4–1.1 $\tau_{exc} = 40$–60 μs	≈ 0.11
$Yb_3Al_5O_{12}$ O_h^{10}–Ia3d $D_2(Yb^{3+})$	Ho^{3+} 1	2.0960 ^{c}	77	P	143	0.4–2 Excitation into Ho^{3+} and Yb^{3+} bands	–
$Yb_3Al_5O_{12}$ O_h^{10}–Ia3d $D_2(Yb^{3+})$	Er^{3+} ≈ 3	1.6615	77	P	75	0.4–1.6 Excitation into Er^{3+} and Yb^{3+} bands	–
$(Yb,Lu)_3Al_5O_{12}$ O_h^{10}–Ia3d $D_2(Yb^{3+}, Lu^{3+})$	Yb^{3+} ≈ 35 ≈ 50	1.0294	77 77	P P	10 ≈ 350	0.9–1.1	0.4 ≈ 1.1 ($\approx 1\%$)
$LuAlO_3$ D_{2h}^{16}–Pbnm (Pnma) $C_s(Lu^{3+})$	Nd^{3+} 1 [112]‖F	1.0671 C 1.0831 A 1.0675* C 1.0759* D 1.0832 A 1.0671*→ →1.0687 C 1.0831→ →1.0834 A	77 120 300 300 300 77→550 120→ →500	P P P P P P P	≈ 1.7 ≈ 55 – – ≈ 8 – –	0.5–0.9	0.16 (0.2%) 0.16
	1 [112]‖F	1.3437	300	P	≈ 15	0.5–0.9	0.16 (0.2%)
$LuAlO_3$ D_{2h}^{16}–Pbnm (Pnma) $C_s(Lu^{3+})$	Ho^{3+} – [112]‖F	2.1348 ^{c}	≈ 90	P	3	0.5–2	–

^{c} The detailed spectroscopic investigations were not carried out.

Lumines-cence linewidth [cm^{-1}]	Laser transition	Energy terminal level [cm^{-1}]	Excitation system, pump source, filter, cooling	Rod length, rod diameter [mm]; resonator cavity parameters, mirrors (coating, transmittance)	Remarks	Reference
	$^4I_{11/2} \to {^4I_{13/2}}$					
≈ 4		≈ 6560	ER with Xe flashtube ISP–250	40; 5.5 $\parallel$, 5", $R = 600$ mm, DM, 1%	RED from Er^{3+}	[655]
	$^5I_7 \to {^5I_8}$					
–		≈ 100	ER with Xe flashtube IFP–400, N$_2$	25; 2.5 $\parallel$, Au		[422]
	$^4I_{13/2} \to {^4I_{15/2}}$					
≈ 12		≈ 500	Low-temp. ER with Xe flashtube, QLP, N$_2$	20; 5 $\parallel$, $R = 600$ mm, DM, 2%	EET to Er^{3+}	[486]
	$^2F_{5/2} \to {^2F_{7/2}}$					
≈ 10		621	Low-temp. ER with Xe flashtube IFP-–400, QLP, N$_2$	25; 5 $\parallel$, 10", $R = 576$ mm, DM, 1%		[86, 546]
	$^4F_{3/2} \to {^4I_{11/2}}$					
≈ 1.7		2022	Low- and high-temp. ER with Xe flash-tube IFP–400, QLP, ZhS–17, N$_2$ and its vapor	18; 5 $\parallel$, 10", $R = 576$ mm, DM, 1%	At ≈ 160 K $E_{thr}(A) =$ $= E_{thr}(C)$ *In CAM laser	[271, 565, 637]
≈ 2.8		2160				
≈ 7.5		2026				
≈ 8.5		2099				
11.5		2161				
1.7$\to$		2026				
3$\to$		2160				
	$^4F_{3/2} \to {^4I_{13/2}}$					
≈ 7.5		3950	ER with Xe flashtube IFP–400, ZhS–17	18; 5 $\parallel$, 10", $R = 600$ mm, DM, 1%	$n_a = 0.91$ $n_b = 1.93$ $n_c = 1.88$ (at $\lambda = 0.6328$ μm)	[271, 565, 637]
	$^5I_7 \to {^5I_8}$					
–		≈ 470	ER with Xe flashtube ISP–250, N$_2$ vapor	–		[571]

^{9} The E_{thr} value depends on τ_{exc}.

(*Continued*)

Table 5.1   (*continued*)

Crystal host, space group, cation-site symmetry	Activator [%]	Emission wavelength [μm]	Temperature [K]	Mode of operation	Threshold energy [J] or power [W]	Excitation spectral range [μm]	Luminescence lifetime [ms]
$LuAlO_3$ D_{2h}^{16}–Pbnm (Pnma) $C_s(Lu^{3+})$	Er^{3+} 1 [112]‖F	1.6675 [c]	≈90	P	8	0.3–0.55	–
$Lu_3Al_5O_{12}$ O_h^{10}–Ia3d $D_2(Lu^{3+})$	Nd^{3+} ≈0.6	0.9473	77	P	40	0.5–0.9	0.245
	≈0.6	1.06425 A	300	P	1	0.5–0.9	0.245
		1.0605 B	77	P	0.5		0.245
		1.06375→ →1.0672 A	120→ →900	P	–		
		1.0610* B	300	P	–		
		1.0535* C	300	P			
		1.06425 A	300	CW	950	0.3–0.9	
	≈0.6	1.3209*	300	P	–	0.5–0.9	0.245
		1.3326*	300	P	–		
		1.3342*	300	P	–		
		1.3382**	300	P	2.5		
		1.3410*	300	P	–		
		1.3532	300	P	4.5		
		1.3319*	77	P	–		0.245
		1.3333*	77	P	–		
		1.3376*	77	P	–		
		1.3499*	77	P	–		
		1.3525	77	P	0.9		
		1.3382	300	CW	2000	0.3–0.9	
$Lu_3Al_5O_{12}$ O_h^{10}–Ia3d $D_2(Lu^{3+})$	Ho^{3+} ≈2	2.1020	77	P	50	0.4–2	4–5
	≈5	2.9460	300	P	60 [g]	0.4–1.2 $\tau_{exc} = 60$– –100 μs	–

[c] The detailed spectroscopic investigations were not carried out.

Lumines-cence linewdth [cm^{-1}]	Laser transition	Energy terminal level [cm^{-1}]	Excitation system, pump source, filter, cooling	Rod length, rod diameter [mm]; resonator cavity parameters, mirrors (coating, transmittance)	Remarks	Reference
	$^4S_{3/2} \rightarrow {}^4I_{9/2}$					
–		≈ 12300	ER with Xe flashtube ISP–250, N$_2$ vapor	25; 5 $\parallel$, $R = 500$ mm, DM, 1%		[571]
	$^4F_{3/2} \rightarrow {}^4I_{9/2}$					
≈ 6		878	Low-temp. ER with Xe flashtube IFP–1000, N$_2$	25; 5 $\parallel$, 5″, $R = 576$ mm, DM, 2%	$n \approx 1.8$	[191, 271]
	$^4F_{3/2} \rightarrow {}^4I_{11/2}$					
5.3		2099	Low- and high-	25; 5	$\sigma_e = 3.5 \times 10^{-19}$	[191, 210,
1.4		2004	temp. ER	$\parallel$, 5″, $R = 576$		271, 278]
$2 \rightarrow$		2100	with Xe flash-tube IFP–400, QLP, ZhS–17,	mm, DM, 1%	At ≈ 177 K $E_{\mathrm{thr}}(A) =$ $= E_{\mathrm{thr}}(B)$	
5		2003	N$_2$ and its		*In CAM laser	
4.6		2003	vapor			
5.3		2099	ER with Xe lamp DKsTV–3000, water		$\delta\lambda/\delta T = 4.9$	
	$^4F_{3/2} \rightarrow {}^4I_{13/2}$					
9		3924	Low- and room-	25; 5	*In CAM laser	[89, 191,
7		3924	temp. ER	$\parallel$, 5″, $R = 600$		222, 271,
7.3		3933	with Xe flash-tube IFP–400,	mm, DM, 1%		591]
9.5		4022	QLP, ZhS–17,		**From this	
10.5		4038	N$_2$		line was obtain-	
10.5		4038			ed the second	
4.3		3926			harmonic using	
4.8		3934			LiIO$_3$ crystal.	
5.2		4026				
5.5		4026				
5.8		4040				
9.5		4022	ER with Xe lamp DKsTV–3000, water			
	$^5I_7 \rightarrow {}^5I_8$					
11		≈ 480	Low-temp. ER with Xe flashtube IFP–400, N$_2$	23; 6 $\parallel$, 10″, $R = 600$ mm, DM, 1%		[143]
	$^5I_6 \rightarrow {}^5I_7$					
≈ 15		≈ 5400	ER with Xe flashtube ISP–250	37; 6 $\parallel$, 10″, $R = 600$ mm, DM, 1%		[546, 547, 565, 655]

[9] The E_{thr} value depends on τ_{exc}.

(Continued)

Table 5.1   (*continued*)

Crystal host, space group, cation-site symmetry	Activator [%]	Emission wavelength [μm]	Temperature [K]	Mode of operation	Threshold energy [J] or power [W]	Excitation spectral range [μm]	Luminescence lifetime [ms]
$Lu_3Al_5O_{12}$ O_h^{10}–Ia3d $D_2(Lu^{3+})$	Er^{3+} ≈ 8 1.5	0.86325 0.8632	77 300	P P	100 [g] 45 [g]	0.35–0.55 $\tau_{exc} = 60$– –60 μs	≈ 0.11 (0.1%)
	2–5	1.6525 1.6630	77 77	P P	100 40	0.4–1.6	6.4
	1.5	1.7762	300	P	30 [g]	0.35–0.55 $\tau_{exc} = 60$ μs	≈ 0.11
	1.5 ≈ 8	2.9408 2.8298	300 300	P P	55 [g] ≈ 25 [g]	0.4–1.2 $\tau_{exc} = 60$ μs	≈ 0.12
		2.9406	300	P	≈ 25		
$Lu_3Al_5O_{12}$ O_h^{10}–Ia3d $D_2(Lu^{3+})$	Tm^{3+} ≈ 2	1.8855 2.0240	77 77	P P	400 160	0.4–2	–
$Lu_3Al_5O_{12}$ O_h^{10}–Ia3d $D_2(Lu^{3+})$	Yb^{3+} ≈ 2	1.0294 1.0297	77 175	P P	1.5 35	0.9–1.1	1 ≈ 1.1
$Lu_3Al_5O_{12}$: :Nd^{3+} O_h^{10}–Ia3d $D_2(Lu^{3+})$	Yb^{3+} ≈ 2	1.0294	77	P	1	0.5–1.1 Excitation into Nd^{3+} and Yb^{3+} bands	≈ 1

[g] The E_{thr} value depends on τ_{exc}.

Luminescence linewidth [cm⁻¹]	Laser transition	Energy terminal level [cm⁻¹]	Excitation system, pump source, filter, cooling	Rod length, rod diameter [mm]; resonator cavity parameters, mirrors (coating, transmittance)	Remarks	Reference
≈ 8.5 ≈ 11	$^4S_{3/2} \to {}^4I_{13/2}$	6824 6818	Low- and room-temp. ER with Xe flash-tube IFP–400, QLP, N_2	35–43; 6–7 $\parallel$, 10″, $R = 600$ mm, DM, 1%		[546, 547, 565]
6.5 8.5	$^4I_{13/2} \to {}^4I_{15/2}$	≈ 530 ≈ 560	Low-temp. ER with Xe flashtube IFP-–400, QLP, N_2	23; 6 $\parallel$, 10″, $R = 600$ mm, DM, 1%		[143]
–	$^4S_{3/2} \to {}^4I_{9/2}$	12772	ER with Xe flashtube ISP–250	43; 7 $\parallel$, 10″, $R = 600$ mm, DM, 1%		[546, 547, 565]
≈ 9 ≈ 8.5 ≈ 9	$^4I_{11/2} \to {}^4I_{13/2}$	6885 ≈ 6885 (6847) ≈ 6885	ER with Xe flashtube ISP–250	43; 7 $\parallel$, 10″, $R = 600$ mm, DM, 1%		[546, 547, 565, 572]
6 10.5	$^3H_4 \to {}^3H_6$	≈ 225 ≈ 580	Low-temp. ER with Xe flashtube IFP-–400, QLP, N_2	23; 6 $\parallel$, 10″, $R = 600$ mm, DM, 1%		[143]
≈ 7 ≈ 21	$^2F_{5/2} \to {}^2F_{7/2}$	621 ≈ 621	Low-temp. ER with Xe flashtube IFP–400, QLP, N_2 and its vapor	30; 6 $\parallel$, 5″, $R = 576$ mm, DM, 2%	$\sigma_e = 1.8 \times 10^{-19}$ $\Delta v_g = 3$ cm⁻¹	[86, 483]
≈ 7	$^2F_{5/2} \to {}^2F_{7/2}$	621	Low-temp. ER with Xe flashtube IFP–400, QLP, N_2	15, 5 $\parallel$, 5″, $R = 576$ mm, DM, 2%	$Nd^{3+} = 1$ at.% $\sigma_e = 1.8 \times 10^{-19}$ $\Delta v_g \approx 1$ cm⁻¹ EET to Yb^{3+}	[86, 483]

(*Continued*)

Table 5.1 (*continued*)

Crystal host, space group, cation-site symmetry	Activator [%]	Emission wavelength [μm]	Temperature [K]	Mode of operation	Threshold energy [J] or power [W]	Excitation spectral range [μm]	Luminescence lifetime [ms]
$Lu_3Al_5O_{12}$: :Nd^{3+}, Cr^{3+} O_h^{10}–Ia3d $D_2(Lu^{3+})$	Yb^{3+} 5	1.0294	77	P	1.5	0.35–1.1 Excitation into Nd^{3+}, Yb^{3+}, and Cr^{3+} bands	≈ 1
$Lu_3Al_5O_{12}$: :Tm^{3+} O_h^{10}–Ia3d $D_2(Lu^{3+})$	Er^{3+} ≈ 10	2.6990	300	P	18 ^{g}	0.4–1.1 $\tau_{exc} = 60$ μs	–
$Lu_3Al_5O_{12}$: :Ho^{3+}, Tm^{3+} O_h^{10}–Ia3d $D_2(Lu^{3+})$	Er^{3+} ≈ 9	2.6990	300	P	25 ^{g}	0.4–1.1 $\tau_{exc} = 60$ μs	–
$Lu_3Al_5O_{12}$: :Er^{3+}, Tm^{3+} O_h^{10}–Ia3d $D_2(Lu^{3+})$	Ho^{3+} ≈ 2	2.1020 [(40)]	77 77	P CW	15 65	0.4–2 Excitation into Ho^{3+}, Er^{3+}, and Tm^{3+} bands	–
$Lu_3Sc_2Al_3O_{12}$ O_h^{10}–Ia3d $D_2(Lu^{3+})$	Nd^{3+} ≈ 1	1.0599 1.0620* 1.0591	300 300 77	P P P	3 – 3	0.5–0.9	≈ 0.28 (0.3%) ≈ 0.28
	≈ 1	1.3360	300	P	5	0.5–0.9	≈ 0.28 (0.3%)

^{g} The E_{thr} value depends on τ_{exc}.
[(40)] Nearly the same spectral and energetic characteristics hold for lasers based on the $Lu_3Al_5O_{12}$:Er^{3+}, Tm^{3+}, Yb^{3+}–Ho^{3+} crystal.

Lumines-cence linewidth [cm^{-1}]	Laser transition	Energy terminal level [cm^{-1}]	Excitation system, pump source, filter, cooling	Rod length, rod diameter [mm]; resonator cavity parameters, mirrors (coating, transmittance)	Remarks	Reference
≈ 7	$^2F_{5/2} \rightarrow {}^2F_{7/2}$	621	Low-temp. ER with Xe flashtube IFP–400, QLP, N$_2$	31; 5 $\parallel$, 7″, $R = 576$ mm, DM, 1%	Nd^{3+} = 0.5%, Cr^{3+} = 0.1% $\sigma_e = 1.8 \times 10^{-19}$ $\Delta\nu_g = 1$ cm^{-1} EET to Yb^{3+}	[483]
≈ 4	$^4I_{11/2} \rightarrow {}^4I_{13/2}$	≈ 6560	ER with Xe flashtube ISP–250	31; 5.5 $\parallel$, 5″, $R = 600$ mm, DM, 1%	Tm^{3+} = 3 at.% RED from Er^{3+}	[655]
≈ 4	$^4I_{11/2} \rightarrow {}^4I_{13/2}$	≈ 6560	ER with Xe flashtube ISP–250	29; 6 $\parallel$, 10″, $R = 600$ mm, DM, 1%	Ho^{3+} = 3 at.%, Tm^{3+} = 1 at.% RED from Er^{3+}	[546, 547, 565]
≈ 11	$^5I_7 \rightarrow {}^5I_8$	≈ 480	Under P conditions, ER with flashtube IFP–400; and under CW conditions, with W lamp, N$_2$	23; 6 $\parallel$, 10″, $R = 600$ mm, DM, 1%	Er^{3+} = Tm^{3+} = = 2 at.% EET to Ho^{3+}	[143]
≈ 10 ≈ 15 ≈ 5	$^4F_{3/2} \rightarrow {}^4I_{11/2}$	1978 2111 1979	Low- and room-temp. ER with Xe flashtube IFP–400, QLP, ZhS–17, N$_2$	25; 5 $\parallel$, 7″, $R = 576$ mm, DM, 1%	*In CAM laser	[86, 488, 529]
≈ 12	$^4F_{3/2} \rightarrow {}^4I_{13/2}$	3931	ER with Xe flashtube IFP––400, ZhS–17	25; 5 $\parallel$, 7″, $R = 600$ mm, DM, 1%		[86, 488, 529]

(Continued)

Table 5.1 (*continued*)

Crystal host, space group, cation-site symmetry	Activator [%]	Emission wavelength [μm]	Temperature [K]	Mode of operation	Threshold energy [J] or power [W]	Excitation spectral range [μm]	Luminescence lifetime [ms]
$Lu_3Sc_2Al_3O_{12}$: :Nd^{3+} O_h^{10}–Ia3d $D_2(Lu^{3+})$	Yb^{3+} 2	1.0299	77	P	5	0.5–1.1 Excitation into Nd^{3+} and Yb^{3+} bands	≈1.2
$Lu_3Ga_5O_{12}$ O_h^{10}–Ia3d $D_2(Lu^3)$	Nd^{3+} ≈1	1.0594* C 1.0609 B 1.0623 A 1.0587 C 1.06025 B 1.0616 A	300 300 300 77 77 77	P P P P P P	– 8 1.5 8 4.5 10	0.5–0.9	≈0.27
	≈1	1.3315	300	P	6	0.5–0.9	≈0.27
$Lu_3Ga_5O_{12}$: : Nd^{3+} O_h^{10}–Ia3d $D_2(Lu^{3+})$	Yb^{3+} 2	1.0230	77	P	1.5	0.5–1.1 Excitation into Nd^{3+} and Yb^{3+} bands	≈1.3
$PbMoO_4$ C_{4h}^6–I4$_1$/a $S_4(Pb^{2+})$	Nd^{3+} –	1.0586	300	P	60 [f]	0.4–0.9	0.13
	1.5 $c \perp F$	1.3340 1.3425* 1.3320 1.3375* 1.3450* 1.3780*	300 300 77 77 77 77	P P P P P P	50 – 100 – – –	0.5–0.9 0.4–0.9	0.13 0.13
$Bi_4Si_3O_{12}$ T_d^6–I4̄3d	Nd^{3+} 1	1.0629 1.0629	77 300	P P	1 7	0.5–0.9	0.23 (0.3%) 0.23

[f] The composition of the stimulated-emission spectrum and the threshold energies depend on the mutual orientation of the geometric axis of the rod F (axis of the laser element) and of the crystallographic directions of the crystal.

Luminescence linewidth [cm^{-1}]	Laser transition	Energy terminal level [cm^{-1}]	Excitation system, pump source, filter, cooling	Rod length, rod diameter [mm]; resonator cavity parameters, mirrors (coating, transmittance)	Remarks	Reference
	$^2F_{5/2} \to {}^2F_{7/2}$				$Nd^{3+} = 1.5\%$	
≈ 13		≈ 600	Low-temp. ER with Xe flashtube IFP–400, QLP, ZhS–17, N$_2$	25; 6 $\parallel$, 7″, $R = 576$ mm, DM, 1%	$\Delta v_g = 1.5$ cm^{-1} EET to Yb^{3+}	[483, 488, 529]
	$^4F_{3/2} \to {}^4I_{11/2}$					
≈ 9		≈ 1990	Low- and room-temp. ER with Xe flash-tube IFP–400, QLP, ZhS–17, N$_2$	23; 5 $\parallel$, 8″, $R = 576$ mm, DM, 1%	*In CAM laser $\Delta v_g < 1$ cm^{-1}	[489, 506]
≈ 9		≈ 2005			$\Delta v_g < 1$ cm^{-1}	
≈ 9		≈ 2060			$\Delta v_g < 0.5$ cm^{-1}	
4		1996			$\Delta v_g < 0.5$ cm^{-1}	
4		2010			$\Delta v_g < 0.5$ cm^{-1}	
4,5		2065				
	$^4F_{3/2} \to {}^4I_{13/2}$					
≈ 13		≈ 3930	ER with Xe flashtube IFP––400, ZhS–17	23; 5 $\parallel$, 8″, $R = 600$ mm, DM, 1%		[489, 506]
	$^2F_{5/2} \to {}^2F_{7/2}$				$Nd^{3+} = 1.5\%$	
≈ 8		≈ 600	Low-temp. ER with Xe flashtube IFP–400, QLP, ZhS–17, N$_2$	25; 5 $\parallel$, 7″, $R = 576$ mm, DM, 1%	$\Delta v_g = 1$ cm^{-1} EET to Yb^{3+}	[86, 483]
	$^4F_{3/2} \to {}^4I_{11/2}$					
–		≈ 2000	Illuminator with Xe flash-tube FT–524	52; 3 R, Ag, 1%		[183, 423]
	$^4F_{3/2} \to {}^4I_{13/2}$					
≈ 35		3950	Low- and room-temp. ER with Xe flash-tube IFP–400, ZhS–17 and without it, N$_2$	21; 5 $\parallel$, $R = 600$ mm, DM, 1%	*In CAM laser	[97]
≈ 40		3970				
5		3947				
6		3920				
6		3962				
14		4140				
	$^4F_{3/2} \to {}^4I_{11/2}$					
≈ 3		1928	Low- and room-temp. ER with Xe flash-tube IFP–400, QLP, ZhS–17, N$_2$	17; 4 $\parallel$, 7″, $R = 576$ mm, DM, 1%	$\sigma_e = 9 \times 10^{-19}$ $\Delta v_g = 0.5$ cm^{-1}	[566]
≈ 20		1924			$\sigma_e = 1.3 \times 10^{-19}$ $\Delta v_g = 1.5$ cm^{-1} $n = 2.02$ at $\lambda = 0.633$ μm	

(*Continued*)

Table 5.1 (*continued*)

Crystal host, space group, cation-site symmetry	Activator [%]	Emission wavelength [μm]	Temperature [K]	Mode of oper- ation	Threshold energy [J] or power [W]	Excitation spectral range [μm]	Lumines- cence lifetime [ms]
	1	1.3407	300	P	7	0.5–0.9	0.23 (0.3%)
$Bi_4Ge_3O_{12}$ T_d^6–I$\bar{4}$3d	Nd^{3+} 1	1.0644	300	P	4.5	0.5–0.9	0.22 (0.3%)
		1.06425	77	P	0.65		0.22
	1	1.3418	300	P	4.5	0.5–0.9	0.22 (0.3%)
						Mixed oxide laser crystals	
$LiLa(MoO_4)_2$ C_{4h}^6–I4$_1$/a $S_4(La^{3+})$	Nd^{3+} 2 $c \perp F$	1.0585 [(41)]	300	P	3	0.35–0.9	$\approx$0.15
		1.0658	77	P	15		$\approx$0.15
	2 $c \perp F$	1.3370	300	P	20	0.35–0.9	$\approx$0.15
		1.3375	77	P	35		$\approx$0.15
		1.3440*	77	P	–		
$Li(Nd,La)P_4O_{12}$ C_{2h}^6–C2/c $C_2(Nd^{3+}, La^{3+})$	Nd^{3+} 50	1.0477	300	CW	–	LP; 0.5145	0.19 0.32 ($\approx$1%)
$Li(Nd,Gd)P_4O_{12}$ C_{2h}^6–C2/c $C_2(Nd^{3+}, Gd^{3+})$	Nd^{3+} 50	1.0477	300	CW	0.003 [b]	LP; 0.5145	0.18 0.32 ($\approx$1%)

[b] The absorbed pump power (energy) at which stimulated emission begins is indicated.

[(41)] The emission wavelength corresponds to the wavelength of the luminescence band maximum, the contour of which at 300 K is characterized by overlap of some luminescence lines connected with Stark components of the $^4F_{3/2}$ manifold. Therefore, the λ_g value given does not correctly correspond to the wavelength of any "Stark" transitions that participate in the formation of this contour.

Luminescence linewidth [cm^{-1}]	Laser transition	Energy terminal level [cm^{-1}]	Excitation system, pump source, filter, cooling	Rod length, rod diameter [mm]; resonator cavity parameters, mirrors (coating, transmittance)	Remarks	Reference
≈ 29	$^4F_{3/2} \to {}^4I_{13/2}$	3873	ER with Xe flashtube IFP–400, ZhS–17	17; 4 $\parallel$, 7″, $R = 600$ mm, DM, 1%	$\sigma_e = 10^{-19}$	[566]
≈ 24	$^4F_{3/2} \to {}^4I_{11/2}$	1932	Low- and room-temp. ER with Xe flashtube IFP–400, QLP, ZhS–17, N_2	20–37; 5–6 $\parallel$, 7″, $R = 600$ mm, DM, 1%	$\sigma_e = 1.3 \times 10^{-19}$	[156, 567]
≈ 3		1931			$\sigma_e = 9 \times 10^{-19}$ $n = 2.04$ $K \approx 60$	
≈ 29	$^4F_{3/2} \to {}^4I_{13/2}$	3875	ER with Xe flashtube ISP–1000, ZhS–17	20–37; 5–6 $\parallel$, 7″, $R = 600$ mm, DM, 1%	$\sigma_e = 10^{-19}$ $n \approx 2$ The second harmonic was obtained using $LiIO_3$ crystal.	[567]

with disordered structure

Luminescence linewidth [cm^{-1}]	Laser transition	Energy terminal level [cm^{-1}]	Excitation system, pump source, filter, cooling	Rod length, rod diameter [mm]; resonator cavity parameters, mirrors (coating, transmittance)	Remarks	Reference
≈ 100	$^4F_{3/2} \to {}^4I_{11/2}$	1955–2000	Low- and room-temp. ER with Xe flash-tube IFP–400, QLP, N_2	25; 5 $\parallel$, $R = 576$ mm, DM, 1%	$\Delta v_g \approx 10$ cm^{-1} $n \approx 2.05$	[424]
≈ 30		2003				
≈ 75	$^4F_{3/2} \to {}^4I_{13/2}$	≈ 3910	Low- and room-temp. ER with Xe flash-tube IFP–400, QLP, N_2	25; 5 $\parallel$, $R = 600$ mm, DM, 1%		[97, 424]
30		3909				
30		3945			*In CAM laser	
–	$^4F_{3/2} \to {}^4I_{11/2}$	≈ 1940	CW Ar ion laser	DM, 0.1%	$\sigma_e = 2 \times 10^{-19}$ $\mathbf{E} \parallel c$ axis	[618]
–	$^4F_{3/2} \to {}^4I_{11/2}$	≈ 1940	CW Ar ion laser	DM, 0.1	$\sigma_e = 2 \times 10^{-19}$ $\mathbf{E} \parallel c$ axis	[618]

(*Continued*)

Table 5.1   *(continued)*

Crystal host, space group, cation-site symmetry	Activator [%]	Emission wavelength [μm]	Temperature [K]	Mode of operation	Threshold energy [J] or power [W]	Excitation spectral range [μm]	Luminescence lifetime [ms]
LiGd(MoO$_4$)$_2$ C_{4h}^6–I4$_1$/a S_4(Gd^{3+})	Nd^{3+} ≈ 2 $c \perp F$	1.0599 [41]	300	P	1.3	0.4–0.9	0.14 (0.3%)
			300	CW	1150		
	≈ 2 $c \perp F$	1.3400 1.3400 1.3455*	300 77 77	P P P	13 30 –	0.4–0.9	0.14 (0.3%)
NaLa(MoO$_4$)$_2$ C_{4h}^6–I4$_1$/a S_4(La^{3+})	Nd^{3+} 1–10	1.0595	300	p	5.5 ^{f}	0.35–0.9	0.14 (0.5%)
		1.0653 1.0653→ →1.0665 1.0653	300 300→ →750 300 300	P P q–CW CW	1.6 ^{f} – – 2600		
	1.5 $\angle cF = = 60°$	1.3380 1.3440 1.3380 1.3430 1.3755* 1.3840*	300 300 77 77 77 77	P P P P P P	4.5 10 4 25 – –	0.35–0.9	0.14 (0.5%) 0.14
NaLa(MoO$_3$)$_2$ C_{4h}^6–I4$_1$/a S_4(La^{3+})	Ho^{3+} 1	2.050	90	P	25 ^{f}	0.35–2	8

---

^{f} The composition of the stimulated-emission spectrum and the threshold energies depend on the mutual orientation of the geometric axis of the rod F (axis of the laser element) and of the crystallographic directions of the crystal.

[41] The emission wavelength corresponds to the wavelength of the luminescence band maximum, the contour of which at 300 K is characterized by overlap of some luminescence lines connected with Stark components of the $^4F_{3/2}$ manifold. Therefore, the λ_g value given does not correctly correspond to the wavelength of any "Stark" transitions that participate in the formation of this contour.

Lumines-cence linewidth [cm^{-1}]	Laser transition	Energy terminal level [cm^{-1}]	Excitation system, pump source, filter, cooling	Rod length, rod diameter [mm]; resonator cavity parameters, mirrors (coating, transmittance)	Remarks	Reference
≈ 80	$^4F_{3/2} \rightarrow {}^4I_{11/2}$	2005	ER with Xe flashtube ISP–1000 ER with Xe lamp DKsTV– –3000, water	20; 4 $\parallel$, 12″, $R = 576$ mm, DM, 0.7% 25; 5 $\parallel$, $>10″$, $R = = 576$ mm, DM, 1%	$n \approx 1.95$	[210, 425]
65 25 28	$^4I_{3/2} \rightarrow {}^4I_{13/2}$	≈ 3980 ≈ 3980 ≈ 4000	Low- and room-temp. ER with Xe flash-tube IFP–400, QLP, N$_2$	25; 5 $\parallel$, $R = 600$ mm, DM, 1%	*In CAM laser	[97]
– 50 50→ 50	$^4F_{3/2} \rightarrow {}^4I_{11/2}$	≈ 1945 1996 1996 1996	Room- and high-temp. ER with Xe flash-tubes IFP type ER with PS ER with Xe lamp, water	35; 4 24; 4.8 $\parallel$, 20″, $R = 576$ mm, DM, 0.7%	$\Delta v_g = 5$ cm^{-1} $K \approx 22$ $n \approx 1.97$ $\delta\lambda/\delta T \approx 3.1$	[32, 67, 68, 218, 243]
45 50 15 18 33 35	$^4F_{3/2} \rightarrow {}^4I_{13/2}$	3900 3940 3900 3937 4113 4158	Low- and room-temp. ER with Xe flash-tube IFP–400, QLP, N$_2$	21; 5 $\parallel$, $R = 600$ mm DM, 1%	*In CAM laser	[97]
≈ 24	$^5I_7 \rightarrow {}^5I_8$	≈ 260	ER with Xe flashtube ISP–250, N$_2$ vapor	20; 5–6 $\parallel$, $R = 500$ mm, DM, 0.5%	$\Delta v_g = 10$ cm^{-1}	[406]

(*Continued*)

Table 5.1 (*continued*)

Crystal host, space group, cation-site symmetry	Activator [%]	Emission wavelength [μm]	Temperature [K]	Mode of operation	Threshold energy [J] or power [W]	Excitation spectral range [μm]	Luminescence lifetime [ms]
$NaLa(MoO_4)_2$: :Er^{3+} C_{4h}^6–$I4_1/a$ $S_4(La^{3+})$	Ho^{3+} 3	2.050	≈ 90	P	32 [f,(42)]	0.35–2 Excitation into Ho^{3+} and Er^{3+} bands	10
$NaLa(WO_4)_2$ C_{4h}^6–$I4_1/a$ $S_4(La^{3+})$	Nd^{3+} 1 $c\parallel F$	1.0635	300	P	–	0.35–0.9	–
	1	1.3355	300	P	50 [f]	0.35–0.9	–
$Na_3Nd(PO_4)_2$ D_{2h}^{11}–Pbcm(?)	Nd^{3+} 100	≈ 1.05 [a]	300	CW	–	LP; 0.58	0.023 0.36* (0.5%)
$Na_5Nd(WO_4)_4$ C_{4h}^6–$I4_1/a$ $S_4(Nd^{3+})$	Nd^{3+} 100	1.063 [a]	300	CW	3×10^{-4} [b]	LP; 0.586	0.09 0.22* ($\approx 1\%$)
$NaGd(WO_4)_2$ C_{4h}^6–$I4_1/a$ $S_4(Gd^{3+})$	Nd^{3+} 10–20	≈ 1.06 [a,d]	300	P	16 [f]	0.4–0.9	≈ 0.18
$Na(Nd,Gd)$ $(WO_4)_2$ C_{4h}^6–$I4_1/a$ $S_4(Nd^{3+}, Gd^{3+})$	Nd^{3+} 20–50	≈ 1.06 [a,d]	77	P	4.5 [f]	0.4–0.9	≈ 0.18

[a] λ_g value requires more-accurate definition.

[b] The absorbed pump power (energy) at which stimulated emission begins is indicated.

[c] The detailed spectroscopic investigations were not carried out.

[d] The spectroscopic data confirming this result are not published.

[f] The composition of the stimulated-emission spectrum and the threshold energies depend on the mutual orientation of the geometric axis of the rod F (axis of the laser element) and of the crystallographic directions of the crystal.

[(42)] Such a high E_{thr} value was due to poor quality of the crystal.

Lumines-cence linewidth [cm^{-1}]	Laser transition	Energy terminal level [cm^{-1}]	Excitation system, pump source, filter, cooling	Rod length, rod diameter [mm]; resonator cavity parameters, mirrors (coating, transmittance)	Remarks	Reference
	$^5I_7 \rightarrow {}^5I_8$				$Er^{3+} = 2.8\%$	
≈ 24		≈ 260	ER with Xe flashtube ISP–250, N_2 vapor	20; 5–6 $\parallel$, $R = 500$ mm, DM, 0.5%	EET to Ho^{3+}	[406]
–	$^4F_{3/2} \rightarrow {}^4I_{11/2}$	≈ 2000	ER with Xe flashtube IFP–800	90; 6 $\parallel$, 10″, DM, 15%, DR	$\Delta v_g = 5$ cm^{-1} Retuning in 80 cm^{-1} range $n \approx 1.95$	[426]
≈ 50	$^4F_{3/2} \rightarrow {}^4I_{13/2}$	≈ 4000	ER with Xe flashtube IFP–400	21; 6 $\parallel$, $R = 600$ mm, DM, 1%		[97]
c	$^4F_{3/2} \rightarrow {}^4I_{11/2}$	≈ 2000	CW dye laser	–	*In isostructural $Na_3La(PO_4)_2$ crystal	[577]
–	$^4F_{3/2} \rightarrow {}^4I_{11/2}$	≈ 2000	CW dye laser	$0.2 \times 0.5 \times 0.5$ $\parallel$, DM	*In isostructural $Na_5La(WO_4)_4$ crystal	[577]
–	$^4F_{3/2} \rightarrow {}^4I_{11/2}$	≈ 2000	Illuminator with Xe flash-tube FT–91	28; 2.5 $R = 70$ mm		[427]
–	$^4F_{3/2} \rightarrow {}^4I_{11/2}$	≈ 2000	Illuminator with Xe flash-tube FT–91, N_2	28; 2.5 $R = 70$ mm		[427]

(*Continued*)

Table 5.1 (*continued*)

Crystal host, space group, cation-site symmetry	Activator [%]	Emission wavelength [μm]	Temperature [K]	Mode of oper-ation	Threshold energy [J] or power [W]	Excitation spectral range [μm]	Lumines-cence lifetime [ms]
$KLa(MoO_4)_2$ (?)	Nd^{3+} ≈ 2	1.0587 1.0585	300 77	P P	3.5 [f] 5	0.4–0.9	0.2 (0.5%)
	≈ 2	1.3350 1.3350	300 77	P P	8 [f] 15	0.4–0.9	0.2 (0.5%)
$K_3Nd(PO_4)_2$ C_{2h}^2–$P2_1/m$ $C_i(Nd^{3+})$	Nd^{3+} 100	≈ 1.06 [a, d]	300	–	–	LP	0.021 0.46* (0.5%)
$K_3(Nd,La)$ $(PO_4)_2$ C_{2h}^2–$P2_1/m$ $C_i(Nd^{3+},La^{3+})$	Nd^{3+} 20–40	≈ 1.06 [a, d]	300	–	–	LP	0.46 (0.5%)
$K_5Nd(MoO_4)_4$ D_{3d}^5–$R\bar{3}m$	Nd^{3+} 100	1.0660	300	P	0.025 [f]	LP; 0.745 and 0.805	0.07 0.215* (0.05%)
$K_5Bi(MoO_4)_4$ D_{3d}^5–$R\bar{3}m$	Nd^{3+} ≈ 10	1.0660	300	P	60 [f]	0.5–0.9	0.215 (0.05%)
$CaY_4(SiO_4)_3O$ C_{6h}^2–$P6_5/m$ $C_s(Y^{3+})$	Nd^{3+} – $c \parallel F$	1.0672	300	P	50	0.4–0.9	0.17

[a] λ_g value requires more-accurate definition.
[d] The spectroscopic data confirming this result are not published.
[f] The composition of the stimulated-emission spectrum and the threshold energies depend on the mutual orientation of the geometric axis of the rod F (axis of the laser element) and of the crystallographic directions of the crystal.

Lumines-cence linewidth [cm^{-1}]	Laser transition	Energy terminal level [cm^{-1}]	Excitation system, pump source, filter, cooling	Rod length, rod diameter [mm]; resonator cavity parameters, mirrors (coating, transmittance)	Remarks	Reference
≈ 48 22	$^4F_{3/2} \rightarrow {}^4I_{11/2}$	≈ 1940 1942	Low- and room-temp. ER with Xe flashtube IFP–400, QLP, N$_2$	40; 6 $\parallel$, $R = 576$ mm, Dm, 1%	$\Delta v_g = 10$ cm^{-1} $\Delta v_g \approx 2$ cm^{-1}	[428]
≈ 65 35	$^4F_{3/2} \rightarrow {}^4I_{13/2}$	≈ 3895 3900	Low- and room-temp. ER with Xe flash-tube IFP–400, QLP, N$_2$	40; 6 $\parallel$, $R = 600$ mm, DM, 1%		[428]
–	$^4F_{3/2} \rightarrow {}^4I_{11/2}$	≈ 2000	–	–	*In isostructural K$_3$La(PO$_4$)$_2$ crystal	[619]
–	$^4F_{3/2} \rightarrow {}^4I_{11/2}$	≈ 2000	–	–		[619]
≈ 90	$^4F_{3/2} \rightarrow {}^4I_{11/2}$	≈ 1980	Benzene Raman laser	0.7 $\parallel$, 5″, $R = 600$ mm, DM, 1%	$\sigma_e = 7 \times 10^{-20}$ $\Delta v_g = 25$ cm^{-1} *In isostructural K$_5$Bi(MoO$_4$)$_4$ crystal $n_o \approx 1.78$	[569]
≈ 90	$^4F_{3/2} \rightarrow {}^4I_{11/2}$	≈ 1980	ER with Xe flashtube ISP–1000, ZhS–17	11; 7 $\parallel$, 5″, $R = 600$ mm, DM, 1%	$\sigma_e = 7 \times 10^{-20}$ $n_o \approx 1.78$	[569]
≈ 53	$^4F_{3/2} \rightarrow {}^4I_{11/2}$	≈ 1975	Illuminator with Xe flash-tube PEK type	24; 5.2 $\parallel$, DM, 1%	$n_o = 1.831$ $n_e = 1.816$ $K \approx 17$	[130, 627]

(*Continued*)

Table 5.1 (*continued*)

Crystal host, space group, cation-site symmetry	Activator [%]	Emission wavelength [μm]	Temperature [K]	Mode of operation	Threshold energy [J] or power [W]	Excitation spectral range [μm]	Luminescence lifetime [ms]
$CaY_4(SiO_4)_3O$: Ho^{3+} Er^{3+}, Tm^{3+} C_{6h}^2–$P6_3/m$ $C_s(Y^{3+})$	2.5	2.060	70–77	P	15 [f]	0.4–2 Excitation into Ho^{3+},	2.5
				P	34	Er^{3+}, and Tm^{3+} bands	
$Ca_{0.25}Ba_{0.75}$ $(NbO_3)_2$ D_{4h}^{14}–$P4_2/mnm$ (?)	Nd^{3+} 0.25- -2	1.062 [a]	295	P	380 [f]	0.4–0.9	–
$CaLa_4(SiO_4)_3O$ C_{6h}^2–$P6_3/m$ $C_s(La^{3+})$	Nd^{3+} ≈ 2 $c \parallel F$	1.0612	300	P	1.6	0.4–0.9	0.24
		1.0612	300	CW	525		
	≈ 2	1.0612	300	Q	10	0.4–0.9 $\tau_{exc} = 60$ μs	
	1.5 $c \approx \parallel F$	1.0612	300	Q	–		
	1	1.3354	300	P	10 [f]	0.5–0.9	0.19

[a] λ_g value requires more-accurate definition.
[f] The composition of the stimulated-emission spectrum and the threshold energies depend on the mutual orientation of the geometric axis of the rod F (axis of the laser element) and of the crystallographic directions of the crystal.

Luminescence linewidth [cm^{-1}]	Laser transition	Energy terminal level [cm^{-1}]	Excitation system, pump source, filter, cooling	Rod length, rod diameter [mm]; resonator cavity parameters, mirrors (coating, transmittance)	Remarks	Reference
	$^5I_7 \rightarrow {}^5I_8$				$Er^{3+} = 37.5\%$, $Tm^{3+} = 3.8\%$	
≈ 55		≈ 300	Illuminator with Xe flashtube PEK type, N$_2$	32; 7 $\parallel$, $R = 1000$ mm, DM, 5% DM, 25%	$\Delta\nu_\text{g} = 32$ cm^{-1} EET to Ho^{3+} $\eta_\text{g} = 1.48$	[130, 429]
	$^4F_{3/2} \rightarrow {}^4I_{11/2}$					
–		≈ 2000	Illuminator with Xe flashtube FT–524	EMDM		[156]
	$^4F_{3/2} \rightarrow {}^4I_{11/2}$					
≈ 40		≈ 1904	Illuminator with Xe flashtube PEK type	46–65; 5–6.4 $\parallel$, DM, 1%	$n_\text{o} = 1.857$ $n_\text{e} = 1.823$ $\eta_\text{g} = 1.9$	[130, 142, 474, 570, 627]
≈ 40		≈ 1904	Illuminator with W lamp, water	50; 3 EDAC (MgF$_2$)	$K \approx 19$ The loss at $\lambda = 1.06$ μm are 0.0014– 0.004 cm^{-1}	
≈ 40		≈ 1904	Cylindrical illuminator with Xe flashtube (l = 71 mm)	76; 6.3 $\parallel$, DM, 55% calcite polarizer and KDP Pockelseffect cell	$E_\text{g} = 0.3$ J $\tau_\text{g} = 20$ ns $f_\text{g} = 10$ Hz	
≈ 40		≈ 1904	ER with Xe flashtube	53; 6.4 $\parallel$, EDAC (MgF$_2$), $R = 10$ m, d$_\text{r} = $ 1 m, DM, 35% Cell with dye 9740 type in dichlorethane	$\sigma_\text{e} = 1.6 \times 10^{-19}$ $\Delta\nu_\text{g} = 22$ cm^{-1} $\tau_\text{g} \approx 11$ ps Mode-locked operating regime	
49	$^4F_{3/2} \rightarrow {}^4I_{13/2}$	3842	ER with Xe flashtube ISP–250, ZhS type filter	65; 5 $\parallel$, 20″, $R = 500$ mm, DM, 0.5%		[627]

(*Continued*)

Table 5.1 (*continued*)

Crystal host, space group, cation-site symmetry	Activator [%]	Emission wavelength [μm]	Temperature [K]	Mode of oper-ation	Threshold energy [J] or power [W]	Excitation spectral range [μm]	Lumines-cence lifetime [ms]
$Ca_4La(PO_4)_3O$ $C_{6h}^2-P6_3/m$	Nd^{3+} – $a \parallel F$	1.0613	300	P	22	0.4–0.9	–
$CaGd_4(SiO_4)_3O$ $C_{6h}^2-P6_3/m$ $C_s(Gd^{3+})$	Nd^{3+} –	≈ 1.06 [a,d]	300	P	–	0.4–0.9	–
$SrY_4(SiO_4)_3O$: :Er^{3+}, Tm^{3+} $C_{6h}^6-P6_3/m$ $C_s(Y^{3+})$	Ho^{3+} –	≈ 2 [a]	77	P	34 [f]	0.4–2 Excitation into Ho^{3+}, Er^{3+}, and Tm^{3+} bands	–
$SrLa_4(SiO_4)_3O$ $C_{6h}^2-P6_3/m$ $C_s(La^{3+})$	Nd^{3+} ≈ 2 $c \parallel F$	1.0586	300	P	3.2	0.4–0.9	0.2
$YScO_3$ T_h^7-Ia3 $C_2(Y^{3+}, Sc^{3+})$, $C_{3i}(Y^{3+}, Sc^{3+})$	Nd^{3+} ≈ 0.5	1.0843 A 1.0770 B	300 77	P P	55 70	0.5–0.9	0.24 0.24
		(44)					
$Y_2Ti_2O_7$ O_h^7-Fd3m	Nd^{3+} 1(?)	1.0618 1.0615	300 77	P P	21 12	–	0.175
$ZrO_2-Y_2O_3$ O_h^5-Fm3m	Nd^{3+} 0.5- -0.7	1.0608 1.0673*	300 300	P P	7–20 –	0.4–0.9 0.5–0.9	0.45

[a] λ_g value requires more-accurate definition.

[d] The spectroscopic data confirming this result are not published.

[f] The composition of the stimulated-emission spectrum and the threshold energies depend on the mutual orientation of the geometric axis of the rod F (axis of the laser element) and of the crystallographic directions of the crystal.

Luminescence linewidth [cm^{-1}]	Laser transition	Energy terminal level [cm^{-1}]	Excitation system, pump source, filter, cooling	Rod length, rod diameter [mm]; resonator cavity parameters, mirrors (coating, transmittance)	Remarks	Reference
≈ 30	$^4F_{3/2} \rightarrow {}^4I_{11/2}$	≈ 2000	Illuminator with Xe flash-tube PEK type	76; 6.4 $\parallel$, DM, 65%	$\eta_g = 1.6$ $n_o = 1.699$ $n_e = 1.690$	[130]
$-$	$^4F_{3/2} \rightarrow {}^4I_{11/2}$	≈ 2000	Illuminator with Xe flash-tube PEK type	$-$	(43)	
$-$	$^5I_7 \rightarrow {}^5I_8$	≈ 300	Illuminator with Xe flash-tube PEK type, N$_2$	24; 6 $\parallel$, DM, 45%	Er^{3+} = 37.5%, Tm^{3+} = 3.8% $\eta_g = 0.31$ EET to Ho^{3+}	[130]
≈ 53	$^4F_{3/2} \rightarrow {}^4I_{11/2}$	≈ 2000	Illuminator with PEK type	75; 4.3 $\parallel$, DM, 1%	$\eta_g = 2.1$ $n_f = 1.792$ $n_e = 1.764$ $K \approx 27$	[130, 142]
≈ 70 ≈ 40	$^4F_{3/2} \rightarrow {}^4I_{11/2}$	1950 1890	Low- and room-temp. ER with Xe flash-tube IFP–400, QLP, N$_2$ and its vapor	18; 6 $\parallel$, 10″, R = 576 mm, DM, 1%	$\sigma_e = 3.3 \times 10^{-21}$ $\sigma_e = 1.4 \times 10^{-20}$ At ≈ 170 K $E_{thr}(A) =$ $= E_{thr}(B)$ $n \approx 2$	[639]
$-$ $-$	$^4F_{3/2} \rightarrow {}^4I_{11/2}$	≈ 2000 ≈ 2000	ER with Xe flashtube IFP–400, N$_2$	$-$	e	[755]
$-$ $-$	$^4F_{3/2} \rightarrow {}^4I_{11/2}$	≈ 2000 ≈ 2000	ER with Xe flashtube IFP type, ZhS–17	17–32; 3–6 $\parallel$, R = 576 mm, DM, 1–3%	Y$_2$O$_3$ = 10 wt.% $\Delta v_g = 20$ cm^{-1} *In CAM laser $n = 2.1$–2.2	[90, 430]

^{e} The detailed data are not available. The results are doubtful and require confirmation or correction.
$^{(43)}$ Result was obtained by Westinghouse workers (*Ivey* private communication).
$^{(44)}$ Stimulated emission of a YScO$_3$–Nd^{3+} crystal at 77 K at 0.9200 and 1.0520 μm there is reported without any description of experimental conditions in [470]. [639] criticizes and indicates mistaken results in [470].

Table 5.1 (*continued*)

Crystal host, space group, cation-site symmetry	Activator [%]	Emission wavelength [μm]	Temperature [K]	Mode of operation	Threshold energy [J] or power [W]	Excitation spectral range [μm]	Luminescence lifetime [ms]
	0.5	1.3320	300	P	42	0.5–0.9	0.45
ZrO_2–Er_2O_3 O_h^5–Fm3m	Ho^{3+} 1	2.115 [a]	77	P	400	0.4–2 Excitation into Ho^{3+} and Er^{3+} bands	6.5
	Er^{3+} 12	1.620 [a]	77	P	800	0.4–1.6	–
ZrO_2–Er_2O_3 O_h^5–Fm3m	Tm^{3+} 1	1.896 [a]	77	P	420	0.4–1.9 Excitation into Er^{3+} and Tm^{3+} bands	2.5
	Er^{3+} 12	1.620 [a]	77	P	800	0.4–1.6	–
$Ba_2NaNb_5O_{15}$ C_{2v}^{11}–Cmm2	Nd^{3+} ≈ 0.1 $c \parallel F$	1.0613	300	P	23	0.5–0.9	0.2
$Ba_2MgGe_2O_7$ D_{2d}^3–P$\bar{4}2_1$m	Nd^{3+} ≈ 2	1.05436	300	P	5.1 [f]	0.3–0.9	0.4

[a] λ_g value requires more-accurate definition.
[c] The detailed spectroscopic investigations were not carried out.
[f] The composition of the stimulated-emission spectrum and the threshold energies depend on the mutual orientation of the geometric axis of the rod F (axis of the laser element) and of the crystallographic directions of the crystal.

Lumines-cence linewidth [cm^{-1}]	Laser transition	Energy terminal level [cm^{-1}]	Excitation system, pump source, filter, cooling	Rod length, rod diameter [mm]; resonator cavity parameters, mirrors (coating, transmittance)	Remarks	Reference
≈ 130	$^4F_{3/2} \to ^4I_{13/2}$	3910	ER with Xe flashtube IFP–400, ZhS–17	17; 5 $\parallel$, $R = 600$ mm, DM, 1%		[90]
–	$^5I_7 \to ^5I_8$	≈ 300	ER with Xe flashtube IFP–2000, TQC, N$_2$	33; 6 $\parallel$, Au, 5%	Er$_2$O$_3$ = 12% EET to Ho^{3+}	[431]
–	$^4I_{13/2} \to ^4I_{15/2}$	≈ 300	ER with Xe flashtube IFP–2000, TQC, N$_2$	33; 6 $\parallel$, Au, 5%	c	[431]
–	$^3H_4 \to ^3H_6$	≈ 300	ER with Xe flashtube IFP–2000, TQC, N$_2$	33; 6 $\parallel$, Au, 5%	Er$_2$O$_3$ = 12% EET to Tm^{3+}	[431]
–	$^4I_{13/2} \to ^4I_{15/2}$	≈ 300	ER with Xe flashtube IFP–2000. TQC, N$_2$	33; 6 $\parallel$, Au, 5%	c	[431]
≈ 55	$^4F_{3/2} \to ^4I_{11/2}$	≈ 2000	ER with Xe flashtube ISP–1000, ZhS–17	21; 3.5 $\parallel$, 15″, $R = 576$ mm, DM, 1%	$\Delta v_g = 6$ cm^{-1} $n = 2.1$–2.2	[585]
≈ 40	$^4F_{3/2} \to ^4I_{11/2}$	1895	Round illuminator with Ag coating, and with Xe flashtube	25; 5 $\parallel$, DM, 5–30%	$\sigma_e = 10^{-19}$ $\Delta v_g = 8$ cm^{-1} ECC–K$^+$ or Na$^+$	[432]

(*Continued*)

Table 5.1   (*continued*)

Crystal host, space group, cation-site symmetry	Activator [%]	Emission wavelength [μm]	Temperature [K]	Mode of operation	Threshold energy [J] or power [W]	Excitation spectral range [μm]	Luminescence lifetime [ms]
$Ba_2ZnGe_2O_7$ $D_{2d}^3 - P\bar{4}2_1m$	Nd^{3+} ≈ 2 $c \parallel F$	1.05437	300	P	4.5	0.3–0.9	0.34
$(Nd,Sc)P_5O_{14}$ $C_{2h}^5 - P2_1/c$ $C_1(Nd^{3+}, Sc^{3+})$	Nd^{3+} 50	1.051 [a]	300	CW	0.004 [b] 0.006 [b] 0.005 [b] 0.005 [b] 0.004 [b]	LP; 0.4727 LP; 0.4765 LP; 0.4965 LP; 0.5017 LP; 0.5145	0.3
$(Nd,In)P_5O_{14}$ $C_{2h}^5 - P2_1/c$ $C_1(Nd^{3+}, In^{3+})$	Nd^{3+} 75	1.051 [a]	300	P	–	LP; 0.4765, 0.5017 and 0.5145	0.1- –0.12
$(Nd,La)P_5O_{14}$ $C_{2h}^5 - P2_1/c$ $C_1(Nd^{3+}, La^{3+})$	Nd^{3+} 50	1.0511	300	P	2×10^{-5} b	LP; 0.58	≈ 0.3
			300	CW	2.5×10^{-5} b	LP; 0.476 and 0.514	
		1.0511	300	CW	0.01 [b]	LP; 0.7525	
$(Nd,Gd)Al_3$ $(BO_3)_4$ $D_3^7 - R32$ $D_3(Nd^{3+},Gd^{3+})$	Nd^{3+} 50	1.0635 [a,d]	300	–	–	LP; 0.58	0.025
$LuScO_3$ $T_h^7 - Ia3$ $C_2(Lu^{3+},Sc^{3+})$, $C_{3i}(Lu^{3+},Sc^{3+})$	Nd^{3+} ≈ 1	1.0785	300	P	35	0.5–0.9	≈ 0.22
$HfO_2 - Y_2O_3$ $O_h^5 - Fm3m$	Nd^{3+} 0.5- –0.7	1.0604 1.0666*	300 300	P P	7–20 –	0.5–0.9	0.45

[a] λ_g value requires more-accurate definition.

[b] The absorbed pump power (energy) at which stimulated emission begins is indicated.

[d] The spectroscopic data confirming this result are not published.

[(13)] Crystallographic faces (natural-growth surfaces) or cleavage planes of these crystals were used as the laser-rod end faces.

Luminescence linewidth [cm^{-1}]	Laser transition	Energy terminal level [cm^{-1}]	Excitation system, pump source, filter, cooling	Rod length, rod diameter [mm]; resonator cavity parameters, mirrors (coating, transmittance)	Remarks	Reference
≈ 48	$^4F_{3/2} \rightarrow {}^4I_{11/2}$	1890	Illuminator with two Xe flashtubes L–268 type	27; 5 $\parallel$, DM, 15%	$\sigma_e \approx 10^{-19}$ ECC–K$^+$ or Na$^+$	[280]
–	$^4F_{3/2} \rightarrow {}^4I_{11/2}$	1965	CW Ar ion laser	1.25 × 0.675 × × 0.4 $\parallel$, $R = 90$ mm, $d_r = 180$ mm, DM, 0.1%	P_g several mW	[533]
–	$^4F_{3/2} \rightarrow {}^4I_{11/2}$	≈ 2000	Multimode P Ar ion laser	Several mm, $R = 100$ mm, 2–2.5%	In$^{3+} = 25\%$ $\mathbf{E} \parallel b$ axis	[615]
–	$^4F_{3/2} \rightarrow {}^4I_{11/2}$	≈ 1964	P rhodamine 6G dye laser CW rhodamine 6G dye laser	0.035 $\parallel$ [13] 1.4 $\parallel$ [13], $R = 37.5$ mm, 1.5%	$\Delta v_g = 16$ cm^{-1} $n \approx 1.62$ $\eta_g = 11$	[475, 505, 658]
–		≈ 1964	CW Kr ion laser, LPG	0.68 × 0.012 × 0.012, $\parallel$, microguide element	$\sigma_e = 1.5 \times 10^{-19}$ $n \approx 1.48$ for external cover of microguide	
–	$^4F_{3/2} \rightarrow {}^4I_{11/2}$	≈ 2000	Dye laser	–		[579]
≈ 30	$^4F_{3/2} \rightarrow {}^4I_{11/2}$	≈ 1900	ER with Xe flashtube IFP– –400, ZhS–17	11; 4.3 $\parallel$, 15″, $R = 576$ mm, DM, 1%	$\Delta v_g = 3$ cm^{-1}	[639]
– –	$^4F_{3/2} \rightarrow {}^4I_{11/2}$	≈ 2000 ≈ 2000	ER with Xe flashtubes IFP– –800 or IFP– –400, ZhS–17	15–32; 3–6 $\parallel$, $R = 576$ mm, DM, 1–3%	$Y_2O_3 = 10$ wt.% $\Delta v_g = 17$ cm^{-1} *In CAM laser $n = 1.98$–2.02	[90, 430]

(Continued)

Table 5.1   (*continued*)

Crystal host, space group, cation-site symmetry	Activator [%]	Emission wavelength [μm]	Temperature [K]	Mode of operation	Threshold energy [J] or power [W]	Excitation spectral range [μm]	Luminescence lifetime [ms]
	0.5	1.3305	300	P	48	0.5–0.9	0.45
$Bi_4(Si,Ge)_3O_{12}$ T_d^6–I43d	Nd^{3+} ≈ 1	1.0635	300	P	100	0.5–0.9	≈ 0.25
							Other
$Ca_5(PO_4)_3F$ C_{6h}^2–P6$_3$/m $C_s(Ca^{2+}$–I), $C_3(Ca^{2+}$–II)	Nd^{3+} ≈ 1 $a \parallel F$	(π) 1.0630	300	P	0.2	0.4–0.9	0.25
			300	P	7.5		
				CW	145		
	≈ 1 $c \parallel F$	1.3347	300	P	2	0.4–0.9	0.25
		1.3345	77	P	8		0.25
$Ca_5(PO_4)_3F$: :Cr^{3+} C_{6h}^2–P6$_3$/m $C_s(Ca^{2+}$–I), $C_3(Ca^{2+}$–II)	Ho^{3+} 3	2.079 [a]	77	P	25 [f]	0.4–2 Excitation into Ho^{3+} and Cr^{3+} bands	0.5
$Sr_5(PO_4)_3F$ C_{6h}^2–P6$_3$/m $C_s(Sr^{2+}$–I), $C_3(Sr^{2+}$–II)	Nd^{3+} ≈ 2 $a \parallel F$	1.0585	300	P	12	0.4–0.9	0.31
La_2O_2S D_{3d}^3–P3m $C_{3v}(La^{3+})$	Nd^{3+} 1	(π) 1.075 [a]	300 300	P CW	0.3 [f]	0.5–0.9	0.095

[a] λ_g value requires more-accurate definition.
[f] The composition of the stimulated-emission spectrum and the threshold energies depend on the mutual orientation of the geometric axis of the rod F (axis of the laser element) and of the crystallographic directions of the crystal.

Lumines-cence linewidth [cm^{-1}]	Laser transition	Energy terminal level [cm^{-1}]	Excitation system, pump source, filter, cooling	Rod length, rod diameter [mm]; resonator cavity parameters, mirrors (coating, transmittance)	Remarks	Reference
≈ 130	$^4F_{3/2} \to {}^4I_{13/2}$	3910	ER with Xe flashtube IFP–400, ZhS–17	15; 5 $\parallel$, $R = 600$ mm, DM, 1%		[90]
≈ 35	$^4F_{3/2} \to {}^4I_{11/2}$	1928	ER with Xe flashtube ISP–1000, ZhS–17	10; 3 $\parallel$, 10″, $R = 576$ mm, DM, 1%	Si = Ge = 50%	[659]
laser crystals						
6	$^4F_{3/2} \to {}^4I_{11/2}$	≈ 1900	ER with Xe flashtube PEK 1–3 ER with W lamp, water	38; 6.4 $\parallel$, R, DM, 1 and 63% 76; 6.4 $\parallel$, R, DM, 65% 250; 6.3	$\sigma_e = 5 \times 10^{-19}$ $n_o = 1.634$ $n_e = 1.631$ $\eta_g = 2.6$ $K \approx 20$	[90, 214, 215, 281, 434]
≈ 4 2	$^4F_{3/2} \to {}^4I_{13/2}$	3819 3820	Low- and room-temp. ER flashtube IFP–400, N$_2$	17; 4–6 $\parallel$, 10″, $R = 600$ mm, DM, 1%		[90]
≈ 60	$^5I_7 \to {}^5I_8$	≈ 355	Illuminator with Xe flash-tube PEK type, N$_2$	22; 6 $\parallel$, $R = 1000$ mm, DM, 5%	Cr^{3+} = 0.3% $\Delta v_g = 8$ cm^{-1} $\eta_g = 0.17$ EET to Ho^{3+}	[129, 130]
10	$^4F_{3/2} \to {}^4I_{11/2}$	≈ 2000	Illuminator with Xe flash-tube PEK 1–3	33; 6.4 $\parallel$, DM, 5%	$\eta_g = 0.55$	[130]
10	$^4F_{3/2} \to {}^4I_{11/2}$	1910	Under P conditions, illuminator with Xe flashtube; and under CW conditions, with W lamp	75, 10 DM, 3%	$\sigma_e = 2 \times 10^{-18}$ $n \approx 2.16$ $K = 50$	[219, 435]

(Continued)

Table 5.1   (*continued*)

Crystal host, space group, cation-site symmetry	Activator [%]	Emission wavelength [μm]	Temperature [K]	Mode of oper- ation	Threshold energy [J] or power [W]	Excitation spectral range [μm]	Lumines- cence lifetime [ms]
$LaCl_3$ C_{6h}^2–$P6_3/m$ $C_{3h}(La^{3+})$	Pr^{3+} 1	0.4892	5.5–14	P	–	LP; 0.488	≈ 0.015
	1	0.5298	35 12	P P	– 4×10^{-7}	LP; 0.474	0.0025
	1	(σ) 0.6164 0.619 [a]	65 8 –	P P P	12×10^{-7} 10^{-6} –	LP; 0.488	≈ 0.015
	1	(π) 0.6452	300 65	P P	10^{-3} 12×10^{-7}	LP; 0.488	≈ 0.012 ≈ 0.015
$LaBr_3$ C_{6h}^2–$P6_3/m$ $C_{3h}(La^{3+})$	Pr^{3+} 1	0.532 [a]	–	P	–	LP	≈ 0.007
	1	0.621 [a]	–	P	–	LP	≈ 0.012
	1	0.632 [a]	–	P	–	LP	2×10^{-5}
	1	0.647 [a]	–	P	–	LP	≈ 0.012
$CeCl_3$ C_{6h}^2–$P6_3/m$ $C_{3h}(Ce^{3+})$	Nd^{3+} 2.8 $a \parallel F$	(π) 1.0647	300	P	0.06	0.75–0.9	0.175
	2.5 $a \parallel F$		300	CW	0.0075	LP; 0.5145	
$PrCl_3$ C_{6h}^2–$P6_3/m$ $C_{3h}(Pr^{3+})$	Pr^{3+} 100	0.489 [a]	5.5–14	P	–	LP; 0.488	≈ 0.012

[a] λ_g value requires more-accurate definition.

Lumines-cence linewidth [cm^{-1}]	Laser transition	Energy terminal level [cm^{-1}]	Excitation system, pump source, filter, cooling	Rod length, rod diameter [mm]; resonator cavity parameters, mirrors (coating, transmittance)	Remarks	Reference
	$^3P_0 \rightarrow {}^3H_4$					
–		≈ 34	P rhodamine 6G dye laser, He vapor (?)	Several tens of μm	$n \approx 1.77$	[116, 564]
	$^3P_1 \rightarrow {}^3H_5$					
–		2137	P rhodamine 6G dye laser, He vapor (?)	Several tens of μm		[116, 564]
	$^3P_0 \rightarrow {}^3H_6$					
–		4230	P rhodamone 6G dye laser,	Several tens of μm		[116, 564]
–		≈ 4200	He vapor (?)			
	$^3P_0 \rightarrow {}^3F_2$					
–		4923	P rhodamine 6G dye laser	Several tens of μm		[116, 564]
	$^3P_1 \rightarrow {}^3H_5$					
–		≈ 2130	P laser	–		[564]
	$^3P_0 \rightarrow {}^3H_6$					
–		≈ 4200	P laser	–		[564]
	$^3P_2 \rightarrow {}^3F_3$					
–		≈ 6300	P laser	–		[564]
	$^3P_0 \rightarrow {}^3F_2$					
–		≈ 4900	P laser	–		[564]
	$^4F_{3/2} \rightarrow {}^4I_{11/2}$					
–		2055	ER with Xe flashtube FT–91, glass CS2–64 filter	5; 2.85 $\parallel$, 40″, $R = 500$ mm, $d_r = 750$ mm, DM, 0.01 and 20%	$\sigma_e \approx 1.6 \times 10^{-18}$	[482, 620]
			CW Ar ion laser, LPG	8.9; 1.5 $\parallel$, 5″, DM, 1%	Pb glass capsule used	
	$^3P_0 \rightarrow {}^3H_4$					
–		≈ 34	P rhodamine 6G dye laser, He vapor (?)	Several tens of μm	$n \approx 1.77$	[116, 564]

(*Continued*)

Table 5.1 (*continued*)

Crystal host, space group, cation-site symmetry	Activator [%]	Emission wavelength [μm]	Temperature [K]	Mode of operation	Threshold energy [J] or power [W]	Excitation spectral range [μm]	Luminescence lifetime [ms]
	100	0.531 [a]	12	P	12×10^{-7}	LP; 0.474	–
	100	(π) 0.617 [a]	65	P	–	LP; 0.488	≈ 0.011
			8	P	10^{-6}		≈ 0.012
		0.620 [a]	–	P	–		
	100	(σ) 0.647 [a]	300	P	10^{-6}	LP; 0.488	≈ 0.0003
			65	P	–		≈ 0.011
$PrBr_3$ C_{6h}^2–$P6_3/m$ $C_{3h}(Pr^{3+})$	Pr^{3+} 100	0.622 [a]	–	P	–	LP	≈ 0.001
	100	0.649 [a]	300	P	–	LP	≈ 0.001
$Pb_5(PO_4)_3F$ (?)	Nd^{3+} 1	1.0551	300	P	12	0.5–0.9	0.225

[a] λ_g value requires more-accurate definition.

Over the last ten years stimulated emission has been excited in a wide range of activated insulating crystals, in particular (aside from data on new laser activators included in Tables 1.1–4): Pr^{3+} ions in $LiYbF_4$, $LiLuF_4$, $BaYb_2F_8$, PrF_3, and $LiPrP_4O_{12}$; Nd^{3+} ions in $LiLuF_4$, KY_3F_{10}, BaY_2F_8, CaF_2–ScF_3, CaF_2–NdF_3, CaF_2–LuF_3, SrF_2–ScF_3, SrF_2–NdF_3, CdF_2–ScF_3, CdF_2–CeF_3, CdF_2–NdF_3, CdF_2–GdF_3, CdF_2–LuF_3, BaF_2–NdF_3, $NaYGeO_4$, $NaGdGeO_4$, $NaLuGeO_4$, $Ca_3Ga_4O_9$, $RbNd(WO_4)_2$, $CsLa(WO_4)_2$, $CsNd(MoO_4)_2$, $BaGd_2(MoO_4)_2$, $LaGaGe_2O_7$, $(La,Lu)_3(Lu,Ga)_2Ga_3O_{12}$, $NdGaGe_2O_7$, $GdGaGe_2O_7$, $Pb_5Ge_3O_{11}$, $NaY(MoO_4)_2$, $NaGd(MoO_4)_2$, $NaBi(MoO_4)_2$, $NaBi(WO_4)_2$, $Ca_2Ga_2SiO_7$, $Ca_3Ga_2Ge_4O_{14}$, $Ca_3(Nb,Ga)_2Ga_3O_{12}$, $Sr_3Ga_2Ge_4O_{14}$, $BaLaGa_3O_7$, $LaMgAl_{11}O_{19}$, $7La_2O_3 \cdot 9SiO_2$, $La_3Ga_5SiO_{14}$, $La_3Ga_5GeO_{14}$, $La_3Ga_{5.5}Nb_{0.5}O_{14}$, $La_3Ga_{5.5}Ta_{0.5}O_{14}$, $LaSr_2Ga_{11}O_{20}$, $Nd_3Ga_5SiO_{14}$, and β''-$Na_{1+x}Mg_xAl_{11-x}O_{17}$; Dy^{3+} ions in $BaYb_2F_8$; Ho^{3+} ions in $LiYbF_4$, $LiLuF_4$, $BaEr_2F_8$, $BaTm_2F_8$, $BaYb_2F_8$, ErF_3, SrF_2–ErF_3, YF_3–SrF_2, BaF_2–

Lumines-cence linewidth [cm^{-1}]	Laser transition	Energy terminal level [cm^{-1}]	Excitation system, pump source, filter, cooling	Rod length, rod diameter [mm]; resonator cavity parameters, mirrors (coating, transmittance)	Remarks	Reference
–	$^3P_1 \rightarrow {}^3H_5$	≈ 2137	P rhodamine 6G dye laser, He vapor (?)	Several tens of μm		[116, 564]
–	$^3P_0 \rightarrow {}^3H_6$	≈ 4230	P rhodamine 6G dye laser,	Several tens of μm	("")	[116, 564]
–		≈ 4230	He vapor (?)			
–	$^3P_0 \rightarrow {}^3F_2$	≈ 4950	P rhodamine 6G dye laser	Several tens of μm		[76, 564]
–	$^3P_0 \rightarrow {}^3H_6$	≈ 4230	P laser	–		[564]
–	$^3P_0 \rightarrow {}^3F_2$	≈ 4950	P laser	–		[76, 564]
≈ 180	$^4F_{3/2} \rightarrow {}^4I_{11/2}$	1955	ER with Xe flashtube ISP type	15; 3–5.5 $\parallel$, R, DM, 0.5%	$\Delta v_g = 6\text{–}7\text{ cm}^{-1}$	[640]

ErF$_3$, CaF$_2$–HoF$_3$–ErF$_3$, and Tm$_3$Al$_5$O$_{12}$; Er^{3+} ions in LiYbF$_4$, LiLuF$_4$, BaEr$_2$F$_8$, BaYb$_2$F$_8$, SrF$_2$–ErF$_3$, CaF$_2$–HoF$_3$–ErF$_3$, Ca$_3$Ga$_2$Ge$_3$O$_{12}$, and (Gd,Er)$_3$Al$_5$O$_{12}$; Tm^{3+} ions in LiYbF$_4$ and BaYb$_2$F$_8$; Sm^{2+} ions in BaMgF$_4$; Ti^{3+} ions in Al$_2$O$_3$ and BeAl$_2$O$_4$; Cr^{3+} ions in LiCaAlF$_6$, LiSrAlF$_6$, Na$_3$Ga$_2$Li$_3$F$_{12}$, KZnF$_3$, SrAlF$_5$, BeAl$_6$O$_{10}$, Be$_3$Al$_2$(SiO$_3$)$_6$, BeScAlO$_4$, Mg$_2$SiO$_4$, Al$_2$(WO$_4$)$_3$, ScBO$_3$, ZnWO$_4$, (La,Lu)$_3$(Lu,Ga)$_2$Ga$_3$O$_{12}$, Ca$_3$Ga$_2$Ge$_4$O$_{14}$, Sr$_3$Ga$_2$Ge$_4$O$_{14}$, La$_3$Ga$_5$SiO$_{14}$, La$_3$Ga$_5$GeO$_{14}$, La$_3$Ga$_{5.5}$Nb$_{0.5}$O$_{14}$, and La$_3$Ga$_{5.5}$Ta$_{0.5}$O$_{14}$; Cr^{4+} (?) ions in Mg$_2$SiO$_4$; Co^{2+} ions in KZnF$_3$; V^{2+} ions in CsCaF$_3$; defect center in Ca$_3$Ga$_2$Ge$_3$O$_{12}$.

Table 5.2. Crystallographic space groups and possible symmetry sites

Space-group symbol		Schönflies site symmetries
Schönflies	International [436]	

Triclinic systems

C_1^1	$P1$	$[C_1]$
C_i^1	$P\bar{1}$	$C_1(2)$, $[8C_i]$

Monoclinic systems

C_2^1	$P2$	$C_2(2)$, $4C_2$
C_2^2	$P2_1$	$C_1(2)$
C_2^3	$B_2(C_2)$	$C_1(4)$, $2C_2(2)$
C_s^1	Pm	$C_1(2)$, $2C_s$
C_s^2	$Pb(Pc)$	$C_1(2)$
C_s^3	$Bm(Cm)$	$C_1(4)$, $C_s(2)$
C_s^4	$Bb(Cc)$	$C_1(4)$
C_{2h}^1	$P2/m$	$C_1(4)$, $2C_s(2)$, $4C_2(2)$, $[8C_{2h}]$
C_{2h}^2	$P2_1/m$	$C_1(4)$, $C_s(2)$, $[4C_i(2)]$
C_{2h}^3	$B2/m(C2/m)$	$C_1(8)$, $C_s(4)$, $2C_2(4)$, $[2C_i(4)]$, $[4C_{2h}(2)]$
C_{2h}^4	$P2/b(P2/c)$	$C_1(4)$, $2C_2(2)$, $[4C_i(2)]$
C_{2h}^5	$P2_1/b(P2_1/c)$	$C_1(4)$, $[4C_i(2)]$
C_{2h}^6	$B2/b(C2/c)$	$C_1(8)$, $C_2(4)$, $[4C_i(4)]$

Orthorhombic systems

D_2^1	$P222$	$C_1(4)$, $12C_2(2)$, $[8D_2]$
D_2^2	$P222_1$	$C_1(4)$, $4C_2(2)$
D_2^3	$P2_12_12$	$C_1(4)$, $2C_2(2)$
D_2^4	$P2_12_12_1$	$C_1(4)$
D_2^5	$C222_1$	$C_1(8)$, $2C_2(4)$
D_2^6	$C222$	$C_1(8)$, $7C_2(4)$, $4D_2(2)$
D_2^7	$F222$	$C_1(16)$, $6C_2(8)$, $[4D_2(4)]$
D_2^8	$I222$	$C_1(8)$, $6C_2(4)$, $[4D_2(2)]$
D_2^9	$I2_12_12_1$	$C_1(8)$, $3C_2(4)$
C_{2v}^1	$Pmm2$	$C_1(4)$, $4C_s(2)$, $4C_{2v}$
C_{2v}^2	$Pmc2_1$	$C_1(4)$, $2C_s(2)$
C_{2v}^3	$Pcc2$	$C_1(4)$, $4C_2(2)$
C_{2v}^4	$Pma2$	$C_1(4)$, $C_s(2)$, $2C_2(2)$
C_{2v}^5	$Pca2_1$	$C_1(4)$
C_{2v}^6	$Pnc2$	$C_1(4)$, $2C_2(2)$
C_{2v}^7	$Pmn2_1$	$C_1(4)$, $C_s(2)$
C_{2v}^8	$Pba2$	$C_1(4)$, $2C_2(2)$
C_{2v}^9	$Pna2_1$	$C_1(4)$
C_{2v}^{10}	$Pnn2$	$C_1(4)$, $2C_2(2)$
C_{2v}^{11}	$Cmm2$	$C_1(8)$, $2C_s(4)$, $C_2(4)$, $2C_{2v}(2)$
C_{2v}^{12}	$Cmc2_1$	$C_1(8)$, $C_s(4)$
C_{2v}^{13}	$Ccc2$	$C_1(8)$, $3C_2(4)$
C_{2v}^{14}	$Amm2$	$C_1(8)$, $3C_s(4)$, $2C_{2v}(2)$
C_{2v}^{15}	$Abm2$	$C_1(8)$, $C_s(4)$, $2C_2(4)$
C_{2v}^{16}	$Ama2$	$C_1(8)$, $C_s(4)$, $C_2(4)$

(Continued)

Table 5.2   (*continued*)

Space-group symbol		Schönflies site symmetries
Schönflies	International [436]	
C_{2v}^{17}	$Aba2$	$C_1(8)$, $C_2(4)$
C_{2v}^{18}	$Fmm2$	$C_1(16)$, $2C_s(8)$, $C_2(8)$, $C_{2v}(4)$
C_{2v}^{19}	$Fdd2$	$C_1(16)$, $C_2(8)$
C_{2v}^{20}	$Imm2$	$C_1(8)$, $2C_s(4)$, $C_{2v}(2)$
C_{2v}^{21}	$Iba2$	$C_1(8)$, $2C_2(4)$
C_{2v}^{22}	$Ima2$	$C_1(8)$, $C_s(4)$, $C_2(4)$
D_{2h}^{1}	$Pmmm$	$C_1(8)$, $6C_s(4)$, $12C_{2v}(2)$, $[8D_{2h}]$
D_{2h}^{2}	$Pnnn$	$C_1(8)$, $6C_2(4)$, $[2C_i(4)]$, $[4D_2(2)]$
D_{2h}^{3}	$Pccm$	$C_1(8)$, $C_s(4)$, $8C_2(4)$, $[4D_2(2)]$, $[4C_{2h}(2)]$
D_{2h}^{4}	$Pban$	$C_1(8)$, $6C_2(4)$, $[2C_i(4)]$, $[4D_2(2)]$
D_{2h}^{5}	$Pmma$	$C_1(8)$, $3C_s(4)$, $2C_2(4)$, $2C_{2v}(2)$, $[4C_{2h}(2)]$
D_{2h}^{6}	$Pnna$	$C_1(8)$, $2C_2(4)$, $[2C_i(4)]$
D_{2h}^{7}	$Pmna$	$C_1(8)$, $C_s(4)$, $3C_2(4)$, $[4C_{2h}(2)]$
D_{2h}^{8}	$Pcca$	$C_1(8)$, $3C_2(4)$, $[2C_i(4)]$
D_{2h}^{9}	$Pbam$	$C_1(8)$, $2C_s(4)$, $2C_2(4)$, $[4C_{2h}(2)]$
D_{2h}^{10}	$Pccn$	$C_1(8)$, $2C_2(4)$, $[2C_i(4)]$
D_{2h}^{11}	$Pbcm$	$C_1(8)$, $C_s(4)$, $C_2(4)$, $[2C_i(4)]$
D_{2h}^{12}	$Pnnm$	$C_1(8)$, $C_s(4)$, $2C_2(4)$, $[4C_{2h}(2)]$
D_{2h}^{13}	$Pmmn$	$C_1(8)$, $2C_s(4)$, $[2C_i(4)]$, $2C_{2v}(4)$
D_{2h}^{14}	$Pbcn$	$C_1(8)$, $C_2(4)$, $[2C_i(4)]$
D_{2h}^{15}	$Pbca$	$C_1(8)$, $[2C_i(4)]$
D_{2h}^{16}	$Pnma$	$C_1(8)$, $C_s(4)$, $[2C_i(4)]$
D_{2h}^{17}	$Cmcm$	$C_1(16)$, $2C_s(8)$, $C_2(8)$, $[C_i(8)]$, $C_{2v}(4)$, $[2C_{2h}(4)]$
D_{2h}^{18}	$Cmca$	$C_1(16)$, $C_s(8)$, $2C_2(8)$, $[C_i(8)]$, $[2C_{2h}(4)]$
D_{2h}^{19}	$Cmmm$	$C_1(16)$, $4C_s(8)$, $C_2(8)$, $6C_{2v}(4)$, $[2C_{2h}(4)]$, $[4D_{2h}(2)]$
D_{2h}^{20}	$Cccm$	$C_1(16)$, $C_s(8)$, $5C_2(8)$, $[4C_{2h}(4)]$, $[2D_2(4)]$
D_{2h}^{21}	$Cmma$	$C_1(16)$, $2C_s(8)$, $5C_2(8)$, $C_{2v}(4)$, $[4C_{2h}(4)]$, $[2D_2(4)]$
D_{2h}^{22}	$Ccca$	$C_1(16)$, $4C_2(8)$, $[2C_i(8)]$, $[2D_2(4)]$
D_{2h}^{23}	$Fmmm$	$C_1(32)$, $3C_s(16)$, $3C_2(16)$, $3C_{2v}(8)$, $[D_2(8)]$, $[3C_{2h}(8)]$, $[2D_{2h}(4)]$
D_{2h}^{24}	$Fddd$	$C_1(32)$, $3C_2(16)$, $[2C_i(16)]$, $[2D_2(8)]$
D_{2h}^{25}	$Immm$	$C_1(16)$, $3C_s(8)$, $[C_i(8)]$, $[6C_{2v}(4)]$, $[4D_{2h}(2)]$
D_{2h}^{26}	$Ibam$	$C_1(16)$, $C_s(8)$, $4C_2(8)$, $[C_i(8)]$, $[2C_{2h}(4)]$, $[2D_2(4)]$
D_{2h}^{27}	$Ibca$	$C_1(16)$, $3C_2(8)$, $[2C_i(8)]$
D_{2h}^{28}	$Imma$	$C_1(16)$, $2C_s(8)$, $2C_2(8)$, $C_{2v}(4)$, $[4C_{2h}(4)]$

Tetragonal systems

C_4^{1}	$P4$	$C_1(4)$, $C_2(2)$, $2C_4$
C_4^{2}	$P4_1$	$C_1(4)$
C_4^{3}	$P4_2$	$C_1(4)$, $3C_2(2)$
C_4^{4}	$P4_3$	$C_1(4)$
C_4^{5}	$I4$	$C_1(8)$, $C_2(4)$, $C_4(2)$
C_4^{6}	$I4_1$	$C_1(8)$, $C_2(4)$
S_4^{1}	$P\bar{4}$	$C_1(4)$, $3C_2(2)$, $[4S_4]$
S_4^{2}	$I\bar{4}$	$C_1(8)$, $2C_2(4)$, $[4S_4(2)]$
C_{4h}^{1}	$P4/m$	$C_1(8)$, $2C_s(4)$, $C_2(4)$, $2C_4(2)$, $[2C_{2h}(2)]$, $[4C_{4h}]$
C_{4h}^{2}	$P4_2/m$	$C_1(8)$, $C_s(4)$, $3C_2(4)$, $[2S_4(2)]$, $[4C_{2h}(2)]$

(*Continued*)

Table 5.2 (*continued*)

Space-group symbol		Schönflies site symmetries
Schönflies	International [436]	
C_{4h}^3	$P4/n$	$C_1(8)$, $C_2(4)$, $[2C_i(4)]$, $C_4(2)$, $[2S_4(2)]$
C_{4h}^4	$P4_2/n$	$C_1(8)$, $2C_2(4)$, $[2C_i(4)]$, $[2S_4(2)]$
C_{4h}^5	$I4/m$	$C_1(16)$, $C_s(8)$, $C_2(8)$, $[C_i(8)]$, $C_4(4)$, $[S_4(4)]$, $[C_{2h}(4)]$, $[2C_{4h}(2)]$
C_{4h}^6	$I4_1/a$	$C_1(16)$, $C_2(8)$, $[2C_i(8)]$, $[2S_4(4)]$
D_4^1	$P442$	$C_1(8)$, $7C_2(4)$, $2C_4(2)$, $[2D_2(2)]$, $[4D_4]$
D_4^2	$P42_12$	$C_1(8)$, $3C_2(4)$, $C_4(2)$, $[2D_2(2)]$
D_4^3	$P4_122$	$C_1(8)$, $3C_2(4)$
D_4^4	$P4_12_12$	$C_1(8)$, $C_2(4)$
D_4^5	$P4_222$	$C_1(8)$, $9C_2(4)$, $[6D_2(2)]$
D_4^6	$P4_22_12$	$C_1(8)$, $4C_2(4)$, $[2D_2(2)]$
D_4^7	$P4_322$	$C_1(8)$, $3C_2(4)$
D_4^8	$P4_32_12$	$C_1(8)$, $C_2(4)$
D_4^9	$I422$	$C_1(16)$, $5C_2(4)$, $C_4(4)$, $[2D_2(4)]$, $[2D_4(2)]$
D_4^{10}	$I4_122$	$C_1(16)$, $4C_2(4)$, $[2D_2(4)]$
C_{4v}^1	$P4mm$	$C_1(8)$, $3C_s(4)$, $C_{2v}(2)$, $2C_{4v}$
C_{4v}^2	$P4bm$	$C_1(8)$, $C_s(4)$, $C_{2v}(2)$, $C_4(2)$
C_{4v}^3	$P4_2cm$	$C_1(8)$, $C_s(4)$, $C_2(4)$, $2C_{2v}(2)$
C_{4v}^4	$P4_2mm$	$C_1(8)$, $C_s(4)$, $C_2(4)$, $C_{2v}(2)$
C_{4v}^5	$P4cc$	$C_1(8)$, $C_2(4)$, $2C_4(2)$
C_{4v}^6	$P4nc$	$C_1(8)$, $C_2(4)$, $C_4(2)$
C_{4v}^7	$P4_2mc$	$C_1(8)$, $2C_s(4)$, $3C_{2v}(2)$
C_{4v}^8	$P4_2bc$	$C_1(8)$, $2C_2(4)$
C_{4v}^9	$I4mm$	$C_1(16)$, $2C_s(8)$, $C_{2v}(4)$, $C_{4v}(2)$
C_{4v}^{10}	$I4cm$	$C_1(16)$, $C_s(8)$, $C_{2v}(4)$, $C_4(4)$
C_{4v}^{11}	$I4_1md$	$C_1(16)$, $C_s(8)$, $C_{2v}(4)$
C_{4v}^{12}	$I4_1cd$	$C_1(16)$, $C_2(8)$
D_{2d}^1	$P\bar{4}2m$	$C_1(8)$, $C_s(4)$, $5C_2(4)$, $2C_{2v}(2)$, $[2D_2(2)]$, $[4D_{2d}]$
D_{2d}^2	$P\bar{4}2c$	$C_1(8)$, $7C_2(4)$, $[2S_4(2)]$, $[4D_2(2)]$
D_{2d}^3	$P\bar{4}2_1m$	$C_1(8)$, $C_2(4)$, $C_s(4)$, $C_{2v}(2)$, $[2S_4(2)]$
D_{2d}^4	$P\bar{4}2_1c$	$C_1(8)$, $2C_2(4)$, $[2S_4(2)]$
D_{2d}^5	$P\bar{4}m2$	$C_1(8)$, $2C_s(4)$, $2C_2(4)$, $3C_{2v}(2)$, $[4D_{2d}]$
D_{2d}^6	$P\bar{4}c2$	$C_1(8)$, $5C_2(4)$, $[2S_4(2)]$, $[2D_2(2)]$
D_{2d}^7	$P\bar{4}b2$	$C_1(8)$, $4C_2(4)$, $[2D_2(2)]$, $[2S_4(2)]$
D_{2d}^8	$P\bar{4}n2$	$C_2(8)$, $4C_2(4)$, $[2D_2(2)]$, $[2S_4(2)]$
D_{2d}^9	$I\bar{4}m2$	$C_1(16)$, $C_s(8)$, $2C_2(8)$, $2C_{2v}(4)$, $[4D_{2d}(2)]$
D_{2d}^{10}	$I\bar{4}c2$	$C_1(16)$, $4C_2(8)$, $[2S_4(4)]$, $[2D_2(4)]$
D_{2d}^{11}	$I\bar{4}2m$	$C_1(16)$, $C_s(8)$, $3C_2(8)$, $C_{2v}(4)$, $[S_4(4)]$, $[D_2(4)]$, $[2D_{2d}(2)]$
D_{2d}^{12}	$I\bar{4}2d$	$C_1(16)$, $2C_2(8)$, $[2S_4(4)]$
D_{4h}^1	$P4/mmm$	$C_1(16)$, $5C_s(8)$, $7C_{2v}(4)$, $2C_{4v}(2)$, $[2D_{2h}(2)]$, $[4D_{4h}]$
D_{4h}^2	$P4/mcc$	$C_1(16)$, $C_s(8)$, $4C_2(8)$, $2C_4(4)$, $[D_2(4)]$, $[C_{2h}(4)]$, $[2C_{4h}(2)]$, $[2D_4(2)]$
D_{4h}^3	$P4/nbm$	$C_1(16)$, $C_s(8)$, $4C_2(8)$, $C_{2v}(4)$, $C_4(4)$, $[2C_{2h}(4)]$, $[2D_{2d}(2)]$, $[2D_4(2)]$
D_{4h}^4	$P4/nnc$	$C_1(16)$, $4C_2(8)$, $[C_i(8)]$, $C_4(4)$, $[S_4(4)]$, $[D_2(4)]$, $[2D_4(2)]$
D_{4h}^5	$P4/mbm$	$C_1(16)$, $3C_s(8)$, $3C_{2v}(4)$, $C_4(4)$,

(*Continued*)

Table 5.2   (*continued*)

Space-group symbol		Schönflies site symmetries
Schönflies	International [436]	

		$[2D_{2h}(2)]$, $[2C_{4h}(2)]$
D_{4h}^6	$P4/mnc$	$C_1(16)$, $C_s(8)$, $2C_2(8)$, $C_4(4)$,
		$[D_2(4)]$, $[C_{2h}(4)]$, $[2C_{4h}(2)]$
D_{4h}^7	$P4/nmm$	$C_1(16)$, $2C_s(8)$, $2C_2(8)$, $C_{2v}(4)$,
		$[2C_{2h}(4)]$, $C_{4v}(2)$, $[2D_{2d}(2)]$
D_{4h}^8	$P4/ncc$	$C_1(16)$, $2C_2(8)$, $[C_i(8)]$, $C_4(4)$, $[S_4(4)]$, $[D_2(4)]$
D_{4h}^9	$P4_2/mmc$	$C_1(16)$, $3C_s(8)$, $C_2(8)$, $7C_{2v}(4)$, $[2D_{2d}(2)]$, $[4D_{2h}(2)]$
D_{4h}^{10}	$P4_2/mcm$	$C_1(16)$, $2C_s(8)$, $3C_2(8)$, $4C_{2v}(4)$, $[C_{2h}(4)]$, $[D_2(4)]$
D_{4h}^{11}	$P4_2/nbc$	$C_1(16)$, $5C_2(8)$, $[C_i(8)]$, $[S_4(4)]$, $[3D_2(4)]$
D_{4h}^{12}	$P4_2/nnm$	$C_1(16)$, $C_s(8)$, $5C_2(8)$, $C_{2v}(4)$,
		$[2C_{2h}(4),]$, $[2D_2(4)]$, $[2D_{2d}(2)]$
D_{4h}^{13}	$P4_2/mbc$	$C_1(16)$, $C_s(8)$, $3C_2(8)$, $[D_2(4)]$, $[S_4(4)]$, $[2C_{2h}(4)]$
D_{4h}^{14}	$P4_2/mnm$	$C_1(16)$, $2C_s(8)$, $C_2(8)$, $3C_{2v}(4)$,
		$[S_4(4)]$, $[C_{2h}(4)]$, $[2D_{2h}(2)]$
D_{4h}^{15}	$P4_2/nmc$	$C_1(16)$, $C_s(8)$, $C_2(8)$, $[C_i(8)]$, $2C_{2v}(4)$, $[2D_{2d}(2)]$
D_{4h}^{16}	$P4_2/ncm$	$C_1(16)$, $C_s(8)$, $3C_2(8)$, $C_{2v}(4)$,
		$[2C_{2h}(4)]$, $[S_4(4)]$, $[D_2(4)]$
D_{4h}^{17}	$I4/mmm$	$C_1(32)$, $3C_s(16)$, $C_2(16)$, $4C_{2v}(8)$,
		$[C_{2h}(8)]$, $C_{4v}(4)$, $[D_{2d}(4)]$, $[D_{2h}(4)]$, $[2D_{4h}(2)]$
D_{4h}^{18}	$I4/mcm$	$C_1(32)$, $2C_s(16)$, $2C_2(16)$, $2C_{2v}(8)$,
		$C_4(8)$, $[C_{2h}(8)]$, $[D_{2h}(4)]$, $[C_{4h}(4)]$, $[D_{2d}(4)]$, $[D_4(4)]$
D_{4h}^{19}	$I4_1/amd$	$C_1(32)$, $C_s(16)$, $2C_2(16)$, $C_{2v}(8)$, $[2C_{2h}(8)]$, $[2D_{2d}(4)]$
D_{4h}^{20}	$I4_1/acd$	$C_1(32)$, $3C_2(16)$, $[C_i(16)]$, $[D_2(8)]$, $[S_4(8)]$

Trigonal systems

C_3^1	$P3$	$C_1(3)$, $3C_3$
C_3^2	$P3_1$	$C_1(3)$
C_3^3	$P3_2$	$C_1(3)$
C_3^4	$R3$	$C_1(3)$, C_3
I_6^1	$P\bar{3}$	$C_1(6)$, $[2C_i(2)]$, $2C_3(2)$, $[2C_{3i}]$
I_6^2	$R\bar{3}$	$C_1(6)$, $[2C_i(3)]$, $C_3(2)$, $[2C_{3i}]$
D_3^1	$P312$	$C_1(6)$, $2C_2(3)$, $3C_3(2)$, $[6D_3]$
D_3^2	$P321$	$C_1(6)$, $2C_2(3)$, $2C_3(2)$, $[2D_3]$
D_3^3	$P3_112$	$C_1(6)$, $2C_2(3)$
D_3^4	$P3_121$	$C_1(6)$, $2C_2(3)$
D_3^5	$P3_212$	$C_1(6)$, $2C_2(3)$
D_3^6	$P3_221$	$C_1(6)$, $2C_2(3)$
D_3^7	$R32$	$C_1(6)$, $2C_2(3)$, $C_3(2)$, $[2D_3]$
C_{3v}^1	$P3m1$	$C_1(6)$, $C_s(3)$, $3C_{3v}$
C_{3v}^2	$P31m$	$C_1(6)$, $C_s(3)$, $C_3(2)$, C_{3v}
C_{3v}^3	$P3c1$	$C_1(6)$, $3C_3(2)$
C_{3v}^4	$P31c$	$C_1(6)$, $2C_3(2)$
C_{3v}^5	$R3m$	$C_1(6)$, $C_s(3)$, C_{3v}
C_{3v}^6	$R3c$	$C_1(6)$, $C_3(2)$
D_{3d}^1	$P\bar{3}1m$	$C_1(12)$, $C_s(6)$, $2C_2(6)$, $C_3(4)$,
		$[2C_{2h}(3)]$, $C_{3v}(2)$, $[2D_3(2)]$, $[2D_{3d}]$
D_{3d}^2	$P\bar{3}1c$	$C_1(12)$, $[C_i(6)]$, $C_2(6)$, $2C_3(4)$, $[C_{3i}(2)]$, $[3D_3(2)]$

(*Continued*)

Table 5.2 (*continued*)

Space-group symbol		Schönflies site symmetries
Schönflies	International [436]	
D_{3d}^3	$P\bar{3}m1$	$C_1(12)$, $C_s(6)$, $2C_2(6)$, $[2C_{2h}(3)]$, $2C_{3v}(2)$, $[2D_{3d}]$
D_{3d}^4	$P\bar{3}c1$	$C_1(12)$, $C_2(6)$, $[C_i(6)]$, $2C_3(4)$, $[C_{3i}(2)]$, $[D_3(2)]$
D_{3d}^5	$R\bar{3}m$	$C_1(12)$, $C_s(6)$, $2C_2(6)$, $[2C_{2h}(3)]$, $C_{3v}(2)$, $[2D_{3d}]$
D_{3d}^6	$R\bar{3}c$	$C_1(12)$, $C_2(6)$, $[C_i(6)]$, $C_3(4)$, $[C_{3i}(2)]$, $[D_3(3)]$

Hexagonal systems

C_6^1	$P6$	$C_1(6)$, $C_2(3)$, $C_3(2)$, C_6
C_6^2	$P6_1$	$C_1(6)$
C_6^3	$P6_5$	$C_1(6)$
C_6^4	$P6_2$	$C_1(6)$, $2C_2(3)$
C_6^5	$P6_4$	$C_1(6)$, $2C_2(3)$
C_6^6	$P6_3$	$C_1(6)$, $2C_3(2)$
C_{3h}^1	$P\bar{6}$	$C_1(6)$, $2C_s(3)$, $3C_3(2)$, $[6C_{3h}]$
C_{6h}^1	$P6/m$	$C_1(12)$, $2C_s(6)$, $C_2(6)$, $C_3(4)$, $[2C_{2h}(3)]$, $C_6(2)$, $[2C_{3h}(2)]$, $[2C_{6h}]$
C_{6h}^2	$P6_3/m$	$C_1(12)$, $C_s(6)$, $[C_i(6)]$, $2C_3(4)$, $[C_{3i}(2)]$, $[3C_{3h}(2)]$
D_6^1	$P622$	$C_1(12)$, $5C_2(6)$, $C_3(4)$, $[2D_2(3)]$, $C_6(2)$, $[2D_3(2)]$, $[2D_6]$
D_6^2	$P6_122$	$C_1(12)$, $2C_2(6)$
D_6^3	$P6_522$	$C_1(12)$, $2C_2(6)$
D_6^4	$P6_222$	$C_1(12)$, $6C_2(6)$, $[4D_2(3)]$
D_6^5	$P6_422$	$C_1(12)$, $6C_2(6)$, $[4D_2(3)]$
D_6^6	$P6_322$	$C_1(12)$, $2C_2(6)$, $2C_3(4)$, $[4D_3(2)]$
C_{6v}^1	$P6mm$	$C_1(12)$, $2C_s(6)$, $C_{2v}(3)$, $C_{3v}(2)$, C_{6v}
C_{6v}^2	$P6cc$	$C_1(12)$, $C_2(6)$, $C_3(4)$, $C_6(2)$
C_{6v}^3	$P6_3cm$	$C_1(12)$, $C_s(6)$, $C_3(4)$, $C_{3v}(2)$
C_{6v}^4	$P6_3mc$	$C_1(12)$, $C_s(6)$, $2C_{3v}(2)$
D_{3h}^1	$P\bar{6}m2$	$C_1(12)$, $3C_s(6)$, $2C_{2v}(3)$, $3C_{3v}(2)$, $[6D_{3h}]$
D_{3h}^2	$P\bar{6}c2$	$C_1(12)$, $C_s(6)$, $C_2(6)$, $3C_3(4)$, $[3C_{3h}(2)]$, $[3D_3(2)]$
D_{3h}^3	$P\bar{6}2m$	$C_1(12)$, $3C_s(6)$, $C_3(4)$, $2C_{2v}(3)$, $C_{3v}(2)$, $[2C_{3h}(2)]$, $[2D_{3h}]$
D_{3h}^4	$P\bar{6}2c$	$C_1(12)$, $C_s(6)$, $C_2(6)$, $2C_3(4)$, $[3C_{3h}(2)]$, $[D_3(2)]$
D_{6h}^1	$P6/mmm$	$C_1(24)$, $4C_s(12)$, $5C_{2v}(6)$, $C_{3v}(4)$, $[2D_{2h}(3)]$, $C_{6v}(2)$, $[2D_{3h}(2)]$, $[2D_{6h}]$
D_{6h}^2	$P6/mcc$	$C_1(24)$, $C_s(12)$, $3C_2(12)$, $C_3(8)$, $[C_{2h}(6)]$, $[D_2(6)]$, $C_6(6)$, $[C_{3h}(4)]$, $[D_3(4)]$, $[C_{6h}(2)]$, $[D_6(2)]$
D_{6h}^3	$P6_3/mcm$	$C_1(24)$, $2C_s(12)$, $C_2(12)$, $C_3(8)$, $C_{2v}(6)$, $[C_{2h}(6)]$, $C_6(4)$, $[D_3(4)]$, $[C_{3h}(4)]$, $[D_{3d}(2)]$, $[D_{3h}(2)]$
D_{6h}^4	$P6_3/mmc$	$C_1(24)$, $2C_s(12)$, $C_2(12)$, $C_{2v}(6)$, $[C_{2h}(6)]$, $2C_{3v}(4)$, $[3D_{3h}(2)]$, $[D_{3d}(2)]$

Cubic systems

T^1	$P23$	$C_1(12)$, $4C_2(6)$, $C_3(4)$, $[2D_2(3)]$, $[2T]$
T^2	$F23$	$C_1(48)$, $2C_2(24)$, $C_3(16)$, $[4T(4)]$
T^3	$I23$	$C_1(24)$, $2C_2(12)$, $C_3(8)$, $[D_2(6)]$, $[T(2)]$

(Continued)

Table 5.2 *(continued)*

Space-group symbol		Schönflies site symmetries
Schönflies	International [436]	
T^4	$P2_13$	$C_1(12)$, $C_3(4)$
T^5	$I2_13$	$C_1(24)$, $C_2(12)$, $C_3(8)$
T_h^1	$Pm3$	$C_1(24)$, $2C_3(12)$, $C_3(8)$, $4C_{2v}(6)$, $[2D_{2h}(3)]$, $[2T_h]$
T_h^2	$Pn3$	$C_1(24)$, $2C_2(12)$, $C_3(8)$, $[D_2(6)]$, $[2C_{3i}(4)]$, $[T(2)]$
T_h^3	$Fm3$	$C_1(96)$, $C_s(48)$, $C_2(48)$, $C_3(32)$,
		$C_{2v}(24)$, $[C_{2h}(24)]$, $[T(8)]$, $[2T_h(4)]$
T_h^4	$Fd3$	$C_1(96)$, $C_2(48)$, $C_3(32)$, $[2C_{3i}(16)]$, $[2T(8)]$
T_h^5	$Im3$	$C_1(48)$, $C_s(24)$, $C_3(16)$, $2C_{2v}(12)$,
		$[C_{3i}(8)]$, $[D_{2h}(6)]$, $[T_h(2)]$
T_h^6	$Pa3$	$C_1(24)$, $C_3(8)$, $[2S_6(4)]$
T_h^7	$Ia3$	$C_1(48)$, $C_2(24)$, $C_3(16)$, $[2C_{3i}(8)]$
O^1	$P432$	$C_1(24)$, $3C_2(12)$, $C_3(8)$, $2C_4(6)$, $[2D_4(3)]$, $[2O]$
O^2	$P4_132$	$C_1(24)$, $5C_2(12)$, $C_3(8)$, $[3D_2(6)]$, $[2D_3(4)]$, $[T(2)]$
O^3	$F432$	$C_1(96)$, $3C_2(48)$, $C_3(32)$, $C_4(24)$,
		$[D_2(24)]$, $[T(8)]$, $[2O(4)]$
O^4	$F4_132$	$C_1(96)$, $2C_2(48)$, $C_3(32)$, $[2D_3(16)]$, $[2T(8)]$
O^5	$I432$	$C_1(48)$, $3C_2(24)$, $C_3(16)$, $C_4(12)$,
		$[D_2(12)]$, $[D_3(8)]$, $[D_4(6)]$, $[O(2)]$
O^6	$P4_332$	$C_1(24)$, $C_2(12)$, $C_3(8)$, $[2D_3(4)]$
O^7	$P4_132$	$C_1(24)$, $C_2(12)$, $C_3(8)$, $[2D_3(4)]$
O^8	$I4_132$	$C_1(48)$, $3C_2(24)$, $C_3(16)$, $[2D_2(12)]$, $[2D_3(8)]$
T_d^1	$P\bar{4}3m$	$C_1(24)$, $C_s(12)$, $C_2(12)$, $2C_{2v}(6)$,
		$C_{3v}(4)$, $[2D_{2d}(3)]$, $[2T_d]$
T_d^2	$F\bar{4}3m$	$C_1(96)$, $C_s(48)$, $2C_{2v}(24)$, $C_{3v}(16)$, $[4T_d(4)]$
T_d^3	$I\bar{4}3m$	$C_1(48)$, $C_s(24)$, $C_2(24)$, $C_{2v}(12)$,
		$[S_4(12)]$, $C_{3v}(8)$, $[D_{2d}(6)]$, $[T_d(2)]$
T_d^4	$P\bar{4}3n$	$C_1(24)$, $3C_2(12)$, $C_3(8)$, $[2S_4(6)]$, $[D_2(6)]$, $[T(2)]$
T_d^5	$F\bar{4}3c$	$C_1(96)$, $2C_2(48)$, $C_3(32)$, $[2S_4(24)]$, $[2T(8)]$
T_d^6	$I\bar{4}3d$	$C_1(48)$, $C_2(24)$, $C_3(16)$, $[2S_4(12)]$
O_h^1	$Pm3m$	$C_1(48)$, $3C_s(24)$, $3C_{2v}(12)$, $C_{3v}(8)$,
		$2C_{4v}(6)$, $[2D_{4h}(3)]$, $[2O_h]$
O_h^2	$Pn3n$	$C_1(48)$, $2C_2(24)$, $C_3(16)$, $C_4(12)$,
		$[S_4(12)]$, $[C_{3i}(8)]$, $[D_4(6)]$, $[O(2)]$
O_h^3	$Pm3n$	$C_1(48)$, $C_s(24)$, $C_2(24)$, $C_3(16)$,
		$3C_{2v}(12)$, $[D_3(8)]$, $[2D_{2d}(6)]$
O_h^4	$Pn3m$	$C_1(48)$, $C_s(24)$, $3C_2(24)$, $C_{2v}(12)$, $[D_2(12)]$, $C_{3v}(8)$
O_h^5	$Fm3m$	$C_1(192)$, $2C_s(96)$, $3C_{2v}(48)$, $C_{3v}(32)$,
		$C_{4v}(24)$, $[D_{2h}(24)]$, $[T_d(8)]$, $[2O_h(4)]$
O_h^6	$Fm3c$	$C_1(192)$, $C_s(96)$, $C_2(96)$, $C_3(64)$, $C_4(48)$,
		$[C_{2v}(48)]$, $[C_{4h}(24)]$, $D_{2d}(24)$, $[T_h(8)]$, $[O(8)]$
O_h^7	$Fd3m$	$C_1(192)$, $C_2(96)$, $C_s(96)$, $C_{2v}(48)$,
		$C_{3v}(32)$, $[2D_{3d}(16)]$, $[2T_d(8)]$
O_h^8	$Fd3c$	$C_1(192)$, $2C_2(96)$, $C_3(64)$, $[S_4(48)]$,
		$[C_{3i}(32)]$, $[D_3(32)]$, $[T(16)]$
O_h^9	$Im3m$	$C_1(96)$, $2C_s(48)$, $C_2(48)$, $2C_{2v}(24)$, $C_{3v}(16)$, $C_{4v}(12)$,
		$[D_{2d}(12)]$, $[D_{3d}(8)]$, $[D_{4h}(6)]$, $[O_h(2)]$
O_h^{10}	$Ia3d$	$C_1(96)$, $2C_2(48)$, $C_3(32)$, $[S_4(24)]$,
		$[D_2(24)]$, $[D_3(16)]$, $[C_{3i}(16)]$

(Continued)

Note: Each symbol indicates the site positions (possible positions of activator ions) in the unit cell restricted to the given space group. For example the D_{2h}^{15}— $Pbca$ space group contains following sites: $C_1(8)$ and $[2C_i(4)]$. These designations show that in the given case there is one set with C_1 symmetry, consisting of eight sites, and two distinct sets of C_i symmetry, consisting of four sites. The particular sets that are the intersection points of all the elements in the given space group are shown in square brackets.

6. Luminescence and Stimulated-Emission Properties of Laser Crystals in the Y_2O_3–Al_2O_3 System

The key to the development of the laser-crystal physics of activated crystals has always been the search for new laser media. During the years after the laser effect was first observed, a great deal of experience was accumulated [6–8, 438–440], which allowed the synthesis of more than 170 insulating crystal matrices with different properties (see Tables 1.1–5). As previously mentioned, among them, up to now, only $Y_3Al_5O_{12}$ and $YAlO_3$ crystals doped with Nd^{3+} ions and ruby (Al_2O_3-Cr^{3+}) fulfil to the greatest extent the requirements of contemporary quantum electronics and its applications. The remarkable feature of these compounds is the successful combination of reasonable spectral and stimulated-emission properties with a number of such essential qualities as high mechanical strength and hardness, considerable heat conductivity, and broad-band transparency. A fact worthy of special attention is that all the enumerated crystals are from one physical-chemical system Y_2O_3–Al_2O_3. A comprehensive comparison of the properties of the crystals formed in this system may reveal the qualities which cause such a unique combination not only of physical but also of spectral and stimulated-emission properties[1].

Because of the general interest in these crystals, we shall consider here briefly the main luminescence and stimulated-emission properties of $Y_3Al_5O_{12}$ and $YAlO_3$ crystals activated by Nd^{3+} ions. The properties of ruby are known from several reviews and monographs (see, for example, [60, 70, 73, 74, 255]), so we shall not examine them here. A complete picture of the physical characteristics of the laser crystals that belong to the Y_2O_2–Al_2O_3 system may be obtained from Table 6.1 and Fig. 6.1 (see p. 322). Some preliminary data on the effective density of phonon states ρ_{eff} in yttrium orthoaluminate and yttrium aluminum garnet have been cited in Sect. 3.11. These data were obtained in studies of the vibrational spectra of Cr^{3+} ions in $YAlO_3$ crystals [593] and Yb^{3+} ions in $Y_3Al_5O_{12}$ crystals [548].

---

[1] Investigation of the spectral and stimulated-emission properties of the crystals formed in one system was shown to be an effective method of searching for new laser media [172]. Several laser crystals have been found by this method, for example, two crystals in the SrF_2–YF_3 system, four crystals in the SrF_2–LaF_3 system and five crystals in the CaF_2–YF_3 system (all doped with Ln^{3+} ions).

Table 6.1. Physical properties of laser crystals belonging to the Y_2O_3–Al_2O_3 system

Property	Crystal			
	Y_2O_3	$YAlO_3$	$Y_3Al_5O_{12}$	Al_2O_3
Component ratio	1 : 0	1 : 1	3 : 5	0 : 1
Space group	T_h^7–Ia3	D_{2h}^{16}–Pbnm	O_h^{10}–Ia3d	D_{3d}^5–R$\bar{3}$c
Lattice constants [Å]				
without activator[a]	10.595	$a_0 = 5.176$	12.000	$a_0 = 4.7628$
		$b_0 = 5.307$		$b_0 = 13.0032$
		$c_0 = 7.355$		
with Nd^{3+} ions (≈ 1 at.)	10.604	—	≈ 12.01	—
Cation-site symmetry	Y_I^{3+}–(C_2)	Y^{3+}–(C_s)	Y^{3+}–(D_2)	Al^{3+}–(C_3)
	Y_{II}^{3+}–(C_{3i})	Al^{3+}–(C_i)	Al_I^{3+}–(C_{3i})	
			Al_{II}^{3+}–(S_4)	
Melting point [°C]	2430 ± 30	1850 ± 30	1930 ± 20	≈ 2040
Boiling point [°C]	≈ 4300	—	—	≈ 3500
Debye temperature [K]	≈ 460	—	700–750	935–1030
Thermal conductivity [$W \cdot cm^{-1} \cdot K^{-1}$]				
without activator	0.27	0.11	0.13	0.33 $\perp c$ axis
				0.35$\Vert c$ axis
with Nd^{3+} ions (≈ 1 at.)	0.13	—	0.11–0.14	—
Specific heat [$cal \cdot g^{-1} \cdot K^{-1}$]	0.109	0.10	0.14–0.15	0.1813
Thermal expansion [$10^{-6} \cdot K^{-1}$]				
without activator[b]	6–7	9.5$\Vert a$ axis	8.2$\Vert$[100]	5.31$\Vert c$ axis
		4.3$\Vert b$ axis	7.7$\Vert$[110]	4.78 $\perp c$ axis
		10.8$\Vert c$ axis	7.8$\Vert$[111]	
with Nd^{3+} ions (≈ 1 at.)	—	3.37$\Vert b$ axis	≈ 6.96	
Thermal diffusion [$cm^2 \cdot s^{-1}$]				
without activator	0.12	0.049	0.046	≈ 0.084
with Nd^{3+} ions (≈ 1 at.)	0.06	—	0.050	—
Molecular weght	225.81	163.88	593.59	101.96
Density [$g \cdot cm^{-3}$]				
without activator	5.04	5.35	4.55	3.98
with Nd^{3+} ions (≈ 1 at.)	5.06	—	4.55	—
with Cr^{3+} ions (≈ 0.05 at.)	—	—	—	3.92
Hardness				
Mohs scale	6.8	8.5–9	8.25–8.5	9
Knoop scale [$kg \cdot mm^{-2}$]	700–750	977$\Vert a$ axis	1320–1380	2100–2400
		1190$\Vert b$ axis		
		1670$\Vert c$ axis		
Sound velocity [10^5 $cm \cdot s^{-1}$]	—	—	$8.5630[001]_l$	$10.92[11\bar{2}0]_l$
			$5.0293[001]_s$	$5.79[1120]_{ss}$
			$8.6016[110]_l$	$6.69[1120]_{fs}$
Elastic stiffness [10^{12} $dyn \cdot cm^{-2}$]	—	2.2	$c_{11} = 3.33$	$c_{11} = 4.968$
			$c_{12} = 1.11$	$c_{33} = 4.981$
			$c_{44} = 1.15$	$c_{44} = 1.474$
				$c_{12} = 1.636$
				$c_{13} = 1.109$
				$c_{14} = -0.235$
Elastic compliance [10^{-13} $cm^2 \cdot dyn^{-1}$] —	—	$s_{11} - 3.59$	$s_{11} = 2.353$
			$s_{12} = -0.90$	$s_{12} = -0.716$
			$s_{44} = 8.69$	$s_{44} = 6.940$
				$s_{13} = -0.364$
				$s_{14} = 0.489$
				$s_{33} = 2.170$

(*Continued*)

Table 6.1 (*continued*)

Property	Crystal			
	Y_3O_3	$YAlO_3$	$Y_3Al_5O_{12}$	Al_2O_3
Elasto-optic constants[e]	—	—	$p_{11} = -0.029$ $p_{12} = 0.0091$ $p_{44} = -0.0615$	$p_{11} = -0.25$ $p_{33} = -0.23$ $p_{12} = -0.038$ $p_{13} = 0.005$ $p_{31} = -0.032$ $p_{44} = -0.1$ $p_{14} = 0.02$ $p_{41} = 0.01$
Breaking strain [kg·cm^{-2}]	—	—	1.75–2.1	—
Optical transparency range [μm]	0.23–8.0	0.22–6.5	0.24–6	0.14–6.5
Refractive index n				
without activator[d]	1.915 (0.57 μm)	$n_z = 1.929$ $n_\beta = 1.943$ $n_\gamma = 1.952$	1.81523 (1.06 μm)	$n_o = 1.763$ $n_e = 1.755$ (0.694 μm)
with Nd^{3+} ions ($\approx$1 at.%)	—	$n_z = 1.930$ $n_\beta = 1.946$ $n_\gamma = 1.954$ (0.589 μm)	1.81633 (1.06 μm)	—
Linear dispersion $\delta n/\delta T$ [10^{-6}·K^{-1}][e]	—	9.7 (n_a) 14.5 (n_b)	9.86 ± 0.04	1.0–1.4 (n_o)
Nonlinear refractive index n_2 [10^{-13} esu]	—	—	4.08	$\approx$1.4
Dielectric constants	—	—	$\varepsilon_0 = 11.7$ $\varepsilon_\infty = 3.5$	$\varepsilon_0 = 10.55 \| c$ axis $\varepsilon_0 = 8.6 \perp c$ axis $\varepsilon_\alpha = 3.1$ 9.8×10^{-5}
Solubility (in 100 g of water) [g]	1.8×10^{-4}	—	—	
Distribution coefficient for Nd^{3+} ions	$\approx$1	$\approx$0.8	$\approx$0.18	0

With a few exceptions all data for 300 K were taken from the reviews [6–8, 438, 450] and handbooks [451, 510].

[a] In [678] the lattice constants of the $YAlO_3$ crystal were found within the framework of the D_{2h}^{16} space group with another setting (Pnma). They are $a_0 = 5.330$, $b_0 = 7.375$, and $c_0 = 5.180$ Å.

[b] According to [676] the thermal expansion of the $Y_3Al_5O_{12}$ crystal in the temperature range 300–1675 K is 8.9×10^{-6} K^{-1}.

[c] The elasto-optic constants of the Al_2O_3–Cr^{3+} crystal (0.05%) are $p_{11} = -0.23$, $p_{13} = 0.02$, $p_{31} = -0.04$, $p_{41} = 0.01$, $p_{12} = -0.03$, $p_{14} = 0.00$, $p_{33} = -0.20$, and $p_{44} = -0.01$ [510].

[d] In [677], the dispersion of the refractive index for the $Y_3Al_5O_{12}$ crystal was investigated in the range 0.4–4.0 μm. The Sellmeier equation obtained by *Witter* and *Keig* appears for $YAlO_3$ in [220] as

$$n^2 = 1 + \frac{A \lambda^2}{\lambda^2 - B},$$

where λ is in micrometers and

A	B
$\alpha = 2.63468$	$\alpha = 0.011592$
$\beta = 2.67792$	$\beta = 0.012282$
$\gamma = 2.70892$	$\gamma = 0.021607$

In [829] the same measurements for the $YAlO_3$ crystal were also carried out.

[e] The data cited are for activated crystals.

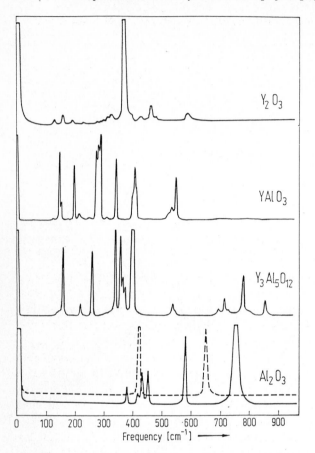

Fig. 6.1. Room-temperature Raman spectra of Y_2O_3–Al_2O_3 crystalline system compounds. The spectra of $Y_3Al_5O_{12}$ and Al_2O_3 crystals are shown for two polarizations of exciting argon-ion laser emission

6.1 Luminescence Properties of $Y_3Al_5O_{12}$-Nd^{3+} Crystals

Detailed investigations of both the luminescence and absorption spectra, as well as the Nd^{3+} ion energy level scheme, in $Y_3Al_5O_{12}$ crystals (see Table 4.3) have been reported in [168, 271, 273–275, 524, 654]. At room and lower temperatures, the optically excited luminescence of $Y_3Al_5O_{12}$-Nd^{3+} crystals occurs in four infrared channels that correspond to transitions from the metastable $^4F_{3/2}$ state to Stark components of manifolds of the ground $^4I_{J'}$ term. As shown in Sect. 4.3, the spectral densities of emission of Nd^{3+} ions in these luminescence channels (the intermanifold branching ratios $\beta_{JJ'}$) are defined by use of the spectroscopic-quality parameter X. According to [514, 662], $X=0.3$ for the $Y_3Al_5O_{12}$-Nd^{3+} crystal. Data

Table 6.2. **Intermanifold branching ratios for the $^4F_{3/2} \rightarrow ^4I_{J'}$ transitions of Nd^{3+} ions in $Y_3Al_5O_{12}$ crystals**[a]

Transition	Luminescence range [μm]	$\beta_{JJ'}$				
		Experimental		Calculated		
		[535]	[541, 750]	[535]	[744]	[514][b]
$^4F_{3/2} \rightarrow ^4I_{9/2}$	0.8690–0.9460	≈ 0.3	0.32	0.32	0.32	≈ 0.28
$^4F_{3/2} \rightarrow ^4I_{11/2}$	1.0510–1.1125	≈ 0.56	0.54	0.53	0.54	0.577
$^4F_{3/2} \rightarrow ^4I_{13/2}$	1.3187–1.4444	≈ 0.14	0.14	0.15	0.14	0.14
$^4F_{3/2} \rightarrow ^4I_{15/2}$	1.7416–2.1285	0.01	—	0.00	0.004	0.007

[a] The coefficients $\beta_{JJ'}$ for $Y_3Al_5O_{12}$-Nd^{3+} crystals have been measured and calculated in [187, 446–448].
[b] Calculated using $\beta_{JJ'}(X)$ plots.

obtained from theoretical estimates and recent measurements of the intermanifold branching ratios $\beta_{JJ'}$ for Nd^{3+} ions in $Y_3Al_5O_{12}$ are summarized in Table 6.2.

The refined schemes of the Stark splitting as well as the infrared luminescence and absorption spectra associated with transitions between the $^4F_{3/2}$ and $^4I_{9/2-15/2}$ manifolds are shown in Fig. 6.2. It should be noted that twofold checking of the Stark-level positions obtained from luminescence and absorption data is particularly desirable in the case of Nd^{3+} ions, since this is the only way to assure sufficient accuracy of such measurements, on which the subsequent identification of the induced transitions largely depends. The same numbering is used for the transitions on the crystal-field splitting scheme (Fig. 6.2) and the corresponding lines of the spectra. Positions of the ground-state Stark levels may also be obtained from investigations of absorption spectra associated with other multiplets, for instance, as shown for the $^4I_{9/2} \rightarrow ^2P_{1/2}$ transition (see Fig. 6.2a).

Analysis of experimental data shows that at low concentrations, Nd^{3+} ions in $Y_3Al_5O_{12}$ crystals form principally one type of activator center [405, 441]. In forming this center, the Nd^{3+} ions replace Y^{3+} ions (six magnetically nonequivalent site types in the unit cell) with the site symmetry D_2 (see Tables 5.2, 6.1). The Stark splitting of the terms that belong to the $4f^N$ configuration in the local coordinate system is described by the following Hamiltonian, which does not take into account the two-particle operators:

$$\mathcal{H} = B_2^0 O_2^0 + B_2^2 O_2^2 + B_4^0 O_4^0 + B_4^2 O_4^2 + B_4^4 O_4^4 + B_6^0 O_6^0$$

$$+ B_6^2 O_6^2 + B_6^4 O_6^4 + B_6^6 O_6^6.$$

Full values of the Stark splitting for terms with $J > 3/2$ are determined mainly by the strong cubic component of the crystal field,

$$\mathcal{H}_{cub} = \frac{B_4^4}{5}(O_4^0 + 5O_4^4) - \frac{B_6^4}{21}(O_6^0 - 21O_6^4),$$

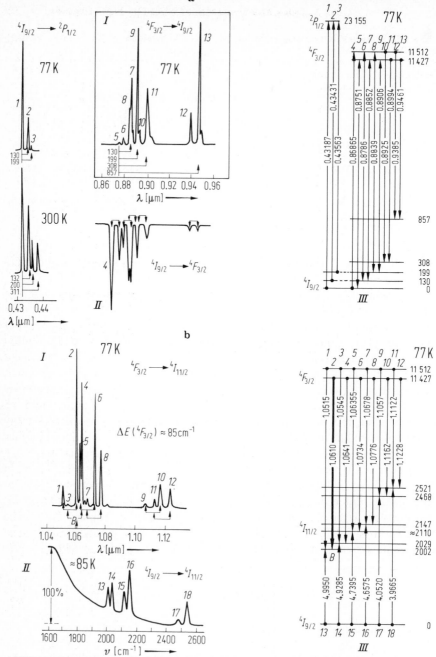

Fig. 6.2. Luminescence (I) and absorption (II) spectra and crystal-field splitting schemes of the $^2P_{1/2}$, $^4F_{3/2}$ and $^4I_{9/2-15/2}$ manifolds (III) of Nd^{3+} ions in $Y_3Al_5O_{12}$. (a) $^4I_{9/2}$ manifold; (b) $^4I_{11/2}$ manifold; (c) $^4I_{13/2}$ manifold; (d)$^4I_{15/2}$ manifold. The arrows in the spectra indicate the ground $^4I_{9/2}$ state splitting (a), and square brackets indicate the $^4F_{9/2}$ state splitting. The other notation is the same as in Fig. 3.22

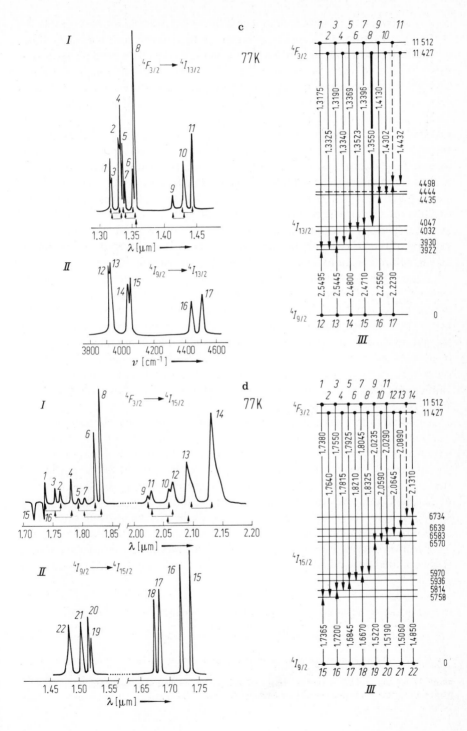

while the axial and rhombic components of the crystal field are responsible for the "fine" structure of the energy spectrum of the impurity ion. The study of the crystal-field parameters B_n^m for Nd^{3+} ions in the $Y_3Al_5O_{12}$ crystal performed in [654] by use of the charge-exchange model determines the main inter-ion interactions responsible for the observed Stark structure of the $^4F_{3/2}$ and $^4I_{9/2-15/2}$ manifolds[2]. Let us discuss some consequences of this study.

The cubic component of the crystal field in the $Y_3Al_5O_{12}$ crystal arises essentially from exchange and Coulomb interactions between $4f$ electrons and the eight nearest neighbors (O^{2-} ions) of the impurity ion. Impurity ions with ionic radius larger than the radius of the Y^{3+} ion (see Table 2.1) push apart the nearest surrounding atoms; the resulting local deformation of the crystal lattice causes the cubic crystal-field parameters B_4^4 and B_6^4 to be almost equal for garnets of different chemical composition. The axial and rhombic crystal-field parameters, especially the quadrupole parameters B_2^0 and B_2^2 and consequently the splitting of the $^4F_{3/2}$ manifold, happen to be most sensitive to the composition and structure constants of the anion sublattice in garnets.

The oxygen ions in garnet lattices that occupy the sites with point-symmetry group C_1 are polarized. Their dipole moments have been calculated in [654], taking into account the change of ionic susceptibility in the crystal and also the overlapping of the wave functions of the nearest neighbors. When summed over the garnet lattice, the contributions to the crystal-field parameters that correspond to the energy of $4f$ electrons in the fields of point charges (B_{2q}^m) and of point dipoles (B_{2D}^m) have opposite signs and approximately equal absolute values[3]. Because of mutual compensation of these contributions, the parameter B_2^2 is close to the parameter B_{2S}^2 (about 300 cm^{-1} for $Y_3Al_5O_{12}$). Since the contribution to B_{2S}^2 from the field of exchange charges is small, the parameter B_2^0 can take on values from $+50$ to -200 cm^{-1}, depending on the structural parameters of the oxygen sub-lattices. Correspondingly, over the homologous row of garnets, the greatest relative changes take place for the Stark splittings of the $^4F_{3/2}$ and $^4S_{3/2}$ manifolds. The noticeable contributions to the Stark splitting of $^4I_{J'}$ manifolds gives interaction between the different terms (the total Stark splittings have the same order of magnitude as the spin–orbit interaction constants for Nd^{3+} ions), as well as the electron–phonon interaction.

For low concentration of activator, the luminescence lifetime of a given state is determined by radiative (spontaneous) and nonradiative transitions [see

---

[2]For completeness and higher reliability of the theoretical results, the crystal field for Nd^{3+} ions in $Y_3Ga_5O_{12}$ and $Y_3Sc_2Al_3O_{12}$ crystals was also studied in [654].

[3]The crystal-field parameters affected by the Ln^{3+} ions in ionic crystals, in the charge-exchange model, are represented as

$$B_n^m = B_{nq}^m + B_{nD}^m + B_{nS}^m,$$

where the B_{nq}^m, B_{nD}^m and B_{nS}^m values determine the $4f$ electron energy in the electric fields of point ions, point dipoles and exchange charges, respectively (see Sect. 4.1).

(2.39)]. The first include both pure-electron and electron-vibrational transitions. In the case of the $Y_3Al_5O_{12}$-Nd^{3+} crystal, electron-vibrational transitions are practically unobserved. Numerous studies of this crystal show [32–34, 169, 187, 272, 344, 399, 402, 408, 446, 539] that the metastable $^4F_{3/2}$ state is predominantly deactivated, due to luminescence transitions,

$$1/\tau_{lum} \approx \sum_i A_i,$$

since $\Sigma_i d_i \ll \Sigma_i A_i$ (here A_i and d_i are the probabilities of spontaneous and nonradiative transitions from the ith level). This is confirmed by the behavior of the temperature dependence $\tau_{lum}(T)$ (see Fig. 6.3) found experimentally for $Y_3Al_5O_{12}$ single crystals in which the amount of Nd^{3+} ions (<0.3 at.%) is low enough that concentration quenching of the luminescence from $^4F_{3/2}$ levels does not occur. The τ_{lum} value (see Fig. 6.3) remains practically constant (within the limits of measurement error) up to about 900 K. This indicates that the inequality above holds and $\tau_{lum} \approx \tau_{rad}$.

For the given laser medium, it is actually difficult to suppose (see Sect. 3.11) the existence of effective multiphonon relaxation through the energy gap of about 4700 cm^{-1} that lies below the $^4F_{3/2}$ levels in the temperature range mentioned, even if this relaxation were due to phonons of the highest energy (see

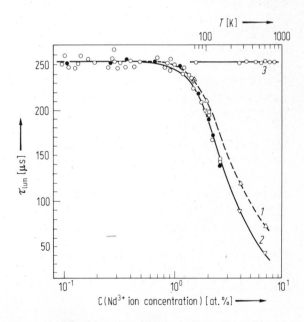

Fig. 6.3. Dependences of $\tau_{lum}(C)$ (curves 1, 2) and $\tau_{lum}(T)$ (curve 3) of the Nd^{3+} ion $^4F_{3/2}$ state in the $Y_3Al_5O_{12}$ crystal. (1) 4.2 K; (2) 77 and 300 K; (3) $C=0.3$ at.%. The triangles correspond to the data of [449]. The open and dark circles correspond to data of the author

Tables 6.3, 4 in which data on infrared reflection and Raman scattering are gathered). Analysis of the $\tau_{lum}(C)$ dependences which are also given in Fig. 6.3, shows that concentrations of Nd^{3+} ions greater than 1 at.% leads to noticeable concentration quenching of the $^4F_{3/2}\to{}^4I_{J'}$ luminescence in $Y_3Al_5O_{12}$ crystals[4]. For such concentrations of Nd^{3+} ions, the τ_{lum} value does not remain constant with increasing temperature [32].

In general, the temperature dependence of τ_{lum} for the $^4F_{3/2}$ state at low neodymium concentration may change under the influence of another factor. Several multiplets ($^4F_{5/2}$, $^2H_{9/2}$ and so on) lie not so far (about 900 cm^{-1}) above this metastable $^4F_{3/2}$ state (see Table 4.3). If their luminescence lifetimes are significantly different from that of the $^4F_{3/2}$ state, then an increase in temperature may cause the static population differences between these manifolds and the $^4F_{3/2}$ state to decrease, and may change the observed luminescence lifetime of that state. The expression for the total probability of spontaneous transitions from a given ith Stark component can be written, taking into account other nearby i levels,

$$A_i' = \frac{\sum_i A_i n_i}{\sum_i n_i} = \frac{\sum_i A_i \exp\left(-\varDelta E_{1i}/kT\right)}{\sum_i \exp\left(-\varDelta E_{1i}/kT\right)},$$

Table 6.3. Peak positions in the infrared reflection spectra of $Y_3Al_5O_{12}$ crystals [cm^{-1}]

Reflection [442]		Transmission [443]	Absorption [444]
Transverse modes (TO)	Longitudinal modes (LO)		
119	125	122	123
165	181	166	163
—	—	178	—
218	227	220	223
—	—	—	255
289	298	292	293
329	341	332	332
370	379	378	373
392	403	390	395
—	—	398	—
429	437	436	430
466	471	465	457
482	505	512	510
521	551	—	530
570	592	569	570
690	707	698	690
737	769	726	722
—	—	789	784
811	921	—	830

[4] For polycrystalline $Y_3Al_5O_{12}$-Nd^{3+} samples, the luminescence lifetime can vary within some range, depending on the method of fabrication. Thus, the luminescence lifetime of a polycrystalline sample with $C \approx 0.01$ at.%Nd^{3+} ions was reported to be about 420 μs at 300 K [716].

Table 6.4. **Peak positions in the Raman-scattering spectra of $Y_3Al_5O_{12}$ crystal [cm^{-1}]**

Peak position [445]	Symmetry of lattice vibrations [442]		
	Fully symmetric, A_{1g}	Tetragonal, E_g	Trigonal, T_{2g}
1	2	3	4
—	—	—	144
162	—	162	—
202	—	—	—
213	—	—	218
—	—	—	243
262	—	—	259
314	—	310	296
339	—	340	—
371	373	—	370[a]
402	—	403	408
441	—	—	436
452	—	—	—
465	—	—	—
481	—	—	—
500	—	—	—
510	—	—	—
536	—	531	530[a]
541	—	537	544
566	561	—	—
—	—	714	690
—	—	—	719
—	—	758	—
781	783	—	—
808	—	—	—
820	—	—	—
835	—	—	—
865	—	—	857

[a] This peak coincides with a line of Ar emission. In [442] Raman scattering was excited with an argon-ion laser.

where n_i is the population of the level i and ΔE_{1i} is the energy gap between the level i ($^4F_{3/2}$, $^4F_{5/2}$ and other higher lying states) and the lowest (the first) Stark component of the $^4F_{3/2}$ manifold. Because the energy gap between the $^4F_{3/2}$ and $^4F_{5/2}$ manifolds is approximately 900 cm^{-1}, the influence of the latter on the τ_{lum} of the metastable state will not be especially noticeable below about 1000 K. Then the temperature dependence of τ_{lum} will be connected only with the difference of spontaneous radiation probabilities for transitions from the components of the $^4F_{3/2}$ state itself. As follows from the two formulas above, τ_{lum} will be reduced, if the total spontaneous probability A_i is larger for the upper level than for the lower component and the lifetime will be increased if the ratio of the A_i values for these levels is opposite. In the case where the spontaneous probabilities A_i for two sublevels of the $^4F_{3/2}$ state are close or equal, τ_{lum} will show either no, or only a slight, temperature dependence, according to the above mentioned factor.

The nature of the concentration quenching of the luminescence of Ln^{3+}, and particularly Nd^{3+}, ions in different media is connected with the formation of quenching centers in the form of pairs or more complex associates and with the presence of resonant or close-to-resonant (phonon-assisted) transitions in absorption and radiation (see Sect. 7.1 for details). Detailed analysis of this phenomenon in crystals containing Nd^{3+} ions has been carried out in a number of investigations, but the most convincing results were reported in [264, 681] for the LaF_3-Nd^{3+} system. There is reason to believe that the conclusions of [264] may be extended also to a number of other crystals with Nd^{3+} ions, and in particular, to $Y_3Al_5O_{12}$-Nd^{3+}. According to [264, 681], the effect responsible for the concentration quenching is cross relaxation between the $^4F_{3/2} \to {}^4I_{15/2}$ and $^4I_{9/2} \to {}^4I_{15/2}$ transitions, with intensity depending on the density of phonons capable of compensation of the energy difference (mismatch) between these transitions, and processes of energy migration, the activity of which increases with decreasing average distance between the Nd^{3+} ions. At helium temperature, the quenching is somewhat weakened even for significant amounts of impurity (see curve 1 in Fig. 6.3). This has its origin in the fact that, in the complex chain of interactions, the energy migration becomes the slowest process. At higher temperatures, the velocity of excitation migration increases and becomes equal to or exceeds the probability of the final interaction that defines the concentration quenching. It should be noted here that for low, and especially for helium temperatures the essential and sometimes decisive (in $YAlO_3$-Nd^{3+} crystals [463]) contribution to the concentration quenching of luminescence in $Y_3Al_5O_{12}$-Nd^{3+} crystals, as well as in other neodymium media [463, 468], may be made by nonresonant cross-relaxation processes which involve the second intermediate $^4I_{13/2}$ state (the cross-relaxing transitions are $^4F_{3/2} \to {}^4I_{13/2}$ and $^4I_{9/2} \to {}^4I_{15/2}$).

The optimum concentration of activator for a given laser medium has special importance because it determines the energetic parameters of laser operation. The optimum content of Nd^{3+} ions in yttrium aluminum garnet may be found by study of the luminescence intensity dependence $I_{lum}(C)$ at 300 K in a series of

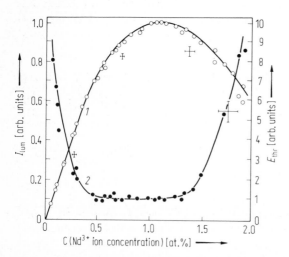

Fig. 6.4. Concentration dependence of luminescence 1.06415 μm line intensity (1) and excitation threshold (2) of the A emission line for a $Y_3Al_5O_{12}$-Nd^{3+} crystal, $^4F_{3/2} \to {}^4I_{11/2}$ transition, 300 K

crystals with different concentrations that have no Tyndall scattering. Most convenient for these purposes is analysis of the intensities for the $^4F_{3/2} \rightarrow {}^4I_{11/2}$ transition or estimation of C_{opt} by the behavior of the 9382 and 9458Å lines due to transitions to the highest Stark component of the $^4I_{9/2}$ state. This component is well spaced 852 cm^{-1} from the ground level and is practically unpopulated at room temperature. The concentration dependence of $I_{lum}(C)$ for the most intense line of the $^4F_{3/2} \rightarrow {}^4I_{11/2}$ transition in $Y_3Al_5O_{12}$-Nd^{3+} crystals, which is a stimulated-emission line at 300 K, is shown in Fig. 6.4. It is seen from this dependence that the maximum of luminescence intensity falls at $C_{opt} \approx 1.1$ at.%. The same concentration series of $Y_3Al_5O_{12}$-Nd^{3+} crystals was used in studying the concentration dependence of the threshold energy $E_{thr}(C)$ [271]. The minimum threshold energy [the middle of the broad extremum of $E_{thr}(C)$] is seen to be in close agreement with the C_{opt} value. On turning to the other spectroscopic characteristics, it should be noted that there is a good correlation between the shapes of $I_{lum}(C)$ and $\tau_{lum}(C)$ dependences. Both of them are sensitive to an increase in the number of pairs or more complex activator centers in $Y_3Al_5O_{12}$-Nd^{3+} crystals.

Investigation of different manifestations of the electron-phonon interaction in $Y_3Al_5O_{12}$-Nd^{3+} crystals [38, 169, 179, 344, 402, 403, 540, 548], study of the temperature dependence of its stimulated-emission characteristics, and measurement of some other parameters and properties [32-34, 169, 170, 187, 408, 446, 539, 679] provide conclusive evidence that the quantum yield of luminescence (in the $^4F_{3/2} \rightarrow {}^4I_{J'}$ bands) is close to unity in a wide temperature range[5]. Another important luminescence characteristic is the branching ratio β_{ij}. If η and τ_{lum} are known, it is possible to determine the values of τ_{rad} and hence A_{ij} (see Sect. 2.2) for a given induced transition. As an example, the experimental values of relative quantum yield versus λ_{exc} for a $Y_3Al_5O_{12}$-Nd^{3+} crystal are shown in Fig. 6.5.

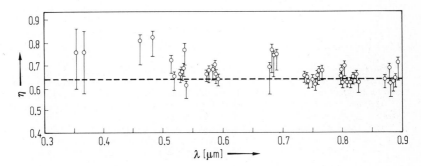

Fig. 6.5. Dependence of the relative quantum yield of Nd^{3+} ions in a $Y_3Al_5O_{12}$ crystal on excitation wavelength [446]. Measurements carried out on the luminescence in the $^4F_{3/2} \rightarrow {}^4I_{11/2}$ transition at an activator concentration of about 1 at.%

---

[5]Subsequent direct precision measurements of the luminescence ($^4F_{3/2} \rightarrow {}^4I_{J'}$) quantum yield for Nd^{3+} ions in $Y_3Al_5O_{12}$ at 300 K undertaken by the authors of [680] show also that $\eta \approx 1$.

Table 6.5. Inter-Stark branching ratios β_{ij} and peak cross sections $\sigma^p_{e,ij}$ for luminescence transitions of Nd^{3+} ions in $Y_3Al_5O_{12}$ crystals measured at 300 K [a]

Transition	λ_{ij} [μm]	β_{ij}	$\Delta\nu_{lum}$ [cm^{-1}]	$\sigma^p_{e,ij}$ [10^{-19} cm^2]
$^4F_{3/2} \rightarrow {}^4I_{9/2}$		$\beta_{JJ'} \approx 0.3$		
$^4F_{3/2} \rightarrow {}^4I_{11/2}$	1.0521 C	0.0382	4.5	0.95
	1.0549	0.0023	4.5	0.06
	1.0615 B	0.0799	3.6	2.5
	1.06415 A	0.1275	≈ 5	3.0
	1.0644 A'	0.0533	4.2	1.45
	1.0682	0.034	6.5	0.6
	1.0737	0.0657	4.6	1.65
	1.0779	0.0463	7.0	0.77
	1.1055	0.0145	11	0.16
	1.1119	0.0297	10.2	0.36
	1.1158	0.0356	10.6	0.42
	1.1225	0.0328	9.9	0.4
		$\beta_{JJ'} \approx 0.56$		
$^4F_{3/2} \rightarrow {}^4I_{13/2}$	1.3187	0.0183	≈ 4	0.95
	1.3203	0.0061	4.6	0.23
	1.3335	0.0073	≈ 3	0.44
	1.3351	0.01	3.3	0.54
	1.3381	0.0243	4.0	1.0
	1.3419	0.0120	≈ 6	0.36
	1.3533	0.0062	≈ 4	0.28
	1.3572	0.0214	≈ 4	0.73
	—	—	—	—
	1.4150	0.0099	8.5	0.2
	1.4271	0.0028	7.0	0.08
	—	—	—	—
	1.4320	0.0066	≈ 10	0.13
	1.4444	0.0128	≈ 9	0.28
		$\beta_{JJ'} \approx 0.14$		
$^4F_{3/2} \rightarrow {}^4I_{15/2}$		$\beta_{JJ'} \approx 0.01$		

[a] The effective cross section σ^{eff}_e for the A transition which gives rise to stimulated emission at the wavelength 1.06415 μm may be considered to be $(3.3 \pm 0.2) \times 10^{-19}$ cm^2. This value results from a sum of corresponding parts of $\sigma^p_{e,ij}$ for two transitions A and A' which have close wavelength (1.06415 and 1.0644 μm, respectively). The given value of $\sigma^{eff}_e = 3.3 \times 10^{-19}$ cm^2 is in excellent agreement with the cross section obtained from stimulated-emission measurements (for example, by gain coefficients). The value of $\sigma_{e,ij}$ calculated by (2.11) by consideration of τ^{ij}_{rad} is $\approx 7.7 \times 10^{-19}$ cm^2, which corresponds to the data given before in [271, 272, 399]. The inter-Stark branching ratios and sections for Nd^{3+} ions in the $Y_3Al_5O_{12}$ crystal at 77 and 300 K have been measured in [187, 271, 401, 447, 541, 750].

According to these results, about 60% of the total luminescence falls into luminescence of the $^4F_{3/2} \to {}^4I_{11/2}$ transition. It should be noted here that precise measurement of the energy distribution over the different luminescent $^4F_{3/2} \to {}^4I_{J'}$ transitions of Nd^{3+} ions, both in the $Y_3Al_5O_{12}$ crystal and in other laser media, is a difficult problem [187, 271, 446, 447, 535, 539, 541].

Knowledge of the cross section (or probability) of luminescence transitions permits calculation of a number of operating characteristics of lasers, for instance, output power, optimum transmittance of optical-resonator mirrors, etc. [60, 70, 72, 73]. From the numerous methods of measuring σ_e for crystals activated by Nd^{3+} ions, the one using the Fuchtbauer-Ladenburg formula receives widest use. For the case of a homogeneously broadened line of Lorentz shape, e.g., for the case of the $Y_3Al_5O_{12}$-Nd^{3+} crystal, this formula has been cited previously. According to (2.11), the determination of σ_e for a given luminescence transition requires knowledge of τ_{rad}^{ij}, that is, the lifetime of the luminescence transition from the ith Stark level of the metastable $^4F_{3/2}$ state to the jth Stark level of one of the $^4I_{J'}$ manifolds. For the case of $Y_3Al_5O_{12}$-Nd^{3+} crystals, τ_{rad}^{ij} may be found from the formula

$$\tau_{rad}^{ij} = \frac{\tau_{rad}}{\beta_{ij}} \frac{b_i}{\sum_i b_i},$$

where $\beta_{ij} = D_{lum}^{ij}/D_{lum}$ is the inter-Stark branching ratio; $b_i = \exp[-\Delta E_{i1}(^4F_{3/2})/kT]$ is the Boltzmann factor; D_{lum}^{ij} is the luminescence density in the $i \to j$ transition; and $D_{lum} = \Sigma_{i,j} D_{lum}^{ij}$ is the total luminescence density for transitions that terminate at the Stark levels of ^4I_J manifolds.

For calculation of the laser characteristics, it is necessary to know the peak value of cross sections ($\sigma_{e,ij}^p$) for the inter-Stark $i \to j$ transition, considering the real (at a given temperature) Boltzmann distribution of population between the (upper) levels of the metastable state. More-accurately defined data on $\sigma_{e,ij}^p$ values at 300 K for a $Y_3Al_5O_{12}$-Nd^{3+} crystal (for the two $^4F_{3/2} \to {}^4I_{11/2,13/2}$ channels) and also on linewidth ($\Delta\nu_{lum}$) and branching ratios, both inter-Stark (β_{ij}) and intermanifold ($\beta_{JJ'}$), are presented in Table 6.5.

6.2 Spectral-Emission Properties of $Y_3Al_5O_{12}$-Nd^{3+} Crystals

Stimulated emission from a $Y_3Al_5O_{12}$-Nd^{3+} crystal was first observed by *Geusic* et al. [9]. They obtained pulsed and CW emission at 77 K as well as at 300 K. Subsequent studies of stimulated emission from this crystal made it possible for the author of the present book to refine the emission wavelength [168] and to show that, for the usual laser scheme with broad-band optical resonator, the stimulated-emission spectrum of $Y_3Al_5O_{12}$-Nd^{3+} at 300 K may consist of two lines, A and B [4, 169]. The main A line with the wavelength $\lambda_g = 1.06415 \ \mu m$

(11507 cm^{-1} $^4F_{3/2} \rightarrow ^4I_{11/2}$ 2110 cm^{-1} transition) has a very low threshold of excitation, while the threshold of the B line with $\lambda_g = 1.0615$ μm (11423 cm^{-1} $^4F_{3/2} \rightarrow ^4I_{11/2}$ 2002 cm^{-1} transition) is approximately 100 times greater (see also Fig. 3.29). At 77 K, under similar experimental conditions, stimulated emission occurs on the B line only, which is related to the transition 11427 cm^{-1} $^4F_{3/2} \rightarrow ^4I_{11/2}$ 2002 cm^{-1}, with $\lambda_g = 1.0610$ μm (see Figs. 3.29, 6.2b).

Using different methods of stimulated-emission spectroscopy (see review [4, 5]), it is possible also to excite stimulated emission from the $Y_3Al_5O_{12}$-Nd^{3+} crystal at other wavelengths. In Fig. 6.6, the crystal-field splitting schemes of the $^4F_{3/2}$ and $^4I_{9/2-15/2}$ manifolds of Nd^{3+} ions in the $Y_3Al_5O_{12}$ crystal are shown, with the identification of all stimulated emission lines, observed in the temperature range from 77 to about 850 K, by various authors (see Table 5.1). In addition, Fig. 6.7 shows the temperature dependence of λ_g for the lines A, B, and C, taken from experiments with both conventional lasers and CAM lasers [4, 169].

Other spectral and generation characteristics measured under different conditions have been considered in Table 5.1. We shall only note here that $Y_3Al_5O_{12}$-Nd^{3+} crystals are very efficient in the CW regime. Their relatively low elasto-optic constants, high mechanical hardness and satisfactory thermal parameters (see Table 6.1) provide very high output power at 300 K for lasers based on these crystals. To obtain an output power of several hundred watts at the main A line (1.06415 μm) is not a problem. A CW laser with $P_g \approx 760$ W is described in [216]. According to the review [437], the output power may exceed 1 kW.

The most important characteristic of a laser is the excitation threshold. It depends on almost all the spectroscopic parameters of the active medium (see Sect. 3.1), and exhibits a pronounced sensitivity to various peculiarities of the interaction of activator ions with the matrix. The threshold energy depends on the operation regime, the excitation conditions, the optical quality of the laser element, and the optical resonator parameters. That is why so much attention has been paid in the literature to analysis of the stimulated-emission threshold energy (see, for example, [60, 69–74]). Unfortunately, in studying the relationship between the threshold energy and the parameters enumerated above, most authors did not sufficiently use the possibilities presented by stimulated-emission studies in a wide temperature range. The methods of the high-temperature spectroscopy of stimulated emission developed in recent years [4, 5, 24, 32, 34, 37, 69], and already mentioned briefly in Chap. 3, significantly enrich the arsenal of methods for spectral studies of activated crystals, and allow determination of many details of the temperature dependences of the stimulated-emission parameters. This is especially true for lasers based on crystals activated by Nd^{3+} ions which are capable of emission up to temperatures much higher than 300 K. The first steps in this direction were made in studies [32–34, 37, 38, 169] devoted to analysis of the laser properties of the $Y_3Al_5O_{12}$-Nd^{3+} crystal. Let us briefly consider the results of these studies.

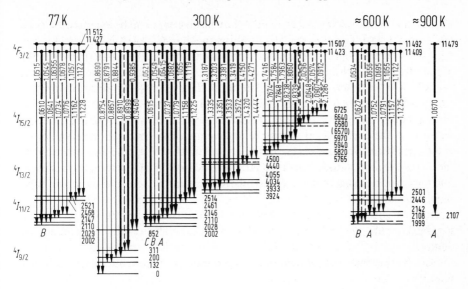

Fig. 6.6. Crystal-field splitting schemes of the $^4F_{3/2}$ and $^4I_{9/2-15/2}$ manifolds of Nd^{3+} ions in the Y$_3$Al$_5$O$_{12}$ crystal with the identification of all registered induced transitions (thick arrows). The position of the levels are given in cm^{-1}, and the wavelengths of transitions between them are given in μm. The other notations are the same as in Fig. 3.30

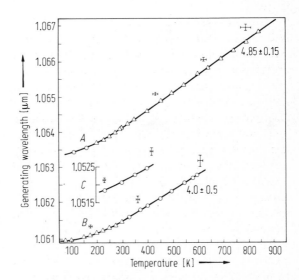

Fig. 6.7. Temperature dependences of emission wavelength λ_g ($^4F_{3/2} \rightarrow {}^4I_{11/2}$ transition) of a Y$_3$Al$_5$O$_{12}$-Nd^{3+} crystal laser [5]. The figures near the curves show values of $(\delta\lambda_g/\delta T)\,10^2\,[\text{Å} \cdot \text{K}^{-1}]$ in the linear parts of $\lambda_g(T)$ dependences. The circles denote experimental points registered in a CAM laser scheme

The temperature dependence of the threshold energy will be examined for the lines of the principal $^4F_{3/2} \rightarrow {}^4I_{11/2}$ emission channel by solving the kinetic equations for the populations of Nd^{3+} ion energy levels that participate directly in the energy balance during excitation and emission. Usually such a treatment is carried out for some generalized situation [60, 69–74]. Application of this treatment to a particular crystal may lead only to some qualitative results. To obtain more-detailed information, it is necessary to take into account the specific features of the energy-level scheme of a given activator ion, and the values of some parameters that characterize the crystal matrix. With the aim of studying the change of stimulated-emission characteristics over a wide temperature range, one may ignore terms that are valid only for certain temperatures in writing the kinetic equations.

In the following analysis, we shall use the conventional energy-levels scheme for Nd^{3+} ions in $Y_3Al_5O_{12}$ given in Fig. 6.8. The orbital degeneracy is completely removed in this crystal. The remaining Kramers degeneracy is the same for all Stark components and need not be taken into account. The

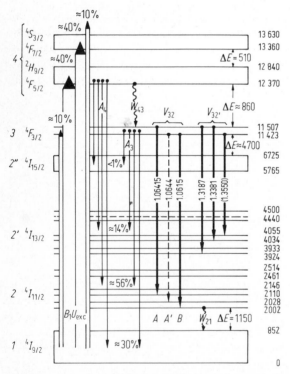

Fig. 6.8. Generalized scheme of Nd^{3+} ion levels participating in the stimulated-emission processes in a $Y_3Al_5O_{12}$ crystal at 300 K [169, 271]. The thick arrows denote induced transitions and the wavy arrows denote nonradiative transitions. The emission line with $\lambda_g = 1.3550$ μm corresponds to the temperature 77 K. The other notations are the same as in Fig. 3.30

multiplets above the $^4F_{7/2}$ and $^4S_{3/2}$ states are practically not involved in the energy balance. In the case of a $Y_3Al_5O_{12}$ crystal excited by a xenon flash lamp, estimates show that they account for only 10% of the total absorbed excitation energy. The $^4I_{15/2}$ and $^4I_{13/2}$ manifolds, between the metastable $^4F_{3/2}$ state and the terminal $^4I_{11/2}$ laser state, are of no significance for the analysis of the behavior of stimulated emission in the principal $^4F_{3/2} \rightarrow {}^4I_{11/2}$ emission channel, for the following reasons. The fraction of luminescence transitions that terminate at these manifolds does not exceed 15% of the total number of radiative transitions from the $^4F_{3/2}$ state (see Table 6.5). The radiative probabilities of the $(^4F_{5/2}, {}^2H_{9/2}) \rightarrow {}^4I_{15/2,13/2}$ transitions are unknown, but these transitions affect the emission in the same manner as do the $(^4F_{5/2}, {}^2H_{9/2}) \rightarrow {}^4I_{11/2}$ transitions. Thus, it is not essential to take them into account separately.

Since the decay of the $^4F_{3/2}$ state is reported to be due to luminescence (see the character of the temperature dependence of τ_{lum} given in Fig. 6.3), we can neglect nonradiative $^4F_{3/2} \rightarrow {}^4I_{15/2}$ transitions. The nonradiative lifetime of the $^4F_{3/2}$ state has been evaluated to be about 30 ms, which is almost 120 times greater than $\tau_{lum} \approx 255$ μs. The rate of nonradiative (direct) transitions between the Stark levels of the same multiplet is much (4–7 orders of magnitude) greater than the rate of intermanifold (multiphonon nonradiative and luminescence) transitions. Therefore, for time intervals of the same order of magnitude as the transient processes that accompany emission, the populations of the Stark levels that belong to one manifold will obey the relation appropriate to thermal equilibrium. At the same time, the total population of each manifold may have the nonequilibrium value defined by the solution of the kinetic equations

$$\dot{N}_i = \sum_j (N_j P_{ji} - N_i P_{ij}),$$

$$\tag{6.1}$$

$$\sum_i N_i = N_0,$$

where $i, j = 1, 2, 3, 4$ are the manifold numbers according to Fig. 6.8, and

$$P_{ij} = A_i \beta_{ij} + V_{ij} + W_{ij}.$$

Here $A_i\beta_{ij}$ and V_{ij} are rates of change of N_i due to spontaneous and stimulated optical transitions, respectively; W_{ij} is the same quantity for nonradiative transitions, and β_{ij} is the branching ratio for radiative transitions from the ith manifold.

The relation between the A, V, and W values and the probabilities of inter-Stark transitions may be expressed as

$$\sum_k \sum_l n_{ik} p_{kl} = P_{ij} N_i. \tag{6.2}$$

Here k and l are the Stark-component indices for ith and jth manifolds respectively; n_{ik} are the populations of these components, and

$$p_{kl} = A_{kl} + \int B_{kl}(v) U_{kl}(v) dv + w_{kl},$$

where A_{kl} is the spontaneous transition probability, $\int B_{kl}(v) U_{kl}(v) dv$ is the induced transition probability, w_{kl} is the nonradiative relaxation probability, and $U_{kl}(v)$ is the volume density of emitted energy per unit frequency range.

Under thermal equilibrium conditions,

$$n_{ik} = b_{ik} n_{i1}$$

and

$$N_i = n_{i1} \sum_k b_{ik},$$

(6.3)

where $b_{ik} = \exp(-\Delta E_{k1}/kT)$ is the Boltzmann factor for the kth Stark level relative to the first Stark level of the ith state.

Substituting (6.2) into (6.3) and introducing the following notation: x for the pair of indices kl; b_x for the Boltzmann factor of the Stark component on which transition x starts; and $b_i = \sum_k b_{ik}$ with summation over the levels of the ith manifold, we have

$$A_i \beta_{ij} = \frac{\sum_x b_x A_x}{b_i},$$

$$V_{ij} = \frac{\sum_x b_x \int B_x(v) U_x(v) dv}{b_i},$$

$$W_{ij} = \frac{\sum_x b_x w_x}{b_i}.$$

Here $A_i^{-1} = \tau_{rad}$ is the radiative lifetime for the ith manifold. This lifetime can vary with temperature because of redistribution of population over the Stark components (see Sect. 6.1).

Let the excitation density U_{exc} be considered constant in the frequency interval that includes all active excitation (absorption) bands. Then for the $1 \rightarrow 4$ transition

$$V_{14} = \frac{U_{exc} \sum_x b_x B_x^{int}}{b_i} = B_1 U_{exc},$$

where B_x^{int} is the Einstein coefficient for induced transition, integrated over the line. The excitation rate is in practice always much lower than the rates of spontaneous and nonradiative deactivation of the upper levels. Therefore, the probability V_{41} is negligible.

As has been pointed out above, stimulated emission in $Y_3Al_5O_{12}$-Nd^{3+} crystals is usually obtained by excitation into two groups of lines connected with the $^4F_{7/2}, ^4S_{3/2}$ and $^4F_{5/2}, ^2H_{9/2}$ states. The energy gap between these two groups of multiplets is considerably less ($\Delta E \approx 510$ cm^{-1}) than that between other states, as shown in Fig. 6.8. Thus, these two groups of manifolds may be considered as being in thermal equilibrium with each other. The probability A_4 of nonradiative deactivation of the state 4 is connected with the probabilities of radiative transitions A_4' from the levels of $^4F_{5/2}, ^2H_{9/2}$ states and A_4'' from the levels of $^4F_{7/2}, ^4S_{3/2}$ states, by the relation

$$A_4 = \frac{A_4' + A_4'' \exp(-\Delta E/kT)}{1 + \exp(-\Delta E/kT)}.$$

Here ΔE is the energy gap between the lowest Stark levels of the states 4' and 4". For the $Y_3Al_5O_{12}$-Nd^{3+} crystal, $\Delta E \approx 900$ cm^{-1}, so up to $T = 1000$ K, we can assume $A_4 \approx A_4'$.

Laser action occurs on transitions between two separate Stark levels 2 and 3. Within the stimulated-emission linewidth Δv_g, the coefficient $B_x(v)$ may be considered to be a constant. Consequently,

$$V_{32} = \frac{b_x B_x(v_g) U_g}{b_3}$$

and

$$V_{23} = \frac{b_x' B_x(v_g) U_g}{b_2},$$

where U_g is the volume density of emitted energy, integrated over the stimulated-emission line and b_x' is the Boltzmann factor of the Stark component that is terminal for induced transition x, this factor being defined just like b_x. In writing the system (6.1) in the final form, we should also take into account the fact that nonradiative relaxation in the channels 4↔1, 4↔2, and 3↔2 is negligibly weak due to the large energy gaps, but, inversely, in the channels 2↔1 and 4↔3, the radiative relaxation is weak. Then the system of equations describing the processes that take place during excitation and stimulated emission is

$$\dot{N}_1 = -N_1(B_1 U_{\text{exc}} + W_{12}) + N_2 W_{21} + N_3 A_3 \beta_{31} + N_4 A_4 \beta_{41},$$

$$\dot{N}_2 = N_1 W_{12} - N_2 \left[W_{21} + \frac{b'_x}{b_2} B_x(v_g) U_g \right]$$

$$+ N_3 \left[A_3 \beta_{32} + W_{32} + \frac{b_x}{b_3} B_x(v_g) U_g \right] + N_4 A_4 \beta_{42},$$

$$\dot{N}_3 = N_2 \frac{b'_x}{b_2} B_x(v_g) U_g \qquad\qquad (6.4)$$

$$- N_3 \left[W_{34} + \frac{1}{\eta_1} A_3 + \frac{b_x}{b_3} B_x(v_g) U_g \right] + N_4 W_{43},$$

$$\dot{N}_4 = N_1 B_1 U_{\text{exc}} + N_3 W_{34} - \frac{1}{\eta} N_4 W_{43},$$

$$N_0 = \sum_{i=1}^{4} N_i,$$

where the designation x is used only for the stimulated-emission transitions; $\eta = W_{43}/(W_{43} + A_4)$ is the excitation efficiency and $\eta_1 = A_3/(A_3 + W_{32})$ is the luminescence quantum yield.

Let us consider the stationary regime of stimulated emission, for which the existence condition is [60]

$$\frac{B_x(v_g)hv_g}{c} (n_{3k} - n_{2l}) = \rho',$$

or, taking into account relations (6.3),

$$N_3 - \frac{b'_x}{b_x} \frac{b_3}{b_2} N_2 = \frac{N_0}{\delta_x}. \qquad\qquad (6.5)$$

Here ρ' is the total loss coefficient, and

$$\delta_x = \frac{N_0 B_x(v_g)hv_g b_x}{c\rho' b_3} = \frac{N_0}{\rho'} \frac{\lambda_g^2}{4\pi^2 c} \frac{b_x}{b_3} \frac{A_x}{\Delta v_{\text{lum}}^x},$$

where c is the velocity of light and Δv_{lum}^x is the luminescence linewidth (Lorentz line shape) for the transition x (both ρ' and Δv_{lum}^x are measured in cm^{-1}). If there are several transitions with coincident frequencies at which stimulated emission occurs, then δ_x in (6.5) is to be replaced by a δ_{eff} equal to the sum of δ_x for appropriate transitions. Solution of (6.4) together with (6.5) gives

$$U_g = \mathcal{T}\frac{\eta(B_1 U_{exc} - B_1 U_{exc}^{thr})}{1 + F\eta\dfrac{B_1 U_{exc}}{W_{21}}},\tag{6.6}$$

where

$$\eta B_1 U_{exc}^{thc} = \frac{A_3}{\delta_x}\left(1 + \frac{b_3}{b_1}\frac{b_x'}{b_x}\delta_x y\right)\cdot\left(1 + \frac{b_y}{b_3}\frac{A_4}{A_3}z\right) = \frac{E_{thr}}{m},$$

$$F = \left(1 + \frac{b_x'}{b_x}\frac{b_3}{b_2}\right) \approx \text{const},\tag{6.7}$$

$$\mathcal{T} = \frac{\delta_x b_3}{B_x(v_g)}.$$

Here m is the coefficient of proportionality between the excitation rate $B_1 U_{exc}$ and the excitation energy E_{exc}; y is the ratio of the Boltzmann factors for the lowest Stark levels of two $^4I_{11/2}$ and $^4I_{9/2}$ states; z is the same ratio for the $^4F_{5/2}$ and $^4F_{3/2}$ states. In deriving (6.7), the values $A_3 \approx 3.93 \times 10^3$ s^{-1}, $\beta_{31} = 0.30$, $\beta_{32} = 0.56$ [271, 447], and $A_4\beta_{41} = 7 \times 10^3$ s^{-1} were used [448]. The results of a number of studies show that the luminescence quantum yield for the $Y_3Al_5O_{12}$-Nd^{3+} crystal is close to unity (see Sect. 6.1). Hence, $W_{43} \gg A_4$ and A_3.

All available experimental data on nonradiative relaxation in crystals doped with Ln^{3+} ions indicate the existence of an approximately exponential dependence of relaxation probability for a given transition on its energy gap ("energy-gap law") [49–53, 703]. This does not contradict the modern theory of the electron–phonon interaction for Ln^{3+} ions [38, 378].

For the $Y_3Al_5O_{12}$-Nd^{3+} crystal, $\Delta E_{21} \approx 1.5\,\Delta E_{43}$ and W_{43} is several times greater than W_{21}. Thus it was accepted that $W_{43} \gg W_{21}$. The inequalities $W_{21} \gg A_3\beta_{32}$, $A_3\beta_{31}$ which are increasingly true with increasing temperature, were also considered. Because of the smallness of z, it was accepted also that $W_{21} \gg z A_4\beta_{41}$. Finally, account was taken of the facts that $\delta_x \gg b_3$; $y, z \ll 1$, and that the values $b_{1,2,3,4}$ are of the order of unity.

The experimental temperature dependences of E_{thr} for the stimulated-emission lines of the $Y_3Al_5O_{12}$-Nd^{3+} crystal are shown in Fig. 6.9 by dashed curves. This figure shows that stimulated emission on the $^4F_{3/2} \rightarrow {}^4I_{11/2}$ transition occurs up to $T \approx 160$ K only on the B line (see the energy-level scheme in Fig. 6.6). Above a temperature of about 160 K, the Boltzmann population of the upper Stark level of the metastable $^4F_{3/2}$ state rises, and stimulated emission becomes possible at the A line also. Since $A_A > A_B$ (see Sect. 3.11), above room temperature, the stimulated-emission condition is fulfilled only for the A line. The switching of stimulated-emission from the A line to the B line is shown in the stimulated-emission spectra (Fig. 3.29).

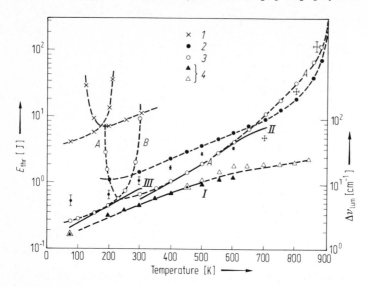

Fig. 6.9. Temperature dependences of Nd^{3+} ion stimulated-emission line excitation energy thresholds E_{thr} in crystals and luminescence linewidth $\Delta\nu_{lum}$ for the A line of a $Y_3Al_5O_{12}$-Nd^{3+} crystal [169, 271]. (1) $E_{thr}(T)$, $^4F_{3/2} \to {}^4I_{11/2}$ transition ($Lu_3Al_5O_{12}$ crystal); (2) $E_{thr}(T)$, $^4F_{3/2} \to {}^4I_{13/2}$ transition ($Y_3Al_5O_{12}$ crystal); (3) $E_{thr}(T)$, $^4F_{3/2} \to {}^4I_{11/2}$ transition ($Y_3Al_5O_{12}$ crystal); (4) $\Delta\nu_{lum}(T)$ of A line ($Y_3Al_5O_{12}$ crystal). Dashed lines are experimental and continuous lines are theoretical dependences of $E_{thr}(T)$. (I) due to $\Delta\nu_{lum}(T)$ changing only; (II) calculated by (6.7), taking into account full resonance between the A and A′ transitions. (III) the same, but without resonance. The points on the curves indicated by open triangles are taken from [169], and the dark triangles are from [33]

In the $Y_3Al_5O_{12}$-Nd^{3+} system, there are two transitions between the levels of the $^4F_{3/2}$ and $^4I_{11/2}$ states that have nearly equal frequencies. These transitions correspond to the A and A′ lines in the low-temperature luminescence spectrum shown in Fig. 6.10, the A′ line arising from the 11427 cm^{-1} $^4F_{3/2} \to {}^4I_{11/2}$ 2029 cm^{-1} transition (see Fig. 6.2b). When the temperature rises above 77 K, the A and A′ lines move closer to each other, due to thermal shifts of the energy levels of the Nd^{3+} ions. For $T \approx 400$ K, the separation between these two lines becomes less than their linewidths (compare the room- and high-temperature spectra shown in Fig. 6.10) and the A′ transition begins to participate in stimulated emission at the frequency of the A line. This is apparent from comparison with experimental data of the theoretical dependences $E_{thr}(T)$ calculated from (6.7). In the temperature range from 230 to about 400 K, the overlapping of the A and A′ lines is not yet complete. Thus the experimental results lie between the two theoretical curves II and III (see Fig. 6.9). For higher temperatures, the dependence $E_{thr}(T)$ is well described by (6.7), if complete resonance (curve II) of the A and A′ transitions is assumed, and if the ratio of probabilities A_4/A_3 is 5. On the basis of that value, one can roughly estimate the value of A_4 to be about 2×10^4 s^{-1}.

Analysis of the experimental data given in Fig. 6.9 by use of (6.7) shows

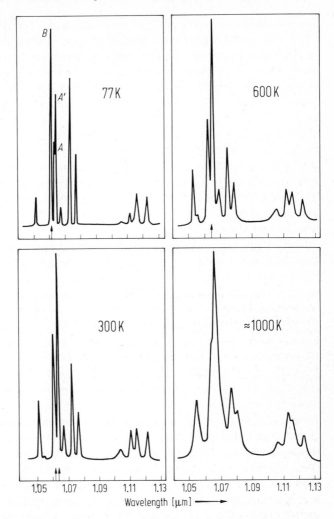

Fig. 6.10. Luminescence spectra of the $^4F_{3/2} \rightarrow {}^4I_{11/2}$ transition of Nd^{3+} ions in a $Y_3Al_5O_{12}$ crystal obtained at different temperatures [169, 271]. The arrows on the wavelength axis indicate the laser lines

that up to $T \approx 400$ K the main influence on the temperature dependence of E_{thr} is the thermal change of luminescence linewidth for an induced transition. The slopes of theoretical and experimental curves are in good agreement up to this temperature. At higher temperatures, the threshold energy is essentially influenced by the thermal increase of population of the $^4I_{11/2}$ terminal laser state[6]

---

[6] As our investigations show, the thermal population factor of the terminal laser state would be especially noticeable in the case of high-concentrated and self-activated neodymium laser media operating in the principal $^4F_{3/2} \rightarrow {}^4I_{11/2}$ transition [795].

and by an increase of the part of the excitation that remains on the levels of the $^4F_{5/2}$, $^2H_{9/2}$ and higher lying manifolds, and is irretrievably lost by 4→1, 2 luminescence transitions. These transitions become appreciable at high temperatures. Detailed investigations show that the increase of terminal laser-state population is already 10% of E_{thr} at about 400 K. The essential role of this influence on the excitation threshold becomes clear when the $E_{thr}(T)$ dependences for the two $^4F_{3/2}→^4I_{11/2}$ and $^4F_{3/2}→^4I_{13/2}$ emission transitions (see Fig. 6.9) are compared in the high-temperature region[7].

The $^4I_{13/2}$ state lies on the energy scale at a value roughly twice that of $^4I_{11/2}$; thus, the corresponding $E_{thr}(T)$ dependence is considerably flatter, that is, it reveals better correlation with the $\Delta\nu_{lum}(T)$ dependence for a given transition. This indicates that lasers based on $Y_3Al_5O_{12}$-Nd^{3+} crystals that emit on the A line of the $^4F_{3/2}→^4I_{11/2}$ transition begin to lose their "four-level" properties above approximately room temperature.

Investigations carried out by the author of the present monograph show that this conclusion is valid also for a number of other media activated by neodymium [271]. Analysis shows that for temperatures higher than about 500 K, luminescence transitions from the $^4F_{5/2}$, $^2H_{9/2}$ and $^4F_{7/2}$, $^4S_{3/2}$ manifolds begin to exert some influence on the value of E_{thr}, which is taken into account by the last factor in (6.7). The excitation leakage in the $Y_3Al_5O_{12}$-Nd^{3+} crystal caused by the thermal population of the Stark components of these states is well illustrated by the high-temperature luminescence spectrum shown in Figure 6.11. It can be seen from this spectrum that for a temperature of about 600 K, with luminescence in the $^4F_{3/2}→^4I_{9/2}$ transition (for comparison, see the low-temperature luminescence spectrum in Fig. 6.2a), some additional lines connected with transitions from the $^4F_{5/2}$, $^2H_{9/2}$ and $^4F_{7/2}$ states become noticeable. These high-temperature luminescence lines have been reliably identified only for the two transitions $^4F_{5/2}→^4I_{9/2}$ and $^2H_{9/2}→^4I_{9/2}$. The figures in brackets denote Stark components of a given state, arranged in increasing order of energy. For example, number (5) indicates that this line is connected with the fifth (the highest) level of the ground $^4I_{9/2}$ state. The dashed brackets in this figure indicate the splitting of the metastable $^4F_{3/2}$ state, which manifests itself in the luminescence spectra of the $^4F_{3/2}→^4I_{9/2}$ transition. The high-temperature leakage of excitation equally affects the $E_{thc}(T)$ dependences for stimulated-emission on both the $^4F_{3/2}→^4I_{11/2}$ and $^4F_{3/2}→^4I_{13/2}$ transitions. For temperatures higher than 700 K, the variation of threshold energy is influenced by factors that are not provided in the commonly accepted model of interaction of activator ions with radiation [169].

The low-temperature dependences of $\Delta\nu_{lum}$ on Nd^{3+} ion concentration for some luminescence lines connected with stimulated-emission on the $^4F_{3/2}→^4I_{11/2}$ and $^4F_{3/2}→^4I_{9/2}$ transitions are shown in Fig. 6.12.

---

[7]The low-temperature part of the $E_{thr}(T)$ dependence for the additional $^4F_{3/2}→^4I_{13/2}$ transition reveals a switching effect, similar to induced-emission switching from the A line to the B line, in the case of the principal $^4F_{3/2}→^4I_{11/2}$ transition.

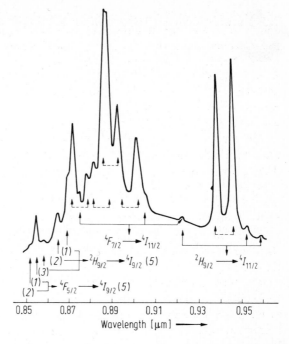

Fig. 6.11. Luminescence spectra of Nd^{3+} ions in a $Y_3Al_5O_{12}$ crystal obtained in the region of 0.9 μm at 600 K [271]. The dashed brackets indicate the splitting of the $^4F_{3/2}$ state. The figures in brackets denote Stark levels inside manifolds in order of increasing energy

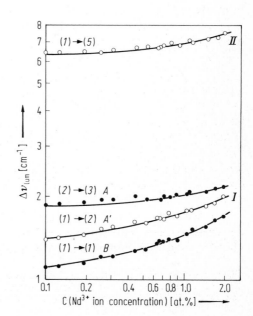

Fig. 6.12. Concentration dependences of luminescence linewidth $\Delta\nu_{lum}$ of Nd^{3+} ions in a $Y_3Al_5O_{12}$ crystal at 77 K. I) Lines connected with the $^4F_{3/2} \rightarrow {}^4I_{11/2}$ transition. II) Lines connected with the $^4F_{3/2} \rightarrow {}^4I_{9/2}$ transition. Other notation as in Fig. 6.11

6.3 Luminescence Properties of $YAlO_3$-Nd^{3+} Crystals

The spectral properties of $YAlO_3$-Nd^{3+} crystals were studied first in [10, 396, 397]. At present, this promising laser medium is being studied by many other research groups [98, 272, 399, 453]. The crystal-field splitting of the $^4F_{3/2}$ and $^4I_{9/2-15/2}$ states, as well as the appropriate luminescence and absorption spectra of this crystal, are shown in Fig. 6.13. Figure 6.14 gives the $\tau_{lum}(C)$ and $\tau_{lum}(T)$ dependences. From the temperature dependence, it is clear that the luminescence quantum yield ($^4F_{3/2} \rightarrow {}^4I_{J'}$ transitions) is close to unity in a wide temperature range.

Unlike the cubic $Y_3Al_5O_{12}$-Nd^{3+} crystal, the anisotropy of orthorhombic $YAlO_3$-Nd^{3+} must be taken into account in determining the spectral density of luminescence from the metastable $^4F_{3/2}$ state to different Stark components of manifolds that belong to the ^4I term. The luminescence of anisotropic crystals has been discussed in Sect. 3.11 where the spectral and stimulated-emission properties of the trigonal LaF_3-Nd^{3+} crystal have been discussed. Now we shall return to this question again.

In connection with the fact that the line intensities in spectra observed in different directions, for example, along the three crystallographic axes of $YAlO_3$, are different then, as was shown in [565, 751], it is reasonable to introduce the concept of orientational cross section $\sigma^{or}_{e,ij}$ for describing the radiative properties of separate Stark $i \rightarrow j$ transitions in this crystal. Then for homogeneously broadened luminescence lines, (2.11) can be written as

$$\sigma^{or}_{e,ij} = \frac{\lambda^2_{ij}}{4\pi^2 n^2 \Delta v_{lum}} \frac{\beta^{or}_{ij} \sum_i b_i}{\tau_{rad} b_i}, \tag{6.8}$$

where $b_i = \exp[-\Delta E_{i1}(^4F_{3/2})]/kT$ is the Boltzmann factor for the ith Stark level of the metastable $^4F_{3/2}$ state calculated with respect to the lowest level of this state ($i=1$), and β^{or}_{ij} is the orientational luminescence inter-Stark branching ratio, describing the relative spectral density of luminescence along the given direction. This characteristic may be determined from the orientational (but nonpolarized) luminescence spectra, the line intensities of which are recalculated taking into account the spectral features of the monochromators and the spectral sensitivity of the photodetectors. The refractive index in (6.8) also depends on orientation, but for the case of the $YAlO_3$ crystal, as follows from Table 6.1, this quantity may be considered to be constant for the wavelength of a given $i \rightarrow j$

Fig. 6.13. Unpolarized luminescence (I) and absorption (II) spectra, and crystal-field splitting ▶ schemes of the $^2P_{1/2}$, $^4F_{3/2}$ and $^4I_{9/2-15/2}$ manifolds (III) of Nd^{3+} ions in the $YAlO_3$ crystal. (a) $^4I_{9/2}$ manifold; (b) $^4I_{11/2}$ manifold; (c) $^4I_{13/2}$ manifold; (d) $^4I_{15/2}$ manifold. The arrows in the spectra indicate the $^4I_{9/2}$ ground-state splitting (a), square brackets denote the splitting for the $^4F_{3/2}$ state, and dashed brackets denote splitting of the first excited Stark component of the $^4I_{9/2}$ state (b). Other notation as in Fig. 3.22

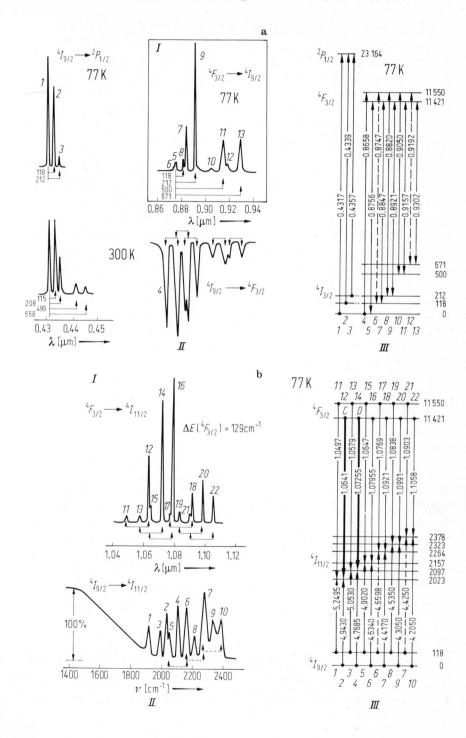

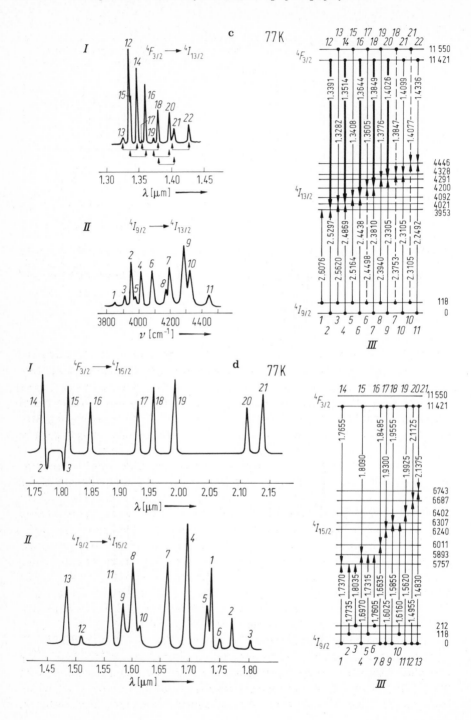

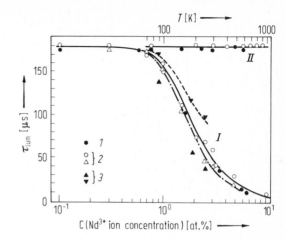

Fig. 6.14. Dependence of $\tau_{\text{lum}}(C)$ (curves 1 and 2) and $\tau_{\text{lum}}(T)$ (curve 3) for Nd^{3+} ions in the YAlO$_3$ crystal. The dashed curve corresponds to 4.2 K, the dash-dot curve corresponds to 77 K, and the continuous curve corresponds to 300 K. Experimental points: (1) [398]; (2) [271]; (3) [463]

transition. It is important to note that the luminescence intensity along the crystallographic directions a, b, and c will also depend on the excitation conditions (isotropic or anisotropic). In real laser devices pumped by flashtubes of elliptical (or other) cross section, some anisotropy is always present in the excitation. Therefore, for anisotropic crystals, such as the YAlO$_3$-Nd^{3+} crystals, optimal orientation of the laser direction relative to the pumping source must be chosen.

Accurate measurements of all the parameters that enter (6.8) and calculations of $\sigma_{\text{e}.ij}^{\text{or};\text{p}}$ for Stark transitions in the two $^4F_{3/2} \rightarrow {}^4I_{11/2,13/2}$ laser channels of the YAlO$_3$-Nd^{3+} crystal, taking into account the Boltzmann population of the metastable $^4F_{3/2}$ state levels at 300 K, have been measured and reported [751]. They are listed in Table 6.6. Here, data are presented for given excitation directions. The values of $\sigma_{\text{e}.ij}^{\text{or};\text{p}}$ for isotropic (cylindrical) pumping can be determined as the arithmetic mean of two corresponding values for anisotropic pumping. For completeness, Table 6.6 includes also data on $\Delta\nu_{\text{lum}}$ and β_{ij}^{or}. (For brevity, the values of β_{ij}^{or} are only given for one orientation of the crystal.) The luminescence orientational spectra obtained at 300 K for the two $^4F_{3/2} \rightarrow {}^4I_{11/2,13/2}$ emission bands are shown in Figs. 6.15, 16.

The intermanifold luminescence branching ratios $\beta_{JJ'}$ for YAlO$_3$-Nd^{3+} crystals were determined using the method considered in Sect. 4.4, that is, by the value of the spectroscopic-quality factor X. This parameter was found to be 0.3 ± 0.05 (taking into account the crystal absorption anisotropy) [751], which leads to the values of intermanifold luminescence branching ratios listed in

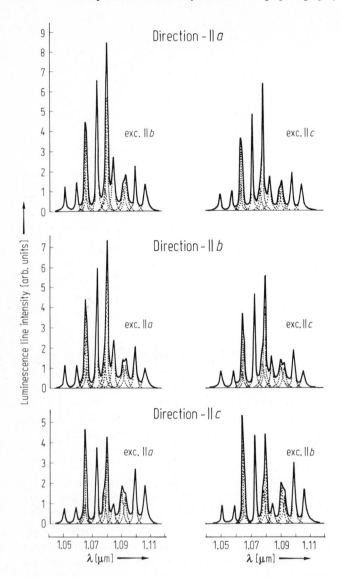

Fig. 6.15. Orientational luminescence spectra ($^4F_{3/2} \to {}^4I_{11/2}$ transition) of Nd^{3+} ions for three crystallographic directions of a $YAlO_3$ crystal at 300 K and excitation directions as shown [751]. The dashed curves show the result of resolving the complex contours of the spectrum. The line intensities are given in relative units

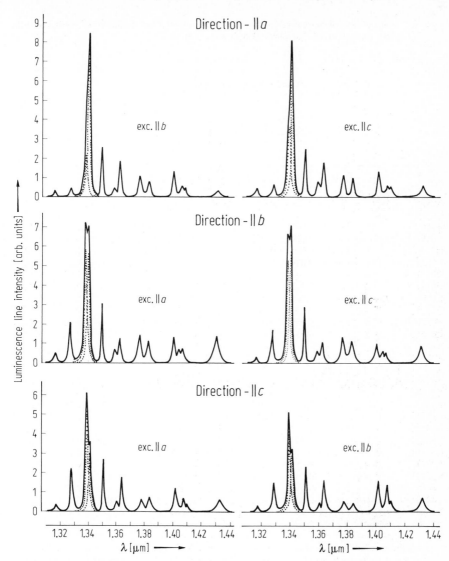

Fig. 6.16. Orientational luminescence spectra ($^4F_{3/2} \rightarrow {}^4I_{13/2}$ transition) of Nd^{3+} ions for three crystallographic directions of a YAlO$_3$ crystal at 300 K and excitation directions as shown [751]. Notation as in Fig. 6.15

Table 6.6. Spectral characteristics and orientational intensities of inter-Stark luminescence transitions for the $^4F_{3/2} \to {}^4I_{11/2,13/2}$ channels of Nd^{3+} ions in $YAlO_3$ crystals at 300 K [751]

Wavelength [μm]	Luminescence linewidth [cm⁻¹]	β_{ij}^{or}		$\sigma_{e,ij}^{or,p}$ [10^{-19} cm²] Crystallographic direction					
				‖a		‖b		‖c	
		Excitation direction							
		‖b	‖c	‖b	‖c	‖a	‖c	‖a	‖b
$^4F_{3/2} \to {}^4I_{11/2}$									
1.0505	9.5 ± 1.0	0.021	0.013	0.034	0.23	0.3	0.23	0.23	0.25
1.0585 B	10.0 ± 1.0	0.026	0.017	0.4	0.28	0.26	0.22	0.25	0.24
1.0644 C	6.5 ± 1.0	0.045	0.037	1.0	0.83	1.07	0.9	1.18	1.38
1.0652 E	11.5 ± 1.5	0.064	0.05	0.82	0.68	0.88	0.55	0.66	0.7
1.0726 D	8.5 ± 1.0	0.095	0.07	1.78	1.3	1.6	1.28	1.05	1.12
1.0775	13.0 ± 1.5	0.024	0.02	0.3	0.25	0.32	0.32	0.4	0.42
1.0796 A	9.8 ± 1.0	0.142	0.1	2.18	1.57	1.83	1.4	1.08	1.1
1.0842 G	14.2 ± 1.5	0.071	0.044	0.75	0.47	0.64	0.43	0.27	0.27
1.0907	17.5 ± 2.0	0.031	0.024	0.3	0.23	0.3	0.33	0.42	0.46
1.0921	12.0 ± 1.5	0.03	0.025	0.4	0.32	0.3	0.26	0.36	0.36
1.0990 F	14.0 ± 1.5	0.053	0.03	0.6	0.52	0.56	0.52	0.78	0.85
1.1058	1.60 ± 1.5	0.038	0.035	0.37	0.33	0.28	0.23	0.54	0.47
$^4F_{3/2} \to {}^4I_{13/2}$									
1.3175	9.5 ± 1.0	0.002	0.003	0.047	0.068	0.068	0.048	0.048	0.048
1.3289	10.5 ± 1.0	0.003	0.004	0.06	0.09	0.29	0.24	0.29	0.2
1.3393	7.5 ± 1.0	0.01	0.017	0.3	0.05	0.8	0.72	0.76	0.62
1.3413	10.5 ± 1.0	0.054	0.05	1.0	0.9	0.78	0.77	0.43	0.39
1.3512	9.0 ± 1.0	0.012	0.011	0.35	0.33	0.42	0.39	0.38	0.32
1.3615	12.0 ± 1.5	0.003	0.006	0.06	0.11	0.1	0.08	0.08	0.6
1.3641	11.0 ± 1.5	0.011	0.01	0.26	0.24	0.17	0.14	0.26	0.23
1.3787	14.0 ± 1.5	0.01	0.01	0.15	0.15	0.19	0.18	0.1	0.08
1.3849	13.0 ± 1.5	0.005	0.006	0.11	0.13	0.16	0.15	0.11	0.07
1.4028	15.0 ± 2.0	0.009	0.009	0.18	0.18	0.17	0.13	0.17	0.21
1.4079	18.0 ± 2.0	0.005	0.005	0.08	0.08	0.1	0.08	0.1	0.18
1.4088[a]	21.0 ± 3.0	—	—	—	—	—	—	—	—
1.4338	20.0 ± 2.0	0.004	0.007	0.05	0.08	0.19	0.12	0.09	0.1

[a] The spectroscopic characteristics of this line, due to low intensity, were determined with an error sometimes exceeding 50%, and so the values are not listed in the table. In the other values, the error is below 20%.

Table 6.7. The orientational values of intermanifold ($\beta_{JJ'}^{or}$) and inter-Stark (β_{ij}^{or}) branching ratios were determined using these data. The relation between these two branching ratios was taken from [188]:

$$\beta_{JJ'} = \frac{1}{3} \sum_{a,b,c} \beta_{JJ'}^{or} = \frac{1}{3} \sum_{i,j} \sum_{a,b,c} \beta_{ij}^{or}. \qquad (6.9)$$

Table 6.7. Intermanifold luminescence branching ratios for the $^4I_{3/2} \rightarrow {}^4I_{J'}$ transitions of Nd^{3+} ions in YAlO$_3$ crystals at 300 K [751]

Transition	Luminescence range [μm]	$\beta_{JJ'}$[a]	$\beta^{or}_{JJ'}$								
			Isotropic cylindrical excitation			Anisotropic excitation					
			Crystallographic direction								
			$\|a$	$\|b$	$\|c$	$\|a$		$\|b$		$\|c$	
						Excitation direction					
						$\|b$	$\|c$	$\|a$	$\|c$	$\|a$	$\|b$
$^4F_{3/2} \rightarrow {}^4I_{9/2}$[b]	0.8662–0.9300	$\approx$0.28	$\approx$0.37	$\approx$0.23	$\approx$0.23	—	—	—	—	—	—
$^4F_{3/2} \rightarrow {}^4I_{11/2}$	1.0505–1.1058	$\approx$0.58	$\approx$0.55	$\approx$0.62	$\approx$0.57	0.64	0.47	0.69	0.55	0.56	0.57
$^4F_{3/2} \rightarrow {}^4I_{13/2}$	1.3175–1.4438	$\approx$0.135	$\approx$0.14	$\approx$0.15	$\approx$0.12	0.13	0.15	0.16	0.14	0.12	0.11
$^4F_{3/2} \rightarrow {}^4I_{15/2}$	1.730 –2.1368	$\approx$0.007	—	—	—	—	—	—	—	—	—

[a] Data on the branching ratios $\beta_{JJ'}$ are found also in [272].
[b] The values of $\beta^{or}_{JJ'}$ for the $^4F_{3/2} \rightarrow {}^4I_{9/2}$ transition are determined by the absorption spectra due to the resonant $^4I_{9/2} \rightarrow {}^4F_{3/2}$ transition.

Table 6.8. Effective cross sections for some intense (laser) luminescence lines of YAlO$_3$-Nd^{3+} crystals [751][a]

Transition	Wavelength [μm]	Temperature [K]	σ_e^{eff}[10^{-19}cm^2]								
			Isotropic cylindrical excitation			Anisotropic excitation					
			Crystallographic direction								
			$\|a$	$\|b$	$\|c$	$\|a$		$\|b$		$\|c$	
						Excitation direction					
						$\|b$	$\|c$	$\|a$	$\|c$	$\|a$	$\|c$
$^4F_{3/2} \rightarrow {}^4I_{9/2}$[b]	0.9300[b]	300	0.38	0.29	0.35	—	—	—	—	—	—
$^4F_{3/2} \rightarrow {}^4I_{11/2}$	1.0641 C	77	4.7	5.85	7.26	4.4	5.05	5.3	6.4	6.95	7.57
	1.07255 D	77	9.15	9.8	6.6	9.5	8.8	10	9.65	6.7	6.64
	1.07955 A	77	9.4	8.95	5.2	10	8.85	9.3	8.6	5.4	5.0
	1.0644 C	300	1.1	1.09	1.38	1.2	1.0	1.18	1.0	1.28	1.48
	1.0726 D	300	1.6	1.54	1.16	1.78	1.3	1.6	1.28	1.05	1.12
	1.0796 A	300	2.05	1.76	1.2	2.4	1.7	1.97	1.55	1.2	1.2
$^4F_{3/2} \rightarrow {}^4I_{13/2}$	1.3390	77	1.64	3.17	3.45	1.47	1.8	3.2	3.14	3.8	3.1
	1.3407	77	0.68	0.46	0.3	0.6	0.76	0.43	0.5	0.26	0.3
	1.3393	300	0.69	0.94	0.78	0.68	0.7	0.98	0.9	0.85	0.7
	1.3413	300	1.13	0.97	0.47	1.16	1.1	0.96	0.97	0.5	0.44

[a] Cross sections for some lines of YAlO$_3$-Nd^{3+} crystals at 300 K have also been measured in [272].
[b] The values of σ_e^{eff} were determined by the absorption spectra.

Here the summation is carried out over all inter-Stark $i \to j$ transitions of a given luminescence band $J \to J'$ and also over the three directions of radiation that coincide with the crystallographic axis a, b, and c. This formula applies to the case of isotropic cylindrical excitation of the luminescence.

As has been repeatedly mentioned, in analyzing the behavior of the excitation threshold for stimulated emission, and in calculating the laser characteristics of a crystal, it is necessary to take into account the amplifying properties of the active medium at the wavelength λ_{ij}, which result from partial or complete overlapping of some radiative (luminescence) lines. This property is conveniently character- ized by some effective cross section σ_e^{eff}, which results from summation of the appropriate parts of σ_e^p for the corresponding transitions (this summation should take into account the polarization of the emission). Numerical values of σ_e^{eff} for the most intense lines on which stimulated emission occurs in $YAlO_3$-Nd^{3+} crystals are given in Table 6.8. Data on the concentration behavior of linewidth for some luminescence lines of Nd^{3+} ions in $YAlO_3$ crystals connected with $^4F_{3/2} \to {}^4I_{9/2}$ and $^4F_{3/2} \to {}^4I_{11/2}$ transitions are given in Fig. 6.17.

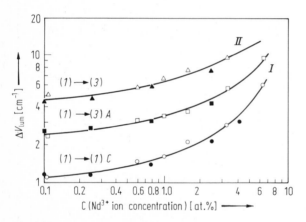

Fig. 6.17. Concentration dependences of Δv_{lum} linewidth for some luminescence lines of Nd^{3+} ions in a $YAlO_3$ crystal at 77 K. (I) The linewidth Δv_{lum} of the $^4F_{3/2} \to {}^4I_{11/2}$ transition lines. (II) The linewidth for the $^4F_{3/2} \to {}^4I_{9/2}$ transition lines. Open dots represent experimental data from [272], dark dots, from [271]. Other notation as in Fig. 6.11

6.4 Spectral-Emission Properties of $YAlO_3$-Nd^{3+} Crystals

Stimulated emission of Nd^{3+} ions in $YAlO_3$ crystals was first reported in [10] by the author of the present monograph with *Bagdasarov* et al. [396, 397], who had found the conditions for growing crystals suitable for laser application. The unique properties of this new active medium permitted realization of both pulse and CW modes of operation at room temperature, even in the first experiments. Some months later, these experiments were repeated [398]. An important property of lasers based on the YAlO$_3$-Nd^{3+} crystal is the high degree of polarization of its radiation (see Sect. 3.11). In comparison with the unpolarized radiation of lasers based on Y$_3$Al$_5$O$_{12}$-Nd^{3+} crystals, this property gives an incontestable advantage both for obtaining harmonics and for modulation of the radiation.

Spectra of the stimulated emission of YAlO$_3$-Nd^{3+} crystals obtained at different temperatures [176] were shown in Fig. 3.33. On the basis of these spectra and also of data obtained from other measurements, the $\lambda_g(T)$ dependences shown in Fig. 6.18 were constructed. More complete information on results of stimulated-emission studies of lasers based on YAlO$_3$-Nd^{3+} crystals can be found in Table 5.1. In this section, we shall confine ourselves to demonstrating

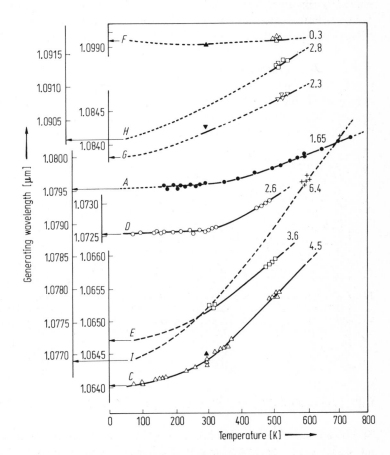

Fig. 6.18. Temperature dependences of generating wavelength λ_g ($^4F_{3/2} \rightarrow {}^4I_{11/2}$ transition) of YAlO$_3$-Nd^{3+} crystal laser. The figures near curves show $(\delta v_g/\delta T)\, 10^2\, [\text{cm}^{-1} \cdot \text{K}^{-1}]$ values in linear parts of $\lambda_g(T)$ dependences. The dark dots are the data from [399]

the crystal-field splitting scheme for $^4F_{3/2}$ and $^4I_{9/2-15/2}$ states. All induced transitions observed to date for this crystal are shown in Fig. 6.19.

Now we shall compare some of the properties of $YAlO_3$ and $Y_3Al_5O_{12}$ crystals activated by Nd^{3+} ions. These crystals are important for creating powerful lasers. Numerous experimental and theoretical studies of lasers based on $Y_3Al_5O_{12}$-Nd^{3+} crystals indicate satisfactory efficiency for generation in the fundamental TEM_{00} mode under reasonably high levels of excitation power can be obtained only with certain difficulties. The highest efficiency reported is about 15% [454]. The main difficulties in increasing it are the induced birefringence and thermal lensing, both of which reduce the useful cross section of the laser rod. It has been shown [455] that the indicatrix of induced refractive indices in the $Y_3Al_5O_{12}$-Nd^{3+} crystal is oriented in the (111) plane and that the value of induced birefringence is connected with the radius of the laser rod by the quadratic relation

$$n_r - n_t = n_{ax}^3 \frac{\alpha P_p}{K} - \frac{(p_{11} - p_{12} + 4p_{44})(v_P + 1)}{48(v_P - 1)} r^2,$$

where n_r and n_t are the induced refractive indices for radial and tangential polarizations; n_{ax} is the refractive index at the axis of the rod; α is the thermal expansion coefficient; K is the thermal conductivity; P_p is the power dissipated as heat per unit volume of crystal; p_{nm} are the elasto-optic constants; v_P is the Poisson ratio; and r is the rod radius.

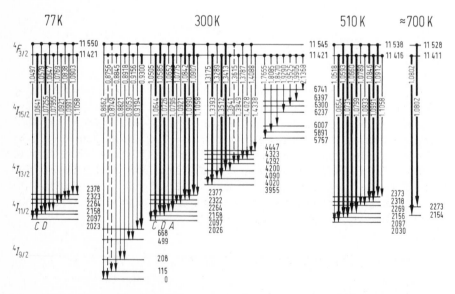

Fig. 6.19. Crystal-field-splitting schemes of the $^4F_{3/2}$ and $^4I_{9/2-15/2}$ manifolds of Nd^{3+} ions in $YAlO_3$ crystal, with the identification of all registered induced transitions (thick arrows). The position of the levels are given in cm^{-1} and the wavelength of transitions between them in μm

Values of $(n_r - n_t)/r^2$ are estimated to be usually of the order of 10^{-4} cm^{-2}. This implies that a beam of linearly polarized light with a Gaussian distribution of density over the cross section will be considerably depolarized after passing through an active element fabricated from a Y$_3$Al$_5$O$_{12}$-Nd^{3+} crystal, which is susceptible to thermal stresses. This is well confirmed by experiment. In spite of the fact that the influence of thermal lensing under low or moderate excitation powers can be sufficiently eliminated by curving the end surfaces of the active element [217, 455, 456], solving this problem in the case of high excitation becomes very difficult. The elliptically polarized (or unpolarized) wave that falls on the active element actually has two components with different polarizations. Thus, the thermal stress in the laser rod leads to formation of two internal thermal lenses with different focal lengths depending on the excitation power. According to [457] this dependence can be expressed as

$$f_1 = \frac{2K\pi r^2}{P_p}\left[\frac{\delta n}{\delta T} - \frac{n_{ax}^3 \alpha c_x''}{48(1 - v_p)} + \frac{2(n_{ax} - 1)\alpha l_0}{L}\right]^{-1}, \qquad (6.10)$$

where L is the crystal length; l_0 is the length of the end section of the rod over which expansion occurs $(l_0 \approx r_0)$; $l_0\alpha = \delta l/\delta T$ and

$$c_x'' = 3(7v_p - 1)(p_{11} + p_{12} + 2p_{44})$$

$$+ (5v_p - 3)(p_{11} + 5p_{12} - 2p_{44})$$

$$+ 8(v_p - 1)(p_{11} + 2p_{12} - 2p_{44}).$$

The first term in the square bracket of (6.10) represents the change of refractive index with temperature. For the Y$_3$Al$_5$O$_{12}$-Nd^{3+} crystal, $\delta n/\delta T = 9.86 \times 10^{-6}$ K^{-1} (for pure Y$_3$Al$_5$O$_{12}$ crystals, $\delta n/\delta T = 7.4 \times 10^{-6}$ K^{-1} [458]). The second term represents the change of refractive index caused by thermal stress that arises in the crystal, and the last term gives the distortion caused by the end-face curvature of the active element. Using (6.10), it is possible to calculate the focal length of the thermal lens that acts on the radiation with radial polarization. The expression for tangentially polarized radiation can be obtained from (6.10), by replacing c_x'' by

$$c_y'' = (7v_p - 1)(p_{11} + 5p_{12} + 2p_{44})$$

$$+ 3(5v_p - 3)(p_{11} + p_{12} + 2p_{44})$$

$$+ 8(v_p - 1)(p_{11} + 2p_{12} - 2p_{44}).$$

Using (6.10) and suitable numerical values of all the parameters that appear, the following simple formulas for focal lengths were derived for a typical $Y_3Al_5O_{12}$-Nd^{3+} CW laser, of length 76 mm and 64 mm in diameter [457]:

$$f_1 = 1.41/P_{exc}$$

and

$$f_2 = 2.0/P_{exr},$$

where $f_{1,2}$ are in meters and P_{exc} is the excitation power (electrical input power into the lamps, $P_{exc} \sim P_p$) in kilowatts.

The effect of thermally induced birefringence can be decreased by special optical compensators introduced into the resonator, or by using $Y_3Al_5O_{12}$-Nd^{3+} crystals with rectangular cross section [459, 460]. The last method is more effective, but unfortunately it is applicable only to lasers with low output power. In this connection, waveguide-type optical resonators are of particular interest. In these resonators, modes can be excited with wavefronts practically undistorted by thermal stress of the active medium[8].

When high pumping powers are used, cooling laser rods of rectangular cross section is a difficult task. In addition, introducing different compensators and polarizers into the optical cavity usually leads to rather significant losses, lowering the total laser efficiency. In fact, the linearly polarized wave from the polarizer, after passing through the crystal, is changed into an elliptically polarized wave. On repeated passages, after reflection, one component of this wave is suppressed by the polarizer, thereby introducing losses into the system.

In this connection, the method of solving the problem of thermally induced birefringence analyzed by *Massey* [454] is of interest. *Massey* proposed concentrating attention on crystals possessing significant natural birefringence. Using uniaxial and biaxial crystals whose spectroscopic, physical and chemical properties are close to those of cubic $Y_3Al_5O_{12}$-Nd^{3+} crystals, it should be possible to develop powerful high-efficiency lasers with linearly polarized radiation. If the active element of circular cross section, which is most favorable for cooling, is fabricated from an anisotropic crystal with such mutual orientation of its optical and geometric axes that thermally induced birefringence is only a small perturbation compared with the natural birefringence (the ratio of these two effects is usually of the order of 10^{-1}–10^{-2}), the injurious depolarization due to high excitation will be decreased to a negligibly low level. *Massey* studied the $YAlO_3$-Nd^{3+} crystal since, so far, it is the only biaxial crystal

---

[8] As has been shown in [527], lasers with waveguide-type resonators produce polarized radiation from isotropic activated media. Another method for obtaining linearly polarized emission from cubic crystals has been proposed in [528]. The authors of this developed an efficient $Y_3Al_5O_{12}$-Nd^{3+} CW laser with polarized emission ($P_g = 40$ W). In this laser, a special prism was used as an output reflector, with its operating face arranged at the Brewster angle to the laser axis.

with properties close to those of Y$_3$Al$_5$O$_{12}$-Nd^{3+}. As follows from [454], the YAlO$_3$-Nd^{3+} crystals grown along the crystal axes a or b have a natural birefringence that significantly exceeds the possible value of that thermally induced.

These conclusions were fully confirmed in subsequent experiments [98]. Comparison of the output powers for CW lasers based on Y$_3$Al$_5$O$_{12}$ and YAlO$_3$ crystals activated with Nd^{3+} ions yielded the results given in Fig. 6.20. These data were obtained with active elements about 75 mm in length and about 6 mm in diameter. The concentration of Nd^{3+} ions in the Y$_3$Al$_5$O$_{12}$ crystal was about 1 at.% and that in the YAlO$_3$ crystal was about 0.9 at.%. Excitation was provided by two krypton-arc lamps placed in a double gold-plated cylinder of elliptic cross section. The transmittance of the output mirror was 4% for the Y$_3$Al$_5$O$_{12}$-Nd^{3+} laser, and 3% for the YAlO$_3$-Nd^{3+} laser.

It should be noted that, to attain high efficiency in YAlO$_3$-Nd^{3+} lasers operating in the $^4F_{3/2} \rightarrow \,^4I_{11/2}$ and $^4F_{3/2} \rightarrow \,^4I_{13/2}$ channels, it is necessary to use active elements with geometric axis F parallel to the crystal axis a or b, because only these orientations allow the highest amplification. Thus, for the A line ($\lambda_g = 1.0796$ μm) of the $^4F_{3/2} \rightarrow \,^4I_{11/2}$ channel, the highest effective cross section occurs with the a orientation (see Table 6.8). For a variety of technological reasons, YAlO$_3$-Nd^{3+} crystals oriented along the b axis are generally used in laser devices. The dependences $P_g(P_{exc})$ given in Fig. 6.20 were obtained under multimode generation in just such a crystal [98]. An attractive feature of YAlO$_3$-Nd^{3+} ($b \| F$) crystal, compared with Y$_3$Al$_5$O$_{12}$-Nd^{3+}, is that it allows generation

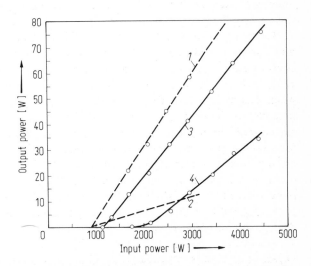

Fig. 6.20. Dependences of $P_g(P_{exc})$ of a CW Y$_3$Al$_5$O$_{12}$-Nd^{3+} crystal laser (curves 1 and 2) and a YAlO$_3$-Nd^{3+} crystal laser (curves 3 and 4) at 300 K [98]. (1) λ_g(A) = 1.06415 μm, unpolarized emission; (2) λ_g(A) = 1.06415 μm, polarized emission; (3) λ_g(A) = 1.0796 μm, polarized emission; (4) λ_g(C) = 1.0644 μm, polarized emission

with high efficiency (about 43% [220]) in a single TEM_{00} mode. Data on $YAlO_3$-Nd^{3+} ($b\|F$) lasers that generate in the TEM_{00} mode at 300 K with 9 W output are given in [220]. For a Q-switched laser, it is reasonable to use $YAlO_3$-Nd^{3+} ($c\|F$) crystals. This orientation is favorable for stimulated emission on C line (1.0644 μm) (see Table 6.8). Use of active elements with $c\|F$ produces higher output power because that laser element has the lowest amplification coefficient (for the $^4F_{3/2} \to {}^4I_{11/2}$ transition); hence, more energy can be stored in the metastable $^4F_{3/2}$ state.

7. Self-Activated Laser Crystals

In this chapter we shall dwell in more detail upon the physics of the concentration quenching of luminescence from laser crystals doped with Ln^{3+} ions, based on results obtained in the last five years. Special emphasis will be placed on the characteristic features of concentration quenching of the Nd^{3+} ion luminescence in high-concentration and self-activated crystals (the latter are sometimes referred to as stoichiometric crystals[1]). We hope that presentation of the current state of this problem will be useful both to specialists in the physics of condensed laser media and to chemical physicists engaged in the search for new high-concentration generating crystals to be used in miniature lasers and laser amplifiers [581, 682]. Development of the latter is of great interest in connection with current investigations in the field of integrated optics.

7.1 Concentration Quenching of Luminescence in Laser Crystals with High Nd^{3+} Ion Content

The problem of concentration quenching of the luminescence of Nd^{3+} ions is not new in laser-crystal physics. However, a new stimulus was imparted to this area by the spectral-emission studies with the self-activated NdP_5O_{14} crystal initiated in 1972 by *Danielmeyer* et al. [84, 505, 507, 534]. With this crystal, they observed anomalously weak concentration quenching of the luminescence from the $^4F_{3/2}$ metastable state, though the Nd^{3+} ion concentration is very high and exceeds the optimal value (about 1 at.%) for the $Y_3Al_5O_{12}$ laser crystal (see Figs. 6.3, 4) by a factor of about thirty. These studies aroused general interest so that in many laboratories throughout the world, much attention was given to the search for new types of neodymium-doped laser crystals with luminescence properties similar to NdP_5O_{14} [533, 569, 576, 577, 579, 610, 618, 656, 658, 772, 782–786] and to detailed exploration of the nature of the weak-concentration-quenching phenomenon [249, 534, 576, 656, 658, 683, 761, 771, 772, 782, 783]. Extensive research in this field has yielded by now about ten self-activated neodymium-doped laser crystals (see Table 7.1[2]) and this number continues to increase. Much

---

[1] The author does not think that this definition is quite correct, because many laser crystals with isomorphic activator ions are also stoichiometric [6–8, 438, 440]. In this respect, a number of disordered media (solid solutions) hardly qualify either.

[2] Table 7.1 does not contain information about high-concentration laser crystals doped with Nd^{3+} ions which have been synthesized as derivatives by "diluting" 100% matrices. Information on these crystals can be found in Tables 1.3, 1.4, and 5.1. Stimulated emission of $K_5NdLi_2F_{10}$ [787] and $Na_2Nd_2Pb_6(PO_4)_6Cl_2$ [788] crystals has been reported.

Table 7.1. Some luminescence characteristics of self-activated neodymium laser crystals [655]

Crystal	Space group	Nd^{3+} ion concentration [10^{21}cm^{-3}]	τ_{rad} [μs]	$\tau_{\text{lum}}^{100\%}$ [μs]	$\tau_{\text{rad}}/\tau_{\text{lum}}^{100\%}$	Quantum yield	Reference
LiNdP$_4$O$_{12}$	C_{2h}^6–$C2/c$	≈ 4.4	325	120	2.71	0.36	[581, 609, 621]
NaNdP$_4$O$_{12}$	C_{2h}^5–$P2_1/c$	4.24	300(?)	110	2.73	0.37	[578]
KNdP$_4$O$_{12}$	C_2^2–$P2_1$	≈ 4.1	≈ 275	105	2.62	0.38	[579, 581]
NdAl$_3$(BO$_3$)$_4$	D_3^7–$R32$	≈ 5.4	50	≈ 19	2.63	0.38	[579, 581, 589]
NdP$_5$O$_{14}$	C_{2h}^5–$P2_1/c$	3.9	300	≈ 110	2.73	0.37	[576, 581, 589, 684–686]
Na$_3$Nd(PO$_4$)$_2$	D_{2h}^{11}–$Pbcm$(?)	—	360	23	15.6	0.06	[577]
Na$_5$Nd(WO$_4$)$_2$	C_{4h}^6–$I4_1/a$	2.6	≈ 220	80	2.75	0.36	[577, 581, 589]
K$_3$Nd(PO$_4$)$_2$	C_{2h}^2–$P2_1/m$	5.0	≈ 460	≈ 21	≈ 22	0.05	[577, 619]
K$_5$Nd(MoO$_4$)$_4$	D_{3d}^5–$R\bar{3}m$	2.3	≈ 215	≈ 70	3.08	0.33	[569, 655]

progress has also been made in understanding of this interesting phenomenon.

Detailed investigations into the luminescence properties of the NdP$_5$O$_{14}$ crystal have revealed the following peculiar features: (i) the luminescence lifetime of the $^4F_{3/2}$ state is independent of temperature over the whole concentration range (the Nd$_x$La$_{1-x}$P$_5$O$_{14}$ system was studied), including that of concentration quenching [534, 683]; (ii) the character of the metastable-state decay is independent of the Nd^{3+} concentration, and is always a single exponential [684]; (iii) the probability of nonradiative deactivation or the concentration quenching constant is

$$K_C = 1/\tau_{\text{lum}} - 1/\tau_{\text{rad}}. \tag{7.1}$$

This behavior is anomalous. A number of studies (see, for example, [576, 683]) indicate that it is a linear function of concentration (Fig. 7.1), i.e., $K_C \sim C$. At the same time, in many other neodymium-doped laser crystals, including Y$_3$Al$_5$O$_{12}$ and YAlO$_3$ which were considered in the previous chapter (see Fig. 7.2), the behavior $K_C \sim C^2$ is observed for a wide range of Nd^{3+} ion concentrations (this more general case will be hereafter called strong concentration quenching; see also [576, 579, 655] and their references). A linear relation of the quenching constant K_C to the luminescent impurity concentration C, as will be shown below, is characteristic of the phenomenon of weak concentration quenching. Anticipating somewhat, we note that in Nd^{3+} doped crystals, the process of weak quenching should in principle always be preceded (at low concentrations) by strong concentration quenching ($K_C \sim C^2$). This point is supported by several recent studies concerned with the $\tau_{\text{lum}}(C)$ dependence and by the analysis of quenching of the Nd^{3+} ion luminescence in [681]. The neodymium concentration at which one quenching process changes to the other will be arbitrarily called the boundary, and designated by C_b.

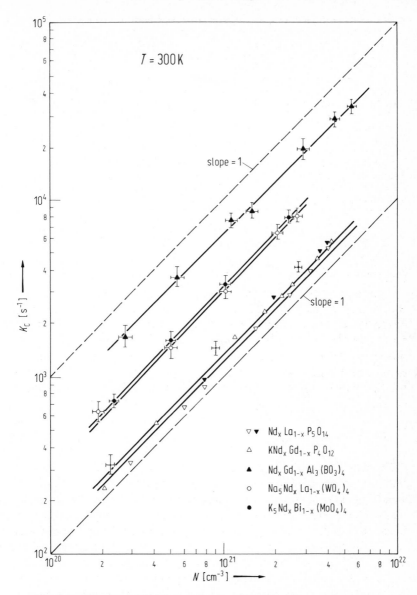

Fig. 7.1. Examples of $K_C(C)$ dependences with slope unity

For the moment, it can be considered a well-established fact that non-radiative de-excitation of the $^4F_{3/2}$ state of Nd^{3+} ions proceeds by energy migration on the donor subsystem [687]. Therefore, to describe this process, two microparameters should be introduced. For electric dipole–dipole interactions, typical of Nd^{3+} ions, the first microparameter,

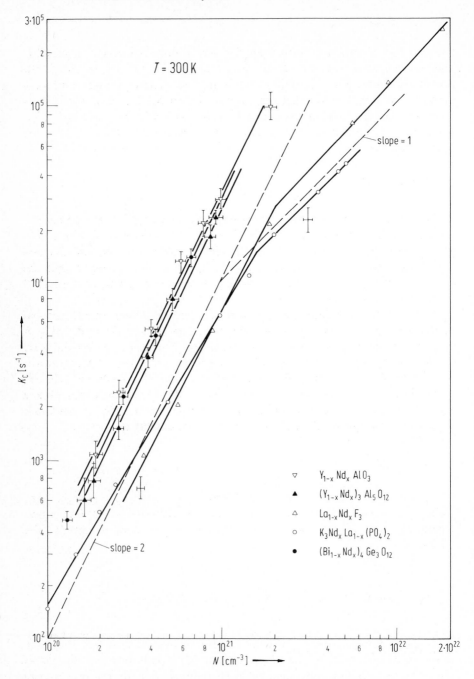

Fig. 7.2. Examples of $K_c(C)$ dependences with the slope of curves equal to two

$$C_{DA} = P_{DA} R^6,$$

describes an elementary quenching for direct donor–acceptor interaction between ions at the distance R (see Sect. 3.4). In this expression, P_{DA} is the excitation–destruction probability, i.e., the number of nonradiative transfers of excitation to acceptors per unit time. In the case of Nd^{3+} ions, this process is stimulated by cross-relaxation channels associated with the $^4I_{15/2}$ and $^4I_{13/2}$ manifolds that lie between the ground and the metastable states. The second microparameter,

$$C_{DD} = P_{DD} R^6,$$

is associated with the migration probability P_{DD} that describes the number of transfers of excitation on donor ions, per unit time, associated with their $^4I_{9/2}$ ground and $^4F_{3/2}$ metastable states. The microparameters C_{DA} and C_{DD} are purely spectroscopic because they depend on oscillator strengths and the corresponding integrals of spectral overlap of interacting transitions (spectral resonance between luminescence and absorption).

Until recently, the diffusion approach was used to explain the processes of nonradiative excitation destruction under conditions of migration [688, 689]. A necessary condition for its validity is a constant coefficient of excitation diffusion over donor ions in the crystal. At low activator concentrations, it breaks down because of fluctuating ion–ion distances [690] and inhomogeneous spectral broadening of the activator-ion lines [691]. These difficulties can be more or less avoided by introducing a quasidiffusion coefficient [690, 691]. A limitation imposed in principle on the diffusion model is the condition

$$C_{DA} \gg C_{DD}. \tag{7.2}$$

This condition can be understood from the following argument. The diffusion approach assumes a gradient of donor excitation density in some region around the acceptor (sphere of radius R_W), whose size is determined by the condition

$$\frac{C_{DA}}{R_W^6} \tau_W \approx 1, \tag{7.3}$$

where τ_W is the time of excitation passage through the region within the sphere. In terms of this model, this time is

$$\tau_W = \frac{d_{jump}^4 R_W^2}{C_{DD}}. \tag{7.4}$$

Here d_{jump} is the characteristic length of one jump. An obvious requirement for the desired gradient to be established would be multiplicity of jumps, i.e., $R_W > d_{jump}$. Thus combination of (7.3) and (7.4) gives the simple expression

$$\frac{C_{DA}}{R_W^4}\frac{d_{jump}^4}{C_{DD}} \approx 1,$$

showing that the condition for applicability of the diffusion approach is the inequality $C_{DA} \gg C_{DD}$.

The authors of [690, 692, 693] have developed a jump theory of luminescence quenching for energy migration on the donor subsystem, which is an alternative to the diffusion approach and fits the case most important for us

$$C_{DD} \gg C_{DA}. \tag{7.5}$$

Unlike the diffusion approach, the jump mechanism interprets electron-excitation destruction as being due to migration when fluctuations in the nonradiative decay rate, caused by this migration, can be described by a Markov process [694].

The next concrete step in developing this alternative approach was a comprehensive study [681] with many experimental results. It was shown in this work, in particular, that for the Nd^{3+} ion case, the donor concentration N_D is equal to the acceptor concentration N_A and to the total N_0 ion concentration ($N_D = N_A = N_0$), i.e., every neodymium ion can act both as an excitation carrier and a quencher. This does not contradict the familiar fact that electron-excitation destruction in Nd^{3+} doped media proceeds by cross-relaxation processes associated with levels of the $^4I_{15/2}$ or $^4I_{13/2}$ and $^4I_{15/2}$ manifolds of the ground $^4I_{J'}$ term [264, 463, 468, 574, 581]. The channel of cross-relaxation quenching associated with the $^4I_{15/2}$ state (type I) operates by the scheme: deexcitation of the acceptor Nd^{3+} ions A to the energy of the $^4I_{15/2}$ manifold ($^4F_{3/2} \rightarrow {}^4I_{15/2}$ transition), accompanied by energy transfer to the donors D, other Nd^{3+} ions ($^4I_{9/2} \rightarrow {}^4I_{15/2}$ transition), with subsequent energy dissipation (or absorption) in the lattice of the host matrix. This channel, as follows from Fig. 7.3, cannot be effective at low temperatures because of the great energy mismatch $\Delta E_{mism}^{(\pm)}$. In particular, for NdP_5O_{14} crystal, the value $\Delta E_{mism}^{(\pm)}$ is negative, i.e., formally for the sequence $^4F_{3/2} \rightarrow {}^4I_{15/2} \rightarrow {}^4I_{9/2}$, the crystalline lattice should impart an energy of about 270 cm^{-1} to the activator subsystem.

The cross-relaxation channel associated with transitions to the $^4I_{13/2}$ and $^4I_{15/2}$ states (type II) functions as follows: de-excitation of the acceptors A to the energy of the $^4I_{13/2}$ manifold ($^4F_{3/2} \rightarrow {}^4I_{13/2}$ transition), energy transfer to the donors D ($^4I_{9/2} \rightarrow {}^4I_{15/2}$ transition), and subsequent phonon emission and energy dissipation through the lattice. As a glance at Fig. 7.3 shows, in this case the mismatch energy is positive. Evidently, the probability for cross relaxation in the channel of type II is almost independent of temperature. It also follows from Fig.

7.3 that the conditions for the cross-relaxation process via channels I and II will be less favorable for the NdP_5O_{14} crystal than for the $Y_3Al_5O_{12}$-Nd^{3+} crystal (the case of strong concentration quenching) because of high values of $\Delta E^{(-)}_{mism}$ and $\Delta E^{(+)}_{mism}$.

We proceed now to the kinetic equation describing the temporal decay of excitation from the metastable $^4F_{3/2}$ state of donor Nd^{3+} ions taking into account their volume distribution over the crystalline lattice sites (including equivalent ones)

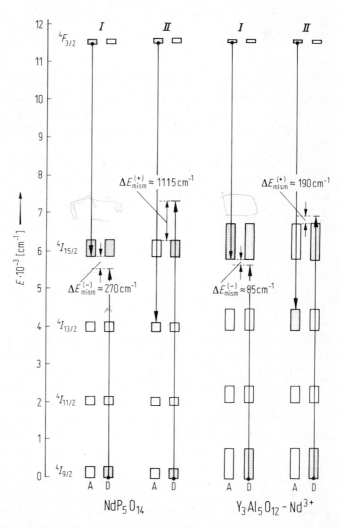

Fig. 7.3. Simplified schemes of Nd^{3+} ion-manifold positions, illustrating cross-relaxation luminescence quenching channels for NdP_5O_{14} and $Y_3Al_5O_{12}$

$$\frac{d\rho(k)}{dt} = -\rho(k)\sum_{l} P_{DD}(R_{kl}) + \sum_{l} P_{DD}(R_{kl})\,\rho(l)$$

$$- \rho(k)\sum_{m} P_{DA}(R_{km}) - \rho(k)/\tau_{rad} \tag{7.6}$$

where $\rho(k)$ and $\rho(l)$ are the densities of excited donor Nd^{3+} ions located in the k and l sites respectively; R_{kl} is the distance between the k and the l donors; and R_{km} is the distance from the m site locations of the acceptor Nd^{3+} ions. The first term of the kinetic equation (7.6) is responsible for excitation flow from the k donors to the l donors; the second term describes energy flow to unexcited k donors; the third term represents quenching due to interaction of the k donors with the m acceptors; and finally, the last term is responsible for radiative de-excitation determined by the spontaneous-emission probability $\Sigma_i\, A_{ji}$ (see Sect. 2.2). As can be seen, (7.6) does not take into account the process of nonlinear luminescence quenching [695] (interaction between excited ions) and excitation leakage by multiphonon nonradiative transitions from the metastable $^4F_{3/2}$ state.

Equation (7.6) will be analyzed for the most interesting case (see Fig. 7.4), where the mean rate of excitation-energy migration ($\bar{\Gamma}$) exceeds the highest rate of its destruction by direct donor-acceptor interaction, $C\Sigma_R P_{DA}(R)$. The quenching process will be, evidently, most active over the shortest possible (for a given crystal) donor–acceptor distance (R_{min}). The case under consideration has been very aptly called superfast migration [690–692]. Its condition is the inequality

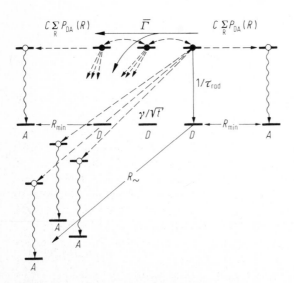

Fig. 7.4. Diagram illustrating donor decay excitation in activated crystals (elementary processes)

$$\frac{C_{\mathrm{DD}}}{\bar{R}^6} > \frac{C_{\mathrm{DA}}}{V^2},\tag{7.7}$$

where $\bar{R} = (V/C)^{1/3}$ is the mean distance between activator ions. In the last expression, C stands for the relative activator concentration and V for some arbitrary unit volume of the crystal, e.g., the volume per activator ion for 100% concentration.

When condition (7.7) is satisfied, the distribution over donor ions is equal, i.e., $\rho(k) = \rho(l) = \rho(k \neq l)$, and the total number of excited donors will be $N_{\mathrm{D}}^* = \Sigma_k \rho(k)$. Thus, for superfast migration, the kinetic equation (7.6) formally describes variations of the number of excited donors with time,

$$\frac{dN_{\mathrm{D}}^*}{dt} = -\rho \sum_{k,m} P_{\mathrm{DA}}(R_{km}) - \frac{N_{\mathrm{D}}^*}{\tau_{\mathrm{rad}}}$$

$$= -CN_{\mathrm{D}}^* \sum_R P_{\mathrm{DA}}(R) - \frac{N_{\mathrm{D}}^*}{\tau_{\mathrm{rad}}}.\tag{7.8}$$

The solution of this differential equation is

$$N_{\mathrm{D}}^* = N_{\mathrm{D}}^*(0) \exp\left\{-\left[C\sum_R P_{\mathrm{DA}}(R) + 1/\tau_{\mathrm{rad}}\right]t\right\}.$$

The established exponential law of excited-donor decay corresponds to the luminescence lifetime,

$$\tau_{\mathrm{lum}} = 1\left/\left[C\sum_R P_{\mathrm{DA}}(R) + 1/\tau_{\mathrm{rad}}\right]\right.$$

or

$$1/\tau_{\mathrm{lum}} - 1/\tau_{\mathrm{rad}} = C\sum_R P_{\mathrm{DA}}(R) = K_C.\tag{7.9}$$

Now it is evident that for superfast migration with $C \geqslant C_{\mathrm{b}}$, condition (7.1) is satisfied, $K_C \sim C$, i.e., the nonradiative de-activation probability of the $^4F_{3/2}$ state of the Nd^{3+} ions becomes linearly dependent on concentration and independent of C_{DD}. At concentrations below C_{b}, the nonradiative decay rate is limited by the migration rate and changes as the square of the concentration. Thus, on the basis of a simple physical consideration, we have established that the character of the quenching process of Nd^{3+} ion luminescence is determined by the value C_{b}, which is constant for a given crystal, and depends, in turn, on its crystalline structure and spectroscopic properties (C_{DA} and C_{DD}). The latter depends upon such spectroscopic features as the character of the crystal-field splitting and the relative positions of activator states $|[S, L]J\rangle$, the oscillator strengths of interacting transitions and the pattern of the impurity-active phonon spectrum (effective phonon density).

Now some words about the value C_b. It automatically follows from (7.7) that for low Nd^{3+} ion concentration,

$$C_b \approx \sqrt{C_{DA}/C_{DD}}, \tag{7.10}$$

i.e., when $C_{DD} \gg C_{DA}$, a single-exponential decay occurs even at $C_b \ll 1$. For other relations between the microparameters C_{DD} and C_{DA} (variations in crystalline matrices) condition (7.7) for superfast migration is valid even at higher Nd^{3+} ion concentrations. This is quite understandable because only the left-hand side of inequality (7.7) depends on relative concentrations.

As pointed out in [681, 683], Nd^{3+} ions always meet the condition $C_{DA} \ll C_{DD}$. This implies that the C_b value cannot be high. In crystals with weak concentration quenching of luminescence, the value of C_b is about, or below, the value at which concentration quenching begins to occur and the decay of the excited $^4F_{3/2}$ state is a single exponential, the rate of nonradiative de-excitation being proportional to C. This situation is most apparent in the case of the $Nd_xLa_{1-x}P_5O_{14}$ system. For the other self-activated neodymium-doped crystals listed in Table 7.1, similar behavior of the $K_C(C)$ plot is observed with crystals (see Fig. 7.1) in which $\tau_{rad}/\tau_{lum}^{100\%} = (2.6-2.8) \approx e$. Thus, here a certain law (the "e" law) manifests itself by reduction of the lifetime of luminescence intensity, and of quantum yield by a factor of approximately e. In the case of crystals of the $K_3Nd_xLa_{1-x}(PO_4)_2$ system, the $K_C(C)$ curve (as seen in Fig. 7.2) has an inflection, i.e., changes from the $K_C \sim C^2$ shape to the $K_C \sim C$ shape, showing very clearly a boundary concentration of Nd^{3+} ions. The inflection point of the $K_C(C)$ curve for the crystals concerned corresponds to a Nd^{3+} ion concentration of about 1.5×10^{21} cm^{-3} (or about 8 at.%) [619]. Such a change from the $K_C \sim C^2$ to the $K_C \sim C$ form occurs also in crystals of the $Na_3Nd_xLa_{1-x}(PO_4)_2$ system. For more complete information, Fig. 7.2 also shows the $K_C(C)$ plot for crystals of the $Nd_xLa_{1-x}F_3$ system. In this case the C_b value corresponds to a Nd^{3+} ion concentration of about 2×10^{21} cm^{-3} (or 7–9 at.%) [681].

The excitation decay from the $^4F_{3/2}$ metastable state of Nd^{3+} ions in crystals of the $Nd_xLa_{1-x}P_5O_{14}$ system [534, 683] is independent of temperature because the temperature-dependent parameter C_{DD} does not enter (7.8) which describes the kinetics of the deactivation process, and the parameter C_{DA} is practically independent of temperature because of the relative positions of the $^4F_{3/2}$, $^4I_{15/2}$, $^4I_{13/2}$, and $^4I_{9/2}$ states.

Thus, all the peculiar features of the concentration quenching of luminescence of $Nd_xLa_{1-x}P_5O_{14}$ crystals are quite well interpreted in microscopic terms. The microscopic approach is also very fruitful for study of specific features of nonradiative excitation destruction in other condensed laser media doped with Nd^{3+} ions.

Now we briefly consider the kinetics of donor decay at $C < C_b$, i.e., at $K_C \sim C^2$. As in other sections, the limited size of the present book permits only a condensed description; therefore those interested in the problem should consult more comprehensive publications [57, 681, 688, 690, 692, 693]. Under conditions of concentration quenching, the luminescence intensity ($^4F_{3/2} \rightarrow {}^4I_J$ transitions) of Nd^{3+} ions in the general case and, particularly, in the $C < C_b$ case, varies with time according to

$$I_{lum} = I_{lum}^0 \exp\{-[t/\tau_{rad} + \Pi(t)]\}. \tag{7.11}$$

The function $\Pi(t)$ in (7.11) which determines the channel of nonradiative losses at various instants of time t after excitation, can take one of the following three forms

$$\Pi(t) = \begin{cases} tC\sum_R P_{DA}(R) & \text{at } t < t_1, \\ \gamma\sqrt{t} & \text{at } t_1 < t < t_2, \\ \bar{\Gamma}t & \text{at } t > t_2, \end{cases} \tag{7.12}$$

(the physical sense of which will be explained below), or it may be determined by the sum of two forms when $t > t_1$. The parameters γ and $\bar{\Gamma}$ in (7.12) are related to the interaction parameters by

$$\gamma = 4/3\pi^{3/2} N_0 C_{DA}^{1/2}$$

and

$$\bar{\Gamma} = \pi(2\pi/3)^{5/2} N_0^2 (C_{DA} C_{DD})^{1/2}.$$

The behavior of the function $\Pi(t)$ will be considered using Fig. 7.4, 5. It is evident from Fig. 7.5 that at $C < C_b$, the donor de-excitation process falls into

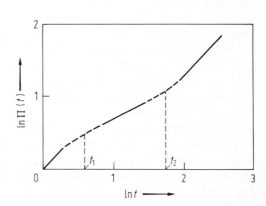

Fig. 7.5. Nonradiative loss kinetics of donor excitations [681]

three stages. The initial exponential portion at $t < t_1$ is determined by direct donor–acceptor interactions over the shortest possible distances $R_{\min}$. The next most probable channel of nonradiative losses for the interval $t_1 < t < t_2$ at $C < C_b$ is also direct donor–acceptor interaction but over different distances $R_{\min}$. *Förster*, who was the first to study this type of donor de-excitation [57], has shown that the decay behavior depends on the square root of the time, $\gamma\sqrt{t}$. In Fig. 7.5, this corresponds to the portion with a slope of 0.5. Finally, the exponential decay in the final quenching stage ($t > t_2$) is determined by excitation migration through donors to acceptors and destruction there [693]. Donor de-excitation in the interval $t < t_1$ is known elsewhere as static "ordered" (interactions over equal distances $R_{\min}$); that at $t_1 < t < t_2$, as static "disordered" (interactions over different distances $R_\sim$); and that at $t > t_2$, as migration decay. As has been discussed before, for $C > C_b$, donor de-excitation, when $\bar{\Gamma} \gg C\Sigma_R P_{DA}(R)$, will be of a superfast-migration character.

To complete this discussion of the selection of a model for quenching of luminescence from levels of the $^4F_{3/2}$ state in Nd^{3+} doped crystals, we briefly mention another limiting case: when $C_{DA} \gg C_{DD}$. According to [681, 693], as a first approximation, the law of decay at any Nd^{3+} ion concentration will be determined by the expression

$$\exp \sum_R \ln\{1 - C + C \exp[-tP_{DA}(R)]\}.$$

It follows from expression (7.10) that for $C_{DA} \gg C_{DD}$ the transition to superfast-migration decay (7.9) at low and intermediate concentrations is physically impossible ($C > 1$). However, when $C \to 1$, it is possible, the C_b value being defined by [681]

$$C_b = 1 - \exp(-C_{DA}/C_{DD}),$$

whereas the decay condition (7.9) at $C \to 1$ is

$$C_{DD} > C_{DA}/|\ln(1 - C)|.$$

These considerations show that in the two limiting cases (7.2) and (7.5), the nonradiative-decay rate may depend linearly on concentration. When $C_{DD} \gg C_{DA}$, this is possible even at rather low C, but when $C_{DA} \gg C_{DD}$ only at $C \to 1$, i.e., according to the diffusion model, weak concentration quenching of luminescence is not practically feasible.

As a continuation of this point, we briefly mention the work of *Danielmeyer* [696], who suggests another explanation of concentration quenching of Ln^{3+} ion luminescence in crystals, including those that are self-activated. He assumes the occurrence of activator formations with somewhat different spectroscopic properties, due to mixing of wave functions of the $4f$ and $5d$ configurations of neighboring impurity Ln^{3+} ions when their spacing in a crystal is of the order of, or less than, 5Å. Recent measurements of oscillator strengths of some transitions of Nd^{3+} ions in the $Nd_xLa_{1-x}P_5O_{14}$ system by *Auzel* [658] have revealed, within the experimental accuracy, that these spectroscopic parameters do not change over the whole range of x. Similar measurements were made by the authors of [681] with crystals in the $Nd_xLa_{1-x}F_3$ system, in which strong concentration quenching of luminescence is observed. They also found the oscillator strengths to be constant for variable x. For this reason, the model of concentration quenching of luminescence suggested in [696] for systems similar to $Nd_xLa_{1-x}P_5O_{14}$ in their properties is not adequate to explain the observed luminescence properties. This idea, however, must not be abandoned without more experiments, since we know many examples of pair and more complex associates of Ln^{3+} ions formed in crystals. To discover manifestations of the mixing effect of wave functions of the $4f$ and $5d$ configurations of neighboring Ln^{3+} ions (the effect may prove very weak) necessitates, as *Auzel* [658] quite rightly stated, precision measurements on crystals characterized by transitions with weak oscillator strengths (say, $LnCl_3$ or $LnBr_3$ crystals). Numerous new interesting ideas, based on up-to-date experimental methods, have been suggested to explain the anomalous weak luminescence concentration quenching of Nd^{3+} ions [761, 771, 772, 782, 783].

We conclude this section by saying that even with numerous reliable experimental results and considerable progress in understanding, the nature of concentration quenching of luminescence is not yet absolutely clear. Many theoretical aspects of this phenomenon, which is extremely important for the laser-crystal physics, do need further clarification and discussion. Unfortunately some workers adopt too-narrow objectives and forget that concentration quenching of luminescence affects practically all spectroscopic properties of a crystal-ion system. Experience shows that interpretation of results on concentration quenching of luminescence of a given Ln^{3+} ion must be based on thorough knowledge of its Stark-level scheme, the phonon properties of the whole system, peculiarities of the fine structure of the matrix and of the activator centers, intensities of radiative and absorption transitions, and other parameters. Any of these points may produce a surprise and radically change the concept of the process. Evidence of this is the production and study of self-activated neodymium-doped laser crystals.

7.2 Erbium and Holmium High-Concentration and Self-Activated Laser Crystals

Though neodymiun-doped self-activated crystals exhibit fairly good spectroscopic characteristics, except for the NdP_5O_{14} crystal [742], they have so far found application only in laser radiation converters. The same applies to most high-concentration Nd^{3+}-doped crystals which were produced by "diluting" self-activated crystals (see Table 3.13). Utilization of such crystals in conventional pulsed and CW lasers with broad-band gas-discharge pumping is still problematic for several reasons. The technological difficulties that prevent production of sufficiently large and optically perfect single crystals have to be overcome and questions associated with weak concentration quenching of luminescence in neodymium-doped laser crystals have to be solved. These questions are very much concerned with their spectral properties [574, 575, 589, 656]. Unfortunately such unique crystalline matrices as $Y_3Al_5O_{12}$ and $YAlO_3$ cannot be employed in lasers with high Nd^{3+} ion concentrations. However, recent investigations, in which the author of the present monograph has taken part [546, 547, 565, 572, 590, 595, 641, 655] (see also [490]) have indicated that these excellent crystals, and related crystals with similar properties (such as the character of the electron-phonon interaction), that contain other Ln^{3+} ions and in particular Ho^{3+} and Er^{3+} ions, have attractive laser characteristics. When excited by conventional lamp pumping, they generate, rather efficiently, stimulated emission at room and higher temperatures over wide spectral ranges, including the 3 μm region, which is extremely important for some applications. Nevertheless, the quantity of "operating" ions in such crystals may be several times the quantity in self-activated neodymium-doped crystals.

Table 7.2 lists all the known high-concentration laser crystals doped with Ho^{3+} and Er^{3+} ions capable of emitting at room temperature around 3 μm.[3] Information concerning other high-concentration and self-activated laser crystals is contained in Tables 3.4 and 5.1.

The above-mentioned studies have shown that the luminescence lifetime of the initial "three-micrometer" laser state ($^4I_{11/2}$ for Er^{3+} ions and $^5I_6$ for Ho^{3+} ions) is practically unchanged in aluminum garnets over a wide activator concentration range, up to $x = 1$ (for the $Er_3Al_5O_{12}$ crystal, $N_0 \approx 1.3 \times 10^{22}$ cm^{-3}, which is much in excess of the Nd^{3+} ion concentration: $\approx 4 \times 10^{21}$ cm^{-3} for the NdP_5O_{14} crystal). Such behavior of $\tau_{lum}^{init}(C)$ is indicative of the absence of cross-relaxation channels of luminescence quenching attributed to transitions from these states. This is also indicated by the different frequencies observed for the $^4I_{11/2} \rightarrow ^4I_{13/2}$ and the $^4I_{15/2} \rightarrow ^4I_{13/2}$ transitions in the case of Er^{3+} ions and for the $^5I_6 \rightarrow ^5I_7$ transitions in the case of Ho^{3+} ions (see the schematic diagrams of the energy states of Ln^{3+} ions in Fig. 1.1). According to [548], the maximum effective phonon density for $Y_3Al_5O_{12}$ crystals (and certainly for $Lu_3Al_5O_{12}$ and other Ln aluminum garnets), which determines the activity of their electron-phonon interaction, is within an energy range from ≈ 350 to ≈ 450 cm^{-1} (for more details, see Sect. 3.11). For Er^{3+} and Ho^{3+} ions in these garnet crystals, the energy mismatch in those transitions exceeds the maximum possible energy of impurity-active phonons. This, again, indicates a minimum probability for a cross-relaxation quenching process. Figure 7.6 shows that, in the case of the

---

[3]Stimulated emission of Ho^{3+} ions on the $^5I_6 \rightarrow ^5I_7$ transition in $LiYF_4$ [778], $BaYb_2F_8$ [791], $KY(WO_4)_2$ [789], $KLu(WO_4)_2$ [790] and $Gd_3Ga_5O_{12}$ [774] crystals and Er^{3+} ions on the $^4I_{11/2} \rightarrow ^4I_{13/2}$ transition in $LiYF_4$ [776], $KEr(WO_4)_2$ [792], $KLu(WO_4)_2$ [793] and $Gd_3Ga_5O_{12}$ [774] crystals has also been reported.

Table 7.2. Room-temperature 3 μm crystal lasers [655]

Crystal	Laser ion concentration [at.%]	Deactivator ions	Laser wavelength [μm]	Threshold energy [J]	Terminal level energy [cm^{-1}]	τ_{lum}^{init} [ms]	τ_{lum}^{term} [ms]	Reference
			Ho^{3+}($^5I_6 \to {}^5I_7$ transition)					
KGd(WO$_4$)$_2$	3–5	—	2.9342	≈25	≈5500	—	—	[641]
YAlO$_3$	2	—	2.9180	7.5	≈5300	0.8–1.0	6.6	[547, 565, 655]
	10	—	3.0132	50	≈5600			[655]
Y$_3$Al$_5$O$_{12}$	≈10	—	2.9403	30	≈5400	0.06	≈7.5	[547, 565]
LaNbO$_4$	2	—	2.8510	50	≈5500	—	—	[607]
Lu$_3$Al$_5$O$_{12}$	5	—	2.9460	60	≈5400	0.06	≈7.5	[546, 547, 565]
			Er^{3+}($^4I_{11/2} \to {}^4I_{13/2}$ transition)					
CaF$_2$	10	—	2.7307	400	≈6500	8.8	≈9	[469]
CaF$_2$–ErF$_3$– –TmF$_3$	12.5	Tm^{3+}(?)	≈2.69	10	≈6500	4.5	—	[111]
KY(WO$_4$)$_2$	≈30	Ho^{3+},Tm^{3+}	2.6887	10	≈6600	—	—	[595]
KGd(WO$_4$)$_2$	3–25	—	2.7222	50	6515	—	—	[641]
			2.7990	15	6710			
YAlO$_3$	2	—	2.7039	40	6642	0.9	≈5	[547, 565]
Y$_3$Al$_5$O$_{12}$	≈33	—	2.7953	≈10	6779	0.1	6.4	[794, 797, 798]
			2.8302	≈10	6818(?)			[572]
			2.9365	≈10	6879			[490, 572]
	≈33	Tm^{3+}	2.6975	≈10	6544	—	—	[794, 797, 798]
			2.7953	≈10	6779			
			2.8302	12	6818(?)			
	≈33	Ho^{3+},Tm^{3+}	2.6975	15	6544	—	—	[794, 797, 798]
Er$_3$Al$_5$O$_{12}$	100	—	2.9367	12	≈6880	0.07	≈2	[572]
Lu$_3$Al$_5$O$_{12}$	≈33	—	2.7987	≈10	6798	0.1	6.5	[794, 797, 798]
			2.8298	≈10	6847(?)			[572, 655]
			2.9395	≈10	6885			[546, 565]
	≈33	Tm^{3+}	2.6990	≈10	6559	—	—	[794, 796–798]
			2.7987	≈11	6798			
			2.8298	15	6847(?)			
	≈33	Ho^{3+},Tm^{3+}	2.6990	15	6559	—	—	[655, 796]

Lu$_3$Al$_5$O$_{12}$-Er^{3+} crystal, the mismatch due to resonant $^4I_{11/2} \to {}^4I_{13/2} \leftarrow {}^4I_{15/2}$ transitions, which is undesirable for the cross-relaxation process, is $\Delta E_{mism}^{cr} \approx 2100$ cm^{-1}. The slight reduction of the value of τ_{lum}^{init} observed with the (Y$_{1-x}$Er$_x$)$_3$Al$_5$O$_{12}$ system (from ≈100 to ≈75 μs, see Fig. 7.7) can be attributed to some ions of uncontrollable impurities in the crystals [697].

The peculiar features of stimulated emission on self-saturating laser transitions were treated in Sect. 3.6. Therefore, we now briefly consider some questions of exciting laser action and improving the laser characteristics for the most familiar crystals of the two systems: (Y$_{1-x}$Er$_x$)$_3$Al$_5$O$_{12}$ and (Er$_x$Lu$_{1-x}$)$_3$Al$_5$O$_{12}$ [547, 572, 590]. The short lifetime of their initial operating state ($\tau_{lum}^{init} \approx 100$ μs) imposes limitations on the pump-pulse duration. It has been shown experimentally that for $\tau_{exc} \lesssim \tau_{lum}^{init}$ the excitation threshold for stimulated emission on the additional $^4I_{11/2} \rightarrow {}^4I_{13/2}$ transition is subject to very slight variations, whereas at $\tau_{exc} \approx 3\tau_{lum}^{init}$ it is much higher. Longer pump pulses fail to excite 3 μm stimulated emission at 300 K. For lasers based on these crystals to find wide application (which there is good reason to expect [490, 628]), another urgent question must be solved—reduction of the undesirable effect of the self-saturation that limits the laser output. The experience of the author, accumulated in the course of cascade-laser studies [25, 106], has shown that this difficult problem can be solved using the mechanism of nonresonant transfer of electron-excitation energy from emitting Er^{3+} ions to others, which must function in this

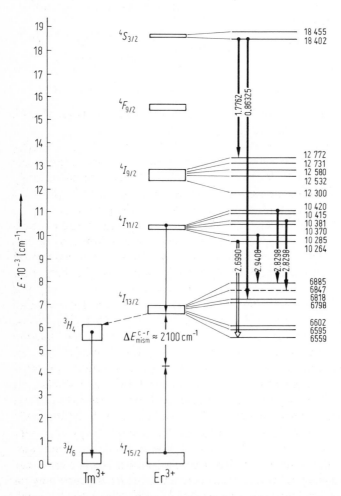

Fig. 7.6. Simplified scheme of $Lu_3Al_5O_{12}$-Er^{3+} crystal-laser operation with deactivating Tm^{3+} ions [547]. The double arrow indicates the short-wavelength laser transition. Other notation as in Figs. 3.4, 15

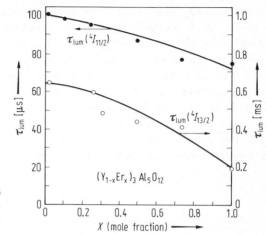

Fig. 7.7. Concentration dependences of $\tau_{lum}(x)$ for the $^4I_{11/2}$ and $^4I_{13/2}$ manifolds of Er^{3+} ions in the $Y_3Al_5O_{12}$ crystal at 300 K [697]

case as deactivators (see also Sect. 3.6). As is shown in Refs. [25, 106, 546, 547, 565], such deactivator ions may be Tm^{3+} and Ho^{3+} ions.

Let us consider in more detail the 3 μm emission of $(Er_xLu_{1-x})_3Al_5O_{12}$ crystals. Insufficiently fast decay of residual excitation from the levels of the terminal $^4I_{13/2}$ laser manifold following $^4I_{11/2} \rightarrow {}^4I_{13/2}$ transitions immediately after a pump pulse leads to greater population of the lower Stark levels of the $^4I_{13/2}$ state. Under these conditions, the longer wavelength $^4I_{11/2} \rightarrow {}^4I_{13/2}$ transitions are preferable for stimulated emission. A laser transition with $\lambda_g = 2.7987\ \mu$m (10370 cm^{-1} $^4I_{11/2} \rightarrow {}^4I_{13/2}$ 6798 cm^{-1}) is not shown in Fig. 7.6. This line is emitted early in the excitation process but then rapidly switches off due to self-saturation effects. It is immediately replaced by two other long-wave transitions. When deactivator Tm^{3+} ions are present in the crystal, this transition occurs in full-power lasing. At lower temperatures a number of other short-wave laser lines are excited with the same behavior as the 2.8787 μm line [794]. This is one of many manifestations of a red shift of stimulated-emission wavelength in lasers. The results in [546, 547, 565, 590] show that this is the case for stimulated emission by Er^{3+} ions in the $Lu_3Al_5O_{12}$ crystal. Similar behavior has been observed with $KGd(WO_4)_2$-Er^{3+} crystals [641]. Figure 7.6 shows that, in the case of the $Lu_3Al_5O_{12}$-Er^{3+} crystal, under such conditions laser action is achieved on the two long-wave transitions 10420 cm^{-1} $^4I_{11/2} \rightarrow {}^4I_{13/2}$ 6885 cm^{-1} (or the 10381 cm^{-1} $\rightarrow$ 6847 cm^{-1} transition of the same frequency) with $\lambda_g = 2.8298\ \mu$m, and the 10285 cm^{-1} $^4I_{11/2} \rightarrow {}^4I_{13/2}$ 6885 cm^{-1} transition, with $\lambda_g = 2.9406\ \mu$m. It should be added that with no deactivator ions, the decay rate of residual excitation from the levels of the $^4I_{13/2}$ state depends only on the probability of luminescence transitions (for the $Lu_3Al_5O_{12}$-Er^{3+} crystal, $\tau_{lum}^{term} = \tau_{rad} = 1/\Sigma\ A_{13/2} \approx 6.4$ ms) because the luminescence quantum yield from this state is close to unity.

If the $(Er_xLu_{1-x})_3Al_5O_{12}$ crystals (at small and intermediate x values) contain Tm^{3+} and Ho^{3+} ions to deactivate the terminal $^4I_{13/2}$ laser state, the spectrum of Er^{3+} ion stimulated emission on the self-saturating $^4I_{11/2} \rightarrow {}^4I_{13/2}$ transition is radically changed. Efficient nonresonant energy transfer to the deactivator ions is due to the migration process (by the $^4I_{13/2} \leftrightarrow {}^4I_{15/2}$ scheme) because the high Er^{3+} ion concentration accelerates excitation decay from the levels of the $^4I_{13/2}$ manifold. It results in a marked reduction of the value τ_{lum}^{term} and a sharp drop (by the same factor) of the luminescence intensity in the $^4I_{13/2} \rightarrow {}^4I_{15/2}$ channel. The stimulated-emission lines are observed to switch over, as is anticipated in such a situation. According to Fig. 7.6, in this case, stimulated-emission connects the lowest Stark levels of the $^4I_{11/2}$ and $^4I_{13/2}$ manifolds (double arrow). By preliminary estimates, deactivation of the terminal $^4I_{13/2}$ laser state of the system is accelerated by a factor of about ten or more times. To improve conditions of excitation and lasing processes occurring on the self-saturating $^4I_{11/2} \rightarrow {}^4I_{13/2}$ transition of Er^{3+} ions, some new laser operating schemes have been suggested and realized in [796–798]. In particular, the scheme called

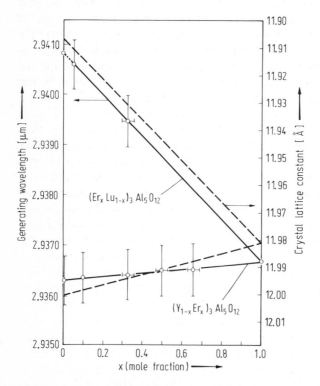

Fig. 7.8.   Concentration dependence of wavelength in the long-wave laser transition ($^4I_{11/2} \rightarrow {}^4I_{13/2}$, 300 K) and crystal-lattice constant (dashed lines) of crystals in the $(Y_{1-x}Er_x)_3Al_5O_{12}$ and $(Er_xLu_{1-x})_3Al_5O_{12}$ systems

"feed-flowing" when applied to $Y_3Al_5O_{12}$ and $Lu_3Al_5O_{12}$ crystals permitted 3 μm emission of Er^{3+} ions with an excitation threshold energy of only a few joules. The extra-feeding of the initial laser $^4I_{11/2}$ state in these crystals was provided by Yb^{3+} and Cr^{3+} sensitizer ions, and acceleration of excitation flowing from the terminal $^4I_{13/2}$ state was provided by Tm^{3+} and Ho^{3+} deactivator ions.

Further problems to be solved along these lines of investigation are establishment of optimal concentrations of stimulated-emission ions for a reasonable choice of deactivator and sensitizer ions and their concentrations. Thus, it is clear now that excessive increase of Ho^{3+} ion concentration results in worse conditions for stimulated emission excitation of Er^{3+} ions on the $^4I_{11/2} \rightarrow ^4I_{13/2}$ transition due to reabsorption occurring from the excited $^5I_7$ state of Ho^{3+} ions. These problems arise also for crystals doped with Ho^{3+} ions with stimulated emission on the self-saturating $^5I_6 \rightarrow ^5I_7$ transition.

Wide variations of Er^{3+} ion concentrations in the $(Y_{1-x}Er_x)_3Al_5O_{12}$ and the $(Er_xLu_{1-x})_3Al_5O_{12}$ crystal systems permit variations, within a narrow spectral range, of the stimulated-emission wavelength for these lasers on the additional $^4I_{11/2} \rightarrow ^4I_{13/2}$ transition, without affecting other emission parameters. This is of importance, because nowadays infrared lasers enjoy wide usage in laser chemistry, nonlinear laser spectroscopy, and other applications that require scanning of the excited line of the substance under study. Figure 7.8 shows the $\lambda_g(C)$ plots obtained for these crystals. This figure shows that the behavior of the $\lambda_g(C)$ curves correlates very well with the crystal-lattice constant $a_0(C)$ of the garnet $(Y_{1-x}Er_x)_3Al_5O_{12} \rightarrow Er_3Al_5O_{12} \leftarrow (Er_xLu_{1-x})_3Al_5O_{12}$ crystals. A change of a_0 slightly changes the parameters of the crystal field that acts on Er^{3+} ions in these crystals, which results in a shift of their energy levels (see Sect. 4.2).

Lasers based on $YAlO_3$ crystals doped with Ho^{3+} and Er^{3+} ions have somewhat better conditions for exciting stimulated emission on self-saturating transitions at $\lambda_g \approx 3$ μm, since the lifetimes τ_{lum} of their initial states are several times those of yttrium and lutetium garnets. Unfortunately, these crystals have not received sufficient study to be recommended for particular infrared-laser designs. It is noteworthy that the $YAlO_3$ crystal doped with Ho^{3+} ions, and emitting on the $^5I_6 \rightarrow ^5I_7$ transition has the lowest excitation threshold [547, 565] and stimulated-emission of the longest-wavelength (3.0132 μm) among crystal-laser media that emit in the 3 μm spectral region at room temperature. A possibility of exciting stimulated emission on the self-saturating $^5I_6 \rightarrow ^5I_7$ transition of Ho^{3+} ions in the $YAlO_3$ crystal was analyzed by *Caird* and *DeShazer* [562].

8. Stimulated-Emission Wavelengths of Lasers Based on Activated Crystals

An index of stimulated-emission wavelengths for all known lasers based on activated crystals has been compiled from the data of Table 5.1. It is presented in Table 8.1 and facilitates selection of a laser with a desired stimulated-emission wavelength; it indicates the type of crystal, and its induced transition and operating temperature. This information will be useful both for constructing conventional lasers with given spectral properties and for choice of CAM lasers [26–30], as well as crystals for parametric generators [469, 490]. These data may provide help in the investigation of various processes that occur during interaction of powerful laser radiation with the medium. These include such problems as producing and investigating high-temperature plasma [231, 495] and also studying the physics of multiphoton transitions [542]. The considerable extension of the spectra of stimulated emission of crystal lasers as far as ≈ 3 μm in recent years [655] encourages the prospect of their wide application in laser chemistry [543, 544]. Because crystal lasers are characterized by high monochromaticity and high-peak-power output, they may be used, together with gas lasers, for selective excitation of molecules that have extremely close vibration frequencies. If the molecules of an excited mixture have different isotopic compositions, then laser-produced dissociation of one of the molecules, with subsequent decomposition, may result in isotope separation.

At present, crystal-laser emission wavelengths cover and sufficiently fill (except for a narrow interval at about 2.2 μm and a band at 3.2–3.9 μm), a wide region from 0.32 to 3.91 μm. In the table, the highest accuracies of the indicated laser wavelengths occur in the visible and near-infrared regions (to about 1.2 μm). Here, for the majority of λ_g values, the accuracy is better than ± 0.0002 μm. In the longer wavelength region, where laser-emission spectra are recorded mainly by photoelectric methods, the dispersion of the λ_g values may reach ± 0.001 μm. In cases in which the wavelength values of stimulated emission need more accurate determination, or confirmation by additional measurements, they are marked with the letter "a". The CaF_2 crystals with oxygen impurity differ from usual pure CaF_2 crystals in their spectroscopic and laser properties; they are marked by an asterisk. For some crystals, the space (crystallographic) group is indicated in parentheses. Stimulated emission of numerous new crystals and many new transitions of known crystals were observed in 1978–1979. This information was included in this book only in Table 8.1, where these new lines are marked by the letter "b". Some new laser data on activated crystals can also be found in Chap. 9 and in footnotes.

For some crystals with Nd^{3+} ions and for ruby, data are available on the thermal change of the stimulated-emission wavelength. These data were listed in Table 3.7.

Table 8.1. Stimulated-emission wavelengths of activated laser crystals

Stimulated-emission wavelength [μm]	Ion	Crystal host	Induced transition	Temperature [K]	Reference
0.3255[b]	Ce^{3+}	$LiYF_4$	$5d \rightarrow {}^2F_{7/2}$	300	[781]
0.4526[b]	Tm^{3+}	$LiYF_4$	${}^1D_2 \rightarrow {}^3F_4({}^3H_4)$	300	[779]
0.479[a]	Pr^{3+}	$LiYF_4$	${}^3P_0 \rightarrow {}^3H_4$	300	[606]
0.489[a]	Pr^{3+}	$PrCl_3$	${}^3P_0 \rightarrow {}^3H_4$	5.5–14	[116]
0.4892	Pr^{3+}	$LaCl_3$	${}^3P_0 \rightarrow {}^3H_4$	5.5–14	[116]
0.5298	Pr^{3+}	$LaCl_3$	${}^3P_1 \rightarrow {}^3H_5$	35	[116]
0.531[a]	Pr^{3+}	$PrCl_3$	${}^3P_1 \rightarrow {}^3H_5$	12	[116]
0.532[a]	Pr^{3+}	$LaBr_3$	${}^3P_1 \rightarrow {}^3H_5$	—	[564]
0.5445	Tb^{3+}	$LiYF_4$	${}^5D_4 \rightarrow {}^7F_5$	300	[83]
0.5512	Ho^{3+}	CaF_2	${}^5S_2 \rightarrow {}^5I_8$	77	[102]
0.5515	Ho^{3+}	$Ba(Y,Yb)_2F_8$	${}^5S_2 \rightarrow {}^5I_8$	77	[31]
0.5540	Er^{3+}	BaY_2F_8	${}^4S_{3/2} \rightarrow {}^4I_{15/2}$	77	[103]
0.5617	Er^{3+}	BaY_2F_8	${}^2H_{9/2} \rightarrow {}^4I_{13/2}$	77	[103]
0.5932[b]	Sm^{3+}	TbF_3	${}^4G_{5/2} \rightarrow {}^4H_{7/2}$	116	[780]
0.5985[a]	Pr^{3+}	LaF_3	${}^3P_0 \rightarrow {}^3H_6$	77	[75]
0.6113	Eu^{3+}	Y_2O_3	${}^5D_0 \rightarrow {}^7F_2$	220	[395]
0.6164	Pr^{3+}	$LaCl_3$	${}^3P_0 \rightarrow {}^3H_6$	65	[116]
0.617[a]	Pr^{3+}	$PrCl_3$	${}^3P_0 \rightarrow {}^3H_6$	8–65	[116]
0.619[a]	Pr^{3+}	$LaCl_3$	${}^3P_0 \rightarrow {}^3H_6$	—	[564]
0.6193[a]	Eu^{3+}	YVO_4	${}^5D_0 \rightarrow {}^7F_2$	90	[409]
0.620[a]	Pr^{3+}	$PrCl_3$	${}^3P_0 \rightarrow {}^3H_6$	—	[564]
0.621[a]	Pr^{3+}	$LaBr_3$	${}^3P_0 \rightarrow {}^3H_6$	—	[564]
0.622[a]	Pr^{3+}	$PrBr_3$	${}^3P_0 \rightarrow {}^3H_6$	—	[564]
0.632[a]	Pr^{3+}	$LaBr_3$	${}^3P_2 \rightarrow {}^3F_3$	—	[564]
0.6374[b]	Pr^{3+}	PrP_5O_{14}	${}^3P_0 \rightarrow {}^3F_2$	300	[799]
0.6452	Pr^{3+}	$LaCl_3$	${}^3P_0 \rightarrow {}^3F_2$	300	[116]
0.647[a]	Pr^{3+}	$LaBr_3$	${}^3P_0 \rightarrow {}^3F_2$	—	[564]
		$PrCl_3$	${}^3P_0 \rightarrow {}^3F_2$	—	[76, 564]
0.649[a]	Pr^{3+}	$PrBr_3$	${}^3P_0 \rightarrow {}^3F_2$	—	[76, 564]
0.6700	Er^{3+}	$Ba(Y,Yb)_2F_8$	${}^4F_{9/2} \rightarrow {}^4I_{15/2}$	77	[31]
0.6709	Er^{3+}	BaY_2F_8	${}^4F_{9/2} \rightarrow {}^4I_{15/2}$	77	[103]
		$Ba(Y,Yb)_2F_8$	${}^4F_{9/2} \rightarrow {}^4I_{15/2}$	77	[31]
0.6799	Cr^{3+}	$BeAl_2O_4$	${}^2E(\bar{E}, 2\bar{A}) \rightarrow {}^4A_2$	≈77	[753]
0.6804[b]	Cr^{3+}	$BeAl_2O_4$	${}^2E(\bar{E}, 2\bar{A}) \rightarrow {}^4A_2$	300	[802]
0.6874	Cr^{3+}	$Y_3Al_5O_{12}$	${}^2E(\bar{E}) \rightarrow {}^4A_2$	≈77	[400]
0.6929 R$_2$	Cr^{3+}	Al_2O_3	${}^2E(2\bar{A}) \rightarrow {}^4A_2$	300	[301]
0.6934 R$_1$	Cr^{3+}	Al_2O_3	${}^2E(\bar{E}) \rightarrow {}^4A_2$	77	[363]

[a] Stimulated-emission wavelength value requires more-accurate definition.
[b] See the explanation on page 380.

(Continued)

Table 8.1 (*continued*)

Stimulated-emission wavelength [μm]	Ion	Crystal host	Induced transition	Temperature [K]	Reference
0.6969	Sm^{2+}	SrF_2	$^5D_0 \rightarrow {^7F_1}$	4.2	[322]
0.701[b,c]	Cr^{3+}	$BeAl_2O_4$	$^2E \rightarrow {^4A_2}$	300	[801, 802]
0.7009	$Cr^{3+}-Cr^{3+}$	Al_2O_3	—	77	[366]
0.7037	Er^{3+}	BaY_2F_8	$^2H_{9/2} \rightarrow {^4I_{11/2}}$	77	[103]
0.7041	$Cr^{3+}-Cr^{3+}$	Al_2O_3	—	77	[366]
0.7083	Sm^{2+}	CaF_2	$5d \rightarrow {^7F_1}$	20	[14, 256]
0.7085	Sm^{2+}	CaF_2*	$5d \rightarrow {^7F_1}$	20	[326]
0.7503[b]	Ho^{3+}	$LiYF_4$	$^5S_2 \rightarrow {^5I_7}$	300	[762]
0.7516[b]	Ho^{3+}	$LiYF_4$	$^5S_2 \rightarrow {^5I_7}$	116	[778]
0.7555[b]	Ho^{3+}	$LiYF_4$	$^5S_2 \rightarrow {^5I_7}$	116	[778]
0.7498	Ho^{3+}	$LiYF_4$	$^5S_2 \rightarrow {^5I_7}$	≈ 90, 300	[568, 778]
0.7670	Cr^{3+}	Al_2O_3	$^2E \rightarrow {^4A_2}$	300	[122]
0.8430	Er^{3+}	CaF_2-YF_3	$^4S_{3/2} \rightarrow {^4I_{13/2}}$	77	[25, 106]
0.8456	Er^{3+}	CaF_2	$^4S_{3/2} \rightarrow {^4I_{13/2}}$	77	[105]
		CaF_2-YF_3	$^4S_{3/2} \rightarrow {^4I_{13/2}}$	77	[25, 106, 546]
		$CaF_2-HoF_3-ErF_3$	$^4S_{3/2} \rightarrow {^4I_{13/2}}$	77	[25]
		$CaF_2-ErF_3-TmF_3$	$^4S_{3/2} \rightarrow {^4I_{13/2}}$	77	[25]
0.8468	Er^{3+}	$KGd(WO_4)_2$	$^4S_{3/2} \rightarrow {^4I_{13/2}}$	300	[641]
0.84965	Er^{3+}	$YAlO_3$	$^4S_{3/2} \rightarrow {^4I_{13/2}}$	77	[547, 565]
0.84975	Er^{3+}	$YAlO_3$	$^4S_{3/2} \rightarrow {^4I_{13/2}}$	300	[547, 565]
0.8500	Er^{3+}	$LiYF_4$	$^4S_{3/2} \rightarrow {^4I_{13/2}}$	300	[104]
0.85[a,b]	Er^{3+}	$KLu(WO_4)_2$	$^4S_{3/2} \rightarrow {^4I_{13/2}}$	300	[793]
0.8503[b]	Er^{3+}	$LiYF_4$	$^4S_{3/2} \rightarrow {^4I_{13/2}}$	110, 300	[776]
0.85165	Er^{3+}	$YAlO_3$	$^4S_{3/2} \rightarrow {^4I_{13/2}}$	77	[547, 565]
0.8535[b]	Er^{3+}	$LiYF_4$	$^4S_{3/2} \rightarrow {^4I_{13/2}}$	110	[776]
0.8548	Er^{3+}	CaF_2	$^4S_{3/2} \rightarrow {^4I_{13/2}}$	77	[105]
0.8594	Er^{3+}	$YAlO_3$	$^4S_{3/2} \rightarrow {^4I_{13/2}}$	300	[547, 565]
0.8621[b]	Er^{3+}	$KY(WO_4)_2$	$^4S_{3/2} \rightarrow {^4I_{13/2}}$	300	[803]
0.86275	Er^{3+}	$Y_3Al_5O_{12}$	$^4S_{3/2} \rightarrow {^4I_{13/2}}$	77, 300	[546, 547]
0.8632	Er^{3+}	$Lu_3Al_5O_{12}$	$^4S_{3/2} \rightarrow {^4I_{13/2}}$	300	[546, 547]
0.853[b]	Er^{3+}	$Bi_4Ge_3O_{12}$	$^4S_{3/2} \rightarrow {^4I_{13/2}}$	77	[800]
0.86325	Er^{3+}	$Lu_3Al_5O_{12}$	$^4S_{3/2} \rightarrow {^4I_{13/2}}$	77	[546, 547]
0.8910	Nd^{3+}	$Y_3Al_5O_{12}$	$^4F_{3/2} \rightarrow {^4I_{9/2}}$	300	[81]
0.8999	Nd^{3+}	$Y_3Al_5O_{12}$	$^4F_{3/2} \rightarrow {^4I_{9/2}}$	300	[81]
0.9137	Nd^{3+}	$KY(WO_4)_2$	$^4F_{3/2} \rightarrow {^4I_{9/2}}$	77	[77]
0.9145	Nd^{3+}	$CaWO_4$	$^4F_{3/2} \rightarrow {^4I_{9/2}}$	77	[78]
0.930	Nd^{3+}	$YAlO_3$	$^4F_{3/2} \rightarrow {^4I_{9/2}}$	300	[79]
0.9385	Nd^{3+}	$Y_3Al_5O_{12}$	$^4F_{3/2} \rightarrow {^4I_{9/2}}$	300	[81]
0.9397[b]	Nd^{3+}	$CaMg_2Y_2Ge_3O_{12}$	$^4F_{3/2} \rightarrow {^4I_{9/2}}$	300	[804, 828]
0.9460	Nd^{3+}	$Y_3Al_5O_{12}$	$^4F_{3/2} \rightarrow {^4I_{9/2}}$	300	[80]
0.9473	Nd^{3+}	$Lu_3Al_5O_{12}$	$^4F_{3/2} \rightarrow {^4I_{9/2}}$	77	[191]
0.979[a]	Ho^{3+}	$LiHoF_4$	$^5F_5 \rightarrow {^5I_7}$	≈ 90	[545]
0.9794	Ho^{3+}	$LiYF_4$	$^5F_5 \rightarrow {^5I_7}$	≈ 90, 300	[568, 778]
1.0143	Ho^{3+}	$LiYF_4$	$^5S_2 \rightarrow {^5I_6}$	≈ 90, 300	[568, 778]
1.0230	Yb^{3+}	$Lu_3Ga_5O_{12}:Nd^{3+}$	$^2F_{5/2} \rightarrow {^2F_{7/2}}$	77	[86, 483]
1.0232	Yb^{3+}	$Gd_3Ga_5O_{12}:Nd^{3+}$	$^2F_{5/2} \rightarrow {^2F_{7/2}}$	77	[483, 486]
1.0233	Yb^{3+}	$Y_3Ga_5O_{12}:Nd^{3+}$	$^2F_{5/2} \rightarrow {^2F_{7/2}}$	77	[483, 486]
1.0293	Yb^{3+}	$Y_3Al_5O_{12}$	$^2F_{5/2} \rightarrow {^2F_{7/2}}$	77	[86, 128]
		$(Y,Yb)_3Al_5O_{12}$	$^2F_{5/2} \rightarrow {^2F_{7/2}}$	77	[546]

[c] Continuous tuning from 0.701 to 0.794 μm was carried out using an optical resonator with dispersive element [801, 802, 824, 825].

(*Continued*)

Table 8.1   (*continued*)

Stimulated-emission wavelength [μm]	Ion	Crystal host	Induced transition	Temperature [K]	Reference
1.0294	Yb^{3+}	$(Yb,Lu)_3Al_5O_{12}$	$^2F_{5/2} \to {}^2F_{7/2}$	77	[546]
		$Lu_3Al_5O_{12}:Nd^{3+},Cr^{3+}$	$^2F_{5/2} \to {}^2F_{7/2}$	77	[86, 483]
1.0299	Yb^{3+}	$Gd_3Sc_2Al_3O_{12}:Nd^{3+}$	$^2F_{5/2} \to {}^2F_{7/2}$	77	[483, 529]
		$Lu_3Sc_2Al_3O_{12}:Nd^{3+}$	$^2F_{5/2} \to {}^2F_{7/2}$	77	[483, 529]
1.0336	Yb^{3+}	$CaF_2:Nd^{3+}$	$^2F_{5/2} \to {}^2F_{7/2}$	120	[152]
1.0369	Nd^{3+}	CaF_2-SrF_2	$^4F_{3/2} \to {}^4I_{11/2}$	300	[347]
1.0370	Nd^{3+}	CaF_2	$^4F_{3/2} \to {}^4I_{11/2}$	300	[32]
		SrF_2	$^4F_{3/2} \to {}^4I_{11/2}$	460	[36]
1.04[a]	Pr^{3+}	$SrMoO_4$	$^1G_4 \to {}^3H_4$	—	[69]
1.0400	Nd^{3+}	LaF_3	$^4F_{3/2} \to {}^4I_{11/2}$	77	[37]
1.0404	Nd^{3+}	CeF_3	$^4F_{3/2} \to {}^4I_{11/2}$	77	[265]
1.04065	Nd^{3+}	LaF_3	$^4F_{3/2} \to {}^4I_{11/2}$	300	[37]
1.0410	Nd^{3+}	CeF_3	$^4F_{3/2} \to {}^4I_{11/2}$	300	[265]
1.0437	Nd^{3+}	SrF_2	$^4F_{3/2} \to {}^4I_{11/2}$	77	[338]
1.0445	Nd^{3+}	SrF_2	$^4F_{3/2} \to {}^4I_{11/2}$	300	[338]
1.0446	Nd^{3+}	SrF_2	$^4F_{3/2} \to {}^4I_{11/2}$	500	[36]
1.0448	Nd^{3+}	CaF_2	$^4F_{3/2} \to {}^4I_{11/2}$	50	[310]
1.0451	Nd^{3+}	LaF_3	$^4F_{3/2} \to {}^4I_{11/2}$	77	[37]
1.0456	Nd^{3+}	CaF_2	$^4F_{3/2} \to {}^4I_{11/2}$	50	[310]
1.0457	Nd^{3+}	CaF_2	$^4F_{3/2} \to {}^4I_{11/2}$	77	[310]
1.0461	Nd^{3+}	CaF_2	$^4F_{3/2} \to {}^4I_{11/2}$	300	[310]
		CaF_2-YF_3	$^4F_{3/2} \to {}^4I_{11/2}$	300	[348]
1.0466	Nd^{3+}	CaF_2	$^4F_{3/2} \to {}^4I_{11/2}$	50	[310]
1.0467	Nd^{3+}	CaF_2	$^4F_{3/2} \to {}^4I_{11/2}$	77	[310]
1.0468	Pr^{3+}	$CaWO_4$	$^1D_2 \to {}^3F_4$	20–90	[385, 386]
1.0471	Nd^{3+}	$LiYF_4$	$^4F_{3/2} \to {}^4I_{11/2}$	300	[261]
1.0477	Nd^{3+}	$LiNdP_4O_{12}$	$^4F_{3/2} \to {}^4I_{11/2}$	300	[608, 609, 805]
		$Li(Nd,La)P_4O_{12}$	$^4F_{3/2} \to {}^4I_{11/2}$	300	[618]
		$Li(Nd,Gd)P_4O_{12}$	$^4F_{3/2} \to {}^4I_{11/2}$	300	[618]
1.0480	Nd^{3+}	CaF_2	$^4F_{3/2} \to {}^4I_{11/2}$	50	[310]
1.048[b]	Nd^{3+}	$K_5NdLi_2F_{10}$	$^4F_{3/2} \to {}^4I_{11/2}$	300	[787]
1.0486	Nd^{3+}	LaF_3-SrF_2	$^4F_{3/2} \to {}^4I_{11/2}$	300	[361]
1.0491	Nd^{3+}	$SrAl_{12}O_{19}$	$^4F_{3/2} \to {}^4I_{11/2}$	300	[96]
1.0493	Nd^{3+}	$Sr_2Y_5F_{19}$	$^4F_{3/2} \to {}^4I_{11/2}$	300	[317]
1.0497	Nd^{3+}	$CaAl_{12}O_{19}$	$^4F_{3/2} \to {}^4I_{11/2}$	300	[96]
1.0498	Nd^{3+}	$Ca_2Y_5F_{19}$	$^4F_{3/2} \to {}^4I_{11/2}$	300	[32, 90]
1.05[a]	Nd^{3+}	$Na_3Nd(PO_4)_2$	$^4F_{3/2} \to {}^4I_{11/2}$	300	[577]
1.0506	Nd^{3+}	$5NaF \cdot 9YF_3$	$^4F_{3/2} \to {}^4I_{11/2}$	300	[66]
1.0507	Nd^{3+}	CaF_2	$^4F_{3/2} \to {}^4I_{11/2}$	50	[310]
1.051[a]	Nd^{3+}	$NaNdP_4O_{12}$	$^4F_{3/2} \to {}^4I_{11/2}$	300	[578]
		$YP_5O_{14} (C_{2h}^5)$	$^4F_{3/2} \to {}^4I_{11/2}$	300	[249, 615]
		LaP_5O_{14}	$^4F_{3/2} \to {}^4I_{11/2}$	300	[576]
		CeP_5O_{14}	$^4F_{3/2} \to {}^4I_{11/2}$	300	[615]
		GdP_5O_{14}	$^4F_{3/2} \to {}^4I_{11/2}$	300	[615]
		$(Nd,Sc)P_5O_{14}$	$^4F_{3/2} \to {}^4I_{11/2}$	300	[533]
		$(Nd,In)P_5O_{14}$	$^4F_{3/2} \to {}^4I_{11/2}$	300	[615]
1.0511	Nd^{3+}	$(Nd,La)P_5O_{14}$	$^4F_{3/2} \to {}^4I_{11/2}$	300	[505]
1.0512	Nd^{3+}	NdP_5O_{14}	$^4F_{3/2} \to {}^4I_{11/2}$	300	[84, 617]
1.0513	Nd^{3+}	NdP_5O_{14}	$^4F_{3/2} \to {}^4I_{11/2}$	300	[617]

(*Continued*)

Table 8.1 (*continued*)

Stimulated-emission wavelength [μm]	Ion	Crystal host	Induced transition	Temperature [K]	Reference
1.0515	Nd^{3+}	$YP_5O_{14}(C_{2h}^5)$	$^4F_{3/2} \to {}^4I_{11/2}$	300	[249, 615]
1.052[a]	Nd^{3+}	$KNdP_4O_{12}$	$^4F_{3/2} \to {}^4I_{11/2}$	300	[579]
		$K_5NdLi_2F_{10}$[b]	$^4F_{3/2} \to {}^4I_{11/2}$	300	[787]
1.0521	Nd^{3+}	BaF_2-YF_3	$^4F_{3/2} \to {}^4I_{11/2}$	300	[487]
		$Y_3Al_5O_{12}$	$^4F_{3/2} \to {}^4I_{11/2}$	300	[167]
		NdP_5O_{14}	$^4F_{3/2} \to {}^4I_{11/2}$	300	[617]
1.0523	Nd^{3+}	BaF_2-LuF_3	$^4F_{3/2} \to {}^4I_{11/2}$	300	[271]
		LaF_3	$^4F_{3/2} \to {}^4I_{11/2}$	77	[37]
1.0525	Nd^{3+}	$YP_5O_{14}(C_{2h}^6)$	$^4F_{3/2} \to {}^4I_{11/2}$	300	[249, 615]
1.0526	Nd^{3+}	BaF_2-GdF_3	$^4F_{3/2} \to {}^4I_{11/2}$	300	[487]
1.0528	Nd^{3+}	SrF_2-GdF_3	$^4F_{3/2} \to {}^4I_{11/2}$	300	[91]
1.0529	Nd^{3+}	NdP_5O_{14}	$^4F_{3/2} \to {}^4I_{11/2}$	300	[617]
1.0530	Nd^{3+}	$LiYF_4$	$^4F_{3/2} \to {}^4I_{11/2}$	300	[261]
1.0534	Nd^{3+}	BaF_2-LaF_3	$^4F_{3/2} \to {}^4I_{11/2}$	300	[22, 92]
1.0535	Nd^{3+}	$CaF_2-SrF_2-BaF_2-YF_3-LaF_3$	$^4F_{3/2} \to {}^4I_{11/2}$	300	[64]
		$Lu_3Al_5O_{12}$	$^4F_{3/2} \to {}^4I_{11/2}$	300	[210]
1.0537	Nd^{3+}	BaF_2-CeF_3	$^4F_{3/2} \to {}^4I_{11/2}$	300	[487]
1.0538	Nd^{3+}	BaF_2-LaF_3	$^4F_{3/2} \to {}^4I_{11/2}$	77	[92]
1.0539	Nd^{3+}	α-$NaCaYF_6$	$^4F_{3/2} \to {}^4I_{11/2}$	300	[202]
1.0540	Nd^{3+}	CaF_2-YF_3	$^4F_{3/2} \to {}^4I_{11/2}$	300	[23, 348]
1.05436	Nd^{3+}	$Ba_2MgGe_2O_7$	$^4F_{3/2} \to {}^4I_{11/2}$	300	[432]
1.05437	Nd^{3+}	$Ba_2ZnGe_2O_7$	$^4F_{3/2} \to {}^4I_{11/2}$	300	[280]
1.0551	Nd^{3+}	$Pb_5(PO_4)_3F$	$^4F_{3/2} \to {}^4I_{11/2}$	300	[640]
1.0554[b]	Nd^{3+}	$LiNdP_4O_{12}$	$^4F_{3/2} \to {}^4I_{11/2}$	300	[805]
1.0560	Nd^{3+}	SrF_2-LuF_3	$^4F_{3/2} \to {}^4I_{11/2}$	300	[487, 536]
1.0566	Nd^{3+}	$SrAl_4O_7$	$^4F_{3/2} \to {}^4I_{11/2}$	77	[88, 96]
1.0567	Nd^{3+}	SrF_2-YF_3	$^4F_{3/2} \to {}^4I_{11/2}$	300	[356]
1.0568	Nd^{3+}	$SrAl_4O_7$	$^4F_{3/2} \to {}^4I_{11/2}$	77	[88, 96]
1.0573	Nd^{3+}	$CaMoO_4$	$^4F_{3/2} \to {}^4I_{11/2}$	295	[211]
1.0574	Nd^{3+}	$SrWO_4$	$^4F_{3/2} \to {}^4I_{11/2}$	77	[183]
1.0575	Nd^{3+}	$Y_3Sc_2Ga_3O_{12}$	$^4F_{3/2} \to {}^4I_{11/2}$	77	[536]
1.05755	Nd^{3+}	$Gd_3Sc_2Ga_3O_{12}$	$^4F_{3/2} \to {}^4I_{11/2}$	77	[636]
1.0576	Nd^{3+}	$SrAl_4O_7$	$^4F_{3/2} \to {}^4I_{11/2}$	300	[88, 96]
		$SrMoO_4$	$^4F_{3/2} \to {}^4I_{11/2}$	295	[183]
1.0580	Nd^{3+}	BaF_2-LaF_3	$^4F_{3/2} \to {}^4I_{11/2}$	77	[22, 92]
		$Gd_3Sc_2Ga_3O_{12}$	$^4F_{3/2} \to {}^4I_{11/2}$	77	[636]
1.0582	Nd^{3+}	$CaWO_4$	$^4F_{3/2} \to {}^4I_{11/2}$	300	[78, 184]
1.0583	Nd^{3+}	LaF_3	$^4F_{3/2} \to {}^4I_{11/2}$	77	[37]
		$Y_3Sc_2Ga_3O_{12}$	$^4F_{3/2} \to {}^4I_{11/2}$	300	[536]
		$Y_3Ga_5O_{12}$	$^4F_{3/2} \to {}^4I_{11/2}$	77	[504, 631]
1.0584	Nd^{3+}	$Gd_3Ga_5O_{12}$	$^4F_{3/2} \to {}^4I_{11/2}$	77	[504]
		$CaY_2Mg_2Ge_3O_{12}$[b]	$^4F_{3/2} \to {}^4I_{11/2}$	300	[828]
1.0585	Nd^{3+}	$YAlO_3$	$^4F_{3/2} \to {}^4I_{11/2}$	300	[10]
		$LiLa(MoO_4)_2$	$^4F_{3/2} \to {}^4I_{11/2}$	300	[424]
		$Na_2Nd_2Pb_6(PO_4)_6Cl_2$[b]	$^4F_{3/2} \to {}^4I_{11/2}$	300	[788]
		$Sr_5(PO_4)_3F$	$^4F_{3/2} \to {}^4I_{11/2}$	300	[130]
		$KLa(MoO_4)_2$	$^4F_{3/2} \to {}^4I_{11/2}$	77	[428]
1.0586	Nd^{3+}	$PbMoO_4$	$^4F_{3/2} \to {}^4I_{11/2}$	300	[183, 423]
		$SrLa_4(SiO_4)_3O$	$^4F_{3/2} \to {}^4I_{11/2}$	300	[183, 423]
1.0587	Nd^{3+}	$CaWO_4$	$^4F_{3/2} \to {}^4I_{11/2}$	300	[184]

(*Continued*)

Table 8.1   *(continued)*

Stimulated-emission wavelength [μm]	Ion	Crystal host	Induced transition	Temperature [K]	Reference
		$KLa(MoO_4)_2$	$^4F_{3/2} \rightarrow {}^4I_{11/2}$	300	[428]
		$Y_3Sc_2Al_3O_{12}$	$^4F_{3/2} \rightarrow {}^4I_{11/2}$	77	[529]
		$Lu_3Ga_5O_{12}$	$^4F_{3/2} \rightarrow {}^4I_{11/2}$	77	[506]
1.0588	Nd^{3+}	$Ca(NbO_3)_2$	$^4F_{3/2} \rightarrow {}^4I_{11/2}$	300	[172]
1.0589	Nd^{3+}	$SrF_2-CeF_3-GdF_3$	$^4F_{3/2} \rightarrow {}^4I_{11/2}$	300	[91, 358]
		$Y_3Ga_5O_{12}$	$^4F_{3/2} \rightarrow {}^4I_{11/2}$	300	[504]
1.05895	Nd^{3+}	$CaAl_4O_7$	$^4F_{3/2} \rightarrow {}^4I_{11/2}$	77	[95]
1.05896	Nd^{3+}	$CaMg_2Y_2Ge_3O_{12}$	$^4F_{3/2} \rightarrow {}^4I_{11/2}$	300	[630]
1.0590	Nd^{3+}	SrF_2-CeF_3	$^4F_{3/2} \rightarrow {}^4I_{11/2}$	300	[271]
		$Ca_2MgSi_2O_7$[b]	$^4F_{3/2} \rightarrow {}^4I_{11/2}$	300	[806]
1.059[a]	Nd^{3+}	$SrMoO_4$	$^4F_{3/2} \rightarrow {}^4I_{11/2}$	77	[183]
		$CaMg_2Y_2Ge_3O_{12}$[b]	$^4F_{3/2} \rightarrow {}^4I_{11/2}$	300	[804]
1.0591	Nd^{3+}	$Gd_3Ga_5O_{12}$	$^4F_{3/2} \rightarrow {}^4I_{11/2}$	300	[504]
		$Lu_3Sc_2Al_3O_{12}$	$^4F_{3/2} \rightarrow {}^4I_{11/2}$	77	[488, 529]
1.05915	Nd^{3+}	$Gd_3Sc_2Al_3O_{12}$	$^4F_{3/2} \rightarrow {}^4I_{11/2}$	77	[529]
1.0594	Nd^{3+}	$Lu_3Ga_5O_{12}$	$^4F_{3/2} \rightarrow {}^4I_{11/2}$	300	[506]
1.0595	Nd^{3+}	$5NaF \cdot 9YF_3$	$^4F_{3/2} \rightarrow {}^4I_{11/2}$	300	[66]
		$Y_3Sc_2Al_3O_{12}$	$^4F_{3/2} \rightarrow {}^4I_{11/2}$	300	[529]
		$NaLa(MoO_4)_2$	$^4F_{3/2} \rightarrow {}^4I_{11/2}$	300	[67, 68]
1.0596	Nd^{3+}	$CaAl_4O_7$	$^4F_{3/2} \rightarrow {}^4I_{11/2}$	300	[95]
1.0597	Nd^{3+}	SrF_2-LaF_3	$^4F_{3/2} \rightarrow {}^4I_{11/2}$	300	[22]
		$Ca_3Ga_2Ge_3O_{12}$[b]	$^4F_{3/2} \rightarrow {}^4I_{11/2}$	300	[807]
1.05975	Nd^{3+}	$Y_3Ga_5O_{12}$	$^4F_{3/2} \rightarrow {}^4I_{11/2}$	77	[631]
1.0599	Nd^{3+}	$Gd_3Ga_5O_{12}$	$^4F_{3/2} \rightarrow {}^4I_{11/2}$	77	[504]
		$Lu_3Sc_2Al_3O_{12}$	$^4F_{3/2} \rightarrow {}^4I_{11/2}$	300	[488, 529]
		$LiGd(MoO_4)_2$	$^4F_{3/2} \rightarrow {}^4I_{11/2}$	300	[425]
1.05995	Nd^{3+}	$Gd_3Sc_2Al_3O_{12}$	$^4F_{3/2} \rightarrow {}^4I_{11/2}$	300	[529]
1.06[a]	Nd^{3+}	BaF_2	$^4F_{3/2} \rightarrow {}^4I_{11/2}$	77	[183]
		$NaGd(WO_4)_2$	$^4F_{3/2} \rightarrow {}^4I_{11/2}$	300	[427]
		$Na(Nd,Gd)(WO_4)_2$	$^4F_{3/2} \rightarrow {}^4I_{11/2}$	77	[427]
		$K_3Nd(PO_4)_2$	$^4F_{3/2} \rightarrow {}^4I_{11/2}$	300	[619]
		$K_3(Nd,La)(PO_4)_2$	$^4F_{3/2} \rightarrow {}^4I_{11/2}$	300	[619]
1.0601	Nd^{3+}	$CaWO_4$	$^4F_{3/2} \rightarrow {}^4I_{11/2}$	77	[184]
1.06025	Nd^{3+}	$Y_3Ga_5O_{12}$	$^4F_{3/2} \rightarrow {}^4I_{11/2}$	77	[504]
		$Lu_3Ga_5O_{12}$	$^4F_{3/2} \rightarrow {}^4I_{11/2}$	77	[506]
1.0603	Nd^{3+}	$Y_3Ga_5O_{12}$	$^4F_{3/2} \rightarrow {}^4I_{11/2}$	300	[504]
1.0604	Nd^{3+}	$HfO_2-Y_2O_3$	$^4F_{3/2} \rightarrow {}^4I_{11/2}$	300	[430]
1.06045	Nd^{3+}	$Gd_3Sc_2Ga_3O_{12}$	$^4F_{3/2} \rightarrow {}^4I_{11/2}$	77	[636]
1.0605	Nd^{3+}	$Lu_3Al_5O_{12}$	$^4F_{3/2} \rightarrow {}^4I_{11/2}$	77	[191, 278]
1.0606	Nd^{3+}	$Gd_3Ga_5O_{12}$	$^4F_{3/2} \rightarrow {}^4I_{11/2}$	300	[504]
		$Gd_2(MoO_4)_3$	$^4F_{3/2} \rightarrow {}^4I_{11/2}$	300	[164]
1.0607	Nd^{3+}	$SrWO_4$	$^4F_{3/2} \rightarrow {}^4I_{11/2}$	77	[183]
1.0608	Nd^{3+}	$(Y,Lu)_3Al_5O_{12}$	$^4F_{3/2} \rightarrow {}^4I_{11/2}$	77	[416]
		$ZrO_2-Y_2O_3$	$^4F_{3/2} \rightarrow {}^4I_{11/2}$	300	[90, 430]
1.0609	Nd^{3+}	$Lu_3Ga_5O_{12}$	$^4F_{3/2} \rightarrow {}^4I_{11/2}$	300	[506]
1.061	Nd^{3+}	$CaMoO_4$	$^4F_{3/2} \rightarrow {}^4I_{11/2}$	300	[183]
1.0610	Nd^{3+}	$Y_3Al_5O_{12}$	$^4F_{3/2} \rightarrow {}^4I_{11/2}$	77	[168]
		$Lu_3Al_5O_{12}$	$^4F_{3/2} \rightarrow {}^4I_{11/2}$	300	[210]
1.0611	Nd^{3+}	$SrMoO_4$	$^4F_{3/2} \rightarrow {}^4I_{11/2}$	77	[183, 384]
1.0612	Nd^{3+}	$Ca(NbO_3)_2$	$^4F_{3/2} \rightarrow {}^4I_{11/2}$	77	[172]

(Continued)

Table 8.1 (*continued*)

Stimulated-emission wavelength [μm]	Ion	Crystal host	Induced transition	Temperature [K]	Reference
		$Gd_3Sc_2Ga_3O_{12}$	$^4F_{3/2} \to {}^4I_{11/2}$	300	[636]
		$CaLa_4(SiO_4)_3O$	$^4F_{3/2} \to {}^4I_{11/2}$	300	[130, 142, 570]
1.0613	Nd^{3+}	$Ca_4La(PO_4)_3O$	$^4F_{3/2} \to {}^4I_{11/2}$	300	[130]
		$Ba_2NaNb_5O_{15}$	$^4F_{3/2} \to {}^4I_{11/2}$	300	[585]
1.0614	Nd^{3+}	$Ca(NbO_3)_2$	$^4F_{3/2} \to {}^4I_{11/2}$	77	[172]
		$Y_3Ga_5O_{12}$	$^4F_{3/2} \to {}^4I_{11/2}$	77	[504, 631]
1.0615	Nd^{3+}	$Ca)NbO_3)_2$	$^4F_{3/2} \to {}^4I_{11/2}$	300	[172]
		$Y_3Al_5O_{12}$	$^4F_{3/2} \to {}^4I_{11/2}$	300	[167]
		$Y_3Sc_2Ga_3O_{12}$	$^4F_{3/2} \to {}^4I_{11/2}$	300	[536]
		$Ba_{0.25}Mg_{2.75}Y_2Ge_3O_{12}$	$^4F_{3/2} \to {}^4I_{11/2}$	300	[433]
1.0616	Nd^{3+}	$Lu_3Ga_5O_{12}$	$^4F_{3/2} \to {}^4I_{11/2}$	77	[506]
1.0618	Nd^{3+}	$LaNbO_4$	$^4F_{3/2} \to {}^4I_{11/2}$	300	[277, 418]
1.0620	Nd^{3+}	$Gd_3Sc_2Al_3O_{12}$	$^4F_{3/2} \to {}^4I_{11/2}$	300	[529]
		$Lu_3Sc_2Al_3O_{12}$	$^4F_{3/2} \to {}^4I_{11/2}$	300	[488, 529]
1.062[a]	Nd^{3+}	$Ca_{0.25}Ba_{0.75}(NbO_3)_2$	$^4F_{3/2} \to {}^4I_{11/2}$	295	[156]
1.06205	Nd^{3+}	$Y_3Ga_5O_{12}$	$^4F_{3/2} \to {}^4I_{11/2}$	300	[504]
1.0621	Nd^{3+}	$SrAl_{12}O_{19}$	$^4F_{3/2} \to {}^4I_{11/2}$	300	[489]
		$Gd_3Ga_5O_{12}$	$^4F_{3/2} \to {}^4I_{11/2}$	300	[504]
1.0622	Nd^{3+}	$Y_3Sc_2Al_3O_{12}$	$^4F_{3/2} \to {}^4I_{11/2}$	300	[529]
1.0623	Nd^{3+}	$CaF_2–SrF_2–BaF_2–YF_3–LaF_3$	$^4F_{3/2} \to {}^4I_{11/2}$	300	[64]
		$Lu_3Ga_5O_{12}$	$^4F_{3/2} \to {}^4I_{11/2}$	300	[506]
1.0625	Nd^{3+}	YVO_4	$^4F_{3/2} \to {}^4I_{11/2}$	300	[167]
1.0626	Nd^{3+}	$Ca(NbO_3)_2$	$^4F_{3/2} \to {}^4I_{11/2}$	77	[172]
1.0627	Nd^{3+}	$SrAl_4O_7$	$^4F_{3/2} \to {}^4I_{11/2}$	77	[88, 96]
		$SrMoO_4$	$^4F_{3/2} \to {}^4I_{11/2}$	77	[183, 384]
		$SrWO_4$	$^4F_{3/2} \to {}^4I_{11/2}$	77	[183]
1.0629	Nd^{3+}	α-$NaCaYF_6$	$^4F_{3/2} \to {}^4I_{11/2}$	300	[202]
		$Bi_4Si_3O_{12}$	$^4F_{3/2} \to {}^4I_{11/2}$	77, 300	[566]
1.0630	Nd^{3+}	$Ca_5(PO_4)_3F$	$^4F_{3/2} \to {}^4I_{11/2}$	300	[90, 214]
1.063[a]	Nd^{3+}	$SrWO_4$	$^4F_{3/2} \to {}^4I_{11/2}$	295	[183]
		$Na_5Nd(WO_4)_4$	$^4F_{3/2} \to {}^4I_{11/2}$	300	[577]
		$NdAl_3(BO_3)_4 : Cr^{3+}$[b]	$^4F_{3/2} \to {}^4I_{11/2}$	300	[808]
1.06305	Nd^{3+}	LaF_3	$^4F_{3/2} \to {}^4I_{11/2}$	77	[344]
1.0632	Nd^{3+}	$CaF_2–YF_3$	$^4F_{3/2} \to {}^4I_{11/2}$	300	[23]
		$CaF_2–YF_3–NdF_3$	$^4F_{3/2} \to {}^4I_{11/2}$	300	[546]
1.06335	Nd^{3+}	LaF_3	$^4F_{3/2} \to {}^4I_{11/2}$	300	[344]
1.0634	Nd^{3+}	$CaWO_4$	$^4F_{3/2} \to {}^4I_{11/2}$	77	[184]
1.0635	Nd^{3+}	$LaF_3–SrF_2$	$^4F_{3/2} \to {}^4I_{11/2}$	300	[361]
		$NdAl_3(BO_3)_4$	$^4F_{3/2} \to {}^4I_{11/2}$	300	[579, 589]
		$NaLa(WO_4)_2$	$^4F_{3/2} \to {}^4I_{11/2}$	300	[426]
		$(Nd,Gd)Al_3(BO_3)_4$	$^4F_{3/2} \to {}^4I_{11/2}$	300	[579]
		$Bi_4(Si,Ge)_3O_{12}$	$^4F_{3/2} \to {}^4I_{11/2}$	300	[659]
1.0636	Nd^{3+}	$(Y,Lu)_3Al_5O_{12}$	$^4F_{3/2} \to {}^4I_{11/2}$	77	[416]
1.0637	Nd^{3+}	$Y_3Al_5O_{12}$	$^4F_{3/2} \to {}^4I_{11/2}$	170	[169]
1.06375	Nd^{3+}	$Lu_3Al_5O_{12}$	$^4F_{3/2} \to {}^4I_{11/2}$	120	[191]
10638	Nd^{3+}	CeF_3	$^4F_{3/2} \to {}^4I_{11/2}$	300	[265]
		$CaAl_4O_7$	$^4F_{3/2} \to {}^4I_{11/2}$	300	[95]
		NdP_5O_{14}	$^4F_{3/2} \to {}^4I_{11/2}$	300	[617]
1.0639	Nd^{3+}	CeF_3	$^4F_{3/2} \to {}^4I_{11/2}$	77	[265]
		$Ca_3Ga_2Ge_3O_{12}$[b]	$^4F_{3/2} \to {}^4I_{11/2}$	300	[807]

(*Continued*)

Table 8.1   (*continued*)

Stimulated-emission wavelength [μm]	Ion	Crystal host	Induced transition	Temperature [K]	Reference
1.0640	Nd^{3+}	$SrMoO_4$	$^4F_{3/2} \rightarrow {}^4I_{11/2}$	77	[183]
1.06405	Nd^{3+}	$YAlO_3$	$^4F_{3/2} \rightarrow {}^4I_{11/2}$	77	[176]
1.0641	Nd^{3+}	YVO_4	$^4F_{3/2} \rightarrow {}^4I_{11/2}$	300	[171]
1.06415	Nd^{3+}	$Y_3Al_5O_{12}$	$^4F_{3/2} \rightarrow {}^4I_{11/2}$	300	[168]
1.0642	Nd^{3+}	$(Y,Lu)_3Al_5O_{12}$	$^4F_{3/2} \rightarrow {}^4I_{11/2}$	295	[416]
1.06425	Nd^{3+}	$Lu_3Al_5O_{12}$	$^4F_{3/2} \rightarrow {}^4I_{11/2}$	300	[210]
		$Bi_4Ge_3O_{12}$	$^4F_{3/2} \rightarrow {}^4I_{11/2}$	77	[567]
1.0644	Nd^{3+}	$YAlO_3$	$^4F_{3/2} \rightarrow {}^4I_{11/2}$	300	[10]
		Y_2SiO_5	$^4F_{3/2} \rightarrow {}^4I_{11/2}$	77	[546]
		$Bi_4Ge_3O_{12}$	$^4F_{3/2} \rightarrow {}^4I_{11/2}$	300	[567]
1.0645	Nd^{3+}	CaF_2-LaF_3	$^4F_{3/2} \rightarrow {}^4I_{11/2}$	300	[536]
		$SrMoO_4$	$^4F_{3/2} \rightarrow {}^4I_{11/2}$	300	[183, 243]
1.0647	Nd^{3+}	$CeCl_3$	$^4F_{3/2} \rightarrow {}^4I_{11/2}$	300	[482]
1.0648	Nd^{3+}	CaF_2	$^4F_{3/2} \rightarrow {}^4I_{11/2}$	50	[310]
		YVO_4	$^4F_{3/2} \rightarrow {}^4I_{11/2}$	300	[171]
1.0649	Nd^{3+}	$CaWO_4$	$^4F_{3/2} \rightarrow {}^4I_{11/2}$	77	[184]
		$CaY_2Mg_2Ge_3O_{12}$[b]	$^4F_{3/2} \rightarrow {}^4I_{11/2}$	300	[828]
1.0650	Nd^{3+}	$CaWO_4$	$^4F_{3/2} \rightarrow {}^4I_{11/2}$	77	[184]
1.065[a]	Nd^{3+}	$CdF_2-YF_3-LaF_3$	$^4F_{3/2} \rightarrow {}^4I_{11/2}$	300	[359]
1.0652	Nd^{3+}	$CaWO_4$	$^4F_{3/2} \rightarrow {}^4I_{11/2}$	300	[78, 184]
		$SrMoO_4$	$^4F_{3/2} \rightarrow {}^4I_{11/2}$	77	[183]
1.0653	Nd^{3+}	α-$NaCaCeF_6$	$^4F_{3/2} \rightarrow {}^4I_{11/2}$	300	[65]
		$NaLa(MoO_4)_2$	$^4F_{3/2} \rightarrow {}^4I_{11/2}$	300	[67, 68]
1.0654	Nd^{3+}	CaF_2-GdF_3	$^4F_{3/2} \rightarrow {}^4I_{11/2}$	300	[536]
1.0656	Nd^{3+}	CdF_2-YF_3	$^4F_{3/2} \rightarrow {}^4I_{11/2}$	300	[359]
1.0657	Nd^{3+}	CaF_2-CeF_3	$^4F_{3/2} \rightarrow {}^4I_{11/2}$	300	[354]
1.0658	Nd^{3+}	$LiLa(MoO_4)_2$	$^4F_{3/2} \rightarrow {}^4I_{11/2}$	300	[424]
1.06585	Nd^{3+}	$CaAl_4O_7$	$^4F_{3/2} \rightarrow {}^4I_{11/2}$	77	[95]
1.0660	Nd^{3+}	$K_5Nd(MoO_4)_4$	$^4F_{3/2} \rightarrow {}^4I_{11/2}$	300	[569]
		$K_5Bi(MoO_4)_4$	$^4F_{3/2} \rightarrow {}^4I_{11/2}$	300	[569]
1.0661	Nd^{3+}	CaF_2	$^4F_{3/2} \rightarrow {}^4I_{11/2}$	120	[152]
1.0664	Nd^{3+}	YVO_4	$^4F_{3/2} \rightarrow {}^4I_{11/2}$	300	[171]
1.0665[a]	Nd^{3+}	CdF_2-LaF_3	$^4F_{3/2} \rightarrow {}^4I_{11/2}$	300	[359]
1.0666	Nd^{3+}	$HfO_2-Y_2O_3$	$^4F_{3/2} \rightarrow {}^4I_{11/2}$	300	[90]
1.0669	Nd^{3+}	$KY(MoO_4)_2$	$^4F_{3/2} \rightarrow {}^4I_{11/2}$	300	[377]
1.067[a]	Nd^{3+}	$Ca_3(VO_4)_2$	$^4F_{3/2} \rightarrow {}^4I_{11/2}$	300	[382]
1.0671	Nd^{3+}	$LuAlO_3$	$^4F_{3/2} \rightarrow {}^4I_{11/2}$	77	[565]
1.0672	Nd^{3+}	$CaY_4(SiO_4)_3O$	$^4F_{3/2} \rightarrow {}^4I_{11/2}$	300	[130]
		$KGd(WO_4)_2$	$^4F_{3/2} \rightarrow {}^4I_{11/2}$	300	[595]
1.0673	Nd^{3+}	$CaMoO_4$	$^4F_{3/2} \rightarrow {}^4I_{11/2}$	77, 300	[183, 211]
		$ZrO_2-Y_2O_3$	$^4F_{3/2} \rightarrow {}^4I_{11/2}$	300	[90]
1.0675	Nd^{3+}	$LuAlO_3$	$^4F_{3/2} \rightarrow {}^4I_{11/2}$	300	[565]
		$Na_2Nd_2Pb_6(PO_4)_6Cl_2$[b]	$^4F_{3/2} \rightarrow {}^4I_{11/2}$	300	[788]
1.0682	Nd^{3+}	$Y_3Al_5O_{12}$	$^4F_{3/2} \rightarrow {}^4I_{11/2}$	300	[601]
1.0687	Nd^{3+}	$KY(WO_4)_2$	$^4F_{3/2} \rightarrow {}^4I_{11/2}$	77	[77]
1.0688	Nd^{3+}	$KY(WO_4)_2$	$^4F_{3/2} \rightarrow {}^4I_{11/2}$	300	[379]
1.0689[a]	Nd^{3+}	$GdAlO_3$	$^4F_{3/2} \rightarrow {}^4I_{11/2}$	77	[471]
1.0690[a]	Nd^{3+}	$GdAlO_3$	$^4F_{3/2} \rightarrow {}^4I_{11/2}$	300	[471]
1.0698	Nd^{3+}	$La_2Be_2O_5$	$^4F_{3/2} \rightarrow {}^4I_{11/2}$	300	[596, 633]
1.0701	Nd^{3+}	$Gd_2(MoO_4)_3$	$^4F_{3/2} \rightarrow {}^4I_{11/2}$	300	[164]
1.0706[b]	Nd^{3+}	$KY(WO_4)_2$	$^4F_{3/2} \rightarrow {}^4I_{11/2}$	300	[809]

(*Continued*)

Table 8.1   (*continued*)

Stimulated-emission wavelength [μm]	Ion	Crystal host	Induced transition	Temperature [K]	Reference
1.0710	Nd^{3+}	Y_2SiO_5	$^4F_{3/2} \to {}^4I_{11/2}$	77	[82]
1.0715	Nd^{3+}	Y_2SiO_5	$^4F_{3/2} \to {}^4I_{11/2}$	300	[82]
1.0720	Nd^{3+}	$CaSc_2O_4$	$^4F_{3/2} \to {}^4I_{11/2}$	300	[488]
1.0721[b]	Nd^{3+}	$KLu(WO_4)_2$	$^4F_{3/2} \to {}^4I_{11/2}$	300	[790]
1.07255	Nd^{3+}	$YAlO_3$	$^4F_{3/2} \to {}^4I_{11/2}$	77	—
1.0726	Nd^{3+}	$YAlO_3$	$^4F_{3/2} \to {}^4I_{11/2}$	300	[10]
		$(Y,Lu)_3Al_5O_{12}$	$^4F_{3/2} \to {}^4I_{11/2}$	77	[416]
1.073[a]	Nd^{3+}	Y_2O_3	$^4F_{3/2} \to {}^4I_{11/2}$	77, 300	[394, 653]
1.0730	Nd^{3+}	$CaSc_2O_4$	$^4F_{3/2} \to {}^4I_{11/2}$	77	[488]
1.0737	Nd^{3+}	$Y_3Al_5O_{12}$	$^4F_{3/2} \to {}^4I_{11/2}$	300	[100, 271]
1.0740	Nd^{3+}	Y_2SiO_5	$^4F_{3/2} \to {}^4I_{11/2}$	77	[546]
1.0741	Nd^{3+}	Gd_2O_3	$^4F_{3/2} \to {}^4I_{11/2}$	300	[419]
1.0742	Nd^{3+}	Y_2SiO_5	$^4F_{3/2} \to {}^4I_{11/2}$	300	[82]
1.075[a]	Nd^{3+}	La_2O_2S	$^4F_{3/2} \to {}^4I_{11/2}$	300	[219]
1.0755	Nd^{3+}	$CaSc_2O_4$	$^4F_{3/2} \to {}^4I_{11/2}$	77	[488]
1.0759[a]	Nd^{3+}	$GdAlO_3$	$^4F_{3/2} \to {}^4I_{11/2}$	77	[471]
		$LuAlO_3$	$^4F_{3/2} \to {}^4I_{11/2}$	300	[565]
1.0760	Nd^{3+}	$GdAlO_3$	$^4F_{3/2} \to {}^4I_{11/2}$	300	[137]
1.07655	Nd^{3+}	$CaAl_4O_7$	$^4F_{3/2} \to {}^4I_{11/2}$	77	[95]
1.0770	Nd^{3+}	$YScO_3$	$^4F_{3/2} \to {}^4I_{11/2}$	77	[639]
1.0772	Nd^{3+}	$CaAl_4O_7$	$^4F_{3/2} \to {}^4I_{11/2}$	77	[95]
1.0776	Nd^{3+}	Gd_2O_3	$^4F_{3/2} \to {}^4I_{11/2}$	77	[419]
1.0779	Nd^{3+}	$Y_3Al_5O_{12}$	$^4F_{3/2} \to {}^4I_{11/2}$	300	[601]
1.078[a]	Nd^{3+}	Y_2O_3	$^4F_{3/2} \to {}^4I_{11/2}$	77	[394]
1.0781	Nd^{3+}	Y_2SiO_5	$^4F_{3/2} \to {}^4I_{11/2}$	77	[82]
1.0782	Nd^{3+}	Y_2SiO_5	$^4F_{3/2} \to {}^4I_{11/2}$	300	[82]
1.0785	Nd^{3+}	$LuScO_3$	$^4F_{3/2} \to {}^4I_{11/2}$	300	[639]
1.0786	Nd^{3+}	$CaAl_4O_7$	$^4F_{3/2} \to {}^4I_{11/2}$	300	[95]
1.0789	Nd^{3+}	Gd_2O_3	$^4F_{3/2} \to {}^4I_{11/2}$	77, 300	[419]
1.079[a]	Nd^{3+}	La_2O_3	$^4F_{3/2} \to {}^4I_{11/2}$	77	[417]
		$La_2Be_2O_5$	$^4F_{3/2} \to {}^4I_{11/2}$	300	[596]
1.07925[b]	Nd^{3+}	Lu_2SiO_5	$^4F_{3/2} \to {}^4I_{11/2}$	300	[810]
1.0796	Nd^{3+}	$YAlO_3$	$^4F_{3/2} \to {}^4I_{11/2}$	300	[10]
1.0804	Nd^{3+}	$LaAlO_3$	$^4F_{3/2} \to {}^4I_{11/2}$	300	[137]
1.08145[b]	Nd^{3+}	Sc_2SiO_5	$^4F_{3/2} \to {}^4I_{11/2}$	300	[810]
1.0828	Nd^{3+}	$SrAl_4O_7$	$^4F_{3/2} \to {}^4I_{11/2}$	300	[88, 96]
1.0831	Nd^{3+}	$LuAlO_3$	$^4F_{3/2} \to {}^4I_{11/2}$	120	[565]
1.0832	Nd^{3+}	$LuAlO_3$	$^4F_{3/2} \to {}^4I_{11/2}$	300	[565]
1.0840	Nd^{3+}	$LiNbO_3$	$^4F_{3/2} \to {}^4I_{11/2}$	77	[156]
		$GdScO_3$	$^4F_{3/2} \to {}^4I_{11/2}$	200	[472]
1.0842	Nd^{3+}	$YAlO_3$	$^4F_{3/2} \to {}^4I_{11/2}$	300	[399]
1.0843	Nd^{3+}	$YScO_3$	$^4F_{3/2} \to {}^4I_{11/2}$	300	[639]
1.0846	Nd^{3+}	$LiNbO_3$	$^4F_{3/2} \to {}^4I_{11/2}$	300	[34, 162]
		$LiNbO_3:Mg^{2+}$	$^4F_{3/2} \to {}^4I_{11/2}$	300	[530]
1.0847	Nd^{3+}	$YAlO_3$	$^4F_{3/2} \to {}^4I_{11/2}$	530	[176]
1.08515	Nd^{3+}	$GdScO_3$	$^4F_{3/2} \to {}^4I_{11/2}$	300	[639]
1.0867	Nd^{3+}	$CaSc_2O_4$	$^4F_{3/2} \to {}^4I_{11/2}$	77	[488]
1.0868	Nd^{3+}	$CaSc_2O_4$	$^4F_{3/2} \to {}^4I_{11/2}$	300	[488]
1.0885	Nd^{3+}	CaF_2*	$^4F_{3/2} \to {}^4I_{11/2}$	300	[318]
		$CaF_2–CeO_2$	$^4F_{3/2} \to {}^4I_{11/2}$	300	[349]

(*Continued*)

Table 8.1   (*continued*)

Stimulated-emission wavelength [μm]	Ion	Crystal host	Induced transition	Temperature [K]	Reference
1.0913	Nd^{3+}	$YAlO_3$	$^4F_{3/2} \to {}^4I_{11/2}$	530	[176]
1.0933	Nd^{3+}	$LiNbO_3$	$^4F_{3/2} \to {}^4I_{11/2}$	300	[163]
		$LiNbO_3:Mg^{2+}$	$^4F_{3/2} \to {}^4I_{11/2}$	300	[530]
1.0990	Nd^{3+}	$YAlO_3$	$^4F_{3/2} \to {}^4I_{11/2}$	300	[399]
1.0991	Nd^{3+}	$YAlO_3$	$^4F_{3/2} \to {}^4I_{11/2}$	500	[176]
1.1119	Nd^{3+}	$Y_3Al_5O_{12}$	$^4F_{3/2} \to {}^4I_{11/2}$	300	[100, 271]
1.1158	Nd^{3+}	$Y_3Al_5O_{12}$	$^4F_{3/2} \to {}^4I_{11/2}$	300	[100, 271]
1.1160	Tm^{2+}	CaF_2	$^2F_{5/2} \to {}^2F_{7/2}$	4.2	[221, 332]
1.1217	V^{2+}	MgF_2	$^4T_2 \to {}^4A_2$	77	[113, 114]
1.1225	Nd^{3+}	$Y_3Al_5O_{12}$	$^4F_{3/2} \to {}^4I_{11/2}$	300	[100, 271]
1.190[b]	Ho^{3+}	$BaYb_2F_8$	$^5I_6 \to {}^5I_8$	300	[791]
1.2085[b]	Ho^{3+}	$Gd_3Ga_5O_{12}$	$^5I_6 \to {}^5I_8$	≈110	[774]
1.2155[b]	Ho^{3+}	$Y_3Al_5O_{12}$	$^5I_6 \to {}^5I_8$	≈110	[775]
		$Y_3Al_5O_{12}:Yb^{3+}$	$^5I_6 \to {}^5I_8$	≈110	[775]
1.2160[b]	Ho^{3+}	$Lu_3Al_5O_{12}$	$^5I_6 \to {}^5I_8$	≈110	[775]
		$Lu_3Al_5O_{12}:Yb^{3+}$	$^5I_6 \to {}^5I_8$	≈110	[775]
1.2308[b]	Er^{3+}	$LiYF_4$	$^4S_{3/2} \to {}^4I_{11/2}$	≈110, 300	[776]
1.26[a]	Er^{3+}	CaF_2^*	$^4S_{3/2} \to {}^4I_{11/2}$	77	[108]
1.3065	Nd^{3+}	$SrAl_{12}O_{19}$	$^4F_{3/2} \to {}^4I_{13/2}$	300	[489]
1.3070	Nd^{3+}	$5NaF \cdot 9YF_3$	$^4F_{3/2} \to {}^4I_{13/2}$	300	[88]
1.3125	Nd^{3+}	LaF_3	$^4F_{3/2} \to {}^4I_{13/2}$	77	[88]
1.3130	Nd^{3+}	CeF_3	$^4F_{3/2} \to {}^4I_{13/2}$	77	[89]
1.3144	Ni^{2+}	MgO	$^3T_2 \to {}^3A_2$	77	[113]
1.3160	Nd^{3+}	SrF_2-LaF_3	$^4F_{3/2} \to {}^4I_{13/2}$	77	[88]
1.3165	Nd^{3+}	CdF_2-YF_3	$^4F_{3/2} \to {}^4I_{13/2}$	77	[89]
		α-$NaCaCeF_6$	$^4F_{3/2} \to {}^4I_{13/2}$	77	[89]
1.3170	Nd^{3+}	CeF_3	$^4F_{3/2} \to {}^4I_{13/2}$	300	[89]
		LaF_3-SrF_2	$^4F_{3/2} \to {}^4I_{13/2}$	77	[93]
1.317[a]	Nd^{3+}	$LiNdP_4O_{12}$	$^4F_{3/2} \to {}^4I_{13/2}$	300	[752]
1.3185	Nd^{3+}	CaF_2-GdF_3	$^4F_{3/2} \to {}^4I_{13/2}$	300	[536]
		BaF_2-LaF_3	$^4F_{3/2} \to {}^4I_{13/2}$	300	[89]
1.3187	Nd^{3+}	$Y_3Al_5O_{12}$	$^4F_{3/2} \to {}^4I_{13/2}$	300	[100, 591]
1.3190	Nd^{3+}	$Ca_2Y_5F_{19}$	$^4F_{3/2} \to {}^4I_{13/2}$	300	[90]
		CaF_2-LaF_3	$^4F_{3/2} \to {}^4I_{13/2}$	300	[536]
		CaF_2-CeF_3	$^4F_{3/2} \to {}^4I_{13/2}$	300	[88]
		$Sr_2Y_5F_{19}$	$^4F_{3/2} \to {}^4I_{13/2}$	300	[317]
		α-$NaCaCeF_6$	$^4F_{3/2} \to {}^4I_{13/2}$	300	[89]
1.3200	Nd^{3+}	$Ca_2Y_5F_{19}$	$^4F_{3/2} \to {}^4I_{13/2}$	77	[90]
		SrF_2-LuF_3	$^4F_{3/2} \to {}^4I_{13/2}$	300	[536]
		BaF_2-YF_3	$^4F_{3/2} \to {}^4I_{13/2}$	300	[487]
		$YP_5O_{14}(C_{2h}^6)^b$	$^4F_{3/2} \to {}^4I_{13/2}$	300	[811]
1.32[a]	Nd^{3+}	$NaNdP_4O_{12}$	$^4F_{3/2} \to {}^4I_{13/2}$	300	[752]
		$KNdP_4O_{12}$	$^4F_{3/2} \to {}^4I_{13/2}$	300	[752]
1.3209	Nd^{3+}	$Lu_3Al_5O_{12}$	$^4F_{3/2} \to {}^4I_{13/2}$	300	[89]
1.323[b]	Nd^{3+}	NdP_5O_{14}	$^4F_{3/2} \to {}^4I_{13/2}$	300	[812]
1.3225	Nd^{3+}	SrF_2-YF_3	$^4F_{3/2} \to {}^4I_{13/2}$	77	[88]
1.3235	Nd^{3+}	LaF_3	$^4F_{3/2} \to {}^4I_{13/2}$	77	[88]
		SrF_2-LaF_3	$^4F_{3/2} \to {}^4I_{13/2}$	77	[88]
1.3240	Nd^{3+}	CeF_3	$^4F_{3/2} \to {}^4I_{13/2}$	77	[89]
1.3245	Nd^{3+}	CdF_2-YF_3	$^4F_{3/2} \to {}^4I_{13/2}$	300	[89]

(*Continued*)

Table 8.1 (*continued*)

Stimulated-emission wavelength [μm]	Ion	Crystal host	Induced transition	Temperature [K]	Reference
1.3250	Nd^{3+}	SrF_2-LaF_3	$^4F_{3/2} \rightarrow {}^4I_{13/2}$	300	[88]
		SrF_2-GdF_3	$^4F_{3/2} \rightarrow {}^4I_{13/2}$	77	[91]
1.3255	Nd^{3+}	CaF_2-YF_3	$^4F_{3/2} \rightarrow {}^4I_{13/2}$	77	[90]
		SrF_2-CeF_3	$^4F_{3/2} \rightarrow {}^4I_{13/2}$	300	[271]
1.3260	Nd^{3+}	SrF_2-GdF_3	$^4F_{3/2} \rightarrow {}^4I_{13/2}$	300	[91]
		α-$NaCaYF_6$	$^4F_{3/2} \rightarrow {}^4I_{13/2}$	77	[89]
1.3270	Nd^{3+}	CaF_2-YF_3	$^4F_{3/2} \rightarrow {}^4I_{13/2}$	300	[90]
1.3275	Nd^{3+}	LaF_3-SrF_2	$^4F_{3/2} \rightarrow {}^4I_{13/2}$	77	[89]
1.3280	Nd^{3+}	BaF_2-LaF_3	$^4F_{3/2} \rightarrow {}^4I_{13/2}$	300	[89]
1.3285	Nd^{3+}	α-$NaCaYF_6$	$^4F_{3/2} \rightarrow {}^4I_{13/2}$	300	[89]
1.3290	Nd^{3+}	BaF_2-LaF_3	$^4F_{3/2} \rightarrow {}^4I_{13/2}$	77	[89]
1.3300	Nd^{3+}	SrF_2-YF_3	$^4F_{3/2} \rightarrow {}^4I_{13/2}$	77	[88]
		$SrMoO_4$	$^4F_{3/2} \rightarrow {}^4I_{13/2}$	77	[97]
1.3305	Nd^{3+}	LaF_3	$^4F_{3/2} \rightarrow {}^4I_{13/2}$	77	[88]
		$Y_3Ga_5O_{12}$	$^4F_{3/2} \rightarrow {}^4I_{13/2}$	300	[504, 631]
		$HfO_2-Y_2O_3$	$^4F_{3/2} \rightarrow {}^4I_{13/2}$	300	[90]
1.3307	Nd^{3+}	$Gd_3Ga_5O_{12}$	$^4F_{3/2} \rightarrow {}^4I_{13/2}$	77	[594]
1.3310	Nd^{3+}	LaF_3	$^4F_{3/2} \rightarrow {}^4I_{13/2}$	300	[88]
		CeF_3	$^4F_{3/2} \rightarrow {}^4I_{13/2}$	77	[89]
		$CaWO_4$	$^4F_{3/2} \rightarrow {}^4I_{13/2}$	77	[93]
		$Y_3Sc_2Ga_3O_{12}$	$^4F_{3/2} \rightarrow {}^4I_{13/2}$	300	[536]
1.3315	Nd^{3+}	LaF_3-SrF_2	$^4F_{3/2} \rightarrow {}^4I_{13/2}$	300	[89]
		$Lu_3Ga_5O_{12}$	$^4F_{3/2} \rightarrow {}^4I_{13/2}$	300	[506]
		$Gd_3Ga_5O_{12}$	$^4F_{3/2} \rightarrow {}^4I_{13/2}$	300	[594]
1.3317[b]	Nd^{3+}	$Ca_3Ga_2Ge_3O_{12}$	$^4F_{3/2} \rightarrow {}^4I_{13/2}$	300	[807]
1.3319	Nd^{3+}	$Lu_3Al_5O_{12}$	$^4F_{3/2} \rightarrow {}^4I_{13/2}$	77	[89]
1.3320	Nd^{3+}	CeF_3	$^4F_{3/2} \rightarrow {}^4I_{13/2}$	300	[89]
		SrF_2-YF_3	$^4F_{3/2} \rightarrow {}^4I_{13/2}$	77	[88]
		$SrAl_4O_7$	$^4F_{3/2} \rightarrow {}^4I_{13/2}$	77	[88, 96]
		$PbMoO_4$	$^4F_{3/2} \rightarrow {}^4I_{13/2}$	77	[97]
		$ZrO_2-Y_2O_3$	$^4F_{3/2} \rightarrow {}^4I_{13/2}$	300	[90]
1.3325	Nd^{3+}	LaF_3-SrF_2	$^4F_{3/2} \rightarrow {}^4I_{13/2}$	77	[89]
		$SrMoO_4$	$^4F_{3/2} \rightarrow {}^4I_{13/2}$	300	[97]
1.3326	Nd^{3+}	$Lu_3Al_5O_{12}$	$^4F_{3/2} \rightarrow {}^4I_{13/2}$	300	[89]
1.3333	Nd^{3+}	$Lu_3Al_5O_{12}$	$^4F_{3/2} \rightarrow {}^4I_{13/2}$	77	[89]
1.3335	Nd^{3+}	$Y_3Al_5O_{12}$	$^4F_{3/2} \rightarrow {}^4I_{13/2}$	300	[100, 271]
1.3340	Nd^{3+}	$CaWO_4$	$^4F_{3/2} \rightarrow {}^4I_{13/2}$	300	[93]
		$PbMoO_4$	$^4F_{3/2} \rightarrow {}^4I_{13/2}$	300	[97]
1.3342	Nd^{3+}	$Lu_3Al_5O_{12}$	$^4F_{3/2} \rightarrow {}^4I_{13/2}$	300	[89]
1.3345	Nd^{3+}	$CaWO_4$	$^4F_{3/2} \rightarrow {}^4I_{13/2}$	77	[93]
		$SrAl_4O_7$	$^4F_{3/2} \rightarrow {}^4I_{13/2}$	300	[88, 96]
		$Ca_5(PO_4)_3F$	$^4F_{3/2} \rightarrow {}^4I_{13/2}$	77	[90]
1.3347	Nd^{3+}	$Ca_5(PO_4)_3F$	$^4F_{3/2} \rightarrow {}^4I_{13/2}$	300	[90]
1.3350	Nd^{3+}	$KLa(MoO_4)_2$	$^4F_{3/2} \rightarrow {}^4I_{13/2}$	77, 300	[428]
1.3351	Nd^{3+}	$Y_3Al_5O_{12}$	$^4F_{3/2} \rightarrow {}^4I_{13/2}$	300	[100, 271]
1.3354	Nd^{3+}	$CaLa_4(SiO_4)_3O$	$^4F_{3/2} \rightarrow {}^4I_{13/2}$	300	[627]
1.3355	Nd^{3+}	SrF_2-LaF_2	$^4F_{3/2} \rightarrow {}^4I_{13/2}$	77	[88]
		$NaLa(WO_4)_2$	$^4F_{3/2} \rightarrow {}^4I_{13/2}$	300	[97]
1.3360	Nd^{3+}	$Y_3Sc_2Al_3O_{12}$	$^4F_{3/2} \rightarrow {}^4I_{13/2}$	300	[529, 536]
		$Gd_3Sc_2Al_3O_{12}$	$^4F_{3/2} \rightarrow {}^4I_{13/2}$	300	[488, 529]

(*Continued*)

Table 8.1   (*continued*)

Stimulated-emission wavelength [μm]	Ion	Crystal host	Induced transition	Temperature [K]	Reference
		$Lu_3Sc_2Al_3O_{12}$	${}^4F_{3/2} \rightarrow {}^4I_{13/2}$	300	[488, 529]
1.3370	Nd^{3+}	CaF_2-YF_3	${}^4F_{3/2} \rightarrow {}^4I_{13/2}$	300	[90]
		$Ca(NbO_3)_2$	${}^4F_{3/2} \rightarrow {}^4I_{13/2}$	77	[93]
		$CaWO_4$	${}^4F_{3/2} \rightarrow {}^4I_{13/2}$	300	[93]
		$LiLa(MoO_4)_2$	${}^4F_{3/2} \rightarrow {}^4I_{13/2}$	300	[97, 424]
1.3372	Nd^{3+}	$CaWO_4$	${}^4F_{3/2} \rightarrow {}^4I_{13/2}$	77	[93]
1.3375	Nd^{3+}	α-$NaCaYF_6$	${}^4F_{3/2} \rightarrow {}^4I_{13/2}$	300	[89]
		$PbMoO_4$	${}^4F_{3/2} \rightarrow {}^4I_{13/2}$	77	[97]
		$LiLa(MoO_4)_2$	${}^4F_{3/2} \rightarrow {}^4I_{13/2}$	77	[97, 424]
1.3376	Nd^{3+}	$Lu_3Al_5O_{12}$	${}^4F_{3/2} \rightarrow {}^4I_{13/2}$	77	[89]
1.3380	Nd^{3+}	CaF_2-YF_3	${}^4F_{3/2} \rightarrow {}^4I_{13/2}$	77	[90]
		$Ca(NbO_3)_2$	${}^4F_{3/2} \rightarrow {}^4I_{13/2}$	300	[93]
		$NaLa(MoO_4)_2$	${}^4F_{3/2} \rightarrow {}^4I_{13/2}$	77, 300	[97]
1.3381	Nd^{3+}	$Y_3Al_5O_{12}$	${}^4F_{3/2} \rightarrow {}^4I_{13/2}$	300	[100, 271]
1.3382	Nd^{3+}	$Lu_3Al_5O_{12}$	${}^4F_{3/2} \rightarrow {}^4I_{13/2}$	300	[89]
1.3390	Nd^{3+}	α-$NaCaYF_6$	${}^4F_{3/2} \rightarrow {}^4I_{13/2}$	77	[89]
		$CaWO_4$	${}^4F_{3/2} \rightarrow {}^4I_{13/2}$	300	[93]
1.3391	Nd^{3+}	$YAlO_3$	${}^4F_{3/2} \rightarrow {}^4I_{13/2}$	77	[93]
1.3393	Nd^{3+}	$YAlO_3$	${}^4F_{3/2} \rightarrow {}^4I_{13/2}$	300	[98, 591]
1.3400	Nd^{3+}	$CaAl_4O_7$	${}^4F_{3/2} \rightarrow {}^4I_{13/2}$	77	[88, 95]
		$LiGd(MoO_4)_2$	${}^4F_{3/2} \rightarrow {}^4I_{13/2}$	77, 300	[97]
1.3407	Nd^{3+}	$Bi_4Si_3O_{12}$	${}^4F_{3/2} \rightarrow {}^4I_{13/2}$	300	[566]
1.3410	Nd^{3+}	$Lu_3Al_5O_{12}$	${}^4F_{3/2} \rightarrow {}^4I_{13/2}$	300	[89]
1.341[b]	Nd^{3+}	$NdAl_3(BO_3)_4$	${}^4F_{3/2} \rightarrow {}^4I_{13/2}$	300	[812]
1.3413	Nd^{3+}	$YAlO_3$	${}^4F_{3/2} \rightarrow {}^4I_{13/2}$	300	[93, 98, 591]
1.3415	Nd^{3+}	$Ca(NbO_3)_2$	${}^4F_{3/2} \rightarrow {}^4I_{13/2}$	77	[93]
		YVO_4	${}^4F_{3/2} \rightarrow {}^4I_{13/2}$	77	[88]
1.3418	Nd^{3+}	$Bi_4Ge_3O_{12}$	${}^4F_{3/2} \rightarrow {}^4I_{13/2}$	300	[567]
1.3420	Nd^{3+}	$CaAl_4O_7$	${}^4F_{3/2} \rightarrow {}^4I_{13/2}$	300	[88, 95]
1.3425	Nd^{3+}	$Ca(NbO_3)_2$	${}^4F_{3/2} \rightarrow {}^4I_{13/2}$	300	[93]
		YVO_4	${}^4F_{3/2} \rightarrow {}^4I_{13/2}$	300	[88]
		$PbMoO_4$	${}^4F_{3/2} \rightarrow {}^4I_{13/2}$	300	[97]
1.3430	Nd^{3+}	$NaLa(MoO_4)_2$	${}^4F_{3/2} \rightarrow {}^4I_{13/2}$	77	[97]
1.3437	Nd^{3+}	$LuAlO_3$	${}^4F_{3/2} \rightarrow {}^4I_{13/2}$	300	[565]
1.3440	Nd^{3+}	$SrMoO_4$	${}^4F_{3/2} \rightarrow {}^4I_{13/2}$	77	[97]
		$LiLa(MoO_4)_2$	${}^4F_{3/2} \rightarrow {}^4I_{13/2}$	77	[97, 424]
		$NaLa(MoO_4)_2$	${}^4F_{3/2} \rightarrow {}^4I_{13/2}$	300	[97]
1.3450	Nd^{3+}	$PbMoO_4$	${}^4F_{3/2} \rightarrow {}^4I_{13/2}$	77	[97]
1.3455	Nd^{3+}	$LiGd(MoO_4)_2$	${}^4F_{3/2} \rightarrow {}^4I_{13/2}$	77	[97]
1.3459	Nd^{3+}	$CaWO_4$	${}^4F_{3/2} \rightarrow {}^4I_{13/2}$	77	[93]
1.3475	Nd^{3+}	$CaWO_4$	${}^4F_{3/2} \rightarrow {}^4I_{13/2}$	300	[93]
1.3482[b]	Nd^{3+}	$KLu(WO_4)_2$	${}^4F_{3/2} \rightarrow {}^4I_{13/2}$	300	[790]
1.3485	Nd^{3+}	$KY(MoO_4)_2$	${}^4F_{3/2} \rightarrow {}^4I_{13/2}$	300	[93]
1.3499	Nd^{3+}	$Lu_3Al_5O_{12}$	${}^4F_{3/2} \rightarrow {}^4I_{13/2}$	77	[89]
1.3510	Nd^{3+}	$KGd(WO_4)_2$	${}^4F_{3/2} \rightarrow {}^4I_{13/2}$	300	[595]
		$La_2Be_2O_5$	${}^4F_{3/2} \rightarrow {}^4I_{13/2}$	300	[633]
1.3512	Nd^{3+}	$YAlO_3$	${}^4F_{3/2} \rightarrow {}^4I_{13/2}$	300	[93]
1.3514	Nd^{3+}	$YAlO_3$	${}^4F_{3/2} \rightarrow {}^4I_{13/2}$	300	[93]
1.3515	Nd^{3+}	$KY(WO_4)_2$	${}^4F_{3/2} \rightarrow {}^4I_{13/2}$	77	[93]

(*Continued*)

Table 8.1 (*continued*)

Stimulated-emission wavelength [μm]	Ion	Crystal host	Induced transition	Temperature [K]	Reference
1.3525	Nd^{3+}	$Ca_2Y_5F_{19}$	$^4F_{3/2} \to {}^4I_{13/2}$	300	[90]
		$KY(WO_4)_2$	$^4F_{3/2} \to {}^4I_{13/2}$	300	[93]
		$Lu_3Al_5O_{12}$	$^4F_{3/2} \to {}^4I_{13/2}$	77	[89]
1.3530	Nd^{3+}	$SrAl_4O_7$	$^4F_{3/2} \to {}^4I_{13/2}$	77	[88, 96]
1.3532	Nd^{3+}	$Lu_3Al_5O_{12}$	$^4F_{3/2} \to {}^4I_{13/2}$	300	[89]
1.3533	Nd^{3+}	$Y_3Al_5O_{12}$	$^4F_{3/2} \to {}^4I_{13/2}$	300	[100, 271]
1.3545	Nd^{3+}	$KY(WO_4)_2$	$^4F_{3/2} \to {}^4I_{13/2}$	77, 300	[93]
1.3550	Nd^{3+}	$LiNbO_3$	$^4F_{3/2} \to {}^4I_{13/2}$	300	[89]
1.3560	Nd^{3+}	$Y_3Al_5O_{12}$	$^4F_{3/2} \to {}^4I_{13/2}$	77	[271]
1.3565	Nd^{3+}	$CaSc_2O_4$	$^4F_{3/2} \to {}^4I_{13/2}$	300	[488]
1.3572	Nd^{3+}	$Y_3Al_5O_{12}$	$^4F_{3/2} \to {}^4I_{13/2}$	300	[100, 271]
1.3575[b]	Nd^{3+}	Lu_2SiO_5	$^4F_{3/2} \to {}^4I_{13/2}$	300	[810]
1.3580	Nd^{3+}	Y_2SiO_5	$^4F_{3/2} \to {}^4I_{13/2}$	77	[82]
1.3585	Nd^{3+}	CaF_2-YF_3	$^4F_{3/2} \to {}^4I_{13/2}$	300	[90]
		Y_2SiO_5	$^4F_{3/2} \to {}^4I_{13/2}$	300	[82]
1.3595	Nd^{3+}	LaF_3	$^4F_{3/2} \to {}^4I_{13/2}$	300	[88]
1.3600	Nd^{3+}	CaF_2-YF_3	$^4F_{3/2} \to {}^4I_{13/2}$	77	[90]
		α-$NaCaYF_6$	$^4F_{3/2} \to {}^4I_{13/2}$	300	[89]
1.3632[b]	Nd^{3+}	Sc_2SiO_5	$^4F_{3/2} \to {}^4I_{13/2}$	300	[810]
1.3644	Nd^{3+}	$YAlO_3$	$^4F_{3/2} \to {}^4I_{13/2}$	77	[93]
1.3665	Nd^{3+}	$SrAl_4O_7$	$^4F_{3/2} \to {}^4I_{13/2}$	300	[88, 96]
1.3670	Nd^{3+}	LaF_3	$^4F_{3/2} \to {}^4I_{13/2}$	77	[88]
1.3675	Nd^{3+}	LaF_3	$^4F_{3/2} \to {}^4I_{13/2}$	300	[88]
		CeF_3	$^4F_{3/2} \to {}^4I_{13/2}$	77	[88, 96]
		$CaAl_4O_7$	$^4F_{3/2} \to {}^4I_{13/2}$	77	[88, 95]
1.3680	Nd^{3+}	$SrAl_4O_7$	$^4F_{3/2} \to {}^4I_{13/2}$	300	[89]
1.3690	Nd^{3+}	CeF_3	$^4F_{3/2} \to {}^4I_{13/2}$	300	[89]
1.3710	Nd^{3+}	$CaAl_4O_7$	$^4F_{3/2} \to {}^4I_{13/2}$	300	[88, 95]
1.3745	Nd^{3+}	$LiNbO_3$	$^4F_{3/2} \to {}^4I_{13/2}$	300	[89]
1.3755	Nd^{3+}	$NaLa(MoO_4)_2$	$^4F_{3/2} \to {}^4I_{13/2}$	77	[97]
1.3780	Nd^{3+}	$PbMoO_4$	$^4F_{3/2} \to {}^4I_{13/2}$	77	[97]
1.3790	Nd^{3+}	$SrMoO_4$	$^4F_{3/2} \to {}^4I_{13/2}$	77	[97]
1.3840	Nd^{3+}	$NaLa(MoO_4)_2$	$^4F_{3/2} \to {}^4I_{13/2}$	77	[97]
1.3849	Nd^{3+}	$YAlO_3$	$^4F_{3/2} \to {}^4I_{13/2}$	77	[93]
1.3870	Nd^{3+}	$LiNbO_3$	$^4F_{3/2} \to {}^4I_{13/2}$	300	[89]
		$LiNbO_3{:}Mg^{2+}$	$^4F_{3/2} \to {}^4I_{13/2}$	300	[530]
		$CaSc_2O_4$	$^4F_{3/2} \to {}^4I_{13/2}$	77	[488]
1.3880	Nd^{3+}	$CaWO_4$	$^4F_{3/2} \to {}^4I_{13/2}$	77	[93]
1.3885	Nd^{3+}	$CaWO_4$	$^4F_{3/2} \to {}^4I_{13/2}$	300	[93]
1.3908[b]	Ho^{3+}	$KY(WO_4)_2$	$^5S_2 \to {}^5I_5$	≈ 110	[775]
1.392[b]	Ho^{3+}	$LiYF_4$	$^5S_2 \to {}^5I_5$	300	[777]
1.3960	Ho^{3+}	$LiYF_4$	$^5S_2 \to {}^5I_5$	116, 300	[568, 778]
1.3982[b]	Ho^{3+}	$KGd(WO_4)_2$	$^5S_2 \to {}^5I_5$	≈ 110	[775]
1.4026	Nd^{3+}	$YAlO_3$	$^4F_{3/2} \to {}^4I_{13/2}$	77	[93]
1.4028[b]	Ho^{3+}	$YAlO_3$	$^5S_2 \to {}^5I_5$	≈ 110	[775]
1.4040[b]	Ho^{3+}	$Gd_3Ga_5O_{12}$	$^5S_2 \to {}^5I_5$	≈ 110	[775]
1.4072[b]	Ho^{3+}	$Y_3Al_5O_{12}$	$^5S_2 \to {}^5I_5$	≈ 110	[775]
1.4085[b]	Ho^{3+}	$Lu_3Al_5O_{12}$	$^5S_2 \to {}^5I_5$	≈ 110	[775]
1.4862[b]	Ho^{3+}	$LiYF_4$	$^5S_2 \to {}^5I_5$	90, 116	[545, 778]
1.5298	Er^{3+}	$CaF_2{*}$	$^4I_{13/2} \to {}^4I_{15/2}$	77	[146]

(*Continued*)

Table 8.1 (*continued*)

Stimulated-emission wavelength [μm]	Ion	Crystal host	Induced transition	Temperature [K]	Reference
1.5308	Er^{3+}	CaF_2*	$^4I_{13/2} \rightarrow {}^4I_{15/2}$	77	[146]
1.547[a]	Er^{3+}	CaF_2-YF_3	$^4I_{13/2} \rightarrow {}^4I_{15/2}$	77	[25, 106]
1.5500	Er^{3+}	$CaAl_4O_7$	$^4I_{13/2} \rightarrow {}^4I_{15/2}$	77	[95]
1.558[b]	Er^{3+}	$Bi_4Ge_3O_{12}$	$^4I_{13/2} \rightarrow {}^4I_{15/2}$	77	[800]
1.5815	Er^{3+}	$CaAl_4O_7$	$^4I_{13/2} \rightarrow {}^4I_{15/2}$	77	[95]
1.6[a]	Ni^{2+}	MnF_2	$^3T_2 \rightarrow {}^3A_2$	77	[113]
1.61[a]	Er^{3+}	$Ca(NbO_3)_2$	$^4I_{13/2} \rightarrow {}^4I_{15/2}$	77	[383]
1.6113	Er^{3+}	LaF_3	$^4I_{13/2} \rightarrow {}^4I_{15/2}$	77	[289]
1.612[a]	Er^{3+}	$CaWO_4$	$^4I_{13/2} \rightarrow {}^4I_{15/2}$	77	[391]
1.617[a]	Er^{3+}	CaF_2	$^4I_{13/2} \rightarrow {}^4I_{15/2}$	77	[320]
1.620[a]	Er^{3+}	$ZrO_2-Er_2O_3:Ho^{3+}$	$^4I_{13/2} \rightarrow {}^4I_{15/2}$	77	[431]
		$ZrO_2-Er_2O_3:Tm^{3+}$	$^4I_{13/2} \rightarrow {}^4I_{15/2}$	77	[431]
1.623[d]	Ni^{2+}	MgF_2	$^3T_2 \rightarrow {}^3A_2$	77	[113]
1.632[a]	Er^{3+}	$Y_3Al_5O_{12}$	$^4I_{13/2} \rightarrow {}^4I_{15/2}$	300	[537]
1.636	Ni^{2+}	MgF_2	$^3T_2 \rightarrow {}^3A_2$	77	[113]
1.6449	Er^{3+}	$Y_3Al_5O_{12}:Yb^{3+}$	$^4I_{13/2} \rightarrow {}^4I_{15/2}$	295	[147]
1.6452	Er^{3+}	$Y_3Al_5O_{12}$	$^4I_{13/2} \rightarrow {}^4I_{15/2}$	77	[128, 135]
1.6459	Er^{3+}	$Y_3Al_5O_{13}: Yb^{3+}$	$^4I_{13/2} \rightarrow {}^4I_{15/2}$	295	[128]
1.6470[b]	Er^{3+}	$LiYF_4$	$^4S_{3/2} \rightarrow {}^4I_{9/2}$	≈ 110	[776]
1.6525	Er^{3+}	$Lu_3Al_5O_{12}$	$^4I_{13/2} \rightarrow {}^4I_{15/2}$	77	[143]
1.6602	Er^{3+}	$Y_3Al_5O_{12}$	$^4I_{13/2} \rightarrow {}^4I_{15/2}$	77	[128, 135]
1.6615	Er^{3+}	$Yb_3Al_5O_{12}$	$^4I_{13/2} \rightarrow {}^4I_{15/2}$	77	[86, 486]
1.6630	Er^{3+}	$Lu_3Al_5O_{12}$	$^4I_{13/2} \rightarrow {}^4I_{15/2}$	77	[143]
1.6632	Er^{3+}	$YAlO_3$	$^4S_{3/2} \rightarrow {}^4I_{9/2}$	300	[109, 110, 547, 565]
1.664[b]	Er^{3+}	$Bi_4Ge_3O_{12}$	$^4S_{3/2} \rightarrow {}^4I_{15/2}$	77	[800]
1.6675	Er^{3+}	$LuAlO_3$	$^4S_{3/2} \rightarrow {}^4I_{9/2}$	≈ 90	[571]
1.673[b]	Ho^{3+}	$LiYF_4$	$^5I_5 \rightarrow {}^5I_7$	300	[777]
1.6734[b]	Ho^{3+}	$LiYF_4$	$^5I_5 \rightarrow {}^5I_7$	116	[778]
1.696[a]	Er^{3+}	CaF_2*	$^4S_{3/2} \rightarrow {}^4I_{9/2}$	77	[108]
1.715[a]	Er^{3+}	CaF_2*	$^4S_{3/2} \rightarrow {}^4I_{9/2}$	77	[108]
1.7155	Er^{3+}	$KGd(WO_4)_2$	$^4S_{3/2} \rightarrow {}^4I_{9/2}$	300	[641]
1.7178[b]	Er^{3+}	$KY(WO_4)_2$	$^4S_{3/2} \rightarrow {}^4I_{9/2}$	300	[803]
1.726[a]	Er^{3+}	CaF_2*	$^4S_{3/2} \rightarrow {}^4I_{9/2}$	77	[108]
1.7320[b]	Er^{3+}	$LiYF_4$	$^4S_{3/2} \rightarrow {}^4I_{9/2}$	110, 300	[776]
1.732[a]	Er^{3+}	$LiErF_4$	$^4S_{3/2} \rightarrow {}^4I_{9/2}$	≈ 90	[545]
1.7325	Er^{3+}	$KGd(WO_4)_2$	$^4S_{3/2} \rightarrow {}^4I_{9/2}$	300	[641]
1.7355[b]	Er^{3+}	$KY(WO_4)_2$	$^4S_{3/2} \rightarrow {}^4I_{9/2}$	300	[803]
1.7383[b]	Er^{3+}	$KLu(WO_4)_2$	$^4S_{3/2} \rightarrow {}^4I_{9/2}$	300	[793]
1.750[a]	Co^{2+}	MgF_2	$^4T_2 \rightarrow {}^4T_1$	77	[113]
1.7757	Er^{3+}	$Y_3Al_5O_{12}$	$^4S_{3/2} \rightarrow {}^4I_{9/2}$	300	[537, 546]
1.7762	Er^{3+}	$Lu_3Al_5O_{12}$	$^4S_{3/2} \rightarrow {}^4I_{9/2}$	300	[546, 547]
1.8035	Co^{2+}	MgF_2	$^4T_2 \rightarrow {}^4T_1$	77	[113]
1.821	Co^{2+}	$KMgF_3$	$^4T_2 \rightarrow {}^4T_1$	77	[113]
1.833[a]	Nd^{3+}	$Y_3Al_5O_{12}$	$^4F_{3/2} \rightarrow {}^4I_{15/2}$	293	[101]
1.8532	Tm^{3+}	$LiNbO_3$	$^3H_4 \rightarrow {}^3H_6$	77	[156]
1.856[a]	Tm^{3+}	$YAlO_3:Cr^{3+}$	$^3H_4 \rightarrow {}^3H_6$	≈ 90	[525]
1.8580	Tm^{3+}	α-$NaCaErF_6$	$^3H_4 \rightarrow {}^3H_6$	150	[134]
1.860	Tm^{3+}	CaF_2-ErF_3	$^3H_4 \rightarrow {}^3H_6$	77	[133, 148]

[d] Continuous tuning from ≈ 1.6 to ≈ 1.8 μm was carried out using an optical resonator with dispersive element [813, 814, 825].

(*Continued*)

Table 8.1   (*continued*)

Stimulated-emission wavelength [μm]	Ion	Crystal host	Induced transition	Temperature [K]	Reference
1.861	Tm^{3+}	$(Y, Er)AlO_3$	$^3H_4 \rightarrow {}^3H_6$	77	[107]
1.865	Ni^{2+}	MnF_2	$^3T_2 \rightarrow {}^3A_2$	20	[113]
1.872[a]	Tm^{3+}	$ErAlO_3$	$^3H_4 \rightarrow {}^3H_6$	77	[225]
1.880	Tm^{3+}	$(Y, Er)_3Al_5O_{12}$	$^3H_4 \rightarrow {}^3H_6$	77	[128, 135]
1.883[a]	Tm^{3+}	$YAlO_3 : Cr^{3+}$	$^3H_4 \rightarrow {}^3H_6$	≈ 90	[525]
1.8834	Tm^{3+}	$Y_3Al_5O_{12}$	$^3H_4 \rightarrow {}^3H_6$	77	[128, 135]
1.884	Tm^{3+}	$(Y, Er)_3Al_5O_{12}$	$^3H_4 \rightarrow {}^3H_6$	77	[128, 135]
1.8845	Tm^{3+}	$(Er, Lu)AlO_3$	$^3H_4 \rightarrow {}^3H_6$	77	[421]
1.8850	Tm^{3+}	$(Er, Yb)_3Al_5O_{12}$	$^3H_4 \rightarrow {}^3H_6$	77	[86, 486]
1.8855	Tm^{3+}	$Lu_3Al_5O_{12}$	$^3H_4 \rightarrow {}^3H_6$	77	[143]
1.8885	Tm^{3+}	$\alpha\text{-}NaCaErF_6$	$^3H_4 \rightarrow {}^3H_6$	150	[134]
1.896[a]	Tm^{3+}	$ZrO_2 - Er_2O_3$	$^3H_4 \rightarrow {}^3H_6$	77	[431]
1.9060	Tm^{3+}	$CaMoO_4 : Er^{3+}$	$^3H_4 \rightarrow {}^3H_6$	77	[21]
1.91[a]	Tm^{3+}	$Ca(NbO_3)_2$	$^3H_4 \rightarrow {}^3H_6$	77	[383]
1.911	Tm^{3+}	$CaWO_4$	$^3H_4 \rightarrow {}^3H_6$	77	[183, 392]
1.9115	Tm^{3+}	$CaMoO_4 : Er^{3+}$	$^3H_4 \rightarrow {}^3H_6$	77	[21]
1.915	Ni^{2+}	MnF_2	$^3T_2 \rightarrow {}^3A_2$	77	[113]
1.916	Tm^{3+}	$CaWO_4$	$^3H_4 \rightarrow {}^3H_6$	77	[183, 392]
1.922	Ni^{2+}	MnF_2	$^3T_2 \rightarrow {}^3A_2$	77	[113]
1.929	Ni^{2+}	MnF_2	$^3T_2 \rightarrow {}^3A_2$	85	[113]
1.933[a]	Tm^{3+}	$YAlO_3 : Cr^{3+}$	$^3H_4 \rightarrow {}^3H_6$	≈ 90	[525]
1.934	Tm^{3+}	Er_2O_3	$^3H_4 \rightarrow {}^3H_6$	77	[149]
1.939	Ni^{2+}	MnF_2	$^3T_2 \rightarrow {}^3A_2$	85	[113]
1.972[a]	Tm^{3+}	SrF_2	$^3H_4 \rightarrow {}^3H_6$	77	[183]
1.99[a]	Co^{2+}	MgF_2	$^4T_2 \rightarrow {}^4T_1$	77	[113]
2.0[a]	Ho^{3+}	$SrY_4(SiO_4)_3O : Er^{3+}, Tm^{3+}$	$^5I_7 \rightarrow {}^5I_8$	77	[130]
	Tm^{3+}	YVO_4	$^3H_4 \rightarrow {}^3H_6$	77	[413]
2.0010	Ho^{3+}	$(Er, Lu)AlO_3$	$^5I_7 \rightarrow {}^5I_8$	77	[421]
2.0132	Tm^{3+}	$Y_3Al_5O_{12}$	$^3H_4 \rightarrow {}^3H_6$	77	[128, 135]
2.014	Tm^{3+}	$(Y, Er)_3Al_5O_{12}$	$^3H_4 \rightarrow {}^3H_6$	85	[128]
2.019	Tm^{3+}	$Y_3Al_5O_{12} : Cr^{3+}$	$^3H_4 \rightarrow {}^3H_6$	295	[128]
2.0195	Tm^{3+}	$(Er, Yb)_3Al_5O_{12}$	$^3H_4 \rightarrow {}^3H_6$	77	[86, 486]
2.0240	Tm^{3+}	$Lu_3Al_5O_{12}$	$^3H_4 \rightarrow {}^3H_6$	77	[143]
2.030	Ho^{3+}	$CaF_2 - ErF_3$	$^5I_7 \rightarrow {}^5I_8$	77	[133]
2.0312	Ho^{3+}	$\alpha\text{-}NaCaErF_6$	$^5I_7 \rightarrow {}^5I_8$	77	[134]
2.0318	Ho^{3+}	$CaF_2 - YF_3$	$^5I_7 \rightarrow {}^5I_8$	77	[134]
2.0345	Ho^{3+}	$\alpha\text{-}NaCaErF_6$	$^5I_7 \rightarrow {}^5I_8$	150	[134]
2.0377	Ho^{3+}	$\alpha\text{-}NaCaErF_6$	$^5I_7 \rightarrow {}^5I_8$	77	[134]
2.0412	Ho^{3+}	$YVO_4 : Er^{3+}, Tm^{3+}$	$^5I_7 \rightarrow {}^5I_8$	77	[660]
2.0416	Ho^{3+}	$ErVO_4 : Tm^{3+}$	$^5I_7 \rightarrow {}^5I_8$	77	[660]
2.046	Ho^{3+}	$CaWO_4$	$^5I_7 \rightarrow {}^5I_8$	77	[183, 390]
2.047	Ho^{3+}	$Ca(NbO_3)_2$	$^5I_7 \rightarrow {}^5I_8$	77	[383]
2.05[a]	Ho^{3+}	$CaF_2 - ErF_3 - TmF_3 - YbF_3$	$^5I_7 \rightarrow {}^5I_8$	100	[111]
	Co^{2+}	MgF_2	$^4T_2 \rightarrow {}^4T_1$	77	[113]
2.050[a]	Ho^{3+}	$NaLa(MoO_4)_2$	$^5I_7 \rightarrow {}^5I_8$	≈ 90	[406]
		$NaLa(MoO_4)_2 : Er^{3+}$	$^5I_7 \rightarrow {}^5I_8$	≈ 90	[406]
2.0505[b]	Ho^{3+}	$LiYF_4$	$^5I_7 \rightarrow {}^5I_8$	300	[815]
2.0515[b]	Ho^{3+}	$LiYF_4$	$^5I_7 \rightarrow {}^5I_8$	300	[815]
2.0534[b]	Ho^{3+}	$LiYF_4$	$^5I_7 \rightarrow {}^5I_8$	300	[815]
2.0555	Ho^{3+}	$BaY_2F_8 : Er^{3+}, Tm^{3+}$	$^5I_7 \rightarrow {}^5I_8$	20	[532]

(*Continued*)

Table 8.1 (*continued*)

Stimulated-emission wavelength [μm]	Ion	Crystal host	Induced transition	Temperature [K]	Reference
2.0556	Ho^{3+}	$CaMoO_4$: Er^{3+}	$^5I_7 \rightarrow {}^5I_8$	77	[21]
2.0563[b]	Ho^{3+}	$BaYb_2F_8$	$^5I_7 \rightarrow {}^5I_8$	300	[791]
2.059	Ho^{3+}	$CaWO_4$	$^5I_7 \rightarrow {}^5I_8$	77	[183, 390]
2.060[a]	Ho^{3+}	CaF_2–ErF_3–TmF_3–YbF_3	$^5I_7 \rightarrow {}^5I_8$	298	[111]
		$CaY_4(SiO_4)_3O$: Er^{3+}, Tm^{3+}	$^5I_7 \rightarrow {}^5I_8$	77	[130, 429]
2.065[a]	Ho^{3+}	BaY_2F_8 : Er^{3+}, Tm^{3+}	$^5I_7 \rightarrow {}^5I_8$	77	[532]
2.0654	Ho^{3+}	$Li(Y,Er)F_4$: Tm^{3+}	$^5I_7 \rightarrow {}^5I_8$	300	[132]
2.0656[b]	Ho^{3+}	$LiYF_4$	$^5I_7 \rightarrow {}^5I_8$	300	[778, 815]
2.066	Ho^{3+}	$LiYF_4$: Er^{3+}	$^5I_7 \rightarrow {}^5I_8$	77	[131]
2.0672	Ho^{3+}	$LiYF_4$	$^5I_7 \rightarrow {}^5I_8$	≈ 90	[568]
2.07[a]	Ho^{3+}	$LaNbO_4$: Er^{3+}	$^5I_7 \rightarrow {}^5I_8$	≈ 90	[406]
2.0707	Ho^{3+}	$CaMoO_4$: Er^{3+}	$^5I_7 \rightarrow {}^5I_8$	77	[21]
2.0720	Ho^{3+}	$K(Y,Er)(WO_4)_2$: Tm^{3+}	$^5I_7 \rightarrow {}^5I_8$	110–220	[595]
2.074[a]	Ho^{3+}	BaY_2F_8 : Er^{3+}, Tm^{3+}	$^5I_7 \rightarrow {}^5I_8$	20	[532]
		$CaMoO_4$: Er^{3+}	$^5I_7 \rightarrow {}^5I_8$	77	[21]
2.0746	Ho^{3+}	BaY_2F_8 : Er^{3+}, Tm^{3+}	$^5I_7 \rightarrow {}^5I_8$	77	[532]
2.0786	Ho^{3+}	$LiNbO_3$	$^5I_7 \rightarrow {}^5I_8$	77	[156]
2.079	Ho^{3+}	$Ca_5(PO_4)_3F$: Cr^{3+}	$^5I_7 \rightarrow {}^5I_8$	77	[129, 130]
2.085[a]	Ho^{3+}	Y_2SiO_5 : Er^{3+}	$^5I_7 \rightarrow {}^5I_8$	≈ 110	[638]
		Y_2SiO_5 : Er^{3+}, Tm^{3+}	$^5I_7 \rightarrow {}^5I_8$	110–300	[638]
		Er_2SiO_5	$^5I_7 \rightarrow {}^5I_8$	≈ 110	[638]
2.086[a]	Ho^{3+}	$Y_3Fe_5O_{12}$: Er^{3+}, Tm^{3+}	$^5I_7 \rightarrow {}^5I_8$	77	[138, 139]
		$Y_3Ga_5O_{12}$	$^5I_7 \rightarrow {}^5I_8$	77	[138, 139]
		$Ho_3Ga_5O_{12}$	$^5I_7 \rightarrow {}^5I_8$	77	[546]
2.0866	Ho^{3+}	BaY_2F_8 : Er^{3+}, Tm^{3+}	$^5I_7 \rightarrow {}^5I_8$	77	[532]
2.087[b]	Ho^{3+}	$Bi_4Ge_3O_{12}$	$^5I_7 \rightarrow {}^5I_8$	77	[800]
2.0885[b]	Ho^{3+}	$Gd_3Ga_5O_{12}$	$^5I_7 \rightarrow {}^5I_8$	≈ 110	[774]
2.089	Ho^{3+}	$Y_3Fe_5O_{12}$: Er^{3+}, Tm^{3+}	$^5I_7 \rightarrow {}^5I_8$	77	[138, 139]
2.090	Ho^{3+}	HoF_3	$^5I_7 \rightarrow {}^5I_8$	77	[346]
2.0914	Ho^{3+}	$Y_3Al_5O_{12}$	$^5I_7 \rightarrow {}^5I_8$	77	[128, 135]
2.0917	Ho^{3+}	$(Y, Er)_3Al_5O_{12}$	$^5I_7 \rightarrow {}^5I_8$	77	[128, 135]
2.092[a]	Ho^{3+}	CaF_2	$^5I_7 \rightarrow {}^5I_8$	77	[183]
		Y_2SiO_2 : Er^{3+}	$^5I_7 \rightarrow {}^5I_8$	≈ 110	[638]
2.0960	Ho^{3+}	$Yb_3Al_5O_{12}$	$^5I_7 \rightarrow {}^5I_8$	77	[422]
2.097[a]	Ho^{3+}	$Ho_3Al_5O_{12}$	$^5I_7 \rightarrow {}^5I_8$	77	[546]
2.0975	Ho^{3+}	$Y_3Al_5O_{12}$	$^5I_7 \rightarrow {}^5I_8$	77	[128, 135]
2.0977	Ho^{3+}	$(Y, Er)_3Al_5O_{12}$: Tm^{3+}	$^5I_7 \rightarrow {}^5I_8$	77	[597]
2.0979	Ho^{3+}	$(Y, Er)_3Al_5O_{12}$	$^5I_7 \rightarrow {}^5I_8$	77	[128, 135]
2.098[a]	Ho^{3+}	$(Y, Er)_3Al_5O_{12}$: Tm^{3+}	$^5I_7 \rightarrow {}^5I_8$	77	[223]
2.0982	Ho^{3+}	$(Y, Er)_3Al_5O_{12}$: Tm^{3+}	$^5I_7 \rightarrow {}^5I_8$	77	[140]
2.0985	Ho^{3+}	$Er_3Sc_2Al_3O_{12}$: Tm^{3+}	$^5I_7 \rightarrow {}^5I_8$	77	[488, 529]
2.0990	Ho^{3+}	$(Y, Er)_3Al_5O_{12}$: Tm^{3+}	$^5I_7 \rightarrow {}^5I_8$	77	[415]
2.1[a]	Ho^{3+}	CaF_2–ErF_3–TmF_3–YbF_3	$^5I_7 \rightarrow {}^5I_8$	77	[145]
		$Y_3Fe_5O_{12}$	$^5I_7 \rightarrow {}^5I_8$	77	[573]
2.1010	Ho^{3+}	$(Er, Tm, Yb)_3Al_5O_{12}$	$^5I_7 \rightarrow {}^5I_8$	77	[86, 486]
2.1020	Ho^{3+}	$Lu_3Al_5O_{12}$	$^5I_7 \rightarrow {}^5I_8$	77	[143]
		$Lu_3Al_5O_{12}$: Er^{3+}, Tm^{3+}	$^5I_7 \rightarrow {}^5I_8$	77	[143]
2.105[a]	Ho^{3+}	Y_2SiO_5 : Er^{3+}	$^5I_7 \rightarrow {}^5I_8$	110–220	[638]
2.1070	Ho^{3+}	$Y_3Fe_5O_{12}$: Er^{3+}, Tm^{3+}	$^5I_7 \rightarrow {}^5I_8$	77	[138, 139]
2.1135	Ho^{3+}	$Ho_3Ga_5O_{12}$	$^5I_7 \rightarrow {}^5I_8$	77	[546]

(*Continued*)

Table 8.1   (*continued*)

Stimulated-emission wavelength [μm]	Ion	Crystal host	Induced transition	Temperature [K]	Reference
2.114	Ho^{3+}	$Y_3Ga_5O_{12}$	$^5I_7 \rightarrow {}^5I_8$	77	[138, 139]
2.115[a]	Ho^{3+}	$ZrO_2-Er_2O_3$	$^5I_7 \rightarrow {}^5I_8$	77	[431]
2.1170	Ho^{3+}	$Ho_3Sc_2Al_3O_{12}$	$^5I_7 \rightarrow {}^5I_8$	77	[546]
2.119	Ho^{3+}	$(Y, Er)AlO_3 : Tm^{3+}$	$^5I_7 \rightarrow {}^5I_8$	300	[141]
2.1205	Ho^{3+}	$ErAlO_3$	$^5I_7 \rightarrow {}^5I_8$	77	[137]
		$(Er, Lu)AlO_3$	$^5I_7 \rightarrow {}^5I_8$	77	[421]
2.121	Ho^{3+}	Er_2O_3	$^5I_7 \rightarrow {}^5I_8$	145	[136]
2.1223	Ho^{3+}	$Y_3Al_5O_{12}$	$^5I_7 \rightarrow {}^5I_8$	77	[128, 135]
2.1224	Ho^{3+}	$Ho_3Al_5O_{12}$	$^5I_7 \rightarrow {}^5I_8$	≈ 90	[526]
2.1227	Ho^{3+}	$(Y, Er)_3Al_5O_{12} : Tm^{3+}$	$^5I_7 \rightarrow {}^5I_8$	85	[140]
		$(Y, Er)_3Al_5O_{12} : Tm^{3+}, Yb^{3+}$	$^5I_7 \rightarrow {}^5I_8$	85	[140]
		$Ho_3Al_5O_{12}$	$^5I_7 \rightarrow {}^5I_8$	77	[546]
2.123	Ho^{3+}	$(Y, Er)AlO_3 : Tm^{3+}$	$^5I_7 \rightarrow {}^5I_8$	300	[107]
		$(Y, Er)_3Al_5O_{12}$	$^5I_7 \rightarrow {}^5I_8$	77	[128, 135]
2.1285	Ho^{3+}	$(Y,Er)_3Al_5O_{12} : Tm^{3+}$	$^5I_7 \rightarrow {}^5I_8$	77	[415]
		$Ho_3Sc_2Al_3O_{12}$	$^5I_7 \rightarrow {}^5I_8$	77	[546]
2.1288	Ho^{3+}	$(Y,Er)_3Al_5O_{12} : Tm^{3+}$	$^5I_7 \rightarrow {}^5I_8$	295	[140]
2.1294	Ho^{3+}	$Ho_3Al_5O_{12}$	$^5I_7 \rightarrow {}^5I_8$	≈ 90	[526]
2.1297	Ho^{3+}	$Ho_3Al_5O_{12}$	$^5I_7 \rightarrow {}^5I_8$	77	[546]
2.13[a]	Ho^{3+}	$(Y,Er)_3Al_5O_{12} : Tm^{3+}$	$^5I_7 \rightarrow {}^5I_8$	300	[224]
2.1348	Ho^{3+}	$LuAlO_3$	$^5I_7 \rightarrow {}^5I_8$	≈ 90	[571]
2.165	Co^{2+}	ZnF_2	$^4T_2 \rightarrow {}^4T_1$	77	[113, 117]
2.171[a]	Ho^{3+}	$BaY_2F_8 : Er^{3+}, Tm^{3+}$	$^5I_7 \rightarrow {}^5I_8$	295	[532]
2.234	U^{3+}	CaF_2	$^4I_{11/2} \rightarrow {}^4I_{9/2}$	77	[333]
2.274[a]	Tm^{3+}	$YAlO_3 : Cr^{3+}$	$^3F_4 \rightarrow {}^3H_5$	300	[563]
2.318[a]	Tm^{3+}	$YAlO_3 : Cr^{3+}$	$^3F_4 \rightarrow {}^3H_5$	300	[563]
2.324[a]	Tm^{3+}	$Y_3Al_5O_{12} : Cr^{3+}$	$^3F_4 \rightarrow {}^3H_5$	300	[563]
2.34[a]	Tm^{3+}	$YAlO_3 : Cr^{3+}$	$^3F_4 \rightarrow {}^3H_5$	90, 300	[525]
2.348[a]	Tm^{3+}	$YAlO_3 : Cr^{3+}$	$^3F_4 \rightarrow {}^3H_5$	300	[112]
2.349[a]	Tm^{3+}	$YAlO_3 : Cr^{3+}$	$^3F_4 \rightarrow {}^3H_5$	300	[112]
2.352[a]	Ho^{3+}	$LiHoF_4$	$^5F_5 \rightarrow {}^5I_5$	≈ 90	[545]
2.3524[b]	Ho^{3+}	$LiYF_4$	$^5F_5 \rightarrow {}^5I_5$	116	[778]
2.353[a]	Tm^{3+}	$YAlO_3 : Cr^{3+}$	$^3F_4 \rightarrow {}^3H_5$	300	[563]
2.354[a]	Tm^{3+}	$YAlO_3 : Cr^{3+}$	$^3F_4 \rightarrow {}^3H_5$	300	[563]
2.355[a]	Tm^{3+}	$YAlO_3 : Cr^{3+}$	$^3F_4 \rightarrow {}^3H_5$	300	[563]
2.35867	Dy^{2+}	CaF_2	$^5I_7 \rightarrow {}^5I_8$	4.2–120	[40, 257]
2.362[a]	Ho^{3+}	BaY_2F_8	$^5F_5 \rightarrow {}^5I_5$	77	[532]
2.363[a]	Ho^{3+}	BaY_2F_8	$^5F_5 \rightarrow {}^5I_5$	77	[532]
2.3659	Dy^{2+}	SrF_2	$^5I_7 \rightarrow {}^5I_8$	20	[339]
2.375[a]	Ho^{3+}	BaY_2F_8	$^5F_5 \rightarrow {}^5I_5$	77	[532]
2.377[a]	Ho^{3+}	BaY_2F_8	$^5F_5 \rightarrow {}^5I_5$	20	[532]
2.407	U^{3+}	SrF_2	$^4I_{11/2} \rightarrow {}^4I_{9/2}$	20–90	[341]
2.439	U^{3+}	CaF_2	$^4I_{11/2} \rightarrow {}^4I_{9/2}$	77	[335]
2.511	U^{3+}	CaF_2	$^4I_{11/2} \rightarrow {}^4I_{9/2}$	77	[334]
2.556	U^{3+}	BaF_2	$^4I_{11/2} \rightarrow {}^4I_{9/2}$	20	[342]
2.571	U^{3+}	CaF_2	$^4I_{11/2} \rightarrow {}^4I_{9/2}$	77	[334]
2.6[a]	U^{3+}	CaF_2	$^4I_{11/2} \rightarrow {}^4I_{9/2}$	4.2	[13]
2.613	U^{3+}	CaF_2	$^4I_{11/2} \rightarrow {}^4I_{9/2}$	77–300	[18]
2.6887	Er^{3+}	$K(Y, Er)(WO_4)_2 : Ho^{3+}, Tm^{3+}$	$^4I_{11/2} \rightarrow {}^4I_{13/2}$	300–350	[595]
2.69[a]	Er^{3+}	$CaF_2-ErF_3-TmF_3$	$^4I_{11/2} \rightarrow {}^4I_{13/2}$	298	[111]

(*Continued*)

Table 8.1   (*continued*)

Stimulated-emission wavelength [μm]	Ion	Crystal host	Induced transition	Temperature [K]	Reference
2.6990	Er^{3+}	$(Er, Lu)_3Al_5O_{12}: Ho^{3+}, Tm^{3+}$	$^4I_{11/2} \rightarrow {}^4I_{13/2}$	300	[655]
		$Lu_3Al_5O_{12}: Tm^{3+}$	$^4I_{11/2} \rightarrow {}^4I_{13/2}$	300	[665]
		$Lu_3Al_5O_{12}: Ho^{3+}, Tm^{3+}$	$^4I_{11/2} \rightarrow {}^4I_{13/2}$	300	[546, 547]
2.7222	Er^{3+}	$KGd(WO_4)_2$	$^4I_{11/2} \rightarrow {}^4I_{13/2}$	300	[641]
2.7307[a]	Er^{3+}	CaF_2-ErF_3	$^4I_{11/2} \rightarrow {}^4I_{13/2}$	300	[469]
2.747[b]	Er^{3+}	$LiYF_4$	$^4I_{11/2} \rightarrow {}^4I_{13/2}$	–	[776]
2.7953[b]	Er^{3+}	$Y_3Al_5O_{12}$	$^4I_{11/2} \rightarrow {}^4I_{13/2}$	300	[794]
2.7987[b]	Er^{3+}	$Lu_3Al_5O_{12}$	$^4I_{11/2} \rightarrow {}^4I_{13/2}$	300	[794]
2.7990	Er^{3+}	$YAlO_3$	$^4I_{11/2} \rightarrow {}^4I_{13/2}$	300	[641]
2.8070[b]	Er^{3+}	$KY(WO_4)_2$	$^4I_{11/2} \rightarrow {}^4I_{13/2}$	300	[803]
		$K(Y, Er)(WO_4)_2$[b]	$^4I_{11/2} \rightarrow {}^4I_{13/2}$	300	[803]
		$KEr(WO_4)_2$[b]	$^4I_{11/2} \rightarrow {}^4I_{13/2}$	300	[792]
2.8092[b]	Er^{3+}	$KLu(WO_4)_2$	$^4I_{11/2} \rightarrow {}^4I_{13/2}$	300	[793]
2.81[a,b]	Er^{3+}	$LiYF_4$	$^4I_{11/2} \rightarrow {}^4I_{13/2}$	300	[816]
2.8218[b]	Er^{3+}	$Gd_3Ga_5O_{12}$	$^4I_{11/2} \rightarrow {}^4I_{13/2}$	300	[774]
2.8298	Er^{3+}	$(Er, Lu)_3Al_5O_{12}$	$^4I_{11/2} \rightarrow {}^4I_{13/2}$	300	[572]
		$Lu_3Al_5O_{12}$	$^4I_{11/2} \rightarrow {}^4I_{13/2}$	300	[572]
2.8302	Er^{3+}	$Y_3Al_5O_{12}$	$^4I_{11/2} \rightarrow {}^4I_{13/2}$	300	[572]
		$(Y, Er)_3Al_5O_{12}$	$^4I_{11/2} \rightarrow {}^4I_{13/2}$	300	[572]
2.850[b]	Ho^{3+}	$LiYF_4$	$^5I_6 \rightarrow {}^5I_7$	300	[778]
2.8510	Ho^{3+}	$LaNbO_4$	$^5I_6 \rightarrow {}^5I_7$	300	[607]
2.870[b]	Er^{3+}	$LiYF_4$	$^4I_{11/2} \rightarrow {}^4I_{13/2}$	110, 300	[776]
2.9[a,b]	Ho^{3+}	$Gd_3Ga_5O_{12}$	$^5I_6 \rightarrow {}^5I_7$	300	[774]
2.9073[b]	Ho^{3+}	$BaYb_2F_8$	$^5I_6 \rightarrow {}^5I_7$	≈ 300	[791]
2.9180	Ho^{3+}	$YAlO_3$	$^5I_6 \rightarrow {}^5I_7$	300	[547, 565]
2.9342	Ho^{3+}	$KGd(WO_4)_2$	$^5I_6 \rightarrow {}^5I_7$	300	[641]
2.9364	Er^{3+}	$Y_3Al_5O_{12}$	$^4I_{11/2} \rightarrow {}^4I_{13/2}$	300	[490, 572]
		$(Y, Er)_3Al_5O_{12}$	$^4I_{11/2} \rightarrow {}^4I_{13/2}$	300	[490, 572]
2.9367	Er^{3+}	$Er_3Al_5O_{12}$	$^4I_{11/2} \rightarrow {}^4I_{13/2}$	300	[572]
2.9395	Er^{3+}	$(Er, Lu)_3Al_5O_{12}$	$^4I_{11/2} \rightarrow {}^4I_{13/2}$	300	[546, 572]
	Ho^{3+}	$KY(WO_4)_2$	$^5I_6 \rightarrow {}^5I_7$	300	[789]
2.9403	Ho^{3+}	$Y_3Al_5O_{12}$	$^5I_6 \rightarrow {}^5I_7$	300	[547]
2.9406	Er^{3+}	$Lu_3Al_5O_{12}$	$^4I_{11/2} \rightarrow {}^4I_{13/2}$	300	[546, 572]
2.9445	Ho^{3+}	$KLu(WO_4)_2$	$^5I_6 \rightarrow {}^5I_7$	300	[790]
2.9460	Ho^{3+}	$Lu_3Al_5O_{12}$	$^5I_6 \rightarrow {}^5I_7$	300	[546, 547]
2.952[b]	Ho^{3+}	$LiYF_4$	$^5I_6 \rightarrow {}^5I_7$	300	[778]
3.0132	Ho^{3+}	$YAlO_3$	$^5I_6 \rightarrow {}^5I_7$	300	[655]
3.022[a]	Dy^{3+}	$Ba(Y, Er)_2F_8$	$^6H_{13/2} \rightarrow {}^6H_{15/2}$	77	[479]
3.914[a,b]	Ho^{3+}	$LiYF_4$	$^5I_5 \rightarrow {}^5I_6$	300	[777]

9. Laser-Crystal Physics

Whereas previous spectroscopic studies of impurity crystalline substances have revealed the possibility of their utilization in lasers, the experience accumulated in studying the parameters of their laser emission has in turn given rise to a new spectroscopic trend, stimulated-emission spectroscopy of activated crystals. In conjunction with such conventional techniques as luminescent and absorption analyses, this new spectroscopic method is at present being widely used in studying the nature of phenomena that occur in excited active media; it is also of great help in solving such a challenging problem as the search for new, more-efficient laser compounds. All the evidence presented in this book indicates convincingly that stimulated-emission spectroscopy provides a substantial contribution to the recognition of new stimulated-emission potentialities of already known laser crystals. The proportions of this contribution, as given in various stages of the development of the physics of laser crystals over nearly twenty years, are illustrated in Fig. 9.1.

Information about the number of activated crystals synthesized annually, the laser effect of which was discovered in the period from 1960 to 1979, is presented in Fig. 9.1a. In construction of the diagram in Fig. 9.1b, account was taken of the discovery of the laser effect for each induced transition between a definite pair of multiplets of the impurity ion in a single crystalline matrix. Information presented in this diagram, to a greater extent, characterizes the achievements of the physics of crystalline lasers, because these data reflect the results of combined studies of both the spectroscopic and stimulated-emission properties of new and already known laser crystals. If we now "subtract" the first diagram from the second, a new diagram is obtained, the time behavior of which reflects the pure contribution from stimulated-emission spectroscopy. Such a diagram is shown in Fig. 9.1c. Thus, the evidence convincingly indicates that the new spectroscopic trend of development has considerably assisted in the progress made in the physics of crystal lasers, with its contribution increasing from year to year, especially since 1965, in addition to sharp rises of research activities in 1973–74 due to generation on the $^4F_{3/2} \to {}^4I_{13/2}$ transition of Nd^{3+} ions and in 1979 due to generation on the $^5S_2 \to {}^5I_5$, $^5I_6 \to {}^5I_8$ and $^5I_6 \to {}^5I_7$ transitions of Ho^{3+} ions and on the $^4I_{11/2} \to {}^4I_{13/2}$ transition of Er^{3+} ions.

Let us return to Fig. 9.1a, from which it is also clear that the whole history of the search for new laser crystals can be provisionally divided into three periods. During the first period, from 1960 to 1964, investigators accumulated data on the

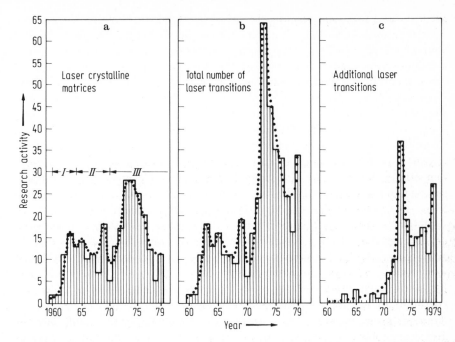

Fig. 9.1. Research activity in laser-crystal spectroscopy from 1960 to 1979. (a) number of laser crystals; (b) taking into account stimulated emission at each induced transition; (c) relative contribution of stimulated-emission spectroscopy

properties of stimulated emission in activated crystals. The second period, from 1965 to 1970, is characterized by the trends directly related with the practically unlimited possibilities that have opened up for crystal lasers in science and applications, together with other types of lasers. The highly useful controversy that arose at that time as to which of the types of lasers was more promising advanced laser efficiency as a major problem. During those years, a great variety of sensitized compounds (the second peak in Figs. 9.1a and 9.1b) and mixed disordered systems were produced, whose utilization in lasers made possible a substantial rise in laser efficiency. During the next (third) period, when each laser type had already found its own sphere of application and when, as a result, the competition between them had become much less acute, a substantial number of studies, including those involving the search for new laser substances, were directed toward gaining a deeper insight into the physics of the processes giving rise to stimulated emission and its excitation in new channels. Our ideas about the processes of excitation-energy transfer between coactivator ions and about the concentration quenching processes have been revised and refined at the level of elementary processes. Greater interest has been shown by investigators (both theorists and experimenters) in the problem of electron-phonon interaction. Such manifestations of this phenomenon as the temperature broadening and shift of the zero-phonon lines, electron-vibrational structure of spectra and multiphonon nonradiative transitions play a major part in the formation of the spectral and kinetic characteristics of crystalline lasers. This enabled us to pass from separate and disconnected studies of the first period to a systematic search for new laser crystals with desired properties, and to realize better their spectroscopic potential by developing new crystal-laser types.

In the years to come, we may expect continuation, on a larger scale, of the intensive search for laser crystals capable of efficient emission at room temperature, both in the ultraviolet and in the visible region of the optical spectrum, as well as in the infrared region, especially for wavelengths from 2 to 5 μm. This would be promoted by the new view of specialists on the acceptable value of luminescence quantum yield of the initial laser states of activated crystals, resulting from the creation of numerous lasers on the basis of self-activated neodymium compounds (with $\eta \approx 0.4$) and crystals with Ho^{3+} and Er^{3+} ions lasing in the three-micrometer region (with $\eta \gtrsim 0.01$) at 300 K. The long-wave limits mentioned above indicate that crystalline compounds with a narrow phonon spectrum (probably, not greater than about 200 cm^{-1} using lamp pumping[1]) should be the proper candidates for such laser media. There is also every reason to believe that, in the next few years, the "mixed garnet" problem will be solved successfully, as indicated by the numerous encouraging results obtained by different groups of researchers [416, 504, 506, 520–523, 529]. As an illustration of the possibility of creating a multicenter disordered crystal that has the structure of the $Y_3Al_5O_{12}$ type, Fig. 9.2 shows the spectra of stimulated emission ($^4F_{3/2} \rightarrow {}^4I_{11/2}$ transition at 77 and 300 K) by ten crystals with the garnet structure doped with Nd^{3+} ions (for details, see Table 5.1). As the cited investigations indicate, some of those compounds are capable of forming solid solutions with each other, while retaining without any significant changes all the physical properties of the $Y_3Al_5O_{12}$ crystal. It is also necessary to mention here the important fact (for the search for new laser crystals) that the three-sublattice structure of these compounds permits incorporation of various ions, thus leading to the formation of a wide variety of garnets.

As is well-known [760], the unit cell of a compound with the garnet structure $A_3B_5O_{12}$ contains eight formula units. The B atoms occupy two types of sites with different oxygen coordination: 16 octahedral a positions (C_{3i}), and 24 tetrahedral d positions (S_4); the a sites form a body-centered-cubic lattice. The 24 A ions lie in dodecahedral (D_2) c positions[2]. The Ln^{3+} ions, which are responsible for the laser properties of the crystals, occupy mainly the c positions

---

[1] Crystals with more extended phonon spectra can be applied when nanosecond and picosecond pulse laser pumping is used for stimulated-emission excitation.

[2] For convenience, the formula of a garnet can be written as $\{A_3\}[B_2](B_3)O_{12}$, where the braces, square brackets, and parentheses mark the cations occupying the c, a, and d crystallographic positions, respectively.

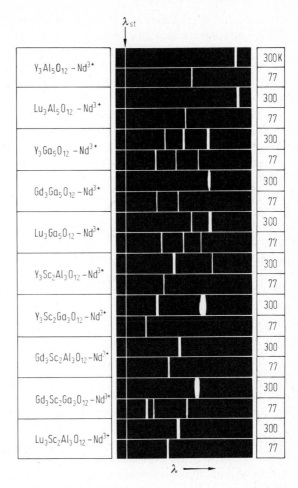

Fig. 9.2. Stimulated-emission spectra ($^4F_{3/2} \rightarrow \, ^4I_{11/2}$ transition) for garnet-structure crystals with Nd^{3+} ions at 77 and 300 K, in the scheme of conventional lasers [524]

(see Fig. 9.3). The oxygen ions lie in 96 general h positions. At the same time, X-ray structural and crystal-chemical investigations have shown that there can exist an exceedingly wide class of compounds with the garnet structure in which the a, c, and d positions can be occupied by ions of a large number of elements of various valences (practically all the groups in the periodic table). Some of these show exceptionally great selectivity and occupy these crystallographic positions completely, i.e., they form their own sublattice in the three-sublattice structure of garnet, and this will exert a specific influence on the cations in the other two sublattices (for example, on the orbital moments of their valence electrons) or the anions (electrical polarization of the oxygen sublattice). Hence follows the possibility of fine control of the intracrystal field of the activator (generation)

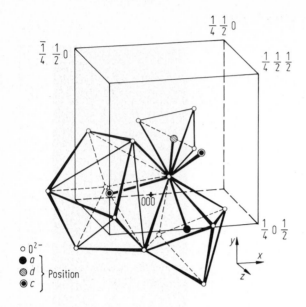

$\frac{1}{4}\,\frac{1}{2}\,0$

$\overline{\frac{1}{4}}\,\frac{1}{2}\,0$

$\frac{1}{4}\,\frac{1}{2}\,\frac{1}{2}$

$\frac{1}{4}\,0\,\frac{1}{2}$

000

$\circ\ 0^{2-}$
$\bullet\ a$
$\oslash\ d$
$\odot\ c$ } Position

y x z

Fig. 9.3. Motif of garnet crystal structure [760]

ions. As well as its importance for applications, this is also of great scientific interest,because it gives wide prospects for posing and solving a number of problems in the theory of the crystal field, in lattice dynamics, in electron–phonon interactions, etc.

To illustrate the structural possibilities of the garnets, we have compiled Table 9.1, in which we list the ions, their degrees of oxidation, and their possible coordinations. This table is based on the results of X-ray structural and crystal-chemical research on both natural and synthetic garnets, on analysis of their magnetic and gamma resonance properties, and on study of their optical and EPR spectra. We must supplement Table 9.1 with the remark that the incorporation in garnets of ions with different valences from the A^{3+} and B^{3+} ions, in order to preserve electrical neutrality, necessitates the introduction of compensator ions of the appropriate valences. Similar considerations must be taken into account concerning the ionic radii of the companion ions in the crystal.

By now, about 30 different crystalline matrices of the garnet-type structure with Ln^{3+} ions have been obtained, among them 13 crystals doped with Nd^{3+} ions (see Table 9,2)[3], which are remarkable for their spectral and laser properties (see Table 5.1). Thirteen emission channels (see Fig. 9.4)[4] of garnet crystals make it possible to produce laser emission over a rather wide spectral region extending

---

[3]The new $Ca_3Ga_2Ge_3O_{12}$-Nd^{3+} garnet laser crystal was synthesized in [807].

[4]Stimulated emission of Ho^{3+} ions in $Y_3Al_5O_{12}$, $Lu_3Al_5O_{12}$ and $Gd_3Ga_5O_{12}$ crystals was also excited on the $^5I_6 \rightarrow {}^5I_8$ transition in [774, 775].

Table 9.1. Structural possibilities of garnets [524, 655]

Crystallographic positions (anion coordination)	Valence						
	1 +	2 +	2 − (1 −)	3 +	4 +	5 +	6 +
a positions (6)	Li,Na	Mg,Ca(?), Mn,Fe,Co, Ni,Cu,Zn,Cd	—	Al,Sc,V,Cr, Mn,Fe,Co, Ga,Y,Ru, Rh(?),In,Pt	Si,Ti,Ge,Zr, Ru(?),Sn,Hf	V(?),Nb, Sb,Ta	Te,W
c positions (8)	Li,Na, K(?), Cu(?)	Mg,Ca,Mn, Fe,Co,Cu, Sr,Cd,Ba, Pb(?)	—	Y,In( )), Ln-ions,Bi	Zr,Hf	—	—
d positions (4)	Li	Mg,Mn,Co, Ni,Zn	—	Al,Fe,Co, Ga	Si,Ti,Fe, Ge,Sn	P,V,As	—
h positions	—	—	O,(F)	—	—	—	—

In certain conditions, some Ln^{3+} ions can occur in the *a* positions, More-detailed information on this and other topics in the crystal chemistry of garnets can be found, for example, in [699].

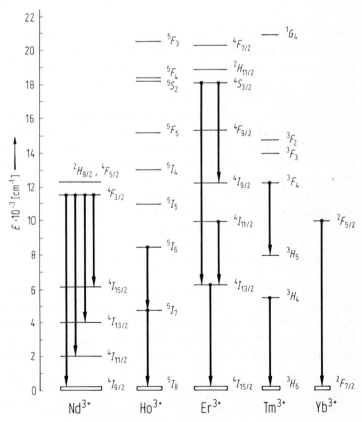

Fig. 9.4. Induced transitions of Ln^{3+} ions in crystals with garnet structure

from 0.86 to about 3 μm. In most cases, stimulated emission is excited at room temperature. In mentioning here the results obtained and the rapid advance in the study of these crystals with their unique structure and properties, we may confidently hope that in the physics of laser crystals and in the spectroscopy of activated media, garnets will play the same prominent role that they have played in the physics of magnetism. Speaking figuratively, the first pages of a new chapter in crystal physics "the physics of laser garnets," have already begun to be filled.

In connection with the development of integrated optics, among the current topics in the physics of activated crystals there are investigation of and search for high-concentration and self-activated laser compounds. A large number of laser media have already been found in which the emitting ions are the main constituents of the crystalline lattice. Evidently, in the investigation of such phenomena as electron-excitation migration and concentration quenching of luminescence in these compounds, much effort will yet be needed, by both experimenters and theorists. The conditions of excitation and study of luminescence and stimulated emission will not always be trivial. For further progress in the search for high-concentration laser crystals, as well as for new efficient laser

Table 9.2. Garnet-structure crystalline laser matrices with Ln^{3+} ions [655]

Garnet	Activator ion				
	Nd^{3+}	Ho^{3+}	Er^{3+}	Tm^{3+}	Yb^{3+}
$CaMg_2Y_2Ge_3O_{12}$	+				
$Y_3Al_5O_{12}$	+	+	+	+	+
$Y_3Sc_2Al_3O_{12}$	+				
$Y_3Sc_2Ga_3O_{12}$	+				
$Y_3Fe_5O_{12}$		+			
$Y_3Ga_5O_{12}$	+	+			
$(Y,Er)_3Al_5O_{12}$		+	+	+	
$(Y,Yb)_3Al_5O_{12}$					+
$(Y,Lu)_3Al_5O_{12}$	+				
$Ba_{0.25}Mg_{2.75}Y_2Ge_3O_{12}$	+				
$Gd_3Sc_2Al_3O_{12}$	+				+
$Gd_3Sc_2Ga_3O_{12}$	+				
$Gd_3Ga_5O_{12}$	+				+
$Ho_3Al_5O_{12}$		+			
$Ho_3Sc_2Al_3O_{12}$		+			
$Ho_3Ga_5O_{12}$		+			
$Er_3Al_5O_{12}$			+	+	
$Er_3Sc_2Al_3O_{12}$		+			
$(Er,Tm,Yb)_3Al_5O_{12}$		+			
$(Er,Yb)_3Al_5O_{12}$				+	
$(Er,Lu)_3Al_5O_{12}$			+		
$Yb_3Al_5O_{12}$		+	+		
$(Yb,Lu)_3Al_5O_{12}$					+
$Lu_3Al_5O_{12}$	+	+	+	+	+
$Lu_3Sc_2Al_3O_{12}$	+				+
$Lu_3Ga_5O_{12}$	+				+

garnets, not the least role will be played, as usual, by crystal chemistry and technology. To emphasize their role in the production of crystalline laser media, it also seems appropriate to point out that polycrystalline materials produced by hot pressing, or by cold pressing followed by sintering, may sometimes be highly promising. Lasers of CaF_2 and SrF_2 polycrystals doped with Dy^{2+} ions were produced by the former technique as far back as 1964 [705]; the latter method has been employed quite recently [706, 707] to synthesize a polycrystalline laser medium, "yttralox" ($Y_2O_3-ThO_2-Nd_2O_3$), which is as good as the ED-2 neodymium glass so far as some of its stimulated-emission parameters are concerned. There is good reason to believe that both these methods will be further developed, especially to obtain large-size active elements for laser amplifiers.

An extremely wide range of laser and spectroscopic properties have been exhibited in recent years by crystals of the $LiYF_4$ type doped with Ln^{3+} ions (see, for example Refs. [83, 123, 132, 261, 545, 568, 606, 762, 777–779, 781]). It is important to note here that among all the known laser crystals, compounds of the $LiYF_4$ type can be used with the largest number of Ln^{3+} ions (see Chap. 1), such as: $Ce^{3+}, Pr^{3+}, Nd^{3+}, Tb^{3+}, Ho^{3+}, Er^{3+}$, and Tm^{3+}. Nineteen channels permit excitation of stimulated emission in the wide region from 0.3225 μm ($5d \rightarrow {}^2F_{7/2}$ transition of Ce^{3+} ions) to 3.914 μm (${}^5I_5 \rightarrow {}^5I_6$ transition of Ho^{3+} ions). Taking into account specific crystal chemical and physical properties of compounds of this type, one can look forward to the broadening of the list of both lasing Ln^{3+} ions and their emission channels. Like garnets, this class of crystals is of great interest for the spectroscopy and the physics of laser crystals, showing itself a model material for the study of different processes occurring in the regime of excitation and stimulated emission. Practical applications are already projected for some of them.

Let us now list the main problems towards the solution of which the physics and stimulated-emission spectroscopy of activated crystals is being directed at present. The study of these problems is contributing considerably to the general development of the physics of solid-state lasers. The most important of these problems are:

1) Search for new activated crystals (with lanthanide and actinide ions) capable of exhibiting stimulated emission.

2) Search for new operating schemes and novel principles of excitation (including nonoptical methods) for crystalline lasers.

3) Search for new channels of stimulated-emission and for more efficient means of exciting stimulated emission in both new and already developed laser crystals.

4) Exact measurement of the frequency (wavelength) of induced transitions.

5) Study of the relation between the main parameter of stimulated-emission (the threshold of excitation of the laser medium) and the most important spectroscopic characteristics of the medium, which should be carried out over a wide temperature range.

6) Study of the temperature behavior of induced transitions; here the mechanisms of switching and "ignition" of stimulated-emission channels are of special interest.

7) Analysis of the processes of energy transfer between unlike impurity ions, which play an important role both in the migration of absorbed energy through the excited states of the crystal and in the phenomena of sensitization (deactivation) and quenching of luminescence.

As the experience of the last few years suggests, use of the methods of stimulated-emission spectroscopy plays a role that is far from unimportant.

8) In combined spectroscopic studies of such manifestations of the electron-phonon interaction as the temperature shift and broadening of zero-phonon lines, as well as the temperature dependence of the nonradiative transition probability.

9) In combined spectroscopic analysis for establishing the exact crystal-field-splitting scheme of activator-ion multiplets in laser crystals.

The successful solution of the problems enumerated above, especially the last two, will also depend on the further development of all branches of the theoretical spectroscopy of activated crystals.

In conclusion, a few words about tunable lasers based on insulating crystals. Even though study of some aspects of them have been discussed in Sects. 3.10, 11, it is necessary to underline the urgency of the practical application of such devices and the extreme importance of the search for active media for such lasers and the study of the physical processes that take place. Attention should be concentrated on crystals with divalent and trivalent lanthanide ions, which are characterized by intense, broad-band luminescence with a high quantum yield in the ultraviolet, visible, and infrared regions of the spectrum. Recent studies have revealed (see, for example, [476, 499, 698, 708, 781, 817–820]) that certain transitions between the states of the $5d$ and $4f$ configurations of such ions as Ce^{3+}, Pr^{3+}, Nd^{3+}, Er^{3+}, Tm^{3+}, and also Sm^{2+}, Eu^{2+}, Yb^{2+} and other ions, in some fluoride and oxygen-containing crystals may be suitable for the development of tunable lasers. Among ions promising in this respect are also those of the actinide group, characterized by intense transitions between the states of the $5f$ electron configuration. Of course, use of these ions will be severely limited because of the radioactivity of the elements of this group. Finally, the attention of the reader should be called to the investigations of Allied Chemical Corporation and Lincoln Laboratory (MIT) workers on the creation of tunable phonon-terminated lasers on the basis of crystals doped with iron-group ions. The recent results of these authors [801, 802, 813, 814, 824—827, 832] allow one to hope that effective tunable crystal lasers generating in the near infrared region will soon be created. The most interesting of these papers [824–826] discuss the creation of a tunable laser (in the region from 0.7 to ≈ 0.8 μm) on the base of alexandrite, $BeAl_2O_4$-Cr^{3+}. This crystal has unique physical properties (for instance, its thermal conductivity is equal to 0.23 W $cm^{-1}K^{-1}$). When conventionally pumped by a lamp, it generates efficiently in the long-pulse, Q-switched, and CW

Table 9.3. Spectral and energetic parameters of some color-center crystal lasers [713, 714]

Parameter \ Crystal, color-center type	KCl–Li$^+$ $F_A(II)^a$	RbCl–Li$^+$ $F_A(II)^b$	KCl–Na$^+$ $F_B(II)$	LiF F_2^-
Threshold excitation power [mW]	13	100	20	0.8 [MW/cm^2]
Power output [mW]	85	6	6	—
Slope efficiency [%]	7.7	2	2	0.75–2
Tuning range [μm]	2.5–2.9	2.75–3.0	2.2–2.65	0.88–1.2

[a] Laser properties were also investigated in [709–712].
[b] Laser properties were also investigated in [710–711].

regimes, both at room and elevated temperatures [824]. In particular, about 25–30 W output power was achieved in the CW regime [826]. This work will give a new impetus to the search for new laser crystals doped with iron-group ions.

Another way of solving the problem of development of the tunable crystalline lasers, which has already proved to be promising, is the utilization of the luminescence properties of color centers in alkali-halide crystals [709], which are due to the nature of their own and impurity point defects. Although in the first studies of lasers based on crystals with color centers [709], broad-band pumping by a pulsed-xenon lamp was used, [710–714] mention the whole arsenal of modern means (for example, laser pumping, and resonators with distributed feedback) that have been used in the development of tunable dye lasers. The accomplishments and the yet-unrealized potentialities of this new class of lasers based on crystals have been vividly demonstrated by the authors of recent papers [713, 759]. Table 9.3 contains some of the most recent results obtained in studies of lasers based on color-center crystals. Recent progress in the field of color-center laser construction and their application is elucidated in [821–823]. While noting the urgency of the problem of development of such lasers, we may hope that the diversity of their active media will be supplemented, aside from the alkali-halide crystals, by other types of crystals, and, in particular, by activated crystals. The combination of the spectroscopic potential of the latter type with the specific properties of color centers may stimulate the development of new, promising types of lasers and operating schemes for solid-state lasers[5]. Attainment of this combination may reasonably be expected to occur in the not-so-distant future. The promise of the field of color-center lasers is revealed by *Feofilov* et al. [763], who obtained laser action, at room temperature, on color centers in CaF_2 and SrF_2 crystals doped with Na^+ ions. Recent progress in the field of color-center laser construction and their application is discussed in [821–823].

---

[5]The sensitizing role of color centers in CaF_2-Er^{3+} laser crystals is already known [146].

References

1 A. Einstein: Strahlungs-Emission und -Absorption nach der Quantentheorie. Verh. Dtsch. Phys. Ges. *18*, 318–323 (1916)
2 N.G. Basov, A.M. Prokhorov: Molecular beams application for radiospectroscopic study of rotational molecular spectra. Zh. Eksp. Teor. Fiz. *27*, 431–438 (1954)
3 J.P. Gordon, H.J. Zeiger, C.H. Townes: Molecular microwave oscillator and new hyperfine structure in the microwave spectrum of NH_3. Phys. Rev. *95*, 282–284 (1954)
4 A.A. Kaminskii: Stimulated emission spectroscopy: A review. Opto-electronics *3*, 19–35 (1971)
5 A.A. Kaminskii: "Investigation of spectroscopic characteristics in the experiments on stimulated emission," in *Spektriskopiya Kristallov*, ed. by P.P. Feofilov (Nauka, Leningrad 1973) pp. 70–93
6 A.A. Kaminskii, V.V. Osiko: Inorganic laser materials with ionic structure. Izv. Akad. Nauk SSSR, Neorg. Mater. *1*, 2049–2087 (1965)
   English transl.: Inorg. Mater. (USSR) *1*, 1853–1887 (1965)
7 A.A. Kaminskii, V.V. Osiko: Inorganic laser materials with ionic structure. Izv. Akad. Nauk SSSR, Neorg. Mater. *3*, 417–463 (1967)
   English transl.: Inorg. Mater. (USSR) *3*, 367–415 (1967)
8 A.A. Kaminskii, V.V. Osiko: Inorganic laser materials with ionic structure. III. Izv. Akad. Nauk SSSR, Neorg. Mater. *6*, 629–696 (1970)
   English transl.: Inorg. Mater. (USSR) *6*, 555–614 (1970)
9 J.E. Geusic, H.M. Marcos, L.G. Van Uitert: Laser oscillations in Nd-doped yttrium aluminum, yttrium gallium and gadolinium garnets. Appl. Phys. Lett. *4*, 182–184 (1964)
10 Kh.S. Bagdasarov, A.A. Kaminskii: $YAlO_3$ with TR^{3+} ion impurity as an active laser medium. Zh. Eksp. I Teor. Fiz. Pis'ma Red. *9*, 501–502 (1969)
   English transl.: JETP Lett. *9*, 303 (1969)
11 A.L. Schawlow, C.H. Townes: Infrared and optical masers. Phys. Rev. *112*, 1940–1949 (1958)
12 T.H. Maiman: Stimulated optical radiation in ruby masers. Nature *187*, 493–494 (1960)
13 P.P. Sorokin, M.J. Stevenson: Stimulated infrared emission from trivalent uranium. Phys. Rev. Lett. *5*, 557–559 (1960)
14 P.P. Sorokin, M.J. Stevenson: Solid-state optical maser using divalent samarium in calcium fluoride. IBM J. Res. Dev. *5*, 56–58 (1961)
15 L.N. Galkin, P.P. Feofilov: The luminescence of trivalent uranium. Dok. Akad. Nauk SSSR *114*, 745–747 (1957)
   English transl.: Sov. Phys. -Dokl. 2, 255–257 (1957)
16 P.P. Feofilov: Absorption and luminescence of divalent rare-earth ions in the crystals of artificial and natural fluorite. Opt. Spektrosk. *1*, 992–999 (1956)
17 L.F. Johnson, K. Nassau: Infrared fluorescence and stimulated emission of Nd^{3+} in $CaWO_4$. Proc. IRE *49*, 1704–1706 (1961)
18 G.D. Boyd, R.J. Collins, S.P.S. Porto, A. Yariv, W.A. Hargreaves: Excitation, relaxation and continuous maser action in the 2.613-micron transition of $CaF_2:U^{3+}$. Phys. Rev. Lett. *8*, 269–272 (1962)
19 F.J. McClung, R.W. Hellwarth: Giant optical pulsations from ruby. J. Appl. Phys. *33*, 828–829
20 L.F. Johnson, R.E. Dietz, H.J. Guggenheim: Optical maser oscillations from Ni^{2+} in MnF_2 involving simultaneous emission of phonons. Phys. Rev. Lett. *11*, 318–320 (1963)
21 L.F. Johnson, L.G. Van Uitert, J.J. Rubin, R.A. Thomas: Energy transfer from Er^{3+} to Tm^{3+} and Ho^{3+} ions in crystals. Phys. Rev. *A 133*, 494–498 (1964)

22　Yu.K. Voronko, A.A. Kaminskii, V.V. Osiko, A.M. Prokhorov: A new type of crystal for optically pumped lasers. Izv. Akad. Nauk SSSR, Neorg. Mater. 2, 1161–1170 (1966)
English transl.: Inorg. Mater (USSR) 2, 991–998 (1966)

23　Kh.S. Bagdasarov, Yu.K. Voronko, A.A. Kaminskii, V.V. Osiko, A.M. Prokhorov: Room-temperature induced emission of yttrofluorite crystals containing Nd^{3+}. Kristallografiya 10, 746–747 (1965)
English transl.: Sov. Phys.-Crystallgr. 5, 626–627 (1966)

24　A.A. Kaminskii: High-temperature effects observed in stimulated emission from CaF_2 and LaF_3 crystals activated with Nd^{3+}. Zh. Eksp. Teor. Fiz. Pis'ma Red. 6, 615–618 (1967)
English transl.: JETP Lett. 6, 115–118 (1967)

25　A.A. Kaminskii: Cascade laser schemes based on activated crystals. Izv. Akad. Nauk SSSR, Neorg. Mater. 7, 904–907 (1971)
English transl.: Inorg. Mater. (USSR) 7, 802–804 (1971)

26　A.A. Kaminskii: Laser with combined active medium. Zh. Eksp. Teor. Fiz. Pis'ma Red. 7, 260–262 (1968)
English transl.: JETP Lett. 7, 201–203 (1968)

27　A.A. Kaminskii: Laser with combined active medium. Dok. Akad. Nauk SSSR 180, 59–62 (1968)
English transl.: Sov. Phys.-Dokl. 13, 413–416 (1968)

28　A.A. Kaminskii: On the methods for investigation of autoresonance energy transfer in active media for lasers. Zh. Eksp. Teor. Fiz. 54, 1659–1674 (1968)
English transl.: Sov. Phys.-JETP 27, 889–896 (1968)

29　L.G. DeShazer: "An investigation of ionic cross relaxation by a quantum electronics technique," in Optical Properties of Ions in Crystals, ed. by H.M. Crosswhite and H.W. Moos (Wiley, New York 1967) pp. 507–518

30　L.G. DeShazer, E.A. Maunders: Spectral control of laser oscillations by secondary light sources. IEEE J. Quantum Electron. 642–644 (1968)

31　L.F. Johnson, H.J. Guggenheim: Infrared-pumped visible laser. Appl. Phys. Lett. 19, 44–47 (1971)

32　A.A. Kaminskii: High-temperature spectroscopic studies of the stimulated emission of laser crystals and glasses activated by Nd^{3+} ions. Zh. Eksp. Teor. Fiz. 54, 727–750 (1968)
English transl.: Sov. Phys.-JETP 27, 388–399 (1968)

33　V.A. Sychugov, G.P. Shipulo: Thermal investigation on Nd^{3+} doped yttrium aluminum garnet. Fiz. Tverd. Tela (Leningrad) 10, 2821–2824 (1968)
English transl.: Sov. Phys.-Solid State 10, 2224–2225 (1969)

34　A.A. Kaminskii: High-temperature spectroscopic investigation of stimulated emission from lasers based on crystals, activated with Nd^{3+} ions. Phys. Status Solidi (a) 1, 573–589 (1970)

35　C.K. Jörgensen: Absorption Spectra and Chemical Bonding in Complexes (Pergamon Press, New York 1962)

36　A.A. Kaminskii: New high-temperature induced transition of an optical quantum generator based on SrF_2-Nd^{3+} crystals (type I). Izv. Akad. Nauk SSSR, Neorg. Mater. 5, 615–616 (1969)
English transl.: Inorg. Mater. (USSR) 5, 525–526 (1969)

37　D.N. Vylegzhanin, A.A. Kaminskii: Investigation of electron-phonon interaction phenomena in the LaF_3-Nd^{3+} crystal. Zh. Eksp. Teor. Fiz. 62, 685–700 (1972)
English transl.: Sov. Phys.-JETP 35, 361–369 (1973)

38　I.S. Andriesh, V.Ya. Gamurar, D.N. Vylegzhanin, A.A. Kaminskii, S.I. Klokishner, Yu.E. Perlin: Electron-phonon interaction in $Y_3Al_5O_{12}$-Nd^{3+}. Fiz. Tverd. Tela (Leningrad) 14, 2967–2979 (1972)
English transl.: Sov. Phys. Solid State 14, 2550–2558 (1973)

39　Yu.K. Voronko, V.V. Osiko, A.M. Prokhorov, I.A. Shcherbakov: Some aspects of spectroscopy of laser crystals with ionic structure. Proc. P.N. Lebedev Phys. Inst. Vol. 60, ed. by D.V. Skobeltsin (Consultants Bureau, New York, London 1974) pp. 1–30.

40　Z.J. Kiss, R.C. Duncan: Pulsed and continuous optical maser action in CaF_2:Dy^{2+}. Proc. IRE 50, 1531–1532 (1962)

41　L.F. Johnson: Continuous operation of the CaF_2:Dy^{2+} optical maser. Proc. IRE 50, 1691–1692 (1962)

42　A. Yariv: Continuous operation of a CaF_2:Dy^{2+} optical maser. Proc. IRE 50, 1699–1700 (1962)

43 P.P. Feofilov: *The Physical Basis of Polarized Emission* (Consultants Bureau, New York 1961)

44 A.A. Kaplyanskii: The effect of elastic deformation by uniaxial compression and elongation of crystals on the spectra of local anisotropic centers in a cubic lattice. I. Opt. Spektrosk. 7, 677–690 (1959)
English transl.: Opt. Spectros. (USSR) 7, 406–413 (1959)

45 Yu.K. Voronko, A.A. Kaminskii, V.V. Osiko: Analysis of the optical spectra of CaF_2-Nd^{3+} crystals (type I). Zh. Eksp. Teor. Fiz. 49, 420–428 (1965)
English transl.: Sov. Phys.-JETP 22, 295–300 (1966)

46 G.H. Dieke, H.M. Crosswhite: The spectra of the doubly and triply ionized rare earths. Appl. Opt. 2, 675–686 (1963)

47 G.H. Dieke: "Spectroscopic observations on maser materials," in *Advances in Quantum Electronics*, ed. by J.R. Singer (Columbia Univ. Press, New York 1961) pp. 164–186

48 B.P. Zakharchenya, A.Ya. Ryskin: Zeeman effect in the absorption and luminescence spectra of CaF_2:Sm^{2+} and SrF_2:Sm^{2+}. Opt. Spektrosk. 13, 875–877 (1962)
English transl.: Opt. Spectros. (USSR) 13, 501–502 (1962)

49 M.J. Weber: Probabilities for radiative and nonradiative decay of Er^{3+} in LaF_3. Phys. Rev. 157, 262–272 (1967)

50 M.J. Weber: Radiative and multiphonon relaxation of rare-earth ions in Y_2O_3. Phys. Rev. 171, 283–291 (1968)

51 L.A. Riseberg, H.W. Moos: Multiphonon orbit–lattice relaxation of excited states of rare earth ions in crystals. Phys. Rev. 714, 429–438 (1968)

52 L.A. Riseberg, H.W. Moos: Multiphonon orbit–lattice relaxation in $LaBr_3$, $LaCl_3$ and LaF_3. Phys. Rev. Lett. 19, 1423–1426 (1967)

53 L.A. Riseberg, H.W. Moos, W.D. Partlow: Multiphonon relaxation in laser materials and applications to the design of quantum electronic devices. IEEE J. Quantum Electron. 4, 609–612 (1968)

54 K.K. Rebane: *Impurity Spectra of Solids* (Plenum Press, New York, London 1970)

55 D.E. McCumber: Theory of phonon-terminated optical masers. Phys. Rev. A 134, 299–306 (1964)

56 D.L. Dexter: A theory of sensitized luminescence in solids. J. Chem. Phys. 21, 836–850 (1953)

57 Th. Förster: Experimentelle und theoretische Untersuchung des zwischenmolekularen Übergangs von Elektronenanregungsenergie. Z. Naturforsch. 4a, 321–327 (1949)

58 M.D. Galanin: The problem of the effect of concentration on the luminescence of solutions. Zh. Eksp. Teor. Fiz. 28, 485–495 (1955)
English transl.: Sov. Phys.-JETP 1, 317–325 (1955)

59 Yu.E. Perlin: Modern methods in the theory of multiphonon processes. Usp. Fiz. Nauk 80, 553–595 (1963)
English transl.: Sov. Phys.-Usp. 6, 542–565 (1964)

60 B.I. Stepanov (ed.): *Metody Rascheta Opticheskikh Kvantovykh Generatorov*, Vol. 1, (Nauka i Tekhnika, Minsk 1966)

61 M.A. Elyashevich: *Spectra of Rare Earths* (Translation USEAC, AEC-tr-4403, Office of Technical Information, Department of Commerce, Washington 1961)

62 H. Walther (ed.): *Laser Spectroscopy of Atoms and Molecules*, Topics in Applied Physics, Vol. 2 (Springer, Berlin, Heidelberg, New York 1976)

63 P.P. Feofilov: Luminescence of tri- and bivalent ions of the rare earths in crystals of fluorite type. Acta Phys. Pol. 26, 331–343 (1964)

64 A.A. Kaminskii, V.V. Osiko, Yu.K. Voronko: A five-component fluoride—a new laser material. Kristallografiya 13, 332 (1968)
English transl.: Sov. Phys.-Crystallogr. 13, 267 (1968)

65 A.A. Kaminskii, Y.E. Lapsker, B.P. Sobolev: Induced emission of $NaCaCeF_6$:Nd^{3+} at room temperature. Phys. Status Solidi 23, K5–K7 (1967)

66 Kh.S. Bagdasarov, A.A. Kaminskii, B.P. Sobolev: Laser action in cubic $5NaF\cdot9YF_3$–Nd^{3+} crystals. Kristallografiya 13, 900–901 (1969)
English transl.: Sov. Phys.-Crystallogr. 13, 779–780 (1969)

67 A.M. Morosov, M.N. Tolstoi, P.P. Feofilov, V.N. Shapovalov: Luminescence and stimulated radiation of neodymium in lanthanum-sodium molybdate crystals. Opt. Spektrosk. 22, 414–419 (1967)
English transl.: Opt. Spectrosc. (USSR) 22, 224–226 (1967)

68 G.M. Zverev, G.Ya. Kolodnyi: Stimulated emission and spectroscopic investigations of double lanthanum-sodium molybdate single crystals with neodymium impurities. Zh. Eksp. Teor. Fiz. *52*, 337–341 (1967)
    English transl.: Sov. Phys.-JETP *25*, 217–220 (1967)
69 A. Yariv, J.P. Gordon: The laser. Proc. IEEE *51*, 4–29 (1963)
70 A.A. Mak, Yu.A. Ananev, B.A. Ermakov: Solid-state lasers. Usp. Fiz. Nauk *92*, 373–426 (1967)
    English transl.: Sov. Phys. Usp. *10*, 419–452 (1968)
71 V.M. Fain, Ya.I. Chanin: *Quantum Radiophysics* (Sov. Radio, Moscow 1965)
72 B. DiBartolo: *Optical Interactions in Solids* (Wiley, New York 1968)
73 A.L. Mikaelyan, M.L. Ter-Mukaelyan, Yu.G. Turkov: *Solid-State Optical Generators* (Sov. Radio, Moscow 1967)
74 A. Yariv: *Quantum Electronics* (Wiley, New York 1975)
75 R. Solomon, L. Mueller: Stimulated emission at 5985 Å from Pr^{3+} in LaF_3. Appl. Phys. Lett. *3*, 135–137 (1963)
76 F. Varsanyi: Surface lasers. Appl. Phys. Lett. *19*, 169–171 (1971)
77 A.A. Kaminskii, P.V. Klevtsov, L. Li, A.A. Pavlyuk: Laser $^4F_{3/2} \rightarrow \, ^4I_{9/2}$ and $^4F_{3/2} \rightarrow \, ^4I_{13/2}$ transitions in $KY(WO_4)_2$:Nd^{3+}. IEEE J. Quantum Electron. *8*, 457–458 (1972)
78 L.F. Johnson, R.A. Thomas: Maser oscillations at 0.9 and 1.35 microns in $CaWO_4$:Nd^{3+}. Phys. Rev. *131*, 2038–2040 (1963)
79 M. Birnbaum, A.W. Tucker: Nd-$YAlO_3$ oscillation at 0.930 μm at 300°K. IEEE J. Quantum Electron. *9*, 46 (1973)
80 R.W. Wallace, S.E. Harris: Oscillation and doubling of the 0.946 μ line in Nd^{3+}:YAG. Appl. Phys. Lett. *15*, 111–112 (1969)
81 M. Birnbaum, A.W. Tucker, P.J. Pomphrey: New Nd:YAG laser transition $^4F_{3/2} \rightarrow \, ^4I_{9/2}$. IEEE J. Quantum Electron. *8*, 501 (1972)
82 Kh.S. Bagdasarov, A.A. Kaminskii, A.M. Kevorkov, A.M. Prokhorov, S.E. Sarkisov, T.A. Tevosyan: Laser properties of Y_2SiO_5-Nd^{3+} crystals irradiated at the $^4F_{3/2} \rightarrow \, ^4I_{11/2}$ and $^4F_{3/2} \rightarrow \, ^4I_{13/2}$ transitions. Dok. Akad. Nauk SSSR *212*, 1326–1327 (1973)
    English transl.: Sov. Phys.-Dokl. *18*, 664 (1973)
83 H.P. Jenssen, D. Castleberry, D. Gabbe, A. Linz: Stimulated emission at 5445 Å in Tb^{3+}:YLF. Digest of Technical Papers CLEA 1973 IEEE/OSA, Washington, May-June 1973, p. 47
84 H.P. Weber, T.C. Damen, H.G. Danielmeyer, B.C. Tofield: Nd-ultraphosphate laser. Appl. Phys. Lett. *22*, 534–536 (1973)
85 A.A. Kaminskii, A.A. Pavlyuk, P.V. Klevtsov: Spectroscopic properties of $KY(MoO_4)_2$ monocrystals activated by Nd^{3+} ions. Opt. Spektrosk. *28*, 292–296 (1970)
    English transl.: Opt. Spectrosc. (USSR) *28*, 157–160 (1970)
86 Kh.S. Bagdasarov, G.A. Bogomolova, D.N. Vylegzhanin, A.A. Kaminskii A.M. Kevorkov, A.G. Petrosyan, A.M. Prohkorov: Luminescence and stimulated emission of Yb^{3+} ions in aluminum garnets. Dok. Akad. Nauk SSSR *216*, 1247–1249 (1974)
    English transl.: Sov. Phys.-Dokl. *19*, 358–359 (1974)
87 B.Ya. Zabokritskii, A.D. Manuilskii, S.G. Odulov, M.S. Soskin: Generation of Nd^{3+} in crystals and glasses at 1.3 μm wavelength and simultaneous generation at 1.06 and 1.3 μm. Ukr. Fiz. Zh. (Russ. Ed.) *17*, 501–503 (1972)
88 A.A. Kaminskii, S.E. Sarkisov, Kh.S. Bagdasarov: Stimulated emission by Nd^{3+} ions in crystals, due to the $^4F_{3/2} \rightarrow \, ^4I_{13/2}$ transition. II. Izv. Akad. Nauk SSSR, Neorg. Mater. *9*, 509–511 (1973)
    English transl.: Inorg. Mater. (USSR) *9*, 457–459 (1973)
89 A.A. Kaminskii, S.E. Sarkisov: Stimulated emission by Nd^{3+} ions in crystals, due to the $^4F_{3/2} \rightarrow \, ^4I_{13/2}$ transition. I. Izv. Akad. Nauk SSSR, Neorg. Mater. *9*, 505–508 (1973)
    English transl.: Inorg. Mater. (USSR) *9*, 453–456 (1973)
90 V.I. Aleksandrov, A.A. Kaminskii, G.V. Maksimova, A.M. Prokhorov, S.E. Sarkisov, A.A. Sobol, V.M. Tatarintsev: Stimulated radiation of Nd^{3+} ions in crystals for the $^4F_{3/2} \rightarrow \, ^4I_{13/2}$ transition. Dok. Akad. Nauk SSSR *211*, 567–570 (1973)
    English transl.: Sov. Phys. Dokl. *18*, 495–497 (1974)
91 A.A. Kaminskii, S.E. Sarkisov, K.B. Seiranyan, B.P. Sobolev: Stimulated emission by Nd^{3+} ions in SrF_2-GdF_3. Izv. Akad. Nauk SSSR, Neorg. Mater. *9*, 340 (1973)
    English transl.: Inorg. Mater. (USSR) *9*, 310 (1973)

92   A.A. Kaminskii: On the possibility of investigation of the "Stark" structure of TR^{3+} ion spectra in disordered fluoride crystal systems. Zh. Eksp. Teor. Fiz. *58*, 407–419 (1970) English transl.: Sov. Phys.-JETP *31*, 216–222 (1970)

93   A.A. Kaminskii, S.E. Sarkisov, L. Li: Investigation of stimulated emission in the $^4F_{3/2} \rightarrow {}^4I_{13/2}$ transition of Nd^{3+} ions in crystals (III). Phys. Status Solidi (a) *15*, K141–K144 (1973)

94   A.A. Kaminskii, P.V. Klevtsov, L. Li, A.A. Pavlyuk: Spectroscopic and generation study of a new laser crystal, $KY(WO_4)_2 – Nd^{3+}$. Izv. Akad. Nauk SSSR, Neorg. Mater. *8*, 2153–2163 (1972)
     English transl.: Inorg. Mater. (USSR) *8*, 1896–1904 (1972)

95   A.M. Kevorkov, A.A. Kaminskii, Kh.S. Bagdasarov, T.A. Tevosyan, S.E. Sarkisov: Spectroscopic properties of crystals of $CaAl_4O_7 – Nd^{3+}$. Izv. Akad. Nauk SSSR, Neorg. Mater. *9*, 161 (1973)
     English transl.: Inorg. Mater. (USSR) *9*, 146–147 (1973)

96   A.M. Kevorkov, A.A. Kaminskii, Kh.S. Bagdasarov, T.A. Tevosyan, S.E. Sarkisov: Spectroscopic properties of crystals of $SrAl_4O_7 – Nd^{3+}$. Izv. Akad. Nauk SSSR, Neorg. Mater. *9*, 1839–1840 (1973)
     English transl.: Inorg. Mater. (USSR) *9*, 1637–1638 (1973)

97   A.A. Kaminskii, S.E. Sarkisov: Investigation of the stimulated emission due to $^4F_{3/2} \rightarrow {}^4I_{13/2}$ transition in Nd^{3+} ions in crystals. IV. Kvant. Elektron. (Moscow) *N3*, 106–108 (1974)
     English. transl.: Sov. J. Quantum Electron. *3*, 248–249 (1973)

98   G.A. Massey, J.M. Yarborough: High average power operation and nonlinear optical generation with the Nd:YAlO$_3$ laser. Appl. Phys. Lett. *18*, 576–579 (1971)

99   R.G. Smith: New room temperature CW laser transition in YAlG:Nd. IEEE J. Quantum Electron. *4*, 505–506 (1968)

100 U. DeSerno, D. Röss, G. Zeidler: Quasicontinuous giant pulse emission of $^4F_{3/2} \rightarrow {}^4I_{13/2}$ transition at 1.32 μm in YAG-Nd^{3+}. Phys. Lett. A *28*, 422–423 (1968)

101 R.W. Wallace: Oscillation of the 1.833 μ line in Nd^{3+}:YAG. IEEE J. Quantum Electron. 7, 203–204 (1971)

102 Yu.K. Voronko, A.A. Kaminskii, V.V. Osiko, A.M. Prokhorov: Stimulated emission of Ho^{3+} in CaF$_2$ at $\lambda = 5512$ Å. Zh. Eksp. Teor. Fiz. Pis'ma Red. *1*, 5–9 (1965)
     English transl.: JETP Lett. *1*, 3–5 (1965)

103 L.F. Johnson, H.J. Guggenheim: New laser lines in the visible from Er^{3+} ions in BaY$_2$F$_8$. Appl. Phys. Lett. *20*, 474–476 (1972)

104 E.P. Chicklis, C.S. Naiman, A. Linz: Stimulated emission at 0.85 μm in Er^{3+}:YLF. Digest of Technical Papers VII Intern. Quantum Electron. Conf., May 8–11, 1972, Montreal, p. 17

105 Yu.K Voronko, V.A. Sychugov: The stimulated emission of Er^{3+} ions in CaF$_2$ at $\lambda_1 = 8456$ Å and $\lambda_2 = 8548$ Å. Phys. Status Solidi *25*, K119–K121 (1968)

106 A.A. Kaminskii: Spectroscopic study of stimulated radiation of Er^{3+} activated CaF$_2$–YF$_3$ crystals. Opt. Spektrosk. *31*, 938–943 (1971)
     English transl.: Opt. Spectrosc. (USSR) *31*, 507–510 (1971)

107 M. Weber, M. Bass, T. Varitimos, D. Bua: Laser action from Ho^{3+}, Er^{3+}, and Tm^{3+} in YAlO$_3$. IEEE J. Quantum Electron. *9*, 1079–1·086 (1973)

108 Yu.K. Voronko, G.M. Zverev, A.M. Prokhorov: Stimulated emission from Er^{3+} ions in CaF$_2$. Zh. Eksp. Teor. Fiz. *48*, 1529–1532 (1965)
     English transl.: Sov. Phys.-JETP *21*, 1023–1025 (1965)

109 M.J. Weber, M. Bass, G.A. DeMars: Laser action and spectroscopic properties of Er^{3+} in YAlO$_3$. J. Appl. Phys. *42*, 301–305 (1971)

110 M.J. Weber, M. Bass, G.A. DeMars, K. Andringa, R.R. Monchamp: Stimulated emission at 1.663 μ from Er^{3+} ions in YAlO$_3$. IEEE J. Quantum Electron. *6*, 654 (1970)

111 M. Robinson, D.P. Devor: Thermal switching of laser emission of Er^{3+} at 2.69 μ and Tm^{3+} at 1.86 μ in mixed crystals of CaF$_2$:ErF$_3$:TmF$_3$. Appl. Phys. Lett. *10*, 167–170 (1967)

112 L.M. Hobrock, L.G. DeShazer, W.F. Krupke, G.A. Keig, D.E. Witter: Four-level operation of Tm:Cr:YAlO$_3$ laser at 2.35 μ. Digest of Technical Papers VII Intern. Quantum Electron. Conf., May 8–11, 1972, Montreal, pp. 15–16

113 L.F. Johnson, H.J. Guggenheim, R.A. Thomas: Phonon-terminated optical masers. Phys. Rev. *149*, 179–185 (1966)

114 L.F. Johnson, H.J. Guggenheim: Phonon-terminated coherent emission from V^{2+} ions in MgF$_2$. J. Appl. Phys. *38*, 4837–4839 (1967)

115 L.F. Johnson, R.E. Dietz, H.J. Guggenheim: Exchange splitting of the ground state of Ni^{2+} ions in antiferromagnetic MnF_2, $KMnF_3$, and $RbMnF_3$. Phys. Rev. Lett. *17*, 13–15 (1966)

116 K.R. German, A. Kiel, H. Guggenheim: Stimulated emission from $PrCl_3$. Appl. Phys. Lett. *22*, 87–89 (1973)

117 L.F. Johnson, R.E. Dietz, H.J. Guggenheim: Spontaneous and stimulated emission from Co^{2+} ions in MgF_2 and ZnF_2. Appl. Phys. Lett. *5*, 21–22 (1964)

118 Yu.S. Vagin, V.M. Marchenko, A.M. Prokhorov: Spectrum of a laser based on CaF_2:Sm^{2+} crystal. Zh. Eksp. Teor. Fiz. *55*, 1717–1726 (1968)
     English transl.: Sov. Phys. JETP *28*, 904–909 (1969)

119 K.K. Rebane, O.I. Sild: To the theory of the induced transitions in electron-vibrational sidebands. Proc. Inst. Phys. Est. SSR Acad. Sci., Tartu *23*, 18–21 (1963)

120 Y. Tanabe, S. Sugano: On the absorption spectra of complex ions, I. J. Phys. Soc. Jpn. *9*, 753–766 (1954)

121 Y. Tanabe, S. Sugano: On the absorption spectra of complex ions, II. J. Phys. Soc. Jpn. *9*, 756–779 (1954)

122 E.J. Woodbury, W.K. Ng: Ruby laser operation in the near IR. Proc. IRE *50*, 2367 (1962)

123 R. Allen, L. Esterowitz, M. Kruer, F. Bartoli: Stimulated emission at 0.64 μm in trivalent Pr doped $LiYF_4$. Abstracts of 13-th Rare Earth Research Conf. USA, October 16–19, 1977, Oglebay Park, p. 26

124 H.F. Ivey: Sensitized luminescence and its application in laser materials. Proc. Intern. Conf. on Luminescence (Academiai Kiado Hungary, Budapest 1968) pp. 2027–2049

125 M.N. Tolstoi: "Nonradiative energy transfer among rare-earth ions in crystals and glasses," in *Spektroskopiya Kristallov*, ed. by P.P. Feofilov (Nauka, Moscow 1970) pp. 124–135

126 Z.J. Kiss, R.C. Duncan: Cross-pumped Cr^{3+}–Nd^{3+}:YAG laser system. Appl. Phys. Lett. *5*, 200–202 (1964)

127 M. Bass, M.J. Weber: Nd, Cr–$YAlO_3$ laser tailored for high-energy Q-switched operation. Appl. Phys. Lett. *17*, 395–398 (1970)

128 L.F. Johnson, J.E. Geusic, L.G. Van Uitert: Coherent oscillations from Tm^{3+}, Ho^{3+}, Yb^{3+}, and Er^{3+} ions in yttrium aluminum garnet. Appl. Phys. Lett. *7*, 127–128 (1965)

129 R.H. Hopkins, R.G. Seidensticker: Redistribution of holmium and chromium during Czochralski growth of fluorapatite. J. Cryst. Growth *11*, 97–98 (1971)

130 K.B. Steinbruegge, T. Henningsen, R.H. Hopkins, R. Mazelsky, N.T. Melamed, E.P. Riedel, G.W. Roland: Laser properties of Nd^{+3} and Ho^{+3} doped crystals with the apatite structure. Appl. Opt. *11*, 999–1012 (1972)

131 R.L. Remski, L.T. James, K.H. Gooen, B. DiBartolo, A. Linz: Pulsed laser action in $LiYF_4$:Er^{3+}, Ho^{3+} at 77°K. IEEE J. Quantum Electron. *5*, 214 (1969)

132 E.P. Chicklis, C.S. Naiman, R.C. Folweiler, J.L. Doherty: Stimulated emission in multiply doped Ho^{3+}: YLF and YAG—a comparison. IEEE J. Quantum Electron. *8*, 225–230 (1972)

133 M.V. Dmitruk, A.A. Kaminskii, V.V. Osiko, M.M. Fursikov: Sensitization in optical, quantum generators based on CaF_2–ErF_3–Ho^{3+}. Izv. Akad. Nauk SSSR, Neorg. Mater. *3*. 579–581 (1967)
     English transl.: Inorg. Mater. (USSR) *3*, 516–518 (1967)

134 Kh.S. Bagdasarov, A.A. Kaminskii, B.P. Sobolev: Stimulated laser radiation using α-$NaCaErF_6$–Ho^{3+} and α-$NaCaErF_6$–Tu^{3+} crystals. Izv. Akad. Nauk SSSR, Neorg. Mater. *5*, 617–618 (1969)
     English transl.: Inorg. Mater. (USSR) *5*, 527–528 (1969)

135 L.G. Van Uitert, L.F. Johnson: Energy transfer between rare-earth ions. J. Chem. Phys. *44*, 3514–3521 (1966)

136 R.H. Hoskins, B.H. Soffer: Energy transfer and CW laser action in Ho^{3+}:Er_2O_3. IEEE J. Quantum Electron. *2*, 253–255 (1966)

137 Kh.S. Bagdasarov, G.A. Bogomolova, M.M. Gritsenko, A.A. Kaminskii, A.M. Kevorkov: Spectroscopic study of the laser crystal $LaAlO_3$-Nd^{3+}. Kristallografiya *17*, 415–417 (1972)
     English transl.: Sov. Phys. Crystallogr. *17*, 357–359 (1972)

138 L.F. Johnson, J.E. Dillon, J.P. Remeika: Optical studies of Ho^{3+} ions in YGaG and YIG. J. Appl. Phys. *40*, 1499–1500 (1969)

139 L.F. Johnson, J.F. Dillon, J.P. Remeika: Optical properties of Ho^{3+} ions in yttrium gallium garnet and yttrium iron garnet. Phys. Rev. *B1*, 1935–1946 (1970)

140   L.F. Johnson, J.E. Geusic, L.G. Van Uitert: Efficient, high-power coherent emission from Ho^{3+} ions in yttrium aluminum garnets, assisted by energy transfer. Appl. Phys. Lett. *8*, 200–202 (1966)

141   M.J. Weber, M. Bass, E. Comperchio, L.A. Riseberg: Ho^{3+} laser action in $YAlO_3$ at 2.119 μ. IEEE J. Quantum Electron. *7*, 497–498 (1971)

142   R.H. Hopkins, G.W. Roland, K.B. Steinbruegge, W.D. Partlow: Silicate oxyapatites: new high-energy storage laser host for Nd^{3+}. J. Electrochem. Soc. *118*, 637–639 (1971)

143   A.A. Kaminskii, Kh.S. Bagdasarov, A.G. Petrosyan, S.E. Sarkisov: Investigation of stimulated emission from $Lu_3Al_5O_{12}$ crystals with Ho^{3+}, Er^{3+}, and Tm^{3+} ions. Phys. Status Solidi (a) *18*, K31–K34 (1973)

144   N.C. Chang: Energy levels and crystal-field splitting of Nd^{3+} in yttrium oxide. J. Chem. Phys. *44*, 4044–4050 (1966)

145   Yu.K. Voronko, M.V. Dmitruk, T.M. Murina, V.V. Osiko: Continuous laser action using mixed crystals of the yttrofluorite type. Izv. Akad. Nauk SSSR, Neorg. Mater. *5*, 506–509 (1969)
      English transl.: Inorg. Mater. (USSR) *5*, 422–424 (1969)

146   P.A. Forrester, D.F. Sampson: A new laser line due to energy transfer from colour centers to erbium ions in CaF_2. Proc. Phys. Soc. London *88*, 199–204 (1966)

147   K.O. White, S.A. Schleusener: Coincidence of Er:YAG laser emission with methane absorption at 1645.1 nm. Appl. Phys. Lett. *21*, 419–420 (1972)

148   A.A. Kaminskii, V.V. Osiko: Sensitization in optical quantum generators based on CaF_2-ErF_3-Tm^{3+}. Izv. Akad. Nauk SSSR, Neorg. Mater. *3*, 582–583 (1967)
      English transl.: Inorg. Mater. (USSR) *3*, 519–520 (1967)

149   B.H. Soffer, R.H. Hoskins: Energy transfer and CW laser action in Tm^{3+}:Er_2O_3. Appl. Phys. Lett. *6*, 200 (1965)

150   L.G. Van Uitert, W.H. Grodkiewicz, E.F. Dearborn: Growth of large optical-quality yttrium and rare-earth aluminum garnets. J. Am. Ceram. Soc. *48*, 105–108 (1965)

151   S.A. Pollack: Radiative and radiationless transitions from $^4F_{5/2}$ state of the Nd^{3+} ions. J. Chem. Phys. *54*, 291–293 (1971)

152   M. Robinson, C.K. Asawa: Stimulated emission from Nd^{3+} and Yb^{3+} in noncubic sites of neodymium- and ytterbium-doped CaF_2. J. Appl. Phys. *38*, 4495–4501 (1967)

153   P.P. Feofilov, V.V. Ovsyankin: Cooperative luminescence of solids. Appl. Opt. *6*, 1828–1833 (1967)

154   V.V. Ovsyankin, P.P. Feofilov: Cooperative sensitization of luminescence in crystals activated with rare earth ions. Zh. Eksp. Teor. Fiz. Pis'ma. Red. *4*, 471–474 (1966)
      English transl.: JETP Lett. *4*, 317–318 (1966)

155   V.V. Ovsyankin, P.P. Feofilov: "Cooperative luminescence in crystals with rare earth activators," in *Spektroskopiya Kristallov*, ed. by P.P. Feofilov (Nauka, Moscow 1970) pp. 135–143

156   L.F. Johnson, A.A. Ballman: Coherent emission from rare earth in electro-optic crystals. J. Appl. Phys. *40*, 297–302 (1969)

157   S.A. Akhmanov, R.V. Khokhlov: *Problemy Nelineinoi Optiki* (USSR Information Institute, Moscow 1964)

158   N. Bloembergen: *Nonlinear Optics* (Benjamin, New York, Amsterdam 1965)

159   R.W. Minck, R.W. Terhune, C.C. Wang: Nonlinear optics. Appl. Opt. *5*, 1595–1612 (1966)

160   I.S. Rez: Crystals with nonlinear polarizability. Usp. Fiz. Nauk *93*, 633–674 (1967)
      English transl.: Sov. Phys. Usp. *10*, 759–782 (1967)

161   J.F. Nye: *Physical Properties of Crystals* (Clarendon Press, Oxford 1964)

162   N.F. Evlanova, A.S. Kovalev, V.A. Koptsik, L.S. Kornienko, A.M. Prokhorov, L.N. Rashkovich: Stimulated emission of $LiNbO_3$ crystals with neodymium impurity. Zh. Eksp. Teor. Fiz. Pis'ma. Red. *5*, 351–352 (1967)
      English transl.: JETP Lett. *5*, 291–292 (1967)

163   L.I. Ivleva, A.A. Kaminskii, Yu.S. Kusminov, V.N. Shpakov: Absorption, luminescence, and induced emission of $LiNbO_3$-Nd^{3+} crystals. Dok. Akad. Nauk SSSR *183*, 1068–1071 (1968)
      English transl.: Sov. Phys. Dokl. *13*, 1185–1187 (1969)

164   A.A. Kaminskii: Laser and spectroscopic properties of activated ferroelectrics. Kristallografiya *17*, 231–246 (1972)
      English transl.: Sov. Phys. Crystallogr. *17*, 194–207 (1972)

165   A.A. Kaminskii: Laser und spectroskopische Eigenschaften von aktivierten Ferroelektrika. Laser- und Elektro-Optik, *N1*, 30–38 (1972)

166   Kh.S. Bagdasarov, G.A. Bogomolova, A.A. Kaminskii, A.M. Prokhorov, T.M. Prokhortseva: Laser and spectroscopic properties of $Gd_2(MoO_4)_3 - Nd^{3+}$ crystals. Dok. Akad. Nauk SSSR *197*, 557–559 (1971)
      English transl.: Sov. Phys. Dokl. *16*, 216–218 (1971)

167   A.A. Kaminskii: An analysis of the Stark structure of TR^{3+} ion spectra by the stimulated radiation spectra technique. Zh. Eksp. Teor. Fiz. *56*, 83–87 (1969)
      English transl.: Sov. Phys. JETP *29*, 46–48 (1969)

168   A.A. Kaminskii: Stimulated radiation from $Y_3Al_5O_{12} - Nd^{3+}$ crystals. Zh. Eksp. Teor. Fiz. *51*, 49–58 (1966)
      English transl.: Sov. Phys. JETP *24*, 33–39 (1967)

169   A.A. Kaminskii, D.N. Vylegzhanin: Stimulated emission investigation of effects of electron-phonon interaction in crystals activated with Nd^{3+} ions. IEEE J. Quantum Electron. *7*, 329–338 (1971)

170   T. Kushida: Linewidth and thermal shifts of spectral lines in neodymium-doped yttrium aluminum garnet and calcium fluorophosphate. Phys. Rev. *185*, 500–508 (1969)

171   A.A. Kaminskii, G.A. Bogomolova, L. Li: Absorption, luminescence, induced radiation, and crystal splitting of the levels of ions of Nd^{3+} in the crystal YVO_4. Izv. Akad. Nauk SSSR, Neorg. Mater. *5*, 673–690 (1969)
      English transl.: Inorg. Mater. (USSR) *5*, 573–586 (1969)

172   A.A. Kaminskii, L. Li: Spectroscopic and laser studies on crystalline compounds in the system $CaO-Nb_2O_5$. $Ca(NbO_3)_2 - Nd^{3+}$ crystals. Izv. Akad. Nauk SSSR, Neorg. Mater. *6*, 294–306 (1970)
      English transl.: Inorg. Mater. (USSR) *6*, 254–263 (1970)

173   A.A. Kaminskii, G.I. Rogov, Kh.S. Bagdasarov: Stimulated radiation from a $Ca(NbO_3)_2 - Nd^{3+}$ crystal laser. Phys. Status Solidi *31*, K87–K90 (1969)

174   R. Pappalardo: Calculated quantum yield for photon-cascade emission (PCE) for Pr^{3+} and Tm^{3+} in fluoride host. J. Lumin. *14*, 159–193 (1976)

175   A.A. Mak, D.S. Prilezhaev, B.M. Sedov, V.I. Utyugov, V.A. Fromzel: "Some peculiarities of generation in activated media with inhomogeneously broadened luminescence line" in *Neodnorodnoe Ushirenie Spektralnykh Linii Aktivnykh Sred OKG* (Instit. of Phys., UkSSR Acad. Sci., Kiev 1969) pp. 134–142

176   A.A. Kaminskii: Temperature pulsations and multifrequency generation of $YAlO_3 - Nd^{3+}$. Zh. Eksp. Teor. Fiz. Pis'ma Red. *14*, 333–337 (1971)
      English transl.: JETP Lett. *14*, 222–225 (1971)

177   I.I. Kondilenko, P.A. Korotkov, O.N. Koshel: Peculiarities of multiple ruby OQG generation kinetics. Phys. Status Solidi (*a*) *4*, 43–52 (1971)

178   N.S. Belokrinitskii, G.S. Litvinov, A.D. Manuilskii, M.S. Soskin: "Investigation of the internal structure of 1.06 μm luminescence band of Nd^{3+} ions in $CaWO_4$ by the induced emission spectra," in *Spektroskopiya Kristallov*, ed. by P.P. Feofilov (Nauka, Moscow 1970) pp. 200–204

179   V.I. Kravchenko, M.S. Soskin, V.B. Timofeev: "Continuous tuning and generation frequency control in solid-state lasers," in *Kvantovaya Elektronika*, (Naukova Dumka, Kiev 1969) pp. 39–53

180   N.S. Belokrinitskii, N.G. Zubrilin, M.T. Shpak: Time development of wide-band-generation spectra of Nd^{3+}-doped phosphate glasses and disordered crystals in resonators with dispersion. Opt. Spektrosk. *31*, 766–768 (1971)
      English transl.: Opt. Spectrosc. (USSR) *31*, 411–412 (1971)

181   W. Low, E.L. Offenbacher: Review of ESR results in rutile, perovskites, spinels, and garnet structures. Solid State Phys. *17*, 135–216 (1965)

182   Yu.K. Voronko, V.V. Osiko: Uniform crystal splitting of Ln^{3+} levels. Izv. Akad. Nauk SSSR, Neorg. Mater. *3*, 413–414 (1967)
      English transl.: Inorg. Mater. (USSR) *3*, 365–366 (1967)

183   L.F. Johnson: Optical maser characteristics of rare-earth ions in crystals. J. Appl. Phys. *34*, 897–909 (1963)

184   A.A. Kaminskii: Spectral composition of laser light from neodymium-doped calcium tungstate crystals. Izv. Akad. Nauk SSSR, Neorg. Mater. *6*, 396–397 (1970)

English transl.: Inorg. Mater. (USSR) 6, 347–349 (1970)

185   L.F. Johnson: Characteristics of the CaWO$_4$:Nd^{3+} optical maser. Proc. 3rd Quantum Electron. Int. Congr., ed. by P. Grivet, N. Bloembergen (Columbia Univ. Press, New York 1964) pp. 1021–1035

186   G.V. Maksimova, A.A. Sobol: Investigation of the optical centers of Nd^{3+} in CaWO$_4$. Izv. Akad. Nauk SSSR, Neorg. Mater. 6, 307–313 (1970)
       English transl. Inorg. Mater. (USSR) 6, 264–268 (1970)

187   T. Kushida, H.M. Marcos, J.E. Geusic: Laser transition cross section and fluorescence branching ratio for Nd^{3+} in yttrium aluminum garnet. Phys. Rev. 167, 289–291 (1968)

188   K.H. Hellwege: Zur Ausmessung atomarer Strahlungsquellen in Kristallen. Z. Phys. 121, 589–603 (1943)

189   A.M. Prokhorov, V.A. Sychugov, G.P. Shipulo: Double frequency generation in neodymium-activated LaF$_3$ crystals. Zh. Eksp. Teor. Fiz. 56, 1806–1814 (1969)
       English transl.: Sov. Phys. JETP 29, 970–974 (1969)

190   J.R. Izatt, R.C. Mitchell, H.A. Daw: Thermal dependence of ruby laser emission. J. Appl. Phys. 37, 1558–1562 (1966)

191   A.A. Kaminskii, G.A. Bogomolova, Kh.S. Bagdasarov, A.G. Petrosyan: Luminescence, absorption and stimulated emission of Lu$_3$Al$_5$O$_{12}$–Nd^{3+} crystals. Opt. Spektrosk. 39, 1119–1125 (1975)
       English transl.: Opt. Spectrosc. (USSR) 39, 643–646 (1975)

192   H.K. Paetzold: Über den Temperatur- und Druckeinfluss auf Elektronenterme in Kristallen. Z. Phys. 129, 123–139 (1951)

193   P. Hill, S. Hüfner: Lineshift and linewidth in optical spectra of europium salts. Z. Phys. 240, 168–184 (1970)

194   B.Z. Malkin: "Theory of spin-lattice relaxation by Kronig-Van Vleck and calculation of zero-phonon linewidth in the optical spectra of paramagnetic crystals," in Paramagnitnyi Resonans, N4 (University Press, Kazan 1968) pp. 3–28

195   F.Z. Gilfanov, B.Z. Malkin, I.K. Nasyrov, A.L. Stolov: Temperature dependence of the widths and shifts of phononless absorption lines of fluoride crystals activated with gadolinium. Fiz. Tverd. Tela (Leningrad) 8, 3070–3074 (1966)
       English transl.: Sov. Phys. Solid State 8, 2449–2452 (1967)

196   S.A. Johnson, H.G. Freie, A.L. Schawlow, W.M. Yen: Thermal shifts in the energy levels of LaF$_3$:Nd^{3+}. J. Opt. Soc. Am. 57, 734–737 (1967)

197   D.E. McCumber, M.D. Sturge: Linewidth and temperature shift of the R lines in ruby. J. Appl. Phys. 34, 1682–1684 (1963)

198   G.F. Imbusch, W.M. Yen, A.L. Schawlow, D.E. McCumber, M.D. Sturge: Temperature dependence of the width and position of the $^2E \rightarrow {}^4A_2$ fluorescence lines of Cr^{3+} and V^{2+} in MgO. Phys. Rev. A 133, 1029–1034 (1964)

199   T. Kushida, M. Kikuchi: R, R' and B absorption linewidths and phonon-induced relaxation in ruby. J. Phys. Soc. Jpn. 23, 1333–1348 (1967)

200   M.G. Blazha, D.N. Vylegzhanin, A.A. Kaminskii, S.I. Klokishner, Yu.E. Perlin: Phonon density determination in crystals by side-bands of zero-phonon lines in the spectra of RE ion luminescence. Izv. Akad. Nauk SSSR, Ser. Fiz. 40, 1851–1856 (1976)
       English transl.: Bull. Acad. Sci. USSR, Phys. Ser. 40, 69–73 (1976)

201   J.T. Gourley: Spectral-linewidth studies of Pr^{3+} in yttrium aluminum garnet. Phys. Rev. B 5, 22–30 (1972)

202   Kh.S. Bagdasarov, A.A. Kaminskii, Ya.E. Lapsker, B.P. Sobolev: Laser with neodymium-activated α-gagarinite. Zh. Eksp. Teor. Fiz. Pis'ma Red. 5, 220–223 (1967)
       English transl.: JETP Lett. 5, 175–178 (1967)

203   D.F. Nelson, W.S. Boyle: A continuously operating ruby optical maser. Appl. Opt. 1, 181–183 (1962)

204   M. Birnbaum, P.H. Wendzikowski, C.L. Fincher: Continuous-wave nonspiking single mode ruby lasers. Appl. Phys. Lett. 16, 436–438 (1970)

205   D. Roess: Analysis of room temperature CW ruby lasers. IEEE J. Quantum Electron. 2, 208–214 (1966)

206   V. Evtuhov, J.K. Neeland: Power output and efficiency of continuous ruby laser. J. Appl. Phys. 38, 4051–4056 (1967)

207   Yu.K. Voronko, M.V. Dmitruk, A.A. Kaminskii, V.V. Osiko, V.N. Shpakov: Continuous

stimulated radiation of a LaF_3-Nd^{3+} laser at room temperature. Zh. Eksp. Teor. Fiz. *54*, 751–753 (1968)
English transl.: Sov. Phys. JETP *27*, 400–401 (1968)

208   L.F. Johnson, G.D. Boyd, K. Nassau, R.R. Soden: Continuous operation solid-state optical maser. Phys. Rev. *126*, 1406–1409 (1962)

209   A.A. Kaminskii, L.S. Kornienko, G.V. Maksimova, V.V. Osiko, A.M. Prokhorov, G.P. Shipulo: A continuous optical quantum generator on $CaWO_4$ with Nd^{3+} operating at room temperature. Zh. Eksp. Teor. Fiz. *49*, 31–35 (1965)
English transl.: Sov. Phys. JETP *22*, 22–25 (1966)

210   A.A. Kaminskii, P.V. Klevtsov, Kh.S. Bagdasarov, A.A. Maier, A.A. Pavlyuk, A.G. Petrosyan, M.V. Provotorov: New CW crystal lasers. Zh. Eksp. Teor. Fiz. Pis'ma Red. *16*, 548–551 (1972)
English transl.: JETP Lett. *16*, 387–389 (1972)

211   R.C. Duncan: Continuous room-temperature $Nd^{3+}:CaMoO_4$ laser. J. Appl. Rhys. *36*, 874–875 (1965)

212   Kh.S. Bagdasarov, M.M. Gritsenko, F.M. Zubkova, A.A. Kaminskii, A.M. Kevorkov, L. Li: A continuous-wave laser based on $Ca(NbO_3)_2-Nd^{3+}$ crystals. Kristallografiya *15*, 380–382 (1970)
English transl.: Sov. Phys. Crystallogr. *15*, 323–324 (1970)

213   J. Aubrée: Tirage du niobate de calcium monocristallin dopé au néodyme. J. Cryst. Growth *13–14*, 675–680 (1972)

214   R.C. Ohlmann, K.B. Steinbruegge, R. Mazelsky: Spectroscopic and laser characteristics of neodymium-doped calcium fluorophosphate. Appl. Opt. *7*, 905–914 (1968)

215   R. Mazelsky, R.C. Ohlmann, K. Steinbruegge: Crystal growth of a new laser material, fluorapatite, J. Electrochem. Soc. *115*, 68–70 (1968)

216   E.G. Erickson: Holobeam reports 760 watts CW from a segmented Nd:YAG system. Laser Focus *6*, N4, 16 (1970)

217   W. Koechner: Multi-hundred watt Nd:YAG continuous laser. Rev. Sci. Instrum. *41*, 1699–1706 (1970)

218   A.A. Kaminskii, G.Ya. Kolodnyi, N.I. Sergeeva: Continuous $LaNa(MoO_4)_2:Nd^{3+}$ laser operating at 300°K. Zh. Prikl. Spektrosk. *9*, 884–886 (1968)
English transl.: J. Appl. Spectrosc. (USSR) *9*, 1275–1276 (1968)

219   R.V. Alves, R.A. Buchanan, K.A. Wickersheim, E.A. Yates: Neodymium-activated lanthanum oxysulfide: a new high-gain laser material. J. Appl. Phys. *42*, 3043–3048 (1971)

220   M. Bass, M.J. Weber: Yalo: robust at age 2. Laser Focus *7*, N9, 34–36 (1971)

221   R.C. Duncan, Z.J. Kiss: Continuously operating $CaF_2:Tm^{2+}$ optical maser. Appl. Phys. Lett. *3*, 23–24 (1963)

222   A.A. Kaminskii, S.E. Sarkisov, Kh.S. Bagdasarov, A.G. Petrosyan P.V. Klevtsov, A.A. Pavlyuk: Investigation of stimulated emission in the $^4F_{3/2} \rightarrow {}^4I_{13/2}$ transition of Nd^{3+} ions in crystals. V. Phys. Status Solidi (*a*) *17*, K75–K77 (1973)

223   D.P. Devor, B.H. Soffer: 2.1-μm laser of 20 W output power and 4-percent efficiency from Ho^{3+} in sensitized YAG, IEEE J. Quantum Electron. *8*, 231–234 (1972)

224   R.L. Remski, D.J. Smith: Temperature dependence of pulsed laser threshold in YAG:Er^{3+}, Tm^{3+}, Ho^{3+}. IEEE J. Quantum Electron. *6*, 750–751 (1970)

225   V.A. Gorbachev, V.I. Zhekov, T.M. Murina, V.V. Osiko, B.P. Starikov, M.I. Timoshechkin: Spectroscopic and laser properties of erbium-aluminate with Tm^{3+} ions impurity. Sb. Kratk. Soobshch. Po Fiz. Akad. Nauk SSSR-Fiz. Inst. Im. P.N. Lebedeva *N4*, 16–22 (1973)

226   Z.J. Kiss, H.R. Lewis, R.C. Duncan: Sun pumped continuous optical maser. Appl. Phys. Lett. *2*, 93–96 (1963)

227   A.A. Kaminskii, L.S. Kornienko, A.M. Prokhorov: Continuous solar laser using Dy^{2+} in CaF_2. Dok. Akad. Nauk SSSR *161*, 1063–1064 (1965)
English transl.: Sov. Phys. Dokl. *10*, 334–335 (1965)

228   C.G. Young: A sun-pumped CW one-watt laser. Appl. Opt. *5*, 993–997 (1966)

229   C.W. Reno: Solar-pumped modulated laser. RCA Rev. *23*, 149–157 (1966)

230   V.N. Chernyshev, A.G. Sheremetev, V.V. Kobsarev: *Lasery v Sistemakh Svyazi*, (Svyaz', Moscow 1966)

231   I.J. Spalding: Lasers and the reactor ignition problem. Kvant. Elektron. (Moscow) *N4*(10), 40–47 (1972)
English transl.: Sov. J. Quantum Electron. *2*, 329–334 (1973)

232 Yu.K. Danileiko, A.A. Manenkov, A.M. Prokhorov, V.Ya. Khaimov-Malkov: Surface damage of ruby crystals by laser radiation. Zh. Eksp. Teor. Fiz. *58*, 33–36 (1970)
English transl.: Sov. Phys. JETP *31*, 18–21 (1970)

233 L.D. Khazov, A.N. Shestov: Surface discharge at the end of an active laser rod. Opt. Spektrosk. *23*, 487–489 (1967)
English transl.: Opt. Spectrosc. (USSR) *23*, 261–262 (1966)

234 L.A. Nesterov, A.A. Poplavskii, I.A. Fersman, L.D. Khazov: Variation of destruction threshold of a transparent dielectric with laser pulse length. Zh. Tekh. Fiz. *40*, 651–653 (1970)
English transl.: Sov. Phys. Tech. Phys. *15*, 505–507 (1970)

235 V.A. Pashkov, G.M. Zverev: Destruction of ruby and leucosapphire crystals by strong laser radiation. Zh. Eksp. Teor. Fiz. *51*, 777–779 (1966)
English transl.: Sov. Phys. JETP *24*, 516–518 (1967)

236 T.P. Belikova, A.N. Savchenko, E.A. Sviredenkov: Optic breakdown in ruby and related effects. Zh. Eksp. Teor. Fiz. *54*, 37–45 (1968)
English transl.: Sov. Phys. JETP *27*, 19–23 (1968)

237 C.R. Guiliano, L.D. Hess: Damage threshold studies in ruby and sapphire. Proc. Symp. on Damage in Laser Materials. Nat. Bur. Stand. USA N341, 76–90 (1970)

238 E.M. Akulenok, Yu.K. Danileiko, A.A. Manenkov, B.S. Nichitailo, A.D. Piskun, V.Ya. Khaimov-Malkov: Mechanics of damage to ruby crystal by laser radiation. Zh. Eksp. Teor. Fiz. Pis'ma Red. *16*, 336–339 (1972)
English transl.: JETP Lett. *16*, 238–240 (1972)

239 G.M. Zverev, E.A. Levchuk. V.A. Pashkov, Yu.D. Poredin: Laser-radiation induced damage to the surface of lithium niobate and tantalate single crystals. Kvant. Elektron. (Moscow) *N2*, 94–96 (1972)
English transl.: Sov. J. Quantum Electron. *2*, 167–169 (1972)

240 G.M. Zverev, E.A. Levchuk, E.K. Maldutis: Destruction of KDP, ADP and $LiNbO_3$ crystals by intense laser radiation. Zh. Eksp. Teor. Fiz. *57*, 730–736 (1969)
English transl.: Sov. Phys. JETP *30*, 400–403 (1970)

241 A.I. Bodretsova, A.A. Kaminskii, S.I. Levikov, V.V. Osiko: A quasicontinuous laser with pyrotechnical excitation. Zh. Prikl. Spektrosk. *6*, 254–256 (1967)
English transl.: J. Appl. Spectrosc. (USSR) *6*, 168–169 (1970)

242 A.I. Bodretsova, Kh.S. Bagdasarov, A.A. Kaminskii, N.N. Kirilova, S.I. Levikov: High-power $Y_3Al_5O_{12} - Nd^{3+}$ laser with an explosion-type lamp. Kvant. Elektron. (Moscow) N2(8), 107–108 (1972)
English transl.: Sov. J. Quantum Electron. *2*, 183–184 (1972)

243 A.A. Kaminskii, A.I. Bodretsova, S.I. Levikov: Pyrotechnically excited quasi-CW laser. Zh. Tekh. Fiz. *39*, 535–542 (1969)
English transl.: Sov. Phys. Tech. Phys. *14*, 396–402 (1969)

244 E.M. Zolotov: Investigation of laser and luminescence properties of calcium and strontium fluoride crystals doped with divalent dysprosium. Proc. P.N. Lebedev Phys. Inst. [Acad. Sci. USSR] Vol. 60, ed. by D.V. Skobeltsin (Consultants Bureau, New York, London 1974) pp. 87–132

245 S.K. Isaev, L.S. Kornienko, E.G. Lariontsev, M.S. Khritankov: Investigation of the delay of stimulated emission from a $CaF_2:Dy^{2+}$ laser relative to pumping pulses. Kvant. Elektron. (Moscow) N4(10), 48–54 (1972)
English transl.: Sov. J. Quantum Electron. *2*, 335–338 (1973)

246 M. Birnbaum, C.L. Fichner: Ruby, Nd^{3+}:glass, and $CaWO_4$ lasers pumped by a pulsed argon ion laser. Proc. IEEE *56*, 1096–1097 (1968)

247 D. Röss, G. Zeidler: Pumpen von Neodymlasern mit Rubin-Laserlicht. Z. Naturforsch. *a 21*, 336–340 (1966)

248 R.B. Allen, S.J. Scalise: Continuous operation of a YAlG:Nd laser by injection luminescent pumping. Appl. Phys. Lett. *14*, 188–190 (1969)

249 W.W. Krühler, G. Huber, H. Danielmeyer: Correlations between site geometries and level energies in the laser system $Nd_{1-x}Y_xP_5O_{14}$. Appl. Phys. *8*, 261–268 (1975)

250 F.W. Ostarmayer, R.B. Allen, E.G. Dierschke: Room-temperature CW operation of a $GaAs_{1-x}P_x$ diode-pumped YAG:Nd laser. Appl. Phys. Lett. *19*, 289–292 (1971)

251 M. Ross: YAG laser operation by semiconductor laser pumping. Proc. IEEE *56*, 196–197 (1968)

252 W.W. Krühler, J.P. Jeser, H.G. Danielmeyer: Properties and laser oscillation of the (Nd, Y) pentaphosphate system. Appl. Phys. 2, 329–333 (1973)

253 V.A. Sychugov, G.P. Shipulo: Rate of relaxation from the lower laser level. Zh. Eksp. Teor. Fiz. 58, 817–820 (1970)
     English transl.: Sov. Phys. JETP 31, 438–439 (1970)

254 A.R. Reinberg, L.A. Riseberg, R.M. Brown, R.W. Wacker, W.C. Holton: GaAs:Si LED pumped Yb-doped YAG laser. Appl. Phys. Lett. 19, 11–13 (1971)

255 S.V. Vonsovskii, S.V. Grum-Grzhimailo, V.U. Cherepanov, A.N. Men, D.T. Sviridov, Yu.F. Smirnov, A.E. Nikiforov: Teoriya Kristallicheskogo Polya i Opticheskie Spektry Primesnykh Ionov s Nesapolnennoi d-Obolochkoi, (Nauka, Moscow 1969)

256 W. Waiser, C.G.B. Garrett, D.L. Wood: Fluorescence and optical maser effects in $CaF_2:Sm^{2+}$. Phys. Rev. 123, 766–776 (1961)

257 R.J. Pressley, J.P. Wittke: $CaF_2:Dy^{2+}$ lasers. IEEE J. Quantum Electron. 3, 116–129 (1967)

258 W.M. Yen, W.C. Scott, A.L. Schawlow: Phonon-induced relaxation in excited optical states of trivalent praseodymium in LaF_3. Phys. Rev. A 136, 271–283 (1964)

259 W.T. Carnall, P.R. Fields, R. Sarup: $^1S_0$ level of Pr^{3+} in crystal matrices and energy-level parameters for the $4f^2$ configuration of Pr^{3+} in LaF_3. J. Chem. Phys. 51, 2587–2591 (1969)

260 G.H. Dieke, R. Sarup: Fluorescence spectrum of $PrCl_3$ and the levels of the Pr^{3+} ions. J. Chem. Phys. 29, 741–745 (1958)

261 A.L. Harmer, A. Linz, D.R. Gabbe: Fluorescence of Nd^{3+} in lithium yttrium fluoride. J. Phys. Chem. Solids 30, 1483–1491 (1969)

262 D. Sengupta, J.O. Artman: Energy-level scheme for Nd^{3+} in $LiYF_4$. J. Chem. Phys. 53, 838–840 (1970)

263 H.H. Caspers, H.E. Rast, R.A. Buchanan: Intermediate coupling energy levels for Nd^{3+} $(4f)^3$ in LaF_3. J. Chem. Phys. 42, 3214–2315 (1965)

264 Yu. K. Voronko, V.V. Osiko, N.V. Savostyanova, V.S. Fedorov, I.A. Shcherbakov: Study of the processes of deactivation of the metastable state of excited Nd^{3+} ions in LaF_3 crystals. Fiz. Tverd. Tela (Leningrad) 14, 2656–2663 (1972)
     English transl.: Sov. Phys. Solid State 14, 2294–2299 (1973)

265 M.V. Dmitruk, A.A. Kaminskii, I.A. Shcherbakov: Spectroscopic investigations of stimulated radiation from a CeF_3-Nd^{3+} laser. Zh. Eksp. Teor. Fiz. 54, 1680–1686 (1968)
     English transl.: Sov. Phys. JETP 27, 900–903 (1968)

266 V.T. Gabrielyan, A.A. Kaminskii, L.Li: Absorption and luminescence spectra and energy levels of Nd^{3+} and Er^{3+} ions in $LiNbO_3$ crystals. Phys. Status Solidi (a) 3, K37–K42 (1970)

267 W.F. Krupke: New rare earth quantum electronics devices: a calculational approach. Proc. IEEE Region VI Conf., April 24–26, 1974, Albuquerque, New Mexico, pp. 17–31

268 A.M. Morozov, M.N. Tolstoi, P.P. Feofilov: Luminescence of neodymium in crystals of the scheelite type. Opt. Spektrosk 22, 258–265 (1967)
     English transl.: Opt. Spectrosc. (USSR) 22, 139–142 (1967)

269 Ya.E. Kariss, A.M. Morozov, P.P. Feofilov: On the luminescence of Nd^{3+} in $CaWO_4$. Opt. Spektrosk. 17, 887–892 (1964)
     English transl.: Opt. Spectrosc. (USSR) 17, 481–484 (1964)

270 G.M. Zverev, G.Ya. Kolodnyi, A.I. Smirnov: Optical spectra of Nd^{3+} in single crystals of scandium and yttrium oxides. Opt. Spektrosk. 23, 606–610 (1967)
     English transl.: Opt. Spectrosc. (USSR) 23, 325–327 (1967)

271 A.A. Kaminskii: "Achievements and problems of stimulated emission spectroscopy of activated crystals," in Spektroskopiya Kristallov, ed. by A.A. Kaminskii, Z.L. Morgenshtern, D.T. Sviridov (Nauka, Moscow 1975) pp. 92–122

272 M.J. Weber, T.E. Varitimos: Optical spectra and intensities of Nd^{3+} in $YAlO_3$. J. Appl. Phys. 42, 4996–5005 (1971)

273 J.A. Koningstein: Crystal-field studies of excited states of trivalent neodymium in yttrium gallium garnet and yttrium aluminum garnet. J. Chem. Phys. 44, 3957–3968 (1966)

274 J.A. Koningstein: Energy levels and crystal field calculations of neodymium in yttrium aluminum garnet. Phys. Rev. A 136, 711–716 (1964)

275 P.P. Feofilov, V.A. Timofeeva, M.N. Tolstoi, L.M. Belyaev: On the luminescence of neodymium and chromium in yttrium aluminum garnet. Opt. Spektrosk. 19, 817–819 (1965)
     English transl.: Opt. Spectrosc. (USSR) 19, 451–452 (1965)

420   References

276   J.R. Henderson, M. Muramoto: Spectrum of Nd^{3+} in lanthanide oxide crystals. J. Chem. Phys. *46*, 2515–2520 (1967)
277   N.A. Godina, M.N. Tolstoi, P.P. Feofilov: Luminescence of neodymium in yttrium and lanthanum niobates and tantalates. Opt. Spektrosk. *23*, 756–761 (1967)
       English transl.: Opt. Spectrosc. (USSR) *23*, 411–413 (1967)
278   Z.T. Azamatov, P.A. Arsenev, M.V. Chukichev: Investigation of stimulated emission and the diagram of the energy levels of Nd^{3+} in single crystals of lutetium-aluminum garnet (LuAG). Kristallografiya *15*, 827–829 (1970)
       English transl.: Sov. Phys. Crystallogr. *15*, 713–714 (1971)
279   V.K. Konyukhov, V.M. Marchenko, A.M. Prokhorov: CaF_2–Sm^{2+} crystal laser excited by a ruby laser. Opt. Spektrosk. *20*, 531–532 (1966)
       English transl.: Opt. Spectrosc. (USSR) *20*, 299–300 (1966)
280   D.J. Horowitz, L.F. Gillespie, J.E. Miller, E.J. Sharp: Laser action of Nd^{3+} in a crystal $Ba_2ZnGe_2O_7$. J. Appl. Phys. *43*, 3527–3530 (1972)
281   A.M. Morozov, L.G. Morozova, A.K. Trofimov, P.P. Feofilov: Spectral and luminescent characteristics of fluoroapatite single crystals activated by rare earth ions. Opt. Spektrosk. *29*, 1106–1118 (1970)
       English transl.: Opt. Spectrosc. (USSR) *29*, 590–596 (1970)
282   G.V. Maksimova, A.A. Sobol: Nd^{3+} optical centers in crystals of calcium and strontium fluorophosphates. Proc. P.N. Lebedev Phys. Inst. [Acad. Sci. USSR], Vol. 60, ed. by D.V. Skobeltsin (Consultants Bureau, New York, London 1974) pp. 59–73
283   G.V. Maksimova, A.A. Sobol: Local compensation of the Nd^{3+} ionic charge in fluorapatite crystals. Izv. Akad. Nauk SSSR, Neorg. Mater. *8*, 1077–1082 (1972)
       English transl.: Inorg. Mater. (USSR) *8*, 945–949 (1972)
284   C. Brecher, H. Samelson, A. Lampicki, R. Riley, T. Peters: Polarized spectra and crystal-field parameters of Eu^{3+} in YVO_4. Phys. Rev. *155*, 178–187 (1967)
285   Z.T. Azamatov, P.A. Arsenev, M.V. Chukichev: Spectra of gadolinium aluminum yttrium garnet single crystals. Opt. Spektrosk. *28*, 289–291 (1970)
       English transl.: Opt. Spectrosc. (USSR) *28*, 156–157 (1970)
286   P.A. Arsenev, K.E. Bienert: Absorption, luminescence and stimulated emission spectra of Tm^{3+} in $GaAlO_3$ crystals. Phys. Status Solidi (*a*) *13*, K125–K129 (1972)
287   M.R. Brown, K.G. Roots, W.A. Shand: Energy levels of Er^{3+} in $LiYF_4$. J. Phys. *C 2*, 593–602 (1969)
288   W.F. Krupke, J.B. Gruber: Absorption and fluorescence spectra of $Er^{3+}(4f^{11})$ in LaF_3. J. Chem. Phys. *39*, 1024–1030 (1963)
289   W.F. Krupke, J.B. Gruber: Energy levels of Er^{3+} in LaF_3 and coherent emission at 1.61 μ. J. Chem. Phys. *41*, 1225–1232 (1964)
290   J.P. Van der Ziel, W.A. Bonner, L. Kopf, L.G. Van Uitert: Coherent emission from Ho^{3+} ions in epitaxially grown thin aluminum garnet films. Phys. Lett. *A 42*, 105–106 (1973)
291   E. Bernal G.: Optical spectrum and magnetic properties of Er^{3+} in $CaWO_4$. J. Chem. Phys. *55*, 2538–2549 (1971)
292   J.A. Koningstein, J.E. Geusic: Energy levels and crystal-field calculations of Er^{3+} in yttrium aluminum garnet. Phys. Rev. *A 136*, 726–728 (1964)
293   J.A. Koningstein: "Crystal field of trivalent rare earth ions in garnet host lattices," in *Optical Properties of Ions in* Crystals," ed. by H.M. Crosswhite, H.W. Moos (Wiley, New York 1967) pp. 105–115
294   P.A. Arsenev, K.E. Bienert: Absorption, luminescence, and stimulated emission spectra of Er^{3+} ions in $GdAlO_3$ crystals. Phys. Status Solidi (*a*) *10*, K85–K88 (1972)
295   G.M. Zverev, G.Ya. Kolodnyi, A.M. Onishchenko: Resonant and nonresonant processes of excitation energy transfer from Tm^{3+} or Ho^{3+} ions to the Er^{3+} ions in $(Y, Er)_3Al_5O_{12}$ crystals. Zh. Eksp. Teor. Fiz. *57*, 794–805 (1969)
       English transl.: Sov. Phys. JETP *30*, 435–440 (1970)
296   K.D. Knoll: Absorption and fluorescence spectra of Tm^{3+} in YVO_4 and YPO_4. Phys. Status Solidi (*b*) *45*, 553–559 (1971)
297   P.A. Arsenev, K.E. Bienert: Spectral properties of Ho^{3+} in $GdAlO_3$ crystals. Phys. Status Solidi (*a*) *13*, K129–K132 (1972)
298   J.A. Koningstein: Energy levels and crystal-field calculations of trivalent ytterbium in yttrium aluminum garnet and yttrium gallium garnet. Theor. Chim. Acta *3*, 271–277 (1965)

299 W.A. Hargreaves: High-resolution measurements of absorption, fluorescence and crystal-field splitting of solutions of divalent, trivalent, and tetravalent uranium ions in fluoride crystals. Phys. Rev. *156*, 331–342 (1967)

300 L.N. Galkin, P.P. Feofilov: Luminescence spectra of trivalent ions of uranium. Opt. Spektrosk. *7*, 840–841 (1959)

301 F.J. McClung, S.E. Schwarz, F.J. Meyers: R_2 line optical maser action in ruby. J. Appl. Phys. *33*, 3139–3140 (1962)

302 T. Kushida: Absorption spectrum of optically pumped ruby. I. Experimental studies of spectrum in excited states. J. Phys. Soc. Jpn. *21*, 1331–1341 (1966)

303 R.S. Mulliken: Electronic structures of polyatomic molecules and valence. IV. Electronic states, quantum theory of the double band. Phys. Rev. *43*, 279–302 (1933)

304 K.F. Tittle: Spectral characteristics for a neodymium laser liquid filter system. Rev. Sci. Instrum. *35*, 522–524 (1964)

305 A.L. Harmer, A. Linz, D.R. Gabbe, L. Gillespie, G.M. Janney, E. Sharp: Fluorescence of Nd^{3+} in YLF ($LiYF_4$). Bull. Am. Phys. Soc. *12*, 1068 (1967)

306 D.R. Gabbe, A.L. Harmer: Scheelite structure fluorides: the growth of pure and rare earth doped $LiYF_4$. J. Cryst. Growth *3*, 544 (1968)

307 R.D. Shannon, C.T. Prewitt: Effective ionic radii in oxides and fluorides. Acta Crystallogr. *B 25*, 925–946 (1969)

308 L.F. Johnson: Optical maser characteristics of Nd^{3+} in CaF_2. J. Appl. Phys. *33*, 756 (1962)

309 A.A. Kaminskii, L.S. Kornienko, L.V. Makarenko, A.M. Prokhorov, M.M. Fursikov: Stimulated emission of Nd^{3+} in CaF_2 at room temperature. Zh. Eksp. Teor. Fiz. *46*, 386–389 (1964)
English transl.: Sov. Phys. JETP *19*, 262–263 (1964)

310 A.A. Kaminskii, L.S. Kornienko, A.M. Prokhorov: A spectral study of induced radiation from Nd^{3+} in CaF_2. Zh. Eksp. Teor. Fiz. *48*, 476–482 (1965)
English transl.: Sov. Phys. JETP *21*, 318–322 (1965)

311 A.A. Kaminskii, L.S. Kornienko, A.M. Prokhorov: Lifetime of the $^4F_{3/2}$ excited state of the Nd^{3+} ion in CaF_2 and $CaWO_4$. Zh. Eksp. Teor. Fiz. *48*, 1262–1266 (1965)
English transl.: Sov. Phys. JETP *21*, 844–847 (1965)

312 A.A. Kaminskii, V.V. Osiko, M.M. Fursikov: The photoreduction $TR^{3+} \rightarrow TR^{2+}$ in fluorite crystals. Zh. Eksp. Teor. Fiz. Pis'ma Red. *4*, 92–96 (1966)
English transl.: JETP Lett. *4*, 62–65 (1966)

313 S.Kh. Batygov, A.A. Kaminskii: Nature of "aging" of fluorite and yttrofluorite crystals activated by Nd^{3+} under stimulated emission conditions. Zh. Eksp. Teor. Fiz. *53* 839–852 (1967)
English transl.: Sov. Phys. JETP *26*, 512–518 (1968)

314 Yu.K. Voronko, A.A. Kaminskii, V.V. Osiko: Effect of hard radiation on the optical centers of TR^{3+} ions in crystals. Zh. Eksp. Teor. Fiz. Pis'ma Red. *2*, 473–478 (1965)
English transl.: JETP Lett. *2*, 294–297 (1965)

315 S.Kh. Batygov, Yu.K. Voronko, M.V. Dmitruk, V.V. Osiko, A.M. Prokhorov, I.A. Shcherbakov: Spectroscopy of optical Nd^{3+} centers in CaF_2 and SrF_2 crystals. Proc. P.N. Lebedev Phys. Inst. [Acad. Sci. USSR], Vol. 60, ed. by D.V. Skobeltsin (Consultants Bureau, New York, London 1974) pp. 31–58

316 Ya.E. Kariss, M.N. Tolstoi, P.P. Feofilov: On the luminescence and absorption of trivalent neodymium in crystals of the fluorite type. Opt. Spektrosk. *18*, 440–445 (1965)
English transl.: Opt. Spectrosc. (USSR) *18*, 247–250 (1965)

317 A.A. Kaminskii, S.E. Sarkisov, K.B. Seiranyan, B.P. Sobolev: Investigation of the stimulated emission from $Sr_2Y_5F_{19}$ crystals doped with Nd^{3+} ions. Kvant. Elektron. (Moscow) *1*, 187–189 (1974)
English transl.: Sov. J. Quantum Electron. *4*, 112–113 (1974)

318 Yu.K. Voronko, A.A. Kaminskii, L.S. Kornienko, V.V. Osiko, A.M. Prokhorov, V.T. Udovenchik: Investigation of stimulated emission of CaF_2-Nd^{3+} (type II) crystals at room temperature. Zh. Eksp. Teor. Fiz. Pis'ma Red. *1*, N2, 3–7 (1965)
English transl.: JETP Lett. *1*, 39–42 (1965)

319 Yu.K. Voronko, A.A. Kaminskii, V.V. Osiko: Optical relaxation of Ho^{3+} and Er^{3+} ions in the CaF_2 lattice (type I) in the visible wavelength ragion. Zh. Eksp. Teor. Fiz. *49*, 1022–1027 (1965)
English transl.: Sov. Phys. JETP *22*, 710–713 (1966)

320 S.A. Pollack: Stimulated emission in $CaF_2:Er^{3+}$. Proc. IEEE *51*, 1793 (1963)
321 R.D. Shannon, C.T. Prewitt: Revised values of effective ionic radii. Acta Crystallogr. *B 26*, 1046–1048 (1970)
322 P.P. Sorokon, M.J. Stevenson, J.R. Lankard, G.D. Pettit: Spectroscopy and optical maser action in SrF_2-Sm^{2+}. Phys. Rev. *127*, 503–508 (1962)
323 V.K. Konyukhov, V.M. Marchenko, A.M. Prokhorov: $CaF_2:Sm^{2+}$ laser with ruby laser excitation. IEEE J. Quantum Electron. *2*, 541–542 (1966)
324 Yu.A. Ananyev, A.K. Grezin, A.A. Mak, B.M. Sedov, Ye.N. Yudina: A fluorite: samarium laser. Opt. Mekh. Promst. *35*, N5, 30–33 (1968)
     English transl.: Sov. J. Opt. Technol. *35*, 313–316 (1968)
325 P.P. Feofilov, A.A. Kaplyanskii: Spectra of divalent rare earth ions in crystals of alkaline earth fluorides. I. Samarium. Opt. Spektrosk. *12*, 493–500 (1962)
     English transl.: Opt. Spectrosc. (USSR) *12*, 272–277 (1962)
326 P.A. Forrester, G.M. Green, D.F. Sampson: The effects of oxygen on the properties of CaF_2 as a laser host. Br. J. Appl. Phys. *16*, 1209–1210 (1965)
327 Z.J. Kiss: The CaF_2-Tm^{2+} and the CaF_2-Dy^{2+} optical maser systems. Proc. 3rd Quantum Electron. Intern. Congr., ed. by P. Grivet, N. Bloembergen (Columbia Univ. Press, New York 1964) pp. 805–815
328 V.K. Konyukhov, B.B. Kostin, L.A. Kulevskii, T.M. Murina, A.M. Prokhorov: CaF_2-Dy^{2+} laser operating in a repetitive giant-pulse mode with continuous pumping. Dok. Akad. Nauk SSSR *165*, 1056–1058 (1965)
     English transl.: Sov. Phys. Dokl. *10*, 1192–1193 (1966)
329 V.V. Kostin, L.A. Kulevsky, T.M. Murina, A.M. Prokhorov, A.A. Tikhonov: CaF_2-Dy^{2+} gaint pulse laser with high repetition rate. IEEE J. Quantum Electron. *2*, 611–612 (1966)
330 N.M. Galaktionova, V.F. Egorova, V.S. Zubkova, A.A. Mak, D.S. Prilezhaev: Effect of the spectral-luminescent properties of an active medium on the characteristics of a solid state laser. Opt. Spektrosk. *23*, 949–953 (1967)
     English transl.: Opt. Spectrosc. (USSR) *23*, 517–519 (1967)
331 A.I. Bodretsova, A.A. Kaminskii, S.I. Levikov, V.V. Osiko: Quasi-continuous laser on $CaF_2:Dy^{2+}$ base with pyrotechnic excitation. Dok. Akad. Nauk SSSR *174*, 337–338 (1967)
     English transl.: Sov. Phys. Dokl. *12*, 507–508 (1967)
332 Z.J. Kiss, R.C. Duncan: Optical maser action in CaF_2-Tm^{2+}. Proc. IRE *50*, 1532–1533 (1962)
333 S.P.S. Porto, A. Yariv: Trigonal sites and 2.24-micron coherent emission of U^{3+} in CaF_2. J. Appl. Phys. *33*, 1620–1621 (1962)
334 J.P. Wittke, Z.J. Kiss, R.C. Duncan, J.J. McCormick: Uranium-doped calcium fluoride as a laser material. Proc. IEEE *51*, 56–62 (1963)
335 S.P.S. Porto, A. Yariv: Low lying energy levels and comparison of laser action of U^{3+} in CaF_2. Proc. 3rd Quantum Electron. Int. Congr., ed. by P. Grivet, N. Bloembergen (Columbia Univ. Press, New York 1964) pp. 717–723
336 Ya.E. Kariss, P.P. Feofilov: The absorption, luminescence, and laser action of neodymium in SrF_2 crystals. Opt. Spektrosk. *14*, 169–172 (1963)
     English transl.: Opt. Spectrosc. (USSR) *14*, 89–90 (1963)
337 A.A. Kaminskii, L. Li: The photoreduction $Nd^{3+} \rightarrow Nd^{2+}$ in SrF_2 (type I) crystals under stimulated emission conditions ("aging"). Phys. Status Solidi *30*, K77–K79 (1968)
338 A.A. Kaminskii, L. Li: Spectroscopic investigations of stimulated emission from a laser based on SrF_2-Nd^{3+}. Zh. Prikl. Spektrosk. *12*, 35–40 (1970)
     English transl.: J. Appl. Spectrosc. (USSR) *12*, 29–34 (1972)
339 E.M. Zolotov, V.V. Osiko, A.M. Prokhorov, G.P. Shipulo: Study of the luminescence and laser properties of SrF_2-Dy^{2+} crystals. Zh. Prikl. Spektrosk. *8*, 1046–1047 (1968)
     English transl.: J. Appl. Spectrosc. (USSR) *8*, 627–628 (1972)
340 S.E. Hatch, W.E. Parson, R.J. Weagley: Hot-pressed polycrystalline CaF_2-Dy^{2+} laser. Appl. Phys. Lett. *5*, 153–154 (1964)
341 S.P.S. Porto, A. Yariv: Excitation, relaxation, and optical maser action at 2.407 micron in $SrF_2:U^{3+}$. Proc. IRE *50*, 1543–1544 (1962)
342 S.P.S. Porto, A. Yariv: Optical maser characteristics of $BaF_2:U^{3+}$. Proc. IRE *50*, 1542–1543 (1962)
343 M.V. Dmitruk, A.A. Kaminskii: Stimulated emission from LaF_3-Nd^{3+} crystal lasers. Zh. Eksp. Teor. Fiz. *53*, 874–881 (1967)

English transl.: Sov. Phys. JETP *26*, 531–534 (1968)

344 A.A. Kaminskii, D.N. Vylegzhanin: Investigation of manifestation of the electron–phonon interaction in crystals activated with Nd^{3+} ions using stimulated emission spectroscopy methods. Vorträge Internat. Tagung Laser und ihre Anwendungen, 10.6–17.6, 1970, Dresden, Teil 7, S. 409–443

345 J.R. O'Conner, W.A. Hargreaves: Lattice energy transfer and stimulated emission from $CeF_3:Nd^{3+}$. Appl. Phys. Lett. *4*, 208–209 (1964)

346 D.P. Devor, B.H. Soffer, M. Robinson: Stimulated emission from Ho^{3+} at 2 μm in HoF_3. Appl. Phys. Lett. *18*, 122–124 (1971)

347 A.A. Kaminskii, R.G. Mikaelyan, I.N. Zigler: Room-temperature induced emission of CaF_2–SrF_2 crystals containing Nd^{3+}. Phys. Status Solidi *31*, K85–K86 (1969)

348 A.A. Kaminskii, V.V. Osiko, A.M. Prokhorov, Yu.K. Voronko: Spectral investigation of the stimulated radiation of Nd^{3+} in CaF_2–YF_3. Phys. Lett. *22*, 419–420 (1966)

349 A.A. Kaminskii, V.V. Osiko, Yu.K. Voronko: Mixed systems on the basis of fluorides as new laser materials for quantum electronics. The optical and emission parameters. Phys. Status Solidi *21*, K17–K22 (1967)

350 A.A. Kaminskii: The nature of "aging" of CaF_2–YF_3–Nd^{3+} crystals (type I) under stimulated emission conditions. Phys. Status Solidi *20*, K51–K54 (1967)

351 Yu.K. Voronko, K.F. Shepilov, I.A. Shcherbakov: Laser action of CaF_2–Nd^{3+}–TR^{3+}. Sb. Kratk. Soobshch. Po Fiz. Akad. Nauk SSSR-Fiz. Insit. Im. P.N. Lebedeva N10, 3–9 (1970)

352 A.A. Kaminskii, N.I. Suchov: Laser with external mirrors for low-temperature experiments. Prib. Tekh. Eksp. N4, 233–234 (1967)
     English transl.: Instrum. Exper. Tech. (USSR) N4, 936–937 (1967)

353 M.V. Dmitruk. A.A. Kaminskii: Laser emission from CaF_2–YF_3 crystals containing Ho^{3+} and Er^{3+}. Kristallografiya *14*, 722–723 (1969)
     English transl.: Sov. Phys. Crystallogr. *14*, 620 (1970)

354 Yu.K. Voronko, A.A. Kaminskii, V.V. Osiko, M.M. Fursikov: Neodymium-doped cerium fluorite in lasers. Kristallografiya *11*, 936–937 (1966)
     English transl.: Sov. Phys. Crystallogr. *11*, 753–755 (1967)

355 A.A. Kaminskii, V.N. Shpakov: Investigations of new crystals for Q-switched lasers. Zh. Eksp. Teor. Fiz. *52*, 103–111 (1967)
     English transl.: Sov. Phys. JETP *25*, 67–71 (1967)

356 L.S. Garashina, A.A. Kaminskii, L. Li, B.P. Sobolev: A laser based on SrF_2–YF_3–Nd^{3+} crystals. Kristallografiya *14*, 925 (1969)
     English transl.: Sov. Phys. Crystallogr. *14*, 799 (1970)

357 A.A. Kaminskii, V.V. Osiko, V.T. Udovenchik: Room-temperature induced emission of neodymium-doped SrF_2–LaF_3 crystals. Zh. Prikl. Spektrosk. *6*, 40–44 (1967)
     English transl.: J. Appl. Spectrosc. (USSR) *6*, 23–25 (1970)

358 A.A. Kaminskii: Exploration of active media for lasers. Dok. Akad. Nauk SSSR *211*, 811–813 (1973)
     English transl.: Sov. Phys. Dokl. *18*, 529–530 (1974)

359 Kh.S. Bagdasarov, O.E. Izotova, A.A. Kaminskii, L. Li, B.P. Sobolev: Optical and generation properties of mixed CdF_2–YF_3 crystals activated with Nd^{3+} ions. Dok. Akad. Nauk SSSR *188*, 1042–1044 (1969)
     English transl.: Sov. Phys. Dokl. *14*, 939–941 (1970)

360 Kh.S. Bagdasarov, Yu.K. Voronko, A.A. Kaminskii, V.V. Osiko: Systems based on fluorides as active materials in quantum electronics. Izv. Akad. Nauk SSSR, Neorg. Mater. *1*, 2088–2092 (1965)
     English transl.: Inorg. Mater. (USSR) *1*, 1888–1891 (1965)

361 M.V. Dmitruk, A.A. Kaminskii, V.V. Osiko, T.A. Tevosyan: Induced emission of hexagonal LaF_3–SrF_2–Nd^{3+} crystals at room temperature. Phys. Status Solidi *25*, K75–K78 (1968)

362 T.H. Maiman: Optical maser action in ruby. Br. Commun. Electron. *7*, 674–675 (1960)

363 R.J. Collins, D.F. Nelson, A.L. Schawlow, W. Bond, C.G.B. Garrett, W. Kaiser: Coherence, narrowing, directionality and relaxation oscillations in the light emission from ruby. Phys. Rev. Lett. *5*, 303–305 (1960)

364 T.H. Maiman: Stimulated optical emission in fluorescent solids. I. Theoretical considerations. Phys. Rev. *123*, 1145–1150 (1961)

365 T.H. Maiman, R.H. Hoskins, I.J.D. Heanens, C.K. Asawa, V. Evtuhov: Stimulated optical

emission in fluorescent solids. II. Spectroscopy and stimulated emission in ruby. Phys. Rev. *123*, 1151–1157 (1961)

366   A.L. Schawlow, G.E. Devlin: Simultaneous optical maser action in two ruby satellite lines. Phys. Rev. Lett. *6*, 96–97 (1961)

367   I. Wieder, L.R. Sarles: Stimulated optical emission from exchange-coupled ions of Cr^{3+} in Al_2O_3. Phys. Rev. Lett. *6*, 95–96 (1961)

368   V. Evtuhov, J.K. Neeland: Continuous operation of a ruby laser at room temperature. Appl. Phys. Lett. *6*, 75–76 (1965)

369   L.W. Riley, M. Bass, F.L. Hahn: Stimulated emission from 4.3% abundant Cr^{50} ions in ruby. Appl. Phys. Lett. *7*, 88–90 (1965)

370   D. Roess, G. Zeidler: Quasicontinuous ruby giant pulse laser using a saturable absorber as a Q switch. Appl. Phys. Lett. *8*, 10–12 (1966)

371   K. Gürs: Ein kontinuierlicher wassergekühlter Rubinlaser. Phys. Lett. *16*, 125–127 (1965)

372   E.V. Sayre, K.M. Sancier, S. Freed: Absorption spectrum and quantum states of the praseodymium ion. I. Single crystals of praseodymium chloride. J. Chem. Phys. *23*, 2060–2065 (1955)

373   T. Kushida: Temperature dependence of absorption linewidths of ruby. J. Phys. Soc. Jpn. *22*, 351 (1967)

374   M.D. Galanin, Z.A. Chizhikova: Luminescence of ruby at high excitation energies and in the laser action. Opt. Spektrosk. *17*, 402–406 (1964)
English transl.: Opt. Spectrosc. (USSR) *17*, 214–215 (1964)

375   R.C. Powell, B. DiBartolo, B. Birang, C.S. Naiman: Fluorescence studies of energy transfer between single and pair Cr^{3+} systems in Al_2O_3. Phys. Rev. *155*, 296–308 (1967)

376   J.Ch. Vienot, J. Pasteur: Some comments on stimulated emission from particular rubies $(Al_2O_3; Cr^{3+}, Eu^{3+})$. Z. Angew. Math. Phys. *16*, 71–72 (1965)

377   A.A. Kaminskii, P.V. Klevtsov, A.A. Pavlyuk. Stimulated emission from $KY(MoO_4)_2 – Nd^{3+}$ crystal laser. Phys. Status Solidi (a) *1*, K91–K94 (1970)

378   T. Mijakawa, D.L. Dexter: Phonon sidebands, multiphonon relaxation of excited states, and phonon-assisted energy transfer between ions in solids. Phys. Rev. *B 1*, 2961–1969 (1970)

379   A.A. Kaminskii, P.V. Klevtsov, L. Li, A.A. Pavlyuk: Stimulated emission from $KY(WO_4)_2 : Nd^{3+}$ crystal laser. Phys. Status Solidi (a) *5*, K79–K81 (1971)

380   A.A. Kaminskii, P.V. Klevtsov, L. Li, A.A. Pavlyuk: Stimulated emission of $KY(WO_4)_2 : Nd^{3+}$ crystals. Kvant. Elektron. (Moscow) N4, 113–116 (1971)
English transl.: Sov. J. Quantum Electron. *1*, 405–407 (1972)

381   L. Huff: Sun-pumped laser. Digest of Technical Papers CLEA 1973 IEEE/OSA, Washington, May-June 1973, p. 48

382   L.H. Brixner, P.A. Flournoy: Calcium orthovanadate $Ca_3(VO_4)_2$—a new laser host crystal. J. Electrochem. Soc. *112*, 303–308 (1965)

383   A.A. Ballman, S.P.S. Porto, A. Yariv: Calcium niobate $Ca(NbO_3)_2$—a new laser host crystal. J. Appl. Phys. *34*, 3155–3156 (1963)

384   P.A. Flournoy, L.H. Brixner: Laser characteristics of niobium compensated $CaMoO_4$ and $SrMoO_4$. J. Electrochem. Soc. *112*, 779–781 (1965)

385   A. Yariv, S.P.S. Porto, K. Nassau: Optical maser emission from trivalent praseodymium in calcium tungstate. J. Appl. Phys. *33*, 2519–2521 (1962)

386   E.G. Reut, A.I. Ryskin: Radiative levels of the Pr^{3+} ion in scheelite type crystals. Opt. Spektrosk. *20*, 172–173 (1966)
English transl.: Opt. Spectrosc. (USSR) *20*, 91 (1966)

387   L.F. Johnson, G.D. Boyd, K. Nassau, R.R. Soden: Continuous operations of the $CaWO_4 – Nd^{3+}$ optical maser. Proc. IRE *50*, 213 (1962)

388   A.A. Kaminskii: High-temperature spectroscopy of the stimulated emission of an optical quantum generator on crystals and glasses activated by Nd^{3+} ions. Electron Technol. (Poland) *2*, 203–208 (1969)

389   N.S. Belokrinitskii, M.T. Shpak: "Luminescent and laser properties of neodymium in some scheelite structures," in *Kvantovaya Elektronika*, (Naukova Dumka, Kiev 1971) pp. 162–227

390   L.F. Johnson, G.D. Boyd, K. Nassau: Optical maser characteristics of Ho^{3+} in $CaWO_4$. Proc. IRE *50*, 87–88 (1962)

391   Z.J. Kiss, R.C. Duncan: Optical maser action in $CaWO_4 – Er^{3+}$. Proc. IRE *50*, 1531 (1962)

392   L.F. Johnson, G.D. Boyd, K. Nassau: Optical maser characteristics of Tm^{3+} in $CaWO_4$. Proc.

IRE *50*, 86–87 (1962)

393   L.F. Johnson, R.R. Soden: Optical maser characteristics of Nd^{3+} in $SrMoO_4$. J. Appl. Phys. *33*, 757 (1962)

394   R.H. Hoskins, B.H. Soffer: Stimulated emission from $Y_2O_3:Nd^{3+}$. Appl. Phys. Lett. *4*, 22–23 (1964)

395   N.C. Chang: Fluorescence and stimulated emission from trivalent europium in yttrium oxide. J. Appl. Phys. *34*, 3500–3504 (1963)

396   Kh.S. Bagdasarov, G.A. Bogomolova, A.A. Kaminskii, A.M. Kevorkov, G.I. Rogov: Yttrium aluminate containing TR^{3+} ions as an active medium for lasers. Kristallografiya *14*, 513–514 (1969)
      English transl.: Sov. Phys. Crystallogr. *14*, 423–424 (1969)

397   Kh.S. Bagdasarov, A.A. Kaminskii, G.I. Rogov: Synthesis and optical properties of crystals of $YAlO_3$ activated by Nd^{3+} ions. Dok. Akad. Nauk SSSR *185*, 1022–1024 (1969)
      English transl.: Sov. Phys. Dokl. *14*, 346–348 (1969)

398   M.J. Weber, M. Bass, K. Andringa, R.R. Monchamp, E. Comperchio: Czochralski growth and properties of $YAlO_3$ laser crystals. Appl. Phys. Lett. *15*, 342–345 (1969)

399   G.A. Keig, L.G. DeShazer: Laserverhalten von Yttrium-Orthoaluminat bei dotierung mit seltenen Erden. Laser und Elektro-Optik *N3*, 45–50 (1972)

400   B.K. Sevastyanov, Kh.S. Bagdasarov, L.B. Pasternak, S.Yu. Volkov, V.P. Orekhova: Lasing on Cr^{3+} ions in yttrium aluminum garnet crystals. Zh. Eksp. Teor. Fiz. Pis'ma Red. *17*, 69–71 (1973)
      English transl.: JETP Lett. *17*, 47–48 (1973)

401   J.E. Geusic, W.B. Bridges, J.I. Pankove: Coherent optical sources for communications. Proc. IEEE *58*, 1419–1439 (1970)

402   A.A. Kaminskii, D.N. Vylegzhanin: A study of electron-phonon processes in the crystal $+Nd^{3+}$ system. Dok. Akad. Nauk SSSR *195*, 827–830 (1970)
      English transl.: Sov. Phys. Dokl. *15*, 1078–1081 (1971)

403   D.N. Vylegzhanin, A.A. Kaminskii: Manifestation of electron–phonon interection in the lasing of a crystal with Nd^{3+}. Zh. Eksp. Teor. Fiz. Pis'ma Red *11*, 569–573 (1970)
      English transl.: JETP Lett. *11*, 393–396 (1970)

404   A.A. Kaminskii: Stimulierte Emissionsspektroskopie. Laser *N2*, 58 (1971)

405   A.A. Kaminskii, Kh.S. Bagdasarov, L.M. Belyaev: Analysis of the optical spectra and stimulated radiation of $Y_3Al_5O_{12}$–Nd^{3+} crystals. Phys. Status Solidi *21*, K23–K29 (1967)

406   A.M. Korovkin, A.M. Morozov, A.M. Tkachuk, A.A. Fedorov, V.A. Fedorov, P.P. Feofilov: "Spontaneous and induced emission of holmium ions in $LaNa(MoO_4)_2$ and $LaNbO_4$ crystals," in *Spektroskopiya Kristallov*, ed by A.A. Kaminskii, Z.L. Morgenshtern, D.T. Sviridov (Nauka, Moscow 1975) pp. 281–287

407   G.M. Zverev, G.Ya. Kolodnyi, A.M. Onishchenko: Nonradiative transitions between levels of trivalent rare-earth ions in yttrium aluminum garnet crystals. Zh. Eksp. Teor. Fiz. *60*, 920–928 (1971)
      English transl.: Sov. Phys. JETP *33*, 497–501 (1971)

408   T. Kushida, J.E. Geusic: Optical refrigeration in Nd-doped $Y_3Al_5O_{12}$. Phys. Rev. Lett. *21*, 1172–1175 (1968)

409   J.R. O'Conner: Optical and laser properties of Nd^{3+} and Eu^{3+} doped YVO_4. Trans. Metall. Soc. AIME *239*, 362–365 (1967)

410   Kh.S. Bagdasarov, G.A. Bogomolova, A.A. Kaminskii, V.I. Popov: Absorption, luminescence, and stimulated emission of YVO_4–Nd^{3+} crystals. Dok. Akad. Nauk SSSR *180*, 1347–1350 (1968)
      English transl.: Sov. Phys. Dokl. *13*, 516–518 (1968)

411   J.R. O'Conner: Unusual crystal-field energy levels and efficient laser properties of $YVO_4:Nd^{3+}$. Appl. Phys. Lett. *9*, 407–409 (1966)

412   Kh.S. Bagdasarov, A.A. Kaminskii, V.S. Krylov, V.I. Popov: Room-temperature induced emission of tetragonal YVO_4 crystals containing Nd^{3+}. Phys. Status Solidi *27*, K1–K3 (1968)

413   J.J. Rubin, L.G. Van Uitert: Growth of large yttrium vanadate single crystals for optical maser studies. J. Appl. Phys. *37*, 2920–2921 (1966)

414   L.F. Johnson, J.P. Remeika, J.F. Dillon: Coherent emission from Ho^{3+} ions in yttrium iron garnet. Phys. Lett. *21*, 37–39 (1966)

415   R.V. Bakradze, G.M. Zverev, G.Ya. Kolodnyi, G.P. Kuznetsova, L.V. Makarenko, A.M.

426   References

Onichchenko, N.I. Sergeeva, Z.I. Tatarov: Sensitized luminescence and stimulated radiation from yttrium aluminum garnet. Zh. Eksp. Teor. Fiz. 53, 490–491 (1967)
English transl.: Sov. Phys. JETP 26, 323 (1968)

416   Yu.K. Voronko, G.V. Maksimova, V.G. Mikhalevich, V.V. Osiko, A.A. Sobol, M.I. Timoshechkin, G.P. Shipulo: Spectral properties and induced emission from crystals of yttrium-lutetium-aluminum garnet with Nd^{3+}. Opt. Spektrosk. 33, 681–688 (1972)
English transl.: Opt. Spectrosc. (USSR) 33, 376–380 (1972)

417   R.H. Hoskins, B.H. Soffer: Fluorescence and stimulated emission from $La_2O_3:Nd^{3+}$. J. Appl. Phys. 36, 323–324 (1965)

418   G.F. Bakhsheeva, V.E. Karapetyan, A.M. Morozov, L.G. Morozova, M.N. Tolstoi, P.P. Feofilov: Optical constants, luminescence and stimulated radiation of single crystals of lanthanum niobate activated by neodymium. Opt. Spektrosk. 28 76–81 (1970)
English transl.: Opt. Spectrosc. (USSR) 28, 38–41 (1970)

419   B.H. Soffer, R.H. Hoskins: Fluorescence and stimulated emission from $Gd_2O_3:Nd^{3+}$ at room temperature and $77°K$. Appl. Phys. Lett. 4, 113–114 (1964)

420   H.J. Borchardt, P.E. Bierstedt: $Gd_2(MoO_4)_3$: a ferroelectric laser host. Appl. Phys. Lett. 8, 50–52 (1966)

421   Kh.S. Bagdasarov, A.A. Kaminskii, A.M. Kevorkov, S.E. Sarkisov, T.A. Tevosyan: Stimulated emission from (Er, Lu)AlO_3 crystals containing Ho^{3+} and Tm^{3+} ions. Kristallografiya 18, 1083–1084 (1973)
English transl.: Sov. Phys. Crystallogr. 18, 681 (1974)

422   P.A. Arsenev: The spectral parameters of trivalent holmium in the lattice of aluminum-ytterbium garnet. Ukr. Fiz. Zh. (Russ. Ed.) 15, 689–692 (1970)

423   Ya.E. Kariss, M.N. Tolstoi, P.P. Feofilov: On the stimulated emission of neodymium in single crystals of lead molybdate. Opt. Spektrosk. 18, 177–179 (1965)
English transl.: Opt. Spectrosc. (USSR) 18, 99–100 (1965)

424   A.A. Kaminskii, A.A. Mayer, M.V. Provotorov, S.E. Sarkisov: Investigation of stimulated emission from $LiLa(MoO_4)_2:Nd^{3+}$ crystal laser. Phys. Status Solidi (a)17, K115–K117 (1973)

425   A.A. Kaminskii, A.A. Mayer, N.S. Nikonova, M.V. Provotorov, S.E. Sarkisov: Stimulated emission from the new $LiGd(MoO_4)_2-Nd^{3+}$ crystal laser. Phys. Status Solidi (a)12, K73–K75 (1972)

426   N.S. Belokrinitskii, N.D. Belousov, V.I. Bonchkovskii, V.A. Kobzar-Zlenko, B.S. Skorobogatov, M.S. Soskin: Investigation of induced radiation of $LaNa(WO_4)_2$ single crystal activated with Nd^{3+}. Ukr. Fiz. Zh. (Russ. Ed.) 14, 1400–1404 (1969)

427   G.E. Peterson, P.M. Bridenbaugh: Laser oscillation at 1.06 μ in the series $Na_{0.5}Gd_{0.5-x}Nd_xWO_4$. Appl. Phys. Lett. 4, 173–175 (1964)

428   A.A. Kaminskii, P.V. Klevtsov, L. Li, S.E. Sarkisov: Stimulated emission of radiation of crystals of $KLa(MoO_4)_2$ with Nd^{3+} ions. Izv. Akad. Nauk SSSR, Neorg. Mater. 9, 2059–2061 (1973)
English transl.: Inorg. Mater. (USSR) 9, 1824–1826 (1973)

429   R.H. Hopkins, N.T. Melamed, T. Nenningsen, G.W. Roland: Crystal growth and properties of $CaY_4(SiO_4)_3O$, a new laser host for Ho^{3+}. J. Cryst. Growth 10, 218–222 (1971)

430   V.I. Aleksandrov, Yu.K. Voronko, V.G. Mikhalevich, V.V. Osiko, A.M. Prokhorov, V.M. Tatarintsev, V.T. Udovenchik, G.P. Shipulo: Spectroscopic properties and emission of Nd^{3+} in ZrO_2 and HfO_2 crystals. Dok. Akad. Nauk SSSR 199, 1282–1283 (1971)
English transl.: Sov. Phys. Dokl. 16, 657–658 (1972)

431   B.I. Aleksandrov, T.M. Murina, V.K. Zhekov, V.M. Tataritsev: Induced emission Tm^{3+}, Ho^{3+} in ZrO_2 crystals. Sb. Kratk. Soobshch. Po Fiz. Akad. Nauk SSSR-Fiz. Inst. Im. P.N. Lebedeva N2, 17–21 (1973)

432   M. Alam, K.H. Gooen, B. DiBartolo, A. Linz, E. Sharp, L. Gillespie, G. Janney: Optical spectra and laser action of neodymium in a crystal $Ba_2MgGe_2O_7$. J. Appl. Phys. 39, 4728–4730 (1968)

433   J.E. Miller, E.J. Sharp, D.J. Horowitz: Optical spectra and laser action of neodymium in a crystal $Ba_{0.25}Mg_{2.75}Y_2Ge_3O_{12}$. J. Appl. Phys. 43, 462–465 (1972)

434   Z.M. Bruk, Yu.K. Voronko, G.V. Maksimova, V.V. Osiko, A.M. Prokhorov, K.F. Shipilov, I.A. Shcherbakov: Optical properties and stimulated emission of Nd^{3+} in fluor-apatite. Zh. Eksp. Teor. Fiz. Pis'ma Red. 8, 357–360 (1968)
English transl.: JETP Lett. 8, 221–223 (1968)

435 L.E. Sobon, K.A. Wickersheim, R.A. Buchanan, R.V. Alves: Growth and properties of lanthanum oxysulfide crystals. J. Appl. Phys. *42*, 3049–3053 (1971)

436 *International Tables for X-Ray Crystallography*, Vol. I (Kynoch Press, Birmingham 1952)

437 M.J. Weber: "Insulating crystal lasers," in *Laser Handbook*, ed. by R.J. Pressley (CRS Press, Cleveland 1971) pp. 371–417

438 K. Nassau: "The chemistry of laser crystals," in *Applied Solid State Science*, Vol. 2 (Academic Press, New York 1971) pp. 173–299

439 K. Nassau: Lasers and laser materials. Mater. Res. Stand. *5*, 3–11 (1965)

440 L.G. Van Uitert: Solid state maser materials. Metallurgy Adv. Electron. Mater. *19*, 305–327 (1962)

441 G.A. Bogomolova, A.A. Kaminskii, V.A. Timofeeva: Nd^{3+} optical centers in $Y_3Al_5O_{12}$ crystals. Phys. Status Solidi *16*, K165–K166 (1966)

442 J.P. Hurrell, S.P.S. Porto, I.F. Chang, S.S. Mitra, R.P. Bauman: Optical phonons of yttrium aluminum garnet. Phys. Rev. *173*, 851–856 (1968)

443 N.T. McDevitt: Infrared lattice spectra of rare-earth aluminum, gallium, and iron garnets. J. Opt. Soc. Am. *59*, 1240–1244 (1969)

444 G.A. Slack, D.W. Oliver, R.M. Chrenko, S. Roberts: Optical absorption of $Y_3Al_5O_{12}$ from 10 to 55000 cm^{-1} wave numbers. Phys. Rev. *177*, 1308–1314 (1969)

445 J.A. Koningstein, O.S. Mortensen: Laser-excited phonon Raman spectrum of garnets. J. Mol. Spectrosc. *27*, 343–350 (1968)

446 R.A. Brandewie, G.L. Telk: Quantum efficiency of Nd^{3+} in glass, calcium tungstate, and yttrium aluminum garnet. J. Opt. Soc. Am. *57*, 1221–1225 (1967)

447 R.K. Watts: Branching ratios for $YAlG:Nd^{3+}$. J. Opt. Soc. Am. *61*, 123–124 (1971)

448 W.F. Krupke: Radiative transition probabilities within the $4f^3$ ground configuration of Nd:YAG. IEEE J. Quantum Electron *7*, 153–159 (1971)

449 I.B. Vasilyev, G.M. Zverev, G.Ya. Kolodnyi, A.M. Onishchenko: Nonresonance excitation energy transfer between impurity rare-earth ions. Zh. Eksp. Teor. Fiz. *56*, 122–133 (1969) English transl.: Sov. Phys. JETP *29*, 69–75 (1969)

450 J.R. Thornton, W.D. Fountain, G.W. Flint, T.G. Crow: Properties of neodymium laser materials. Appl. Opt. *8*, 1087–1102 (1969)

451 E.M. Voronkova, B.N. Grechushnikov, G.I. Distler, I.P. Petrov: *Opticheskie Materialy dlya Infrakrasnoi Tekhniki*, (Nauka, Moscow 1965)

452 V.L. Bakumenko, G.S. Kozina, T.A. Kostynskaya, E.P. Lupachev, E.S. Rvacheva: Stimulated emission of praseodymium in calcium tungstate. Opt. Spektrosk. *19*, 132 (1965) English transl.: Opt. Spectrosc. (USSR) *19*, 68 (1965)

453 P. Mercurio, R.H. Milburn: Hybrid YAG–YAlO laser operation at 1.06 μ. Appl. Opt. *11*, 2097–1200 (1972)

454 G.A. Massey: Criterion for selection of CW laser host materials to increase available power in the fundamental mode. Appl. Phys. Lett. *17*, 213–215 (1970)

455 J.D. Foster, L.M. Osterink: Thermal effects in a Nd:YAG laser. J. Appl. Phys. *41*, 3656–3663 (1970)

456 L.M. Osterink, J.D. Foster: Thermal effects and transverse mode control in a Nd:YAG laser. Appl. Phys. Lett. *12*, 128–131 (1968)

457 W. Koechner: Thermal lensing in a Nd:YAG laser rod. Appl. Opt. *9*, 2548–2553 (1970)

458 J.D. Foster, L.M. Osterink: Index of refraction and expansion thermal coefficients of Nd:YAG. Appl. Opt. *7*, 2428–2429 (1968)

459 W.C. Scott, M. deWit: Birefringence compensation and TEM_{00} mode enhancement in a Nd:YAG laser. Appl. Phys. Lett. *18*, 3–4 (1971)

460 C.M. Stickley: Laser brightness gain and mode control by compensation for thermal distortion. IEEE J. Quantum Electron. *2*, 511–518 (1966)

461 W.B. Jones, L.M. Goldman, J.P. Chernoch, W.S. Martin: The mini-FPL—a face-pumped laser: concept and implemention. Digest of Technical Papers VII Intern. Quantum Electron. Conf., May 8–11, 1972, Montreal, pp. 16–17

462 A.L. Mikaelyan, V.V. Dyachenko: Conservation of wave front in strongly deformed solid media. Zh. Eksp. Teor. Fiz. Pis'ma Red. *16*, 25–29 (1972) English transl.: JETP Lett. *16*, 17–19 (1972)

463 Yu.K. Voronko, T.G. Mamedov, V.V. Osiko, M.I. Timoshechkin, I.A. Shcherbakov: Effect of donor–donor and donor–acceptor interactions on the decay kinetics of the metastable state of

428   References

Nd^{3+} in crystals. Zh. Eksp. Teor. Fiz. *65*, 1141–1156 (1973)
English transl.: Sov. Phys. JETP *38*, 565–572 (1974)

464   A.A. Kaminskii: Multibeam lasers. Izv. Akad. Nauk SSSR, Neorg. Mater. *10*, 2230–2231 (1974)
English transl.: Inorg. Mater. (USSR) *10*, 1911–1912 (1974)

465   N.G. Basov, A.R. Zaritskii, S.D. Zakharov, P.G. Kryukov, Yu.A. Matveets, Yu.V. Senatskii, A.I. Fedosimov, S.V. Chekalin: Generation of high-power light pulses at wavelengths 1.06 and 0.53 μ and their application in plasma heating. II. Neodymium-glass laser with a second-harmonic converter. Kvant. Elektron. (Moscow) N6, 50–55 (1972)
English transl.: Sov. J. Quantum Electron. *2*, 533–535 (1973)

466   F. Auzel: Compteur quantique par transfert d'énergie entre deux ions de terres rares dans un tungstate mixte et dans un verre. C. R. Acad. Sci. Sér. *B 262*, 1016–1019 (1966)

467   V.L. Donlan, A.A. Santiago: Optical spectra and energy levels of erbium-doped yttrium orthoaluminate. J. Chem. Phys. *57*, 4717–4723 (1972)

468   R.V. Bakradze, G.M. Zverev, G.Ya. Kolodnyi, G.P. Kuznetsova, A.M. Onishchenko: Effect of rare-earth impurities on the structure and optical properties of yttrium aluminum garnet. Izv. Akad. Nauk SSSR, Ser. Fiz. *31*, 2070–2073 (1967)
English transl.: Bull. Acad. Sci. USSR, Phys. Ser. *31*, 2113–2116 (1968)

469   S.Kh. Batygov, L.A. Kulevskii, S.A. Lavrukhin, A.M. Prokhorov, V.V. Osiko, A.D. Savelyev, V.V. Smirnov: Laser based on CaF$_2$–ErF$_3$ crystals. Kurzfassungen Internat. Tagung Laser und ihre Anwendungen, 1973, Dresden, Teil 2, K97

470   R.K. Sviridova, P.A. Arsenev: The spectra of ScYO$_3$ crystals containing Nd^{3+} ions. Zh. Prikl. Spektrosk. *17*, 888–890 (1972)
English transl.: J. Appl. Spectrosc. (USSR) *17*, 1482–1483 (1974)

471   P.A. Arsenev, K.E. Binert: Synthesis and optical properties of single crystals of gadolinium aluminate (GdAlO$_3$) doped with neodymium ions. Zh. Prikl. Spektrosk. *17*, 1084–1087 (1972)
English transl.: J. Appl. Spectrosc. (USSR) *17*, 1623–1625 (1974)

472   P.A. Arsenev, K.E. Bienert, R.K. Sviridova: Spectral properties of neodymium ions in the lattice of GdScO$_3$ crystals. Phys. Status Solidi (*a*) 9, K103–K104 (1972)

473   P.A. Arsenev, K.E. Bienert: Growth of yttrium and gadolinium aluminate laser crystals by optical zone-melting techniques and their spectroscopic properties. Kurzfassungen Internat. Tagung Laser und ihre Anwendungen, 4.6–9.6, 1973, Dresden, Teil 1, K 44

474   R.C. Eckardt, J.L. DeRosa, J.P. Letellier: Characteristics of an Nd:CaLaSOAP mode-locked oscillator. IEEE J. Quantum Electron. *10*, 620–622 (1974)

475   H.P. Weber, P.F. Liao, B.C. Tofield, P.M. Bridenbaugh: CW fiber laser of NdLa pentaphosphate. Appl. Phys. Lett. *26*, 692–294 (1975)

476   M.J. Weber: Nonradiative decay from 5d states of rare earths in crystals. Solid State Commun. *12*, 741–744 (1973)

477   M.J. Weber: Multiphonon relaxation of rare earth ions in yttrium orthoaluminate. Phys. Rev. *B 8*, 54–64 (1973)

478   J.P. Van der Ziel, W.A. Bonner, L. Kopf, S. Singh, L.G. Van Uitert: Laser oscillation from Ho^{3+} and Nd^{3+} ions in epitaxially grown thin aluminum garnet films. Appl. Phys. Lett. *22*, 656–657 (1973)

479   L.F. Johnson, H.J. Guggenheim: Laser emission at 3 μ from Dy^{3+} in BaY$_2$F$_8$. Appl. Phys. Lett. *23*, 96–98 (1973)

480   W.A. Bonner: Epitaxial growth of garnets for thin film lasers. J. Electron. Mater. *3*, 193–208 (1974)

481   A.J. Lindop, D.W. Goodwin: EPR and optical spectroscopy of Nd^{3+} ions in calcium aluminate. J. Phys. *C 6*, 1818–1829 (1973)

482   S. Singh, L.G. Van Uitert, J.R. Potopowicz, W.H. Grodkiewicz: Laser emission at 1.065 μm from neodymium-doped anhydrous cerium trichloride at room temperature. Appl. Phys. Lett. *24*, 10–13 (1974)

483   G.A. Bogomolova, D.N. Vylegzhanin, A.A. Kaminskii: Spectral and lasing investigations of garnets with the Yb^{3+} ions. Zh. Eksp. Teor. Fiz. *69*, 860–874 (1975)
English transl.: Sov. Phys. JETP *42*, 440–446 (1976)

484   R.A. Buchanan, K.A. Wickersheim, J.J. Pearson, G.F. Hermann: Energy levels of Yb^{3+} in gallium and aluminum garnets. I. Spectra. Phys. Rev. *159*, 245–251 (1967)

485   I.S. Minhas, K.K. Sharma: Optical absorption and Zeeman spectra of Nd^{3+}-doped PbMoO$_4$.

Phys. Rev. *B 8*, 385–392 (1973)

486 Kh.S. Bagdasarov, A.A. Kaminskii, A.M. Kevorkov, A.M. Prokhorov, S.E. Sarkisov, T.A. Tevosyan: Stimulated emission of TR^{3+} ions in ytterbium-aluminum garnet crystals. Dok. Akad. Nauk SSSR *218*, 550–551 (1974)
English transl.: Sov. Phys. Dokl. *19*, 592 (1975)

487 A.A. Kaminskii, B.P. Sobolev, Kh.S. Bagdasarov, N.L. Tkachenko, S.E. Sarkisov, K.B. Seyranian: Investigation of stimulated emission from crystals with Nd^{3+} ions. Phys. Status Solidi (a) 23, K135–K136 (1974)

488 Kh.S. Bagdasarov, A.A. Kaminskii, A.M. Kevorkov, A.M. Prokhorov: Investigation of the stimulated radiation emitted by Nd^{3+} ions in $CaSc_2O_4$ crystals. Kvant. Elektron. (Moscow)*1*, 1666–1668 (1974)
English transl.: Sov. J. Quantum Electron. *4*, 927–928 (1975)

489 Kh.S. Bagdasarov, A.A. Kaminskii, A.M. Kevorkov, L. Li, A.M. Prokhorov, S.E. Sarkisov, T.A. Tevosyan: Stimulated emission of Nd^{3+} ions in an $SrAl_{12}O_{19}$ crystal at the transitions $^4F_{3/2} \rightarrow {}^4I_{11/2}$ and $^4F_{3/2} \rightarrow {}^4I_{13/2}$. Dok. Akad. Nauk SSSR *216*, 767–768 (1974)
English transl.: Sov. Phys. Dokl. *19*, 350 (1974)

490 E.V. Zharikov, V.I. Zhekov, L.A. Kulebskii, T.M. Murina, V.V. Osiko, A.M. Prokhorov, A.D. Savelev, V.V. Smirnov, B.P. Starikov, M.I. Timoshechkin: Stimulated emission from Er^{3+} ions in yttrium aluminum garnet crystals at $\lambda = 2.94\ \mu$. Kvant. Elektron. (Moscow)*1*, 1867–1869 (1974)
English transl.: Sov. J. Quantum Electron. *4*, 1039–1040 (1975)

491 L.F. Johnson, K.A. Ingersoll: Elimination of degradation in the laser output from Ho^{3+} in sensitized YAG. J. Appl. Phys. *44*, 5444–5446 (1973)

492 I.J. Barton, D.W. Goodwin: Continuous laser action in $YAlO_3$. J. Phys. *D 5*, 228–234 (1972)

493 D.E. Wortman: Ground term energy states for Nd^{3+} in $LiYF_4$. J. Phys. Chem. Solids *33*, 311–318 (1972)

494 J.F.B. Hawkes, M.J.M. Leask: Spectroscopic study of $Er^{3+}-Gd^{3+}$ interaction in $Er:GdAlO_3$. J. Phys. *C 5*, 1705–1715 (1972)

495 B.B. Kadomtsev (ed.): *Lasery i Termoyadernaya Problema*, (Atomizdat, Moscow 1973)

496 J.C. Toledano: Optical spectrum of Nd^{3+} in oxidized calcium fluoride. J. Chem. Phys. *57*, 1046–1050 (1972)

497 P. Görlich, H. Karras, G. Kötitz, R. Lehmann: *Spectroscopic Properties of Activated Laser Crystals*, (Akademie-Verlag, Berlin 1965)

498 F.E. Auzel: Materials and devices using double-pumped phosphors with energy transfer. Proc. IEEE *61*, 758–786 (1973)

499 K.H. Yang, J.A. DeLuca: UV fluorescence of cerium-doped lutetium and lanthanum trifluorides, potential tunable coherent sources from 2760 to 3220 Å. Appl. Phys. Lett. *31*, 594–596 (1977)

500 S.A. Altshuler, B.M. Kozyrev: *Electron Paramagnetic Resonance*, (Academic Press, New York 1964)

501 C. Kane-Maguire, J.A. Koningstein: On the vibroelectronic Raman effect. J. Chem. Phys. *59*, 1899–1904 (1973)

502 H. Bethe: Termaufspaltung in Kristallen. Ann. Physik (Leipzig) *3*, 133–208 (1929)

503 C.J. Ballhausen: *Introduction to Ligand Field Theory*, (McGraw-Hill, New York 1963)

504 Kh.S. Bagdasarov, G.A. Bogomolova, M.M. Grotsenko, A.A. Kaminskii, A.M. Kevorkov, A.M. Prokhorov, S.E. Sarkisov: Spectroscopy of the stimulated radiation of $Gd_3Ga_5O_{12}-Nd^{3+}$ crystals. Dok. Akad Nauk SSSR *216*, 1018–1021 (1974)
English transl.: Sov. Phys. Dokl. *19*, 353–355 (1974)

505 T.C. Damen, H.P. Weber, B.C. Tofield: NdLa pentaphosphate laser performance. Appl. Phys. Lett. *23*, 519–520 (1973)

506 Kh.S. Bagdasarov, G.A. Bogomolova, A.A. Kaminskii, A.M. Kevorkov L. Li, A.M. Prokhorov, S.E. Sarkisov: Study of the stimulated emission of $Lu_3Al_5O_{12}$ crystals containing Nd^{3+} ions at the transitions $^4F_{3/2} \rightarrow {}^4I_{11/2}$ and $^4F_{3/2} \rightarrow {}^4I_{13/2}$. Dok. Akad. Nauk SSSR *218*, 316–319 (1974)
English transl.: Sov. Phys. Dokl. *19*, 584–585 (1975)

507 H.G. Danielmeyer, H.P. Weber: Fluorescence in neodymium ultraphosphate. IEEE J. Quantum Electron. *8*, 805–808 (1972)

508 Yu.E. Perlin, E.I. Perepilitsa, A.M. Tkachuk, L.S. Kharchenko, B.S. Tsukerblat: Tempera-

ture broadening and shift of the R line of Cr^{3+} ion in lanthanum aluminate. Opt. Spektrosk. *33*, 1121–1128 (1972)
English transl.: Opt. Spectrosc. (USSR) *33*, 614–617 (1972)

509   P.A. Arsenev, L.N. Raiskaya, R.K. Sviridova: Spectral properties of neodymium ions in the lattice of Y_2SiO_5 crystals. Phys. Status Solidi (*a*) *13*, K45–K47 (1972)

510   R.J. Pressley (ed.): *Handbook of Lasers*, (CRS Press, Cleveland 1971)

511   N.A. Kozlov, A.A. Mak, B.M. Sedov: Solid state sun-pumped lasers. Opt.-Mekh. Prom. N11, 25–29 (1966)
English transl.: Sov. J. Opt. Technol. *33*, 549–553 (1966)

512   B.Ya. Zabokritskii, V.N. Nikitin, M.S. Soskin, A.I. Khizhnyak: The investigations of 1.06 μm luminescence band of the neodymium glass by giant pulse generation. Digest of Technical Papers IV All-Union Conf. on Spectroscopy of Crystals, September 25–29, 1973, Sverdlovsk, pp. 86–87

513   V.R. Belan, Ch.M. Briskina, V.V. Grigoryants, M.E. Zhabotinskii: Energy transfer between neodymium ions in glass. Zh. Eksp. Teor. Fiz. *57*, 1148–1159 (1969)
English transl.: Sov. Phys. JETP *30*, 627–632 (1969)

514   A.A. Kaminskii, L. Li: Analysis of spectral line intensities of TR^{3+} ions in crystal systems. Phys. Status Solidi (*a*) *26*, 593–598 (1974)

515   Kh.S. Bagdasarov, Yu.K. Voronko, A.A. Kaminskii, L.V. Krotova, V.V. Osiko: Modification of the optical properties of $CaF_2 - TR^{3+}$ crystals by yttrium impurities. Phys. Status Solidi *12*, 905–912 (1965)

516   V.N. Baksheeva, B.I. Maksakov, M.N. Tolstoi, P.P. Feofilov, V.N. Shapovalov: "Spectral and luminescence properties of neodymium and ytterbium in the crystals of triple mixed fluorides," in *Spektroskopiya Kristallov*, ed. by P.P. Feofilov (Nauka, Moscow 1970) pp. 156–159

517   G.B. Bokii, V.B. Kravchenko: "Crystallochemical problems of activation," in *Spektroskopiya Kristallov*, ed. by P.P. Feofilov, (Nauka, Leningrad 1973) pp. 7–15

518   V.B. Aleksandrov, L.S. Garashina: New data on the structure of $CaF_2 - TRF_3$ solid solutions. Dok. Akad. Nauk SSSR *189*, 307–310 (1969)
English transl.: Sov. Phys. Dokl. *14*, 1040–1043 (1970)

519   A.K. Cheetham, B.E.F. Fender, D. Steele, R.I. Taylor, B.T.M. Willis: Defect structure of fluorite compounds containing excess anions. Solid State Commun. *8*, 171–173 (1970)

520   R.V. Bakradze, I.V. Vasilev, G.M. Zverev, G.Ya. Kolodnyi, G.P. Kuznetsova, A.M. Onishchenko: "Investigations of $Y_3(Al, Ga)_5O_{12}$ crystals doped with neodymium," in *Spektroskopiya Kristallov*, ed. by P.P. Feofilov (Nauka, Moscow 1970) pp. 184–187

521   L.G. Morozova, P.P. Feofilov: Luminescence and X-ray study of the $3Y_2O_3 - (5-x)Ga_2O_3 - xSc_2O_3$ system. Izv. Akad. Nauk SSSR, Neorg. Mater. *4*, 1738–1742 (1968)
English transl.: Inorg. Mater. (USSR) *4*, 1516–1519 (1968)

522   B. Cockayne, D.B. Gasson, D. Findlay, D.W. Goodwin, R.A. Clay: The growth and laser characteristics of yttrium-gadolinium-aluminum garnet single crystals. J. Phys. Chem. *29*, 905–910 (1968)

523   L.A. Riseberg, R.W. Brown, W.C. Holton: New class of intermediate-gain laser materials: mixed garnets. Appl. Phys. Lett. *23*, 127–129 (1973)

524   A.A. Kaminskii, G.A. Bogomolova, A.M. Kevorkov: Spectroscopic investigation of crystals with the garnet structure. Izv. Akad. Nauk SSSR, Neorg. Mater. *11*, 884–889 (1975)
English transl.: Inorg. Mater. (USSR) *11*, 757–761 (1975)

525   A.O. Ivanov, I.V. Mochalov, A.M. Tkachuk, V.A. Fedorov, P.P. Feofilov: Spectral characteristics of the thulium ion and cascade generation of stimulated radiation in a $YAlO_3:Tm^{3+};Cr^{3+}$ crystal. Kvant. Elektron. (Moscow) *2*, 188–190 (1975)
English transl.: Sov. J. Quantum Electron. *5*, 117–118 (1975)

526   A.O. Ivanov, I.V. Mochalov, A.M. Tkachuk, V.A. Fedorov, P.P. Feofilov: Emission of $\lambda \approx 2 \mu$ stimulated radiation by holmium in aluminum holmium garnet crystals. Kvant. Elektron. (Moscow) *2*, 186–188 (1975)
English transl.: Sov. J. Quantum Electron. *5*, 115–116 (1975)

527   A.L. Mikaelyan, V.V. Dyachenko: Lasers with waveguide resonators. Kvant. Elektron. (Moscow) *1*, 937–949 (1974)
English transl.: Sov. J. Quantum Electron. *4*, 514–520 (1974)

528   V.G. Mikhalevich, G.P. Shipulo: High-power YAG:Nd laser emitting polarized light. Kvant. Elektron. (Moscow) *1*, 455–456 (1974)

English transl.: Sov. J. Quantum Electron. *4*, 262 (1974)

529   Kh.S. S. Bagdasarov, A.A. Kaminskii, A.M. Kevorkov, A.M. Prokhorov: Rare earth scandium-aluminum garnets with impurity of TR^{3+} ions as active media for solid state lasers. Dok. Akad. Nauk SSSR *218*, 810–813 (1974)
English transl.: Sov. Phys. Dokl. *19*, 671–673 (1975)

530   K.G. Belabaev, A.A. Kaminskii, S.E. Sarkisov: Stimulated emission from ferroelectric $LiNbO_3$ crystals containing Nd^{3+} and Mg^{2+} ions. Phys. Status Solidi (a) *28*, K17–K20 (1975)

531   C.D. Brandle, J.C. Vanderleeden: Growth, optical properties, and CW laser action of neodymium-doped gadolinium scandium aluminum garnet. IEEE J. Quantum Electron. *10*, 67–71 (1974)

532   L.F. Johnson, H.J. Guggenheim: Electronic- and phonon-terminated laser emission from Ho^{3+} in BaY_2F_8. IEEE J. Quantum Electron. *10*, 442–449 (1974)

533   H.G. Danielmeyer, G. Huber, W.W. Krühler, J.P. Jeser: Continuous oscillation of a (Sc, Nd) pentaphosphate laser with 4 milliwatts pump threshold. Appl. Phys. *2*, 335–338 (1973)

534   M. Blatte, H.G. Danielmeyer, R. Urlich: Energy transfer and the complete level system of NdUP. Appl. Phys. *1*, 275–278 (1973)

535   A.A. Kaminskii, L. Li: Analysis of spectral line intensities of TR^{3+} ions in the disordered crystal systems. Phys. Status Solidi (a) *26*, K21–K26 (1974)

536   A.A. Kaminskii, B.P. Sobolev, Kh.S. Bagdasarov, A.M. Kevorkov, P.P. Fedorov, S.E. Sarkisov: Investigation of stimulated emission in the $^4F_{3/2} \rightarrow {}^4I_{13/2}$ transition of Nd^{3+} ions in crystals (VII). Phys. Status Solidi (a) *26*, K63–K65 (1974)

537   G.M. Zverev, V.M. Garmash, A.M. Onishchenko, V.A. Pashkov, A.A. Semenov, Yu.M. Kolbatskov, A.I. Smirnov: Induced emission by trivalent erbium ions in crystals of yttrium-aluminum garnet. Zh. Prikl. Spektrosk. *21*, 820–823 (1974)
English transl.: J. Appl. Spectrosc. (USSR) *21*, 1467–1469 (1974)

538   L.G. DeShazer, M. Bass, U. Ranon, J.K. Guha, E.D. Reed, J.W. Strozyk, L. Rothrock: Laser performance of Nd^{3+} and Ho^{3+} in YVO_4, and Nd^{3+} in gadolinium gallium garnet (GGG). Digest of Technical Papers VIII Intern. Quantum Electron. Conf., June 10–13, 1974, San Francisco, pp. 7–8

539   T.T. Basiev, E.M. Dianov, A.M. Prokhorov, I.A. Shcherbakov: Quantum yield of the luminescence radiation emitted from the metastable state of Nd^{3+} in silicate glasses and $Y_3Al_5O_{12}$. Dok. Akad. Nauk SSSR *216*, 297–299 (1974)
English transl.: Sov. Phys. Dokl. *19*, 288–289 (1974)

540   I.S. Angriesh, V.Ya. Gamurar, D.N. Vylegzhanin, A.A. Kaminskii, S.I. Klokishner, Yu.E. Perlin: Calculation of the thermal broadening of the A phononless line in the $^4F_{3/2} \rightarrow {}^4I_{11/2}$ band of the Nd^{3+} ion in yttrium-aluminum garnet. Kvant. Elektron. (Moscow) *2*, 287–293 (1975)
English transl.: Sov. J. Quantum Electron. *5*, 162–165 (1975)

541   E.M. Dianov, A.M. Prokhorov, V.P. Samoilov, I.A. Shcherbakov: Measurement of the probabilities of the radiative transitions from the metastable level of the Nd^{3+} in silicate glass and a garnet crystal. Dok. Akad. Nauk SSSR *215*, 1341–1344 (1974)
English transl.: Sov. Phys. Dokl. *19*, 219–221 (1974)

542   V.A. Kovarskii: *Mnogokvantovye Perekhody*, (Shtiintsa, Kishinev 1974)

543   N.V. Karlov, Yu.N. Petrov, A.M. Prokhorov, O.M. Stelmakh: Dissociation of boron trichloride molecules by CO_2-laser radiation. Zh. Eksp. Teor. Fiz. Pis'ma Red. *11*, 220–222 (1970)
English transl.: JETP Lett. *11*, 135–137 (1970)

544   N.G. Basov, E.M. Belenov, E.P. Markin, A.N. Oraevskii, A.V. Pankratov: Stimulation of chemical reactions by laser radiation. Zh. Eksp. Teor. Fiz. *64*, 485–487 (1973)
English transl.: Sov. Phys. JETP *37*, 247–252 (1973)

545   A.M. Morozov, I.G. Pogkolzina, A.M. Tkachuk, V.A. Fedorov, P.P. Feofilov: Luminescence and induced emission of lithium-erbium and lithium-holmium binary fluorides. Opt. Spektrosk. *39*, 605–607 (1975)
English transl.: Opt. Spectrosc. (USSR) *39*, 338–339 (1975)

546   A.A. Kaminskii, T.I. Butaeva, A.M. Kevorkov, V.A. Fedorov, A.G. Petrosyan, M.M. Gritsenko: New data on stimulated emission by crystals with high concentrations of Ln^{3+} ions. Izv. Akad. Nauk SSSR, Neorg. Mater. *12*, 1508–1511 (1976)
English transl.: Inorg. Mater. (USSR) *12*, 1238–1241 (1976)

547   A.A. Kaminskii, T.I. Butaeva, A.O. Ivanov, I.V. Mochalov, A.G. Petrosyan, G.I. Rogov, V.A. Fedorov: New data on stimulated emission of crystals containing Er^{3+} and Ho^{3+} ions. Zh. Tekh. Fiz. Pis'ma 2, 787–793 (1976)
      English transl.: Sov. Tech. Phys. Lett. 2, 308–310 (1976)
548   D.R. Tallant, D.S. Moore, J.C. Wright: Defect equilibria in fluorite structure crystals. J. Chem. Phys. 67, 2897–2907 (1977)
549   G.H. Dieke: Spectra and Energy Levels of Rare Earth Ions in Crystals, (Wiley, New York 1968)
550   A. Szabo: Laser-induced fluorescence-line narrowing in ruby. Phys. Rev. Lett. 25, 924–926 (1970)
551   L.A. Riseberg: Laser-induced fluorescence-line-narrowing spectroscopy of glass: Nd^{3+}. Phys. Rev. A 7, 671–678 (1973)
552   C. Delsart, N. Pelletier-Allard, R. Pelletier: Affinement d'une raie de fluorescence de Pr^{3+}: $LaAlO_3$ par excitation laser. Opt. Commun. 11, 84–88 (1974)
553   L.E. Erickson: Fluorescence line narrowing of trivalent praseodymium in lanthanum trifluoride single crystal—phonon-induced relaxation. Phys. Rev. B 11, 77–81 (1975)
554   R. Flach, D.S. Hamilton, P.M. Selzer, W.M. Yen: Time-resolved fluorescence line-narrowing studies in LaF_3:Pr^{3+}. Phys. Rev. Lett. 35, 1034–1037 (1975)
555   L.A. Riseberg, W.C. Holton: Laser-selective excitation of Nd ions in mixed-crystal systems. Opt. Commun. 9, 298–299 (1973)
556   C.L. Tang, J.M. Telle: Laser modulation spectroscopy of solids. J. Appl. Phys. 45, 4503–4505 (1974)
557   J.A. Koningstein: Advances in ionic and molecular electronic Raman spectroscopy (to be published)
558   B.G. Wybourne: Spectroscopic Properties of Rare Earths, (Wiley, New York 1965)
559   B.R. Judd: Optical absorption intensities of rare-earth ions. Phys. Rev. 127, 750–761 (1962)
560   G.S. Ofelt: Intensities of crystal spectra of rare-earth ions. J. Chem. Phys. 37, 511–520 (1962)
561   W.F. Krupke: Optical absorption and fluorescence intensities in several rare-earth-doped Y_2O_3 and LaF_3 single crystals. Phys. Rev. 145, 325–337 (1966)
562   J.A. Caird, L.G. DeShazer: Analysis of laser emission in Ho^{3+}-doped materials. IEEE J. Quantum Electron. 11, 97–99 (1975)
563   J.A. Caird, L.G. DeShazer, J. Nella: Characteristics of room-temperature 2.3-μm laser emission from Tm^{3+} in YAG and $YAlO_3$. IEEE J. Quantum Electron. 11, 874–881 (1975)
564   K.R. German, A. Kiel, H. Guggenheim: Radiative and nonradiative transitions of Pr^{3+} in trichloride and tribromide hosts. Phys. Rev. B 11, 2436–2442 (1975)
565   A.A. Kaminskii, A.O. Ivanov, S.E. Sarkisov, I.V. Mochalov, V.A. Fedorov, L. Li: Comprehensive investigation of the spectral and lasing characteristics of the $LuAlO_3$ crystal doped with Nd^{3+}. Zh. Eksp. Teor. Fiz. 71, 984–1002 (1976)
      English transl.: Sov. Phys. JETP 44, 516–524 (1976)
566   A.A. Kaminskii, S.E. Sarkisov, A.A. Maier, V.A. Lomonov, V.A. Balashov: Eulytine with TR^{3+} ions as a laser medium. Zh. Tekh. Fiz. Pis'ma 2, 156–161 (1976)
      English transl.: Sov. Techn. Phys. Lett. 2, 59–60 (1976)
567   A.A. Kaminskii, D. Schultze, B. Hermoneit, S.E. Sarkisov, L. Li, J. Bohm, P. Reiche, R. Ehlert, A.A. Mayer, V.A. Lomonov, V.A. Balashov: Spectroscopic properties and stimulated emission in the $^4F_{3/2} \rightarrow {}^4I_{11/2}$ and $^4F_{3/2} \rightarrow {}^4I_{13/2}$ transitions of Nd^{3+} ions from cubic $Bi_4Ge_3O_{12}$. Phys. Status Solidi (a) 33, 737–753 (1976)
568   I.G. Podkolozina, A.M. Tkachuk, V.A. Fedorov, P.P. Feofilov: Multifrequency generation of stimulated emission of Ho^{3+} ion in $LiYF_4$ crystals. Opt. Spektrosk. 40, 196–199 (1976)
      English transl.: Opt. Spectrosc. (USSR) 40, 111–112 (1976)
569   A.A. Kaminskii, S.E. Sarkisov, J. Bohm, P. Reiche, D. Schultze, R. Ueeker: Growing, spectroscopic and laser properties of crystals in $K_5Bi_{1-x}Nd_x(MoO_4)_4$ system. Phys. Status Solidi (a) 43, 71–79 (1977)
570   K.B. Steinbruegge, G.D. Baldwin: Evaluation of $CaLaSOAP$:Nd for high-power flash-pumped Q-switched lasers. Appl. Phys. Lett. 25, 220–222 (1974)
571   A.O. Ivanov, I.V. Mochalov, M.V. Petrov, A.M. Tkachuk, P.P. Feofilov: Spectroscopic properties of rare-earth garnet and orthoaluminate single crystals doped with Ho^{3+}, Er^{3+}, and Tm^{3+} ions. Digest of Technical Papers V All-Union Conf. on Spectroscopy of Crystals, June 2–6, 1976, Kazan, p. 102
572   A.M. Prokhorov, A.A. Kaminskii, V.V. Osiko, M.I. Timoshechkin, E.V. Zharikov, T.I.

Butaeva, S.E. Sarkisov, A.G. Petrosyan, V.A. Fedorov: Investigations of 3-μm stimulated emission from Er^{3+} ions in aluminum garnets at room temperature. Phys. Status Solidi (a) 40, K69–K72 (1977)

573   R.G. Stafford, H. Masui, R.L. Farraw, R.K. Chang, L.G. Van Uitert: Coherent emission from Ho^{3+} ions by pumping into the YAG absorption band. J. Appl. Phys. 47, 2483–2485 (1976)

574   H.G. Danielmeyer: "Stoichiometric laser materials," in Festkörperprobleme XV-Advances in Solid State Physics, ed. by H.J. Queisser, (Pergamon, Vieweg, Braunschweig 1975) pp. 253–277

575   H.P. Weber: Review of Nd pentaphosphate lasers. Opt. Quantum Electron. 7, 431–442 (1975)

576   S. Singh, D.C. Miller, J.R. Potopowicz, L.K. Shick: Emission cross section and fluorescence quenching of Nd^{3+} lanthanum pentaphosphate. J. Appl. Phys. 46, 1191–1196 (1975)

577   S.R. Chinn, H.Y-P. Hong: Fluorescence and lasing properties of $NdNa_5(WO_4)_4$, $K_3Nd(PO_4)_2$ and $Na_3Nd(PO_4)_2$. Digest of Technical Papers IX Intern. Quantum Electron. Conf., June 14–18 1976, Amsterdam, pp. 87–88

578   J. Nakano, K. Otsuka, T. Yamada: Fluorescence and laser-emission cross section in $NaNdP_4O_{12}$. J. Appl. Phys. 47, 2749–2750 (1976)

579   S.R. Chinn, H.Y-P. Hong: CW laser action in acentric $NdAl_3(BO_3)_4$ and $KNdP_4O_{12}$. Opt. Commun. 15, 345–350 (1975)

580   M. Sarawatari, T. Kimura, K. Otsuka: Miniaturized CW $LiNdP_4O_{12}$ laser pumped with a semiconductor laser. Appl. Phys. Lett. 29, 291–293 (1976)

581   S.R. Chinn, H.Y-P. Hong, J. Pierce: Minilasers of neodymium compounds. Laser Focus N5, 63–69 (1976)

582   L. Esterowitz, J. Noonan, J. Bahler: Enhancement in a $Ho^{3+}-Yb^{3+}$ quantum counter by energy transfer. Appl. Phys. Lett. 10, 126–127 (1967)

583   T. Kushida, M. Tamatani: Conversion of infrared into visible light. Jpn. Soc. Appl. Phys. (Supplement) 39, 241–247 (1970)

584   I.P. Kaminov, L.W. Stulz: $Nd:LiNbO_3$ laser. IEEE J. Quantum Electron. 11, 306–308 (1975)

585   A.A. Kaminskii, V.A. Koptsik, Yu.A. Maskaev, I.I. Naumova, L.N. Rashkovich, S.E. Sarkisov: Stimulated emission from Nd^{3+} ions in ferroelectric $Ba_2NaNb_5O_{15}$ crystals ("Banana"). Phys. Status Solidi (a) 28, K5–K10 (1975)

586   V.G. Dmitriev, V.A. Zenkin: Amplification and generation of the second optical harmonic in a nonlinear active medium. Kvant. Elektron. (Moscow) 3, 1811–1813 (1976)
      English transl.: Sov. J. Quantum Electron. 6, 984–986 (1976)

587   V.G. Dmitriev, V.A. Zenkin, N.E. Kornienko, A.I. Ryzhkov, V.L. Strizhevskii: Lasers with nonlinear active medium. Kvant. Elektron. (Moscow) 4, 2416–2427 (1978)
      English transl.: Sov. J. Quantum Electron. 8, 1356–1361 (1978)

588   K. Nassau: "Lithium niobate—a new type of ferroelectric: growth, structure, and properties," in Ferroelectricity, ed. by E.E. Weller, (Elsevier, Amsterdam 1967) pp. 259–268

589   H.Y-P. Hong, K. Dwight: Crystal structure and fluorescence lifetime of $NdAl_3(BO_3)_4$, a promising laser material. Mater. Res. Bull. 9, 1661–1666 (1974)

590   A.A. Kaminskii, T.I. Butaeva, V.A. Fedorov, Kh.S. Bagdasarov, A.G. Petrosyan: Absorption, luminescence, and stimulated emission investigations in $Lu_3Al_5O_{12}-Er^{3+}$ crystals. Phys. Status Solidi (a) 39, 541–548 (1977)

591   A.A. Kaminskii, N.V. Karlov, S.E. Sarkisov, O.M. Stelmakh, V.E. Tukish: Precision measurement of the stimulated emission wavelength and continuous tuning of $YAlO_3:Nd^{3+}$ laser radiation due to $^4F_{3/2} \rightarrow {}^4I_{13/2}$ transition. Kvant. Elektron. (Moscow) 3, 2497–2499 (1976)
      English transl.: Sov. J. Quantum Electron. 6, 1371–1373 (1976)

592   E.I. Perepilitsa: Theory of temperature shift and broadening of the ruby R-line. Fiz. Tverd. Tela (Leningrad) 17, 2490–2492 (1975)
      English transl.: Sov. Phys. Solid State 17, 1660–1661 (1976)

593   Yu.E. Perlin, A.A. Kaminskii, S.I. Klokishner, V.N. Enakii, Kh.S. Bagdasarov, G.A. Bogomolova, D.N. Vylegzhanin: Thermal broadening and shift of the energy levels of TR^{3+} ions in crystals. Phys. Status Solidi (a) 40, 643–653 (1977)

594   A.A. Kaminskii, V.V. Osiko, S.E. Sarkisov, M.I. Timoshechkin, E.V. Zhekov, J. Bohm, P. Reiche, D. Schultze. Growth, spectroscopic investigations, and some new stimulated emission data of $Gd_3Ga_5O_{12}-Nd^{3+}$ single crystals. Phys. Status Solidi (a), 49, 305–311 (1978)

595   A.A. Kaminskii, A.A. Pavlyuk, P.V. Klevtsov, I.F. Balashov, V.A. Berenberg, S.E. Sarkisov, V.A. Fedorov, M.V. Petrov, V.V. Lubchenko: Stimulated radiation of monoclinic crystals of

KY(WO$_4$)$_2$ and KGd(WO$_4$)$_2$ with Ln^{3+} ions. Izv. Akad. Nauk SSSR, Neorg. Mater. *13*, 582–583 (1977)
English transl.: Inorg. Mater. (USSR) *13*, 482–483 (1977)

596 R.C. Morris, C.F. Ceine, R.F. Begley, M. Dutoit, P.J. Harget, H.P. Jenssen, T.S. LaFrance, R. Webb: Lanthanum beryllate: a new rare-earth ion laser host. Appl. Phys. Lett. *27*, 444–445 (1975)

597 R. Beck, K. Gūrs: Ho laser with 50-W output and 6.5% slope efficiency. J. Appl. Phys. *46*, 5224–5225 (1975)

598 J. Falk, L. Huff, J.D. Taynai: Solar-pumped mode-locked, frequency-doubled Nd:YAG laser. Digest of Technical Papers 1975 IEEE/OSA CLEA, May 28–30 1975, Washington, pp. 14–15

599 M.J. Weber: "Optical materials for neodymium fusion lasers," in *Critical Materials Problems in Energy Production*, (Academic Press, New York 1976) pp. 261–279

600 A. Glass: "Refractive-index nonlinearity," in *Laser Program Annual Report UCRL-50021-74*, (Lawrence Livermore Laboratory 1975) pp. 256–262

601 V.I. Marin, V.I. Nikitin, M.S. Soskin, A.I. Khizhnyak: Superluminescence emitted by YAG:Nd^{3+} crystals and stimulated emission due to weak transitions. Kvant. Elektron. (Moscow) *2*, 1340–1343 (1975)
English transl.: Sov. J. Quantum Electron. *5*, 732–734 (1975)

602 P. Labudde, W. Seka, H.P. Weber: Gain increase in laser amplifiers by suppression of parasitic oscillations. Appl. Phys. Lett. *29*, 732–734 (1976)

603 A.S. Epifanov, A.A. Manenkov, A.M. Prokhorov: Frequency and temperature dependence of avalanche ionization in solids under the influence of an electromagnetic field. Zh. Eksp. Teor. Fiz. Pis'ma Red. *21*, 483–489 (1975)
English transl.: JETP Lett. *21*, 223–224 (1975)

604 W. Low: *Paramagnetic Resonance in Solids* (Academic Press, New York 1960)

605 P.P. Yaney, L.G. DeShazer: Spectroscopic studies and analysis of the laser states of Nd^{3+} in YVO$_4$. J. Opt. Soc. Am. *66*, 1405–1414 (1976)

606 L. Esterowitz, R. Allen, M. Kruer, F. Bartoli, L.S. Goldberg, H.P. Jenssen, A. Linz, V.O. Nicolai: Blue light emission by a Pr:LiYF$_4$-laser operated at room temperature. J. Appl. Phys. *48*, 650–652 (1977)

607 A.A. Kaminskii, V.A. Fedorov, Ngoc Chan: Three-micron stimulated emission by Ho^{3+} ions in LaNbO$_4$ crystal. Izv. Akad. Nauk SSSR. Neorg. Mater. *14*, 1357 (1978)
English transl.: Inorg. Mater. (USSR) *14*, 1061 (1978)

608 K. Otsuka, T. Yamada: Transversely pumped LNP laser performance. Appl. Phys. Lett. *26*, 311–313 (1975)

609 K. Otsuka, T. Yamada, M. Saruwatari, T. Kimura: Spectroscopy and laser oscillation properties of lithium neodymium tetraphosphate. IEEE J. Quantum Electron. *11*, 330–335 (1975)

610 S.R. Chinn, H.Y-P. Hong: Low-threshold CW LiNdP$_4$O$_{12}$ laser. Appl. Phys. Lett. *26*, 649–651 (1975)

611 K. Otsuka, T. Yamada: Continuous oscillation of a lithium neodymium tetraphosphate laser with 200-μW pump threshold. IEEE J. Quantum Electron. *11*, 845–846 (1975)

612 H.P. Weber, B.C. Tofield: Heating in a CW Nd-pentaphosphate laser. IEEE J. Quantum Electron. *QE-11*, 368–370 (1975)

613 H.Y-P. Hong: Crystal structure of potassium neodymium metaphosphate, KNdP$_4$O$_{12}$, a new acentric laser material. Mater. Res. Bull. *10*, 1105–1110 (1975)

614 A.W. Tucker, M. Birnbaum, C.L. Fincher, L.G. DeShazer: Continuous-wave operation of Nd:YVO$_4$ at 1.06 and 1.34 μ. J. Appl. Phys. *47*, 232–234 (1976)

615 J.G. Gualtieri, T.R. Aucoin: Laser performance of large Nd-pentaphosphate crystals. Appl. Phys. Lett. *28*, 189–192 (1976)

616 S.R. Chinn, J.W. Pierce, H. Heckscher: Low-threshold transversely excited NdP$_5$O$_{14}$ laser. Appl. Opt. *15*, 1444–1449 (1976)

617 G. Winzer, P.G. Möckeé, R. Oberbacher, L. Vite: Laser emission from polished NdP$_5$O$_{14}$ crystals with directly applied mirrors. Appl. Phys. *11*, 121–130 (1976)

618 K. Otsuka, J. Nakano, T. Yamada: Laser emission cross section of the system LiNd$_{0.5}$M$_{0.5}$P$_4$O$_{12}$ (M = Gd, La). J. Appl. Phys. *46*, 5297–5299 (1975)

619 H. Y-P. Hong, S.R. Chinn: Crystal structure and fluorescence lifetime of potassium neodymium orthophosphate, K$_3$Nd(PO$_4$)$_2$, a new laser material. Mater. Res. Bull. *11*, 421–428 (1976)

620 S. Singh, R.B. Chesler, W.H. Grodkiewicz, J.R. Potopowicz, L.G. Van Uitert: Room-temperature CW Nd^{3+}:$CeCl_3$ laser. J. Appl. Phys. *46*, 436–438 (1975)

621 M. Saruwatari, T. Kimura: LED pumped lithium neodymium tetraphosphate lasers. IEEE J. Quantum Electron. *12*, 584–591 (1976)

622 K. Washio, K. Iwamoto, K. Inoue, I. Hino, S. Matsumoto, F. Saito: Room-temperature CW operation of an efficient miniaturized Nd:YAG laser end-pumped by a superluminescent diode. Appl. Phys. Lett. *29*, 720–722 (1976)

623 J. Stone, C.A. Burrus, A.G. Dentai, B.I. Miller: Nd:YAG single-crystal fiber laser: room-temperature CW operation using a single LED as an end pump. Appl. Phys. Lett. *29*, 37–39 (1976)

624 S.R. Chinn, H.Y-P. Hong, J.W. Pierce: Spiking oscillations in diode-pumped NdP_5O_{14} and $NdAl_3(BO_3)_4$ lasers. IEEE J. Quantum Electron. *12*, 189–193 (1976)

625 N.P. Barnes: Diode-pumped solid-state lasers. J. Appl. Phys. *44*, 230–237 (1973)

626 R. Flach, D.S. Hamilton, P.M. Selzer, W.M. Yen: Laser-induced fluorescence-line-narrowing studies of impurity-ion systems: LaF_3:Pr^{3+}. Phys. Rev. *B 15*, 1248–1260 (1977)

627 A.O. Ivanov, L.G. Morozova, I.V. Mochalov, V.A. Fedorov: Spectra of a neodymium ion in Ca, LaSOAP and Ca, YSOAP crystals and stimulated emission in Ca, LaSOAP–Nd crystals. Opt. Spektrosk. *42*, 556–559 (1977)
   English transl.: Opt. Spectrosc. (USSR) *42*, 311–313 (1977)

628 Kh.S. Bagdasarov, V.P. Danilov, V.I. Zhekov, T.M. Murina, A.A. Manenkov, M.I. Timoshechkin, A.M. Prokhorov: Pulsed-periodic $Y_3Al_5O_{12}$:Er^{3+} laser with high activator concentration. Kvant. Elektron. (Moscow) *5*, 150–152 (1978)
   English transl.: Sov. J. Quantum Electron. *8*, 83–85 (1978)

629 J. Marling: 1.05–1.44 μm tunability and performance of the CW Nd^{3+}:YAG laser. IEEE J. Quantum Electron. *14*, 56–62 (1978)

630 E.J. Sharp, J.E. Miller, D.J. Horowitz, A. Linz, V. Belruss: Spectra and laser action in Nd^{3+}-doped $CaY_2Mg_2Ge_3O_{12}$. J. Appl. Phys. *45*, 4974–4979 (1974)

631 A.A. Kaminskii, G.A. Bogomolova, D.N. Vylegzhanin, Kh.S. Bagdasarov, A.M. Kevorkov, M.M. Gritsenko: Spectroscopic properties of Nd^{3+} ions in the garnet compounds forming in the Y_2O_3–Gd_2O_3 system. Phys. Status Solidi (*a*) *38*, 409–422 (1976)

632 H.P. Jenssen, R.F. Begley, R. Webb, R.C. Morris: Spectroscopic properties and laser performance of Nd^{3+} in lanthanum beryllate. J. Appl. Phys. *47*, 1496–1500 (1976)

633 V.N. Matrosov, M.I. Timoshechkin, E.G. Tsvetkov, S.E. Sarkisov, A.A. Kaminskii: Investigations of crystallization conditions of the lanthanum-beryllate crystals. Digest of Technical Papers V All-Union Conf. on Crystal Growth, September 16–19, 1977, Tbilisi, Vol. 2, pp. 167–168

634 N. Karayianis, C.A. Morrison, D.E. Wortman: Analysis of the ground term energy levels for triply ionized neodymium in yttrium orthovanadate. J. Chem. Phys. *62*, 4125–4129 (1975)

635 W. Krühler: Energieniveau und Lasereigenschaften im (Nd, Y)-Pentaphosphat, Ph. D. Thesis (Universität Stuttgart 1974)

636 A.A. Kaminskii, Kh.S. Bagdasarov, G.A. Bogomolova, M.M. Gritsenko, A.M. Kevorkov, S.E. Sarkisov: Luminescence and stimulated emission of Nd^{3+} ions in $Gd_3Sc_2Ga_3O_{12}$ crystals. Phys. Status Solidi (*a*) *34*, K109–K114 (1976)

637 A.O. Ivanov, L.G. Morozova, I.V. Mochalov, P.P. Feofilov: Luminescence of neodymium ions in single crystals of lutetium orthoaluminate. Opt. Spektrosk. *38*, 405–407 (1975)
   English transl.: Opt. Spectrosc. (USSR) *38*, 230–232 (1975)

638 A.M. Morozov, M.V. Petrov, V.R. Startsev, A.M. Tkachuk, P.P. Feofilov: Luminescence and stimulated emission of holmium in yttrium- and erbium-oxyortho-silicate single crystals. Opt. Spektrosk. *41*, 1086–1089 (1976)
   English transl.: Opt. Spectrosc. (USSR) *41*, 541–542 (1976)

639 Kh.S. Bagdasarov, A.A. Kaminskii, A.M. Kevorkov, L. Li, A.M. Prokhorov, T.A. Tevosyan, S.E. Sarkisov: Investigation of the stimulated emission of cubic crystals of $YScO_3$ with Nd^{3+} ions. Dok. Akad. Nauk SSSR *224*, 798–801 (1975)
   English transl.: Sov. Phys. Dokl. *20*, 681–683 (1975)

640 A.M. Morozov, L.G. Morozova, V.A. Fedorov, P.P. Feofilov: Spontaneous and stimulated emission of neodymium in lead fluorophosphate crystals. Opt. Spektrosk. *39*, 612–614 (1975)
   English transl.: Opt. Spectrosc. (USSR) *39*, 343–344 (1975)

641   A.A. Kaminskii, A.A. Pavlyuk, T.I. Butaeva, V.A. Fedorov, I.F. Balashov, V.A. Berenberg,
      V.V. Lubchenko: Stimulated emission by subsidiary transitions of Ho^{3+} and Er^{3+} ions in
      $KGd(WO_4)_2$ crystals. Izv. Akad. Nauk SSSR Neorg. Mater. *13*, 1541–1542 (1977)
      English transl.: Inorg. Mater. (USSR) *13*, 1251–1252 (1977)
642   J. Heber, K.H. Hellwege, U. Köbler, H. Murmann: Energy levels and interaction between
      Eu^{3+}-ions at lattice sites of symmetry C_2 and symmetry C_{3i} in Y_2O_3. Z. Physik *237*, 189–204
      (1970)
643   H. Forest, G. Ban: Random substitution of Eu^{3+} for Y^{3+} in Y_2O_3:Eu^{3+}. J. Electrochem. Soc.
      118, 1999–2001 (1971)
644   M.I. Gaiduk, V.F. Zolin, L.S. Gaigerova: *Spektry Lyuminestsentsii Evropiya* (Nauka, Moscow
      1974) p. 122
645   N. Karayianis, D.E. Wortmann, H.P. Jenssen: Analysis of the optical spectrum of Ho^{3+} in
      $LiYF_4$. J. Phys. Chem. Solids *37*, 675–682 (1976)
646   M. Bass, L.G. DeShazer, U. Ranon: Evaluation of Nd:YVO_4 and Ho:Er:Tm:YVO_4 as pulse
      pumped   Q-switched   lasers.   Research   and   Development   Technical   Report,
      ECOM–74–0104–1, 1974
647   S.M. Kulpa: Optical and magnetic properties of Er^{3+} in $LiYF_4$. J. Phys. Chem. Solids *36*,
      1317–1321 (1975)
648   K.H. Hellwege, S. Hüfner, M. Schinkmann, H. Schmidt: Optical absorption spectrum and
      crystal field of erbium aluminum garnet (ErAlG). Phys. Kondens. Mater. *4*, 397–403 (1966)
649   D.E. Wortman, C.A. Morrison, R.P. Leavitt: Analysis of the ground configuration of Tm^{3+} in
      $CaWO_4$. Phys. Rev. *B 12*, 4780–4789 (1975)
650   L.M. Hobrock: Spectra of thulium in yttrium orthoaluminate crystals and its four level laser
      operation in the middle infrared, Ph. D. Thesis (University of Southern California 1972)
651   V.A. Antonov, P.A. Arsenev, K.E. Bienert, A.V. Potemkin: Spectral properties of rare-earth
      ions in $YAlO_3$ crystals. Phys. Status Solidi (*a*) *19*, 289–299 (1973)
652   J.M. O'Hare, V.L. Donlan: Crystal-field determination for trivalent thulium in yttrium
      orthoaluminate. Phys. Rev. *B 14*, 3732–3743 (1976)
653   W.W. Holloway, M. Kestigian, F.F. Wang, G.F. Sullivan: Temperature-dependent Nd
      fluorescence parameters and laser thresholds. J. Opt. Soc. Am. *56*, 1409–1410 (1966)
654   G.A. Bogomolova, L.A. Bumagina, A.A. Kaminskii, B.Z. Malkin: Crystal field in laser garnets
      with TR^{3+} ions in the exchange charge model. Fiz. Tverd. Tela (Leningrad) *19*, 1439–1452
      (1977)
      English transl.: Sov. Phys. Solid State *19*, 1428–1435 (1977)
655   A.A. Kaminskii: Moderne Probleme der Laserkristall Physik, Vortrag auf der Jahreshaupt-
      tagung der Physikalischen Gesellschaft DDR 1977, Dresden
656   H.Y-P. Hong, S.R. Chinn: Influence of local-site symmetry on fluorescence lifetime in high-
      Nd-concentration laser materials. Mater. Res. Bull. *11*, 461–468 (1976)
657   L.S. Goldberg, J.N. Bradford: Passive mode locking and picosecond pulse generation in
      Nd:lanthanum beryllate. Appl. Phys. Lett. *29*, 585–588 (1976)
658   F. Auzel: Oscillator strengths of Nd^{3+} in $Nd_xLa_{1-x}P_5O_{14}$ and concentration quenching in
      stoichiometric rare-earth laser materials. IEEE J. Quantum Electron. *12*, 258–259 (1976)
659   S.E. Sarkisov, V.A. Lomonov, A.A. Kaminskii, A.A. Mayer, D. Schultze, J. Bohm:
      Spectroscopic investigation of stable crystalline compounds in the $Bi_4(Ge_{1-x}Si_x)_3O_{12}$ doped
      with Nd^{3+} ions. Digest of Technical Papers V all-Union Conf. on Spectroscopy of Crystals,
      June 2–6, 1976, Kazan, p. 195
660   J.A. Wunderlich, J.G. Sliney, L.G. DeShazer: Stimulated emission at 2.04 μm in Ho^{3+}-doped
      $ErVO_4$ and YVO_4. IEEE J. Quantum Electron. *13*, 69 (1977)
661   W.G. Fateley, N.T. McDevitt, F.F. Bentley: Infrared and Raman selection rules for lattice
      vibrations: the correlation method. Appl. Spectrosc. *25*, 155–173 (1971)
662   A.A. Kaminskii, L. Li: Spectroscopic quality of laser media with Nd^{3+} and Pm^{3+} ions. Zh.
      Tekh. Fiz. Pis'ma *1*, 567–571 (1975)
      English transl.: Sov. Tech. Phys. Lett. *1*, 256–258 (1975)
663   A. Abragam, B. Bleaney: *Electron Paramagnetic Resonance of Transition Ions* (Clarendon
      Press, Oxford 1970)
664   J.S. Griffith: *The Theory of Transition-Metal Ions* (Cambridge University Press, Cambridge
      1961)
665   S. Sugano, Y. Tanabe, H. Kamimura: *Multiplets of Transition Metal Ions in Crystals*

(Academic Press, New York, London 1970)

666 M.V. Eremin, A.A. Kornienko: The superposition model in crystal-field theory. Phys. Status Solidi (*b*) *79*, 775–785 (1977)

667 F. Anisimov, R. Dagys: Electronic structure and spectrum of the $(TmF_8)^6$-cluster. Phys. Status Solidi (*b*) *53*, 85–92 (1972)

668 D.J. Newman: Theory of lanthanide crystal fields. Adv. Phys. *20*, 197–256 (1971)

669 I.V. Aizenberg, B.Z. Malkin, A.L. Stolov: Cubic centers of the Er^{3+} ion in crystal of the fluorite type. Fiz. Tverd. Tela (Leningrad) *13*, 2566–2570 (1971)
English transl.: Sov. Phys. Solid State *13*, 2155–2158 (1972)

670 I.V. Aizenberg, M.P. Davydova, B.Z. Malkin, A.I. Smirnov, A.L. Stolov: Trigonal fluorine centers formed Er^{3+} ions in fluorite-type single crystals. Fiz. Tverd. Tela (Leningrad) *15*, 1345–1352 (1973)
English transl.: Sov. Phys. Solid State *15*, 914–918 (1973)

671 R.M. Sternheimer, M. Blume, R.F. Peieris: Shielding of crystal fields at rare-earth ions. Phys. Rev. *173*, 376–389 (1968)

672 M.T. Hutchings, D.K. Ray: Investigation into the origin of crystalline electric field effects on rare earth ions. I. Contribution from neighbouring induced moments. Proc. Phys. Soc. *81*, 663–676 (1963)

673 D.J. Newman, S.S. Bishton, M.M. Curtis, C.D. Taylor: Configuration interaction and lanthanide crystal fields. J. Phys. C *4*, 3234–3248 (1971)

674 A. Edgar, D.J. Newman: Local distortion effects on the spin-hamiltonian parameters of Gd^{3+} substituted into the fluorites. J. Phys. C *8*, 4023–4025 (1975)

675 J.M. O'Hare, V.L. Donlan: Crystal-field determination for trivalent erbium in yttrium orthoaluminate. Phys. Rev. B *15*, 10–16 (1977)

676 T.K. Gupta, J. Valentich: Thermal expansion of yttrium aluminum garnet. J. Am. Ceram. Soc. *54*, 355–356 (1971)

677 W.L. Bond: Measurement of the refractive indices of several crystals. J. Appl. Phys. *36*, 1674–1677 (1965)

678 R. Diehl, G. Brandt: Crystal structure refinement of $YAlO_3$, a promising laser material. Mater. Res. Bull. *10*, 85–90 (1975)

679 P.F. Liao, H.P. Weber: Fluorescence quenching of the $^4F_{3/2}$ state in Nd-doped yttrium aluminum garnet (YAG) by multiphonon relaxation. J. Appl. Phys. *45*, 2931–2934 (1974)

680 E.M. Dianov, A.Ya. Karasik, V.B. Neustruev, A.M. Prokhorov, I.A. Shcherbakov: Direct measurements of fluorescent quantum yield from the metastable $^4F_{3/2}$ state of Nd^{3+} in $Y_3Al_5O_{12}$ crystals. Dok. Akad. Nauk SSSR *224*, 64–67 (1975)
English transl.: Sov. Phys. Dokl. *20*, 622–624 (1976)

681 Yu.K. Voronko, T.G. Mamedov, V.V. Osiko, A.M. Prokhorov, B.P. Sakun, I.A. Shcherbakov: Nature of nonradiative excitation energy relaxation in condensed media with a high activator concentration. Zh. Eksp. Teor. Fiz. *71*, 478–497 (1976)
English transl.: Sov. Phys. JETP *44*, 251–261 (1976)

682 H.G. Danielmeyer: Oscillators and amplifiers for integrated optics, Proc. Laser 75 Optoelectronics Conf., 1975, pp. 20–22

683 I.A. Bondar, B.I. Denker, A.I. Domanskii, T.G. Mamedov, P.P. Mezentseva, V.V. Osiko, I.A. Shcherbakov: Investigation of anomalously weak quenching of Nd^{3+} ion luminescence in $La_{1-x}Nd_xP_5O_{14}$. Kvant. Elektron. (Moscow) *4*, 302–309 (1977)
English transl.: Sov. J. Quantum Electron. 7, 167–171 (1977)

684 I.A. Bondar, T.G. Mamedov, L.P. Mezentseva, I.A. Shcherbakov, A.I. Domanskii: Synthesis and investigation of $Nd_xLa_{1-x}P_5O_{14}$ single crystals. Kvant. Elektron. (Moscow) *1*, 2625–2628 (1974)
English transl.: Sov. J. Quantum Electron. *4*, 1463–1464 (1975)

685 H.Y-P. Hong, K. Dwight: Crystal structure and fluorescence lifetime of a laser material $NdNa_5(WO_4)_4$. Mater. Res. Bull. *9*, 775–780 (1974)

686 H.P. Weber, P.F. Liao, B.C. Tofield: Emission cross-section and fluorescence efficiency of Nd-pentaphosphate. IEEE J. Quantum Electron. *10*, 563–567 (1974)

687 T.T. Basiev, Yu.K. Voronko, T.G. Mamedov, V.V. Osiko, I.A. Shcherbakov:" Relaxation processes of excitation from metastable levels of rare-earth ions in crystals," in *Spektroskopiya Kristallov*, ed. by A.A. Kaminskii, Z.L. Morgenshtern, D.T. Sviridov (Nauka, Moscow 1975) pp. 155–183

688   M. Yokota, O. Tanimoto: Effects of diffusion on energy transfer by resonance. J. Phys. Soc. Jpn. *22*, 779–784 (1967)

689   V.M. Agranovich: *Teoriya Éksitonov* (Nauka, Moscow 1968)

690   M.V. Artamonova, Ch.M. Briskina, A.I. Burshtein, L.D. Zusaman, A.G. Sklezkov: Time variation of Nd^{3+} ion luminescence and an estimation of electron excitation migration along the ions in glass. Zh. Eksp. Teor. Fiz. *62*, 863–871 (1972)
English transl.: Sov. Phys. JETP *35*, 457–461 (1972)

691   T.T. Basiev, Yu.K. Voronko, T.G. Mamedov, I.A. Shcherbakov: Migration of energy between Yb^{3+} ions in garnet crystals. Kvant. Elektron. (Moscow) *2*, 2172–2182 (1975)
English transl.: Sov. J. Quantum Electron. *5*, 1182–1188 (1976)

692   A.I. Burstein: Hopping mechanism of energy transfer. Zh. Eksp. Teor. Fiz. *62*, 1695–1701 (1972)
English transl.: Sov. Phys. JETP *35*, 882–885 (1972)

693   V.P. Sakun: Kinetics of energy transfer in crystals. Fiz. Tverd. Tela (Leningrad) *14*, 2199–2210 (1972)
English transl.: Sov. Phys. Solid State *14*, 1906–1914 (1973)

694   W. Feller: *An Introduction to Probability Theory and Its Applications*, 2nd ed. (Wiley, New York 1966)

695   I.Ya. Gerlovin, A.P. Abramov, N.A. Tolstoi: UV luminescence and nonlinear quenching in ruby. Opt. Spektrosk. *34*, 128–132 (1973)
English transl.: Opt. Spectrosc. (USSR) *34*, 69–71 (1973)

696   H.G. Danielmeyer: Efficiency and fluorescence quenching of stoichiometric rare earth laser materials. J. Lumin. *12/13*, 179–186 (1976)

697   T.T. Basiev, E.V. Zharikov, V.I. Zhekov, T.M. Murina, V.V. Osiko, A.M. Prokhorov, B.P. Starikov, M.I. Timoshechkin, I.A. Shcherbakov: Radiative and nonradiative transitions exhibited by Er^{3+} ions in mixed yttrium-erbium aluminum garnets. Kvant. Elektron. (Moscow) *3*, 1471–1477 (1976)
English transl.: Sov. J. Quantum Electron. *6*, 796–799 (1976)

698   K.H. Yang, J.A. DeLuca: VUV fluorescence of Nd^{3+}-, Er^{3+}-, and Tm^{3+}-doped trifluorides and tunable coherent sources from 1650 to 2600 Å. Appl. Phys. Lett. *29*, 499–501 (1976)

699   S. Geller: Crystal chemistry of the garnets. Z. Kristallogr. *125*, 1–47 (1967)

700   G.A. Massey: Measurement of device parameters for $Nd:YAlO_3$ lasers. IEEE J. Quantum Electron. *8*, 669–674 (1972)

701   W.A. Wall, J.T. Karpick, B. DiBartolo: Temperature dependence of the vibronic spectrum and fluorescence lifetime of $YAG:Cr^{3+}$. J. Phys. C *4*, 3258–3264 (1971)

702   M.O. Henry, J.P. Larkin, G.F. Imbusch: Luminescence from chromium doped yttrium aluminum garnet. Proc. R. Ir. Acad. Sect. A: *75*, 97–106 (1975)

703   F. Auzel: Multiphonon interaction of excited luminescent centers in the weak coupling limit: nonradiative decay and multiphonon side-bands. CNET Note Technique NT/PEC/RPM/288, Juin 1977, Bagneux-Paris

704   L.M. Goldman, W.B. Jones, J.P. Chernoch, W.S. Martin: GE's "mini" YAG and glass. Laser Focus *8*, 31, 33 (1972)

705   S.E. Hatch, W.F. Parsons, R.J. Weagley: Hot-pressed polycrystalline $CaF_2:Dy^{2+}$ laser. Appl. Phys. Lett. *5*, 153–154 (1964)

706   C. Greskovich, J.P. Chernoch: Polycrystalline ceramic lasers. J. Appl. Phys. *44*, 4599–4606 (1973)

707   C. Greskovich, J.P. Chernoch: Improved polycrystalline ceramic lasers. J. Appl. Phys. *45*, 4495–4502 (1974)

708   H.V. Lauer, F.K. Fong: Coupling strength in the theory of radiationless transitions: $f \rightarrow f$ and $d \rightarrow f$ relaxation of rare-earth ions in $YAlO_3$ and $Y_3Al_5O_{12}$. J. Chem. Phys. *60*, 274–280 (1974)

709   B. Fritz, E. Menke: Laser effect in KCl with $F_A(Li)$ centers. Solid State Commun. *3*, 61–63 (1965)

710   L.F. Mollenauer, D.H. Olson: A broadly tunable CW laser using color centers. Appl. Phys. Lett. *24*, 386–388 (1974)

711   L.F. Mollenauer, D.H. Olson: Broadly tunable lasers using color centers. J. Appl. Phys. *46*, 3109–3118 (1975)

712   G.C. Bjorklung, L.F. Mollenauer, W.J. Tomlinson: Distributed-feedback color centers lasers in the 2.5–3.0 μm region. Appl. Phys. Lett. *29*, 116–118 (1976)

713 G. Litfin, R. Beigang, H. Welling: Durchstimmbare kontinuierliche Farbzentren-Laser im Wellenlängenbereich von 1 μm bis 4 μm, 3. Internat. Tagung Laser und ihre Anwendungen (1977, Dresden, DDR)

714 Yu.L. Gusev, S.I. Marennikov, V.N. Chebotaev: Laser action on F_2^+ and F_2^- color centers in a LiF crystal in the range 0.88–1.2 μ. Zh. Tekh. Fiz., Pis'ma 3, 305–307 (1977) English transl.: Sov. Tech. Phys. Lett. 3, 124 (1977)

715 J.F.B. Hawkes, M.J.M. Leask: Spectroscopic study of $Er^{3+}-Gd^{3+}$ interaction in Er:GdAlO$_3$. J. Phys. C 5, 1705–1715 (1972)

716 S. Singh, W.A. Bonner, W.H. Grodkiewicz, M. Grasso, L.G. Van Uitert: Nd-doped yttrium aluminum garnet with improved fluorescent lifetime of the $^4F_{3/2}$ state. Appl. Phys. Lett. 29, 343–345 (1976)

717 D.J. Newman, G.E. Stedman: Interpretation of crystal-field parameters in the rare-earth-substituted garnets. J. Chem. Phys. 51, 3013–3023 (1969)

718 B.Z. Malkin, Z.I. Ivanenko, I.B. Aizenberg: Crystal field in uniaxially compressed MeF$_2$:Ln crystals. Fiz. Tverd. Tela (Leningrad) 12, 1873–1880 (1970) English transl.: Sov. Phys. Solid State 12, 1491–1496 (1970)

719 A.L. Larionov, B.Z. Malkin: Effective hamiltonian of valence electrons of rare-earth elements in ionic crystals. Opt. Spektrosk. 39, 1109–1113 (1975) English transl.: Opt. Spectrosc. (USSR) 39, 637–639 (1975)

720 Z.I. Ivanenko, B.Z. Malkin: Spin–phonon interaction in doped fluorite-type crystals. Fiz. Tverd. Tela (Leningrad) 14, 153–156 (1972) English transl.: Sov. Phys. Solid State 14, 122–124 (1972)

721 B.G. Dick, A.W. Overhauser: Theory of the dielectric constants of alkali halide crystals. Phys. Rev. 112, 90–103 (1958)

722 F. Gaume (ed.): Spectroscopie des Eléments de Transition et des Eléments Lourds Dans les Solides, Proc. Internat. Conf. Lyon, 1976 (Editions C.N.R.S. N255, Paris 1977)

723 J.H. Van Vleck: The puzzle of rare-earth spectra in solids. J. Phys. Chem. 41, 67–80 (1937)

724 L.J.F. Broer, C.J. Gorter, J. Hoogschagen: On the intensities and the multipole character in the spectra of the rare earth ions. Physica 11, 231–250 (1945)

725 A.A. Kaminskii, L. Li: "Intensities of TR^{3+}-ion transitions in laser crystals," in Spektroskopiya Kristallov, ed. by P.P. Feofilov (Nauka, Leningrad 1978) pp. 100–120

726 E.U. Condon, G.H. Shortley: The Theory of Atomic Spectra (Cambridge University Press, Cambridge 1957)

727 J.L. Prather: Atomic Energy Levels in Crystals, NBS Monograph, Vol. 19, 1961 (U.S. govt. prtg. office)

728 W.T. Carnall, P.R. Fields, K. Rajnak: Electronic energy levels in the trivalent lanthanide aquo ions. I. Pr^{3+}, Nd^{3+}, Pm^{3+}, Sm^{3+}, Dy^{3+}, Ho^{3+}, Er^{3+}, and Tm^{3+}. J. Chem. Phys. 49, 4424–4442 (1968)

729 W.T. Carnall, P.R. Fields, K. Rajnak: Electronic energy levels of the trivalent lanthanide aquo ions. IV. Eu^{3+}. J. Chem. Phys. 49, 4450–4446 (1968)

730 W.T. Carnall, P.R. Fields, K. Rajnak: Electronic energy levels of the trivalent lanthanide aquo ions. III. Tb^{3+}. J. Chem. Phys. 49, 4447–4449 (1968)

731 M.J. Weber: Spontaneous emission probabilities and quantum efficiencies for excited states of Pr^{3+} in LaF$_3$. J. Chem. Phys. 48, 4474–4780 (1968)

732 W.F. Krupke: Induced-emission cross sections in neodymium laser glasses. IEEE J. Quantum Electron. 10, 450–457 (1974)

733 S. Kuboniwa, T. Hoshina: Luminescent properties of Tb^{3+} in oxygen-dominated compounds. J. Phys. Soc. Jpn. 32, 1059–1068 (1972)

734 M.J. Weber, B.H. Matsinger, V.L. Donlan, G.T. Surratt: Optical transition probabilities for trivalent holmium in LaF$_3$ and YAlO$_3$. J. Chem. Phys. 57, 562–567 (1972)

735 M.J. Weber: "Relaxation processes for excited states of Eu^{3+} in LaF$_3$," in Optical Properties of Ions in Crystals, ed. by H.M. Crosswhite, H.W. Moos (Wiley, New York 1967) pp. 467–484

736 A.A. Kaminskii, L. Li: Analysis of spectral line intensities of TR^{3+} ions in disordered crystal systems. Phys. Status Solidi (a) 26, K21–K26 (1974)

737 W.F. Krupke, J.B. Gruber: Optical-absorption intensities of rare-earth ions in crystals: the absorption spectrum of thulium ethyl sulfate. Phys. Rev. A 139, 2008–2016 (1965)

738 W.F. Krupke: Assessment of a promethium YAG laser. IEEE J. Quantum Electron. 8, 725–726 (1972)

440   References

739   J.D. Axe: Radiative transition probabilities within $4f^n$ configurations: the fluorescence spectrum of europium ethylsulfate. J. Chem. Phys. *39*, 1154–1160 (1963)

740   C. Delsart, N. Pelletieer-Allard: Intensités des raies d'absorption optique des ions Pr^{3+} dans $LaAlO_3$. J. Phys. (Paris) *32*, 507–515 (1971)

741   M.J. Weber, T.E. Varitimos, B.H. Matsinger: Optical intensities of rare earth ions in yttrium orthoaluminate. Phys. Rev. *B 8*, 47–53 (1973)

742   S.R. Chinn, W.K. Zwicker: Flash-lamp-excited NdP_5O_{14} laser. Appl. Phys. Lett. *31*, 178–181 (1977)

743   J.R. Chamberlain, A.C. Everitt, J.W. Orton: Optical absorption intensities and quantum counter action of Er^{3+} in yttrium gallium garnet. J. Phys. *C 1*, 157–164 (1968)

744   W.F. Krupke: Transition probabilities in Nd:GGG. Opt. Commun. *12*, 210–212 (1974)

745   W.T. Carnall, P.R. Fields, K. Rajnak: Spectral intensities of the trivalent lanthanides and actinides in solution. II. Pm^{3+}, Sm^{3+}, Eu^{3+}, Gd^{3+}, Tb^{3+}, Dy^{3+}, and Ho^{3+}. J. Chem. Phys. *49*, 4412–4423 (1968)

746   O.K.F. Deutschbein: A simplified method for determination of the induced-emission cross section of neodymium doped glasses. IEEE J. Quantum Electron. *12*, 551–554 (1976)

747   R.R. Jacobs, M.J. Weber: Dependence of the $^4F_{3/2} \rightarrow {}^4I_{11/2}$ induced emission cross section for Nd^{3+} on glass composition. IEEE J. Quantum Electron. *12*, 102–111 (1976)

748   F. Auzel, J.-C. Michel: Détermination de la force d'oscillateur de la transition $^4I_{11/2} \rightarrow {}^4F_{3/2}$ de Nd^{3+} dans la fluoroapatite (FAP), le grenat d'yttrium d'aluminium (YAG) et l'ultraphosphate de néodyme (NdUP). C. R. Acad. Sci. *B 279*, 187–190 (1974)

749   I.S. Minhas, K.K. Sharma: Optical-absorption intensities of neodymium-doped lead molybdate. Phys. Rev. *B 14*, 4124–4130 (1976)

750   S. Singh, R.G. Smith, L.G. Van Uitert: Stimulated-emission cross section and fluorescent quantum efficiency of Nd^{3+} in yttrium garnet at room temperature. Phys. Rev. *B 10*, 2566–2572 (1974)

751   A.A. Kaminskii, S.E. Sarkisov, I.V. Mochalov, L.K. Aminov, A.O. Ivanov: Anisotropy of spectroscopic characteristics in the biaxial $YAlO_3$-Nd^{3+} laser crystals. Phys. Status Solidi (a) *51*, 509–520 (1978)

752   K. Otsuka, S. Miyazawa, T. Yamada, H. Iwasaki, J. Nakano: CW laser oscillations in $MeNdP_4O_{12}$ (Me = Li, Na, K) at 1.32 μm. J Appl. Phys. *48*, 2099–2101 (1977)

753   G.V. Bukin, S.Yu. Volkov, V.N. Matrosov, B.K. Sevactyanov, M.I. Timoshechkin: Laser action of chrysoberyll $BeAl_2O_4$–Cr^{3+} crystal. Kvant. Elektron. (Moscow) *5*, 1168–1169 (1978) English transl.: Sov. J. Quantum Electron. *8*, 671–672 (1978)

754   V.A. Antonov, P.A. Arsenev, Sh.A. Vakhidov, E.M. Ibragimova, D.S. Petrova: Spectroscopic properties of neodymium ions in $LiYO_2$ monocrystals. Phys. Status Solidi (a) *41*, 45–50 (1977)

755   V.A. Antonov, P.A. Arsenev, D.S. Petrova: Spectroscopic properties of the Nd^{3+} ion in $Y_2Ti_2O_7$ and $Gd_2Ti_2O_7$ monocrystals. Phys. Status Solidi (a) *41*, K127–K131 (1977)

756   R.H. Hopkins, K.B. Steinbruegge, A.M. Stewart: Laser damage-initiating inclusions in $CaLa_4(SiO_4)_3$:Nd crystals. J. Cryst. Growth *38*, 255–261 (1977)

757   E. Orlich, S. Hüfner, P. Grünberg: Crystal field interaction in erbium garnets. Z. Phys. *231*, 144–153 (1970)

758   H.W. Moos: Spectroscopic relaxation processes of rare earth ions in crystals. J. Lumin. *1, 2*, 106–121 (1970)

759   L.F. Mollenauer: "Color center lasers," in *Method of Experimental Physics*, ed. by C.L. Tang (Academic Press, New York, London 1978) –

760   S.C. Abrahams, S. Geller: Refinement of the structure of a grossularite garnet. Acta Crystallogr. *11*, 437–441 (1958)

761   W. Strek, C. Szafranski, E. Lukowiak, Z. Mazurak, B. Jezowska-Trzebiatowska: Fluorescence quenching in neodymium pentaphosphate. Phys. Status Solidi (a) *41*, 547–553 (1977)

762   E.P. Chicklis, C.S. Naiman, L. Esterowitz, R. Allen: Deep red laser emission in Ho:YLF. IEEE J. Quantum Electron. *13*, 893–895 (1977)

763   B.A. Arkhangelskaya, A.A. Fedorov, P.P. Feofilov: Spontaneous and induced emission of color centers in MeF_2–Na crystals. Opt. Spektrosk. *44*, 409–411 (1978) English transl.: Opt. Spectrosc. (USSR) *44*, 240–241 (1978)

764   M. Saruwatari, K. Otsuka, S. Miyazawa, T. Yamada, T. Kimura: Fluorescence and oscillation characteristics of $LiNdP_4O_{12}$ lasers at 1.317 μm. IEEE J. Quantum Electron. *13*, 836–842 (1977)

765 D. Milam, M.J. Weber, A.J. Glass: Nonlinear refractive index of fluoride crystals. Appl. Phys. Lett. *31*, 822–825 (1977)

766 P.A. Arsenev, A.V. Potemkin, V.V. Fenin, I. Senff: Investigation of stimulated emission of Er^{3+} ions in mixed crystals with perovskite structure. Phys. Status Solidi (*a*) *43*, K15–K18 (1977)

767 E.Y. Wong, I. Richman: Absorption and fluorescence spectra of Pr^{3+} in $LaBr_3$. J. Chem. Phys. *36*, 1889–1892 (1962).

768 H.M. Crosswhite, H.Crosswhite, F.W. Kaseta, R. Sarup: The spectrum of Nd^{3+}: $LaCl_3$. J. Chem. Phys. *64*, 1981–1985 (1976)

769 T.S. Lomheim, L.G. DeShazer: New procedure of determining neodymium fluorescence branching ratios as applied to 25 crystals and glass hosts. Opt. Commun. *24*, 89–94 (1978)

770 J. Stone, C.A. Burrus: $Nd:Y_2O_3$ single-crystal fiber laser: room temperature CW operation at 1.07- and 1.35- μm wavelength. J. Appl. Phys. *49*, 2281–2287 (1978)

771 F. Auzel: L'auto-extinction de Nd^{3+}: son mécanisme fondamental et un critère prédictif simple pour les matériaux minilaser. Mater. Res. Bull. *14*, 223–231 (1979)

772 J.M. Flaherty, R. Powell: Concentration quenching in $Nd_xY_{1-x}P_5O_{14}$ crystals. Phys. Rev. *B19*, 32–42 (1979)

773 L.D. Markle, R. Powell: Energy transfer among Nd^{3+} ions in garnet crystals. Phys. Rev. *B20*, 75–84 (1979)

774 A.A. Kaminskii, V.A. Fedorov, S.E. Sarkisov, J. Bohm, P. Reiche, D. Schultze: Stimulated emission of Ho^{3+} and Er^{3+} ions in $Gd_3Ga_5O_{12}$ crystals and cascade laser action of Ho^{3+} ions over $^5S_2 \rightarrow {}^5I_5 \rightarrow {}^5I_6 \rightarrow {}^5I_8$. Phys. Status Solidi (a), *53*, K219–K222 (1979)

775 A.A. Kaminskii, V.A. Fedorov, A.G. Petrosyan, A.A. Pavlyuk, J. Bohm, P. Reiche, D. Schultze: Specificities of stimulated radiation of Ho^{3+} ions in oxygen-containing crystals at low temperature. Izv. Akad. Nauk SSSR, Neorg. Mater. *15*, 1494–1495 (1979)
     English transl.: Inorg. Mater. (USSR) *15*, 1180–1181 (1979)

776 M.V. Petrov, A.M. Tkachuk: Optical spectra and multifrequency generation of induced emission of $LiYF_4$-Er^{3+} crystals. Opt. Spektrosk. *45*, 147–155 (1978)
     English transl.: Opt. Spectrosc. (USSR) *45*, 81–85 (1978)

777 L. Esterowitz, R.C. Eckardt, R.E. Allen: Long-wavelength stimulated emission via cascade laser action in Ho:YLF. Appl. Phys. Lett. *35*, 236–239 (1979)

778 Sh.N. Gifeisman, A.M. Tkachuk, V.V. Prizmak: Optical spectra of a Ho^{3+} ion in $LiYF_4$ crystals. Opt. Spektrosk. *44*, 120–126 (1978)
     English transl.: Opt. Spectrosc. (USSR) *44*, 68–71 (1978)

779 J.W. Baer, M.G. Knights, E.P. Chiklis, H.P. Jenssen: XeF pumped laser operation of Tm^{3+} at 453 nm. Proc. Internat. Conf. Laser'78, December 11–15, 1978, Orlando (STS Press, McLean, Va, 1979), pp. 770–771

780 B.N. Kasakov, M.S. Orlov, M.V. Petrov, A.L. Stolov, A.M. Tkachuk: Induced emission of Sm^{3+} in visible. Opt. Spektrosk. *47*, 1217–1219 (1979)
     English transl.: Opt. Spectrosc. (USSR) *47*, 676–677 (1979)

781 D.J. Ehrlich, P.F. Moulton, R.M. Osgood: Ultraviolet solid-state Ce:YLF laser at 325 nm. Optics Letters, *4*, 184–186 (1979)

782 D. Fay, G. Huber, W. Lenth: Linear concentration quenching of luminescence in rare earth laser materials. Optics Commun. *28*, 117–122 (1979)

783 A. Lempicki: Concentration quenching in Nd^{3+} stoichiometric materials. Optics Commun. *23*, 376–380 (1977)

784 G. Winzer, L. Vite, W. Krühler, R. Plattner, P. Möckel, H. Pink: Miniature neodymium lasers (MNL) as possible transmitters for fiber-optic-communication systems. Part 1. Stoichiometric materials. Siemens Forsch.-u. Entwickl.-Ber., Bd5, 287–295 (1976)

785 J.-P. Budin, M. Neubauer, M. Rondot: On the design of neodymium miniature lasers. IEEE J. Quantum Electron. *14*, 831–839 (1978)

786 M.M. Choy, W.K. Zwicker, S.R. Chinn: Emission cross section and flashlamp-excited NdP_5O_{14} laser at 1.32 μm. Appl. Phys. Lett. *34*, 387–388 (1979)

787 A. Lempicki, B. McCollum, S.R. Chinn, H.Y.-P. Hong: Lasing and fluorescence in $K_5NdLi_2F_{10}$. Digest of Technical Papers 10th Internat. Quantum Electron. Conf., May 29–June 1, 1978, Atlanta (USA), p. 630

788 J.-P. Budin, J.-C. Michel, F. Auzel: Oscillator strengths and laser effect in $Na_2Nd_2Pb_6(PO_4)_6Cl_2$ (chloroapatite), a new high-Nd-concentration laser material. J. Appl. Phys. *50*, 641–646 (1979)

789   A.A. Kaminskii, A.A. Pavlyuk, Chan Ngok, L.I. Bobovich, V.A. Fedorov, V.V. Lyubchenko:
      3μ stimulated emission by Ho^{3+} ions in $KY(WO_4)_2$ crystals at 300 K. Dok. Akad. Nauk SSSR
      245, 575–576 (1979)
      English transl.: Sov. Phys. Dokl. *24*, 201–202 (1979)
790   A.A. Kaminskii, A.A. Pavlyuk, N.P. Agamalyan, S.E. Sarkisov, L.I. Bobovich, A.V. Lukin,
      V.V. Lyubchenko:Stimulated emission of Nd^{3+} and Ho^{3+} ions in $KLu(WO_4)_2$ monoclinic
      crystals at room temperature. Izv. Akad. Nauk SSSR. Neorg. Mater. *15*, 2092 (1979)
      English transl.: Inorg. Mater. (USSR) *15*, 16 (1979)
791   B.M. Antipenko, A.A. Mak, B.E. Sinitsin, T.V. Uvarova:Laser converter on the base of
      $BaYb_2F_8$:Ho^{3+}. Digest of Technical Papers VI All-Union Conf. on Spectroscopy of Crystals,
      September 21–25, 1979, Krasnodar, p. 30
792   A.A. Kaminskii, A.A. Pavlyuk, T.I. Butaeva, L.I. Bobovich, I.F. Balashov, V.A. Berenberg,
      V.V. Lyubchenko: Stimulated emission in the 2,8 μm band by a self-activated crystal of
      $KEr(WO_4)_2$. Izv. Akad. Nauk SSSR. Neorg. Mater. *15*, 541–542 (1979)
      English transl.: Inorg. Mater. (USSR) *15*, 424–425 (1979)
793   A.A. Kaminskii, A.A. Pavlyuk, N.P. Agamalyan, L.I. Bobovich, A.V. Lukin, V.V.
      Lyubchenko:Stimulated radiation of $KLu(WO_4)_2$-Er^{3+} crystals at room temperature. Izv.
      Akad. Nauk SSSR. Neorg. Mater. *15*, 1496–1497 (1979)
      English transl.: Inorg. Mater. (USSR) *15*, 1182–1183 (1979)
794   A.A. Kaminskii: Three-micron crystal lasers. Digest of Technical Papers VI All-Union Conf.
      on Spectroscopy of Crystals, September 21–25, 1979, Krasnodar, p. 141
795   A.A. Kaminskii, S.E. Sarkisov, Tran Ngoc, B.I. Denker, V.V. Osiko, A.M.
      Prokhorov:Stimulated emission spectroscopy of concentrated lithium-neodymium phosphate
      glasses in $^4F_{3/2} \rightarrow {}^4I_{11/2}$ and $^4F_{3/2} \rightarrow {}^4I_{13/2}$ transitions. Phys. Status Solidi (a), *50*, 745–750 (1978)
796   A.A. Kaminskii, A.G. Petrosyan: Sensitized stimulated emission from self-saturating 3-μm
      transitions of Ho^{3+} and Er^{3+} ions in $Lu_3Al_5O_{12}$ crystals. Izv. Akad. Nauk SSSR. Neorg.
      Mater. *15*, 543–544 (1979)
      English transl.: Inorg. Mater. (USSR) *15*, 425–427 (1979)
797   A.A. Kaminskii, A.G. Petrosyan: New functional scheme for 3-μ crystal lasers. Dok. Akad.
      Nauk SSSR *246*, 63–65 (1979)
      English transl.: Sov. Phys. Dokl. *24*, 363–364 (1979)
798   A.A. Kaminskii:New room-temperature operating three-micron crystal lasers. Proc. Laser-79
      Opto-Electronics Conf. Ed. W. Waidelich (IPC Science and Technology Press, Guildford,
      England, 1979), pp. 109–116
799   M. Szymanski, J. Karolczak, F. Kaczmarek:Laser properties of praseodymium pentaphos-
      phate single crystals. Appl. Phys. *19*, 345–351 (1979)
800   A.A. Kaminskii, S.E. Sarkisov, T.I. Butaeva, G.A. Denisenko, B. Hermoneit, J. Bohm, W.
      Grosskreutz, D. Schultze:Growth, spectroscopy, and stimulated emission of cubic $Bi_4Ge_3O_{12}$
      crystals doped with Dy^{3+}, Ho^{3+}, Er^{3+}, Tm^{3+}, or Yb^{3+} ions. Phys. Status Solidi (a), *55*,
      725–736 (1979)
801   J.C. Walling, H.P. Jenssen, R.C. Morris, E.W. O'Dell, O.G. Peterson:Broad band tuning of
      solid state alexandrite lasers. Opt. Soc. Amer. Meeting, October 31-November 3, San
      Francisco, 1978
802   J.C. Walling, H.P. Jenssen, R.C. Morris, E.W. O'Dell, O.G. Peterson: Tunable-laser
      performance in $BeAl_2O_4$:Cr^{3+}. Opt. Lett. *4*, 182–186 (1979)
803   A.A. Kaminskii, A.A. Pavlyuk, I.F. Balashov, V.A. Berenberg, V.V. Lyubchenko, V.A.
      Fedorov, T.I. Butaeva:Stimulated radiation at 0.85, 1.73 and 2.8 μm waves at 300°K of
      crystals $KY(WO_4)_2$-Er^{3+}. Izv. Akad. Nauk SSSR. Neorg. Mater. *14*, 2256–2258 (1978)
      English transl.: Inorg. Mater. (USSR) *14*, 1765–1767 (1978)
804   M. Birnbaum, A.W. Tucker, C.L. Fincher:CW room-temperature laser operation of
      Nd:CAMGAR at 0.941 and 1.059 μm. J. Appl. Phys. *49*, 2984–2985 (1978)
805   K. Kubodera, K. Otsuka, S. Miyazawa:Stable $LiNdP_4O_{12}$ miniature laser. Appl. Optics *18*,
      884–890 (1979)
806   T. Harig, G. Huber, J. Libertz, I. Shcherbakov: Appl. Phys. (to be published)
807   N.A. Eskov, V.V. Osiko, A.A. Sobol, M.I. Timoshechkin, T.I. Butaeva, Chan Ngoc, A.A.
      Kaminskii:A new laser garnet, $Ca_3Ga_2Ge_3O_{12}$-Nd^{3+}. Izv. Akad. Nauk SSSR. Neorg. Mater.
      14, 2254–2255 (1978)
      English transl.: Inorg. Mater. (USSR) *14*, 1764–1765 (1978)

808 H.-D. Hattendortt, G. Huber, F. Lutz:CW laser action in Nd(Al, Cr)$_3$(BO$_3$)$_4$. Appl. Phys. Lett. *34*, 437–439 (1979)

809 A.A. Kaminskii, S.E. Sarkisov, A.A. Pavlyuk, V.V. Lyubchenko:Anisotropy of luminescent properties of KGd(WO$_4$)$_2$ and KY(WO$_4$)$_2$ laser crystals doped with Nd^{3+} ions. Izv. Akad. Nauk SSSR. Neorg. Mater. *16*, 721–728 (1980)
      English transl.: Inorg. Mater. (USSR) *16*, 501–507 (1980)

810 A.M. Korovkin, L.G. Morozova, M.V. Petrov, A.M. Tkachuk, P.P. Feofilov: Spontaneous and induced emission of neodymium in the crystals of yttrium silicates and rare-earth silicates. Digest of Technical Papers VI All-Union Conf. on Spectroscopy Crystals, September 21–25, 1979, Krasnodar, p. 156

811 W.W. Krühler, R.D. Plättner:Laser emission of (Nd, Y)-pentaphosphate at 1.32 μm. Optics Communications *28*, 217–220 (1979)

812 G. Huber, H.G. Danielmeyer:NdP$_5$O$_{14}$ and NdAl$_3$(BO$_3$)$_4$ lasers at 1.3 μm. Appl. Phys. *18*, 77–80 (1979)

813 P.F Moulton, A. Mooradian, T.B. Reed:Efficient CW optically pumped Ni:MgF$_2$ laser. Digest of Technical Papers, 10-th Internat. Quantum Electron. Conf. May 29-June 1, Washington, 1978, pp. 630–631

814 P.F. Moulton, A. Mooradian.:Continuously tunable CW Ni:MgF$_2$ laser. Digest of Technical Papers IEEE/OSA CLEA, May 30-June 1, Washington, 1979, p. 87

815 P.S. Gillespie, R.L. Armstrong, K.O. White:Spectral characteristics and atmospheric CO$_2$ absorption of the Ho^{+3}:YLF laser at 2.05 μm. Appl. Opt. *15*, 865–867 (1976)

816 E. Chicklis, L. Esterowitz, R. Allen, M. Kruer:Stimulated emission at 2.8 μm in Er^{3+}:YLF. Proc. Internat. Conf. Laser'78, December 11–15, Orlando, 1978 (STS Press, McLean, Va, 1979), pp. 172–175

817 L.R. Elias, Wm.S. Heaps. W.M. Yen:Excitation of UV fluorescence in LaF$_3$ doped with trivalent cerium and praseodymium. Phys. Rev. *B8*, 4989–4995 (1973)

818 R.R. Jacobs, W.F. Krupke, M.J. Weber:Measurement of excited-state-absorption loss for Ce^{3+} in Y$_3$Al$_5$O$_{12}$ and implications for tunable 5d→4f rare-earth lasers. Appl. Phys. Lett. *33*, 410–412 (1978)

819 W.J. Miniscalco, J.M. Pellegrimo, W.M. Yen:Measurements of excited-state absorption in Ce^{3+}:YAG. J. Appl. Phys. *49*, 6109–6111 (1978)

820 K.H. Yang, J.A. DeLuca:Energy transfer between 5d electronic states of trivalent rare-earth ions. Appl. Phys. Lett. *35*, 301–302 (1979)

821 L.F. Mollenauer:Advances in color center lasers and their application. Digest of Technical Papers, IEEE/OSA CLEA, May 30-June 1, Washington, 1979, p. 85

822 G. Liffin, K.-P. Koch, R. Beigang, H. Welling:Progress in color center lasers. Digest of Technical Papers, IEEE/OSA CLEA, May 30-June 1, Washington, 1979, p. 85

823 L.F. Mollenauer, D.M. Bloom:Color center laser generates picosecond pulses and several watts CW over the range 1.24–1.45 μm. Opt. Lett. *4*, 247–249 (1979)

824 H.P. Jenssen, J.C. Walling, O.G. Peterson, R.C. Morris:Broadly tunable laser performance in alexandrite at elevated temperature. Post-Dead-line paper, Internat. Conf. Laser-78, December 11–15, Orlando (USA), 1978, L-3

825 J.C. Waling, O.G. Peterson:High-gain laser performance in alexandrite. Digest of Technical Papers, IEEE/OSA CLEA, May 30-June 1, Washington, 1979, pp. 86–87

826 H.P. Jenssen (private communications)

827 P.F. Moulton, A. Mooradian, T.B. Reed:Efficient CW optically pumped Ni:MgF$_2$ laser. Opt. Lett. *3*, 164–166 (1978)

828 A.W. Tucker, M. Birnbaum:Energy levels and laser action in Nd:CaY$_2$Mg$_2$Ge$_3$O$_{12}$ (CAMGAR). Proc. Internat. Conf. Laser'78, December 11–15, Orlando (STS Press, McLean, Va. 1979), pp. 168–170

829 Y. Kuwano: Refractive indices of YAlO$_3$:Nd. J. Appl. Phys. *49*, 4223–4224 (1978)

830 M.Kh. Ashurov, Yu.K. Voronko, E.V. Zharikov, A.A. Kaminskii, V.V. Osiko, A.A. Sobol, M.I. Timoshechkin, V.A. Fedorov, A.A. Shabaltai: Structural specificities, spectroscopy and stimulated radiation of yttrium-holmium-aluminum garnet crystals. Izv. Akad. Nauk SSSR. Neorg. Mater. *15*, 1250–1255 (1979)
      English transl.: Inorg. Mater. (USSR) *15*, 979–983 (1979)

831 W. Lee Smith, J.H. Bechtel:Laser-induced breakdown and nonlinear refractive index measurements in phosphate glasses, lanthanum beryllate, and Al$_2$O$_3$. Appl. Phys. Lett. *28*, 606 (1976)

444   References

832   P.F. Moulton, A. Mooradian: Broadly tunable CW operation of $Ni:MgF_2$ and $Co:MgF_2$ lasers. Appl. Phys. Lett. *35*, 838 (1979)

833   R.C. Eckardt, L. Esterowitz, I.D. Abella: Digest of Technical Papers CLEO '82 (Anaheim, CA 1982) p. 160

834   B.M. Antipenko, A.A. Mak, O.B. Raba, K.B. Seyranyan, T.V. Uvarova: Kvant. Elektron. (Moscow) *10*, 889 (1983)

835   B.M. Antipenko, S.P. Voronin, T.A. Privalova: Opt. Spektrosk. (USSR) *67*, 1203 (1989)

836   D.J. Ehrlich, P.E. Moulton, R.M. Osgood: Opt. Lett. *5*, 339 (1980)

837   Y.S. Bai, W.R. Babbitt, A.G. Yodh: Proc. 1st Intl Laser Sci. Conf. (American Institute of Physics, New York 1986) p. 417

838   A.A. Kaminskii: Dok. Akad. Nauk SSSR *271*, 1357 (1983)

839   A.A. Kaminskii, K. Kurbanov, K.L. Ovanesyan, A.G. Petrosyan: Phys. Status Solidi (a) *77*, K173 (1983)

840   A.A. Kaminskii: Izv. Akad. Nauk SSSR, Neorg. Mater. *17*, 185 (1981)

841   A.A. Kaminskii, K. Kurbanov, A.V. Pelevin, Yu.A. Polyakova, T.V. Uvarova: Izv. Akad. Nauk SSSR, Neorg. Mater. *23*, 1934 (1987)

842   A.A. Kaminskii, K. Kurbanov, K.L. Ovanesyan, A.G. Petrosyan: Phys. Status Solidi (a) *105*, K155 (1988)

843   A.A. Kaminskii, K. Kurbanov, T.V. Uvarova: Izv. Akad. Nauk SSSR, Neorg. Mater. *23*, 1046 (1987)

844   R.W. Waynant: Appl. Phys. *B 28*, 205 (1982)

845   R.M. Macfarlane, F. Tong, A.J. Silversmith, W. Lenth: Appl. Phys. Lett. *52*, 1300 (1988)

846   A.A. Kaminskii: *Proc. Intl Conf. on Lasers-80*, ed. by C.B. Collins (STS Press, McLean, Va 1981) p. 328

847   R.M. Macfarlane, F. Tong, W. Lenth: IQEC-88 (Japan Society of Applied Physics, Tokyo, 1988) Techn. Digest, p. 570

848   A.A. Kaminskii: Izv. Akad. Nauk SSSR, Ser. Fiz. *45*, 348 (1981)

849   A.A. Kaminskii, B.P. Sobolev, S.E. Sarkisov, G.A. Denisenko, V.V. Ryabchenkov, V.A. Fedorov, T.V. Uvarova: Izv. Akad. Nauk SSSR, Neorg. Mater. *18*, 482 (1982)

850   G. Kintz, L. Esterowitz, R. Allen: Topical Meeting on Tunable Solid State Lasers, Technical Digest Series, Vol. 20 (OSA, Washington 1987) p. 215

851   A.A. Kaminskii: Stimulated emission spectroscopy of activated crystals, Preprint of Institute of Crystallography, USSR Academy of Sciences, Moscow (1989)

852   J.C. Walling, H.P. Jenssen, R.C. Morris, E.W. O'Dell, O.G. Peterson: Opt. Lett. *4*, 182 (1979)

853   P. Moulton: Opt. News *8*, 9 (1982)

854   N.P. Barnes, L. Esterowitz, R.E. Allen: CLEO '84, Anaheim (1984) p. WA5

855   T.J. Kane, R.C. Eckhart, R.L. Byer: IEEE J. QE-*19*, 1351 (1983)

856   R.L. Byer, E.K. Gustafson, R. Tribino (eds.): *Tunable Solid State Lasers for Remote Sensing*, Springer Ser. Opt. Sci., Vol. 51 (Springer, Berlin, Heidelberg 1985) p. 87

857   S.E. Stokowski, M.H. Randles, R.C. Morris: IEEE J. QE-*24*, 934 (1988)

858   K. Yoshida, H. Yoshida, Y. Kato: IEEE J. QE-*24*, 1188 (1988)

859   W. Koechner: *Solid-State Laser Engineering*, 2nd ed., Springer Ser. Opt. Sci., Vol. 1 (Springer, Berlin, Heidelberg 1988) p. 388

860   C.A. Morrison: *Angular Momentum Theory Applied to Interactions in Solids*, Lecture Notes Chem., Vol. 47 (Springer, Berlin, Heidelberg 1988)

861   A.A. Kaminskii: Phys. Status Solidi (a) *102*, 389 (1987)

Subject Index[a]

	Page	Reference

[a]Following the stimulated-emission activator ion, sensitizing or deactivating ions are enclosed in parentheses. References are in square brackets. Crystals CaF_2^* contained oxygen.

Simple oxide laser crystals with ordered structure

	Page	Reference
$Ba_2ZnGe_2O_7$		
Nd^{3+}	7, 137, 304, 384	[280]
$(Nd, Sc)P_5O_{14}$		
Nd^{3+}	7, 107, 304, 383	[533]
$(Nd, In)P_5O_{14}$		
Nd^{3+}	7, 107, 304, 383	[615]
$(Nd, La)P_5O_{14}$		
Nd^{3+}	7, 107, 110, 304, 362, 363, 370, 373, 383	[475, 505, 658, 782, 783, 785]
$(Nd, Gd)Al_3(BO_3)_4$		
Nd^{3+}	7, 107, 304, 362, 386	[579]
$LuScO_3$		
Nd^{3+}	7, 304, 388	[639]
$HfO_2-Y_2O_3$		
Nd^{3+}	7, 33, 165, 304–307, 385, 387, 390	[90, 430, 514]
$Bi_4(Si, Ge)_3O_{12}$		
Nd^{3+}	7, 306, 386	[659]

Other laser crystals

	Page	Reference
$Na_2Nd_2Pb_6(PO_4)_6Cl_2$		
Nd^{3+}	2, 361, 384, 387	[771, 788]
$Ca_5(PO_4)_3F$	24	[8, 438]
Nd^{3+}	7, 33, 99, 137, 306, 386, 390	[90, 129, 130, 170, 214, 215, 281–283, 434, 769, 771]
$Ho^{3+}(Cr^{3+})$	46, 306, 395	[129, 130]
$Sr_5(PO_4)_3F$		
Nd^{3+}	7, 138, 306, 384	[130, 282]
La_2O_2S		
Nd^{3+}	7, 100, 138, 306, 388	[219, 435, 771]
$LaCl_3$		[46, 47, 758]
Pr^{3+}	7, 10, 32, 107, 122, 308, 381	[116, 372, 564]
$LaBr_3$		[758]
Pr^{3+}	7, 32, 107, 308, 381	[564, 767]
$CeCl_3$		[46, 47]
Nd^{3+}	7, 107, 308, 387	[482, 620]
$PrCl_3$		[46, 47, 672]
Pr^{3+}	6, 7, 32, 52, 107, 122, 308–311, 381	[76, 116, 260, 564]
$PrBr_3$		[46, 47, 672]
Pr^{3+}	6, 7, 32, 52, 107, 122, 310, 381	[76, 564]
$Pb_5(PO_4)_3F$		
Nd^{3+}	7, 138, 310, 384	[640]